BIOTECHNOLOGY OF VA MYCORRHIZA: INDIAN SCENARIO

BIOTECHNOLOGY OF VA MYCORRHIZA INDIAN SCENARIO

Dr. Sudhir Chandra
(Emeritus Professor)

and

Dr. Harbans Kaur Kehri
(Reader, Deptt. of Botany)
UGC Centre of Special Assistance & COSIST Support Programme
University of Allahabad, Allahabad - 211 002
Uttar Pradesh (INDIA)

2006

NIPA

New India Publishing Agency
Pitam Pura, New. Delhi- 110 088

ISBN 81-89422-22-7

Typeset at:

Laxmi Art Creations
Shiva Mkt., Pitam Pura, New Delhi
Mobile : 9811482328

Printed at:

Jai Bharat Printing Press
Rohtas Nagar, Shahdara, Delhi

Published by:

New India Publishing Agency
101, Vikas Surya Plaza, CU Block, L.S.C. Mkt.,
Pitam Pura, New Delhi- 110 088, (INDIA)
Phone: 011-27341717, Fax: 011-27341616
E-mail: spjain_niph@rediffmail.com
Web: www.bookfactoryindia.com

Preface

Because of the advantage of their universal presence, minute size and astonishingly rich physiological qualities, microbes play the key role in maintaining the biological equilibrium on earth and in their absence; life on the planet will be extinct. Different forms of interrelationship between plants and microbes have developed over million of years. On the extreme end, destruction of one by the other takes place. On the other hand, in the struggle for existence between the two efficient opponents, various forms of symbiotic relationships have emerged. The possible beneficial role of a symbiotic association between plant roots and fungi was brought out for the first time in 1842 by Vittadini and the name "Mycorrhiza" was given to it by Frank in 1885. Since then our knowledge on mycorrhiza-plant relationship has grown somewhat with slow pace until about 1970 when there was a sudden upsurge of interest on a specialized type of endomycorrhiza-vesicular arbuscular mycorrhiza (VA Mycorrhiza). The prodigious research made during last three decades clearly established its widespread occurrence in various plant species and under different agroclimatic conditions covering broad ecological range including deserts, forests and mangroves. It was also established that this symbiotic association benefits the plants through enhanced nutrient uptake, biological control of root pathogens, and synergistic interaction with nitrogen fixing microorganisms, hormone production and drought resistance. In view of its utility to plants, this biotool has now attracted the attention of microbiologists, agronomists, horticulturists and foresters at the global level.

Microbiologists, plant pathologists, ecologists, biochemists and biotechnologists of the country have been involved in the researches on VA mycorrhiza and have gathered valuable information on the exploitation of this biotool for the welfare of the plant communities. However, no effort has been made to bring all such informations together. In view of this, we made a proposal to the Department of Science and Technology, Government of India, expressing our desire to take up this task. Our proposal was readily accepted for full support and assistance. The present volume **"Biotechnology of VA Mycorrhiza: Indian Scenario"** is the outcome of our proposal. We have taken utmost care to include all the relevant informations in the volume; however, any left over information should be taken as a slip on our part for which we express our sincere regret.

We would like to express our sincere gratitude to the Department of Science and Technology, Government of India for providing financial support to our proposal. We liberally utilized the library and other facilities from our Department for which we wish to thank our Head of the Department for his kind gesture. Our senior colleagues of the Department including Prof. B.K. Verma, Prof. G.K. Srivastava, Prof. Bihari Lal, Prof. G.L. Tiwari, Prof. D.R. Misra and Prof. P.K. Khare extended their ungrudging help and suggestions for which we express our profound appreciation. Our former and present associates including Dr. K.K. Khanna, Dr. Tanushree Chatterji, Dr. Shika Maheswari, Dr. S.S. Kehri, Dr. S.M.H. Zaidi, Dr. Ratna Dwivedi, Dr. Rudal Dubey, Dr. Nidhi Srivastava,Nidhi Sharma, Raghvendra Pratap Narayan, Varun Khare, Rollie Verma, Sudha Singh, Manoj Dubey and Ansu Tiwari provided their valuable assistance from time to time for which we are grateful to them.

We received valuable suggestions from time to time from some distinguished mycorrhizasts of the country including Prof. D.J. Bagyaraj, Late Prof. A. Mahadevan, Prof. K.G. Mukerji, Prof. Ajit Varma, Prof. B.L. Jalali, Prof. B.N. Johri, Prof. C. Manoharachary and Prof. K.V.B.R. Tilak. We gratefully acknowledge their kind gesture. On our request mycorrhizasts of the country readily provided informations about their research contributions on VA Mycorrhiza for which we sincerely express our gratitude to them.

We gratefully appreciate Mr. Jagdish Chandra for painstaking work of composing the chapters. We are also thankful to our family members who spared us to devote our time to this task. It could have not been possible to bring out the Volume without their kindness and encouragement.

December, 2005

Sudhir Chandra
H.K. Kehri

Prelude

Mycorrhiza research is relatively recent. Although it was in 1885 that A.B. Frank made reference for the first time to the "Mycorrhiza", it took him nine more years to establish the symbiotic relationship between the fungus and its host plant. Research carried out between 1917 and 1950 confirmed that "mycorrhizal colonization is a beneficial symbiosis between tree and fungus".

During the last four decades research on mycorrhiza made prodigious progress. This was possible because of the rapid developments in scientific instrumentation (hardware) and to the increasing number of scientists becoming involved in the mycorrhizal research.

The number of mycorrhiza researchers increased rapidly over the years, particularly in North America and as there was a need for communication between them, it was decided to hold, on a periodic basis, a North American Conference on Mycorrhiza (NACOM). Later on, with the growing number of scientists in Europe, Europe decided to have its own forum (ESOM) to enable more European researchers to play an active role.

In 1987, India alone had over 150 mycorrhiza researchers (North America 105, Europe 250, China 200). But only few of them had the opportunity to attend the NACOM or ESOM or even to meet at National or regional forum. This isolation of researcher and lack of communication, instead of promising complementary work and creating synergistic interactions, led to repeated duplications and limited achievements.

To overcome this problem IDRC (Canada), the International Development Research Centre sponsored the first National Workshop on Mycorrhizae in India in March 1987 at J.N.U., New Delhi.

In the same year (1987), in May, Prof. K.G. Mukerji and Prof. A. Mahadevan, while sipping coffee at the poolside of Holiday Inn of America in Gainesville, Florida, where the 7th NACOM was in progress, discussed about the need to organize a get-together of Asian Mycorrhizasts. After discussing the matter for about an hour, both Prof. Mukerji and Prof. Mahadevan requested Prof. N.C. Schenck to make the announcement. The Asian Conference on Mycorrhiza (ACOM) was thus borne during the NACOM in Florida.

Prof. Mahadevan, took up the organizational work soon after returning from Florida. The first Asian Conference was organized at University of Madras, Madras and sponsored by IDRC, Canada. Prof. D.H. Marx, Prof. J.M. Trappe, Prof. R.K. Dixon from U.S.A., Dr. P. Masson from U.K., Mr. Karim Oka and Prof. A. Fortin from Canada encouraged and spend their valuable time in organization and planning. Mycorrhizasts from Bangladesh (1), Canada (1), China (4), Indonesia (3), Korea (2), Malaysia (3), Nepal (2), Sri Lanka (3), Thailand (3), U.K. (1), U.S.A. (4), Vietnam (1) and India (100) attended the conference.

Prof. G. Rangaswami, the doyen of Agricultural Microbiology in India, inaugurated the conference and Mr. Karim Oka, Programme Officer (Forestry), IDRC (Canada) delivered the key-note address.

All the participants expressed the need for continued opportunities to meet and share information and experiences. Meanwhile it became obvious that workshops, symposia and conferences can only constitute one way of communication and ad-hoc-cooperation among scientists at regional, national and international levels. If continuous and effective interaction was to be ensured, more elaborate mechanisms were necessary and concrete means had to be provided to the researchers to make co-operation a reality.

In this context and based on the recommendations of the majority of the scientists attending the above gatherings, it was decided to establish a "Mycorrhiza Network" funded (in its initial stage) by IDRC and coordinated by the Tata Energy Research Institute (TERI) based in New Delhi, India.

The general objective of the **Mycorrhiza Network** was to strengthen research, encourage co-operation, promote exchange of information and germplasm, and facilitate transfer of technology to the field through the establishment of mycorrhiza research network.

The specific objectives of the Network were:

(i) To promote communication between researchers through meetings and workshops.

(ii) To facilitate the exchange of germ plasm.

(iii) To provide advisory services on research needs, methodology and planning and

(iv) To establish a Mycorrhiza Information Centre (MIC).

MIC has been set up at TERI, 7 Jor Bagh, New Delhi (now at TERI, Darbari Seth Block, Habitat Place, New Delhi) with the specific objective of establishing an information network in the region to make available to researchers the growing literature on mycorrhiza, as well as to facilitate information sharing among the members. Various activities of the MIC include collection and organization of resources on mycorrhiza, information dissemination through computerized databases, services and generation of relevant information products like directories, bibliographies and quarterly newsletter.

The Technical Advisory Committee of Mycorrhiza Network Asia Project at TERI, in its very first meeting held in January, 1989, decided to organize the "**First National Conference on Mycorrhiza**" which was held at Haryana Agricultural University, Hisar from 14-16 February, 1990. Over 110 delegates from India and other Asian countries took part in the deliberations of the conference.

Dr. A.H. Chulan from Malaysia, Dr. Yenchai Vasuvat from Thailand and Prof. J.A. Fortin from U.S.A. also participated in the Conference deliberations.

The papers on Ecology, Physiology, Biochemistry, Biological interactions, Soil-Plant Relationships and Biocides were presented at H.A.U., Hisar.

At Hisar, it was decided to hold the Second National Conference on Mycorrhiza at Bangalore. It was held at Bangalore Agriculture University in November, 1990 and was attended by nearly 100 scientists from India, Dr. Barbara Mosse from U.K., Dr. Hiroshi Hirata from Japan, Dr. Kuo Shiu Chien from China and Dr. Yenchai Vasuvat from Thailand.

Prof. Ajit Varma, conducted the International Training Workshop on "**Basics of Mycorrhizae and its Biotechnology**" from 7-21 December, 1989 at the School of Life Sciences, J.N.U. This was attended by 70 delegates from 28 countries. The course dealt with basic and modern techniques and their application to mycorrhiza - a fungus-root symbiont.

The Third National Conference on Mycorrhiza, sponsored by Mycorrhiza Network (a DBT Project) and organized by TERI, New Delhi was held in Teen Murti Bhavan Auditorium, New Delhi, 13-15 March, 1995. Over 120 delegates from India and Nepal participated in the Conference. 129 papers were received from the delegates, out of which 33 were presented orally while many as posters. The deliberations in the conference were held in seven sessions, namely :

* Distribution of mycorrhizal fungi under varying edapho-climatic conditions
* Management of pests, mycorrhizae and biocides for increased plant productivity
* Soil/water pollution and mycorrhizae
* Tripartite biological interactions involving rhizosphere flora, mycorrhizal fungi and plant roots
* Plant growth responses and mycorrhizal dependency
* Physiology and biochemistry and role of mycorrhizae in difficult/disturbed sites.

The Fourth National Conference on Mycorrhiza was held at Barkatulla University from 5 to 7 March 1999. It was organized under the auspices of the Mycorrhiza Network, TERI, at the Institute of Microbiology and Biotechnology, Barkatulla University, Bhopal. Over 75 delegates from India and one each from Sri Lanka, Italy and Zambia participated in the conference. The deliberations in the conference were held under five sessions :

1. Association and distribution of mycorrhizal fungi under varying edapho-climatic conditions
2. Plant growth responses and mycorrhizal dependency
3. Physiology and Biochemistry
4. Interaction between mycorrhiza and other rhizosphere flora and fauna
5. Maintenance of mycorrhizal cultures and application of mycorrhizal technology

In all, 28 papers were presented orally while 54 as posters.

The Fifth National Conference on Mycorrhiza on a theme "Relevance of mycorrhiza in dry land agriculture and forestry" was held at Central Arid Zone Research Institute, Jodhpur from 16-18 November, 1999. It was organized under the auspices of Mycorrhiza Network, TERI, New Delhi. The main focus was on the following themes :

* Biology, ecology, management and establishment of mycorrhizae and mycorrhizal dependency of plant species under different edaphic conditions.
* Inoculum production and biotechnology, mycorrhizal biodiversity and conservation and maintenance
* Rehabilitation of waste lands including coastal sands, sand dunes, mine spoils and saline-alkaline soils

* Use of molecular tools in taxonomy and other aspects of mycorrhiza research
* Interactions among soil microflora, microfauna and mycorrhizal fungi.

During last fifteen years a number of active centres of mycorrhizal research have cropped up in the country. Presently a large number of workers are engaged in tackling various aspects of problems related to mycorrhiza and exploring the possibility of employing this important tool in improving the health of the plants and their productivity not only under normal conditions but also under adverse conditions.

Realizing the importance of plant resources under present situations in the country and the significance of the role which the mycorrhiza as a biotool may play to improve them, Department of Science and Technology, Govt. of India agreed to finance our proposal to prepare a State-of-Art Report on **" Researches on VA mycorrhiza in India : Current status and future prospects"**. The present report is based on publications on VA mycorrhiza related to India which appeared in Proceedings of different conferences, journals, including abstracting journals, books and personal communications. For convenience, we have included a short account (VAM-Resume) of the state of art on VA mycorrhiza at the global level. It is followed by a detailed account of researches on different aspects of VA mycorrhiza carried out in the country by the mycorrhizasts. Based on this account we have attempted to chalk out the future strategies of work on mycorrhiza which should be given priority while planning the work on this important biotool for plant health. In addition, certain recommendations have been made which include not only our thoughts but also of those which have a serious concern about VA mycorrhiza and plant communities.

Contents

Preface v

Prelude vii

CHAPTER 1
VA MYCORRHIZA: RESUME **1-21**

CHAPTER 2
VA MYCORRHIZA: DISTRIBUTION **23-70**

CHAPTER 3
EFFECT OF VA MYCORRHIZA ON GROWTH **71-142**

CHAPTER 4
VA MYCORRHIZA: PHYSIOLOGY **143-161**

CHAPTER 5
VA MYCORRHIZA *VS* OTHER MICROBES **163-204**

CHAPTER 6
FACTORS AFFECTING VA MYCORRHIZA **205-239**

CHAPTER 7
VA MYCORRHIZA AS BIOCONTROL AGENT **241-264**

CHAPTER 8
VA MYCORRHIZA IN WASTELAND DEVELOPMENT **265-279**

CHAPTER 9
VA MYCORRHIZA: INOCULUM PRODUCTION **281-289**

CHAPTER 10
MISCELLANY **291-297**

CHAPTER 11
FUTURE PROSPECTS AND STRATEGIES **299-317**

REFERENCES **319-413**

1

VA MYCORRHIZA : RESUME

Mycorrhiza research is relatively recent. Although, it was in 1885 that A.B. Frank made reference for the first time to the term "**Mycorrhiza**", it took him nine more years to establish the symbiotic relationship between the fungus and its host plant. It took a much longer time to convince the rest of the scientific community of the validity of Frank's theory. Researches carried out between 1917 to 1950 confirmed that "**Mycorrhizal colonization is a beneficial symbiosis between tree and fungus**". It is an unique partnership in nature, where the higher plants share their photosynthates with their fungal partners and provide them shelter, the fungal partners in return provide protection to the root system and assist it in performing its functions especially anchorage and absorption of nutrients / water from soil more efficiently.

The ectomycorrhizal fungi have been in use as a biological tool in improving the performance of forest trees since long. Better performance of forest trees is ensured by making them mycorrhizal, employing efficient strains of ectomycorrhizal fungi.

The endomycorrhizal fungi especially vesicular arbuscular mycorrhizal fungi (VAM fungi) have gained importance as a biological tool in improving the performance of not only forest trees but also a variety of other plants under a variety of habitats and conditions, only during past three decades.

Credit of identifying the VA mycorrhizal fungi as a biological tool in improving the performance of plants goes to Barbara Mosse (1959) who for the first time experimentally produced VA Mycorrhiza and confirmed its involvement in plant nutrition and water uptake. Since then a lot of information on VAM fungi and its importance as a biological tool in improving the nutrition and water uptake in plants and their resistance to various stresses and root pathogens has accumulated as a result of extensive researches throughout the globe. They have been reported to be associated with all groups of plants growing under all types of habitats. They have been found in almost all the tropical soils, but land disturbances due to various reasons have been shown to disturb their native population. Attempts are being made at various research centres of the world to explore the possibility of ensuring the cultivation of different crops including legumes, cereals, plantation crops, industrial crops, horticultural crops as well as ornamentals and medicinal plants less expensive thus more profitable employing efficient strains of VA mycorrhizal fungi. They are also being exploited as biocontrol agents against the root pathogens of serious nature. In addition, attempts are being made to gather information about their ultrastructure, cytology, genetics, physiology as well as their relationship with the host and their specificity.

At a time when economic and environmental considerations are restricting the use of chemical fertilizers (high energy production costs, and soil and water polluting effects), the interest attached to mycorrhiza as an alternative solution appears legitimate. Mycorrhizae, acting as bio-fertilizer and produced with little energy with no polluting effect, are playing a key role as a partial substitute to chemicals.

Traditionally, Frank (1887) is considered the first author to have distinguished between ecto- and endotrophic mycorrhizae. Ectotrophic mycorrhizae (now known as ectomycorrhizae) are characterized by the presence of fungal mantle or sheath covering the root surface, while endotrophic mycorrhizae (now known as endomycorrhizae) do not possess such a fungal sheath and have extensive intracellular fungal penetrations. Frank reached to these conclusions early in 1887 and was able to formulate the distinction between ecto- and endotrophic mycorrhizae on the basis of his detailed studies on Cupuliferae for the former and ericales and orchids for the latter.

However, it was not until the works of Schlicht (1889), Janse (1896-97), Dangeard (1900), Petri (1903) and Gallaud (1905) that endomycorrhizal fungi found in plants other than ericales and orchids (i.e. those now called VAM) were described in detail. This early research was followed by many other important studies, including those of Peyronel (1924a), Mclennan (1926) and Butler (1939). They established the following fundamental points which became the basis of our knowledge of VAM up to the 1970s. i.e.

(1) The vast range of plants is colonized by VAM fungi; Schlicht (1889) described 71 species of herbaceous angiosperms, Janse (1896-97) studied tropical plants from Java (Bryophytes, Gymnosperms, woody dicotyledons and Pteridophytes) while Peyronel (1924b) listed about 150 mycorrhizal species from the Alpine areas.

(2) During its complex life cycle in contact with the root, the fungal symbiont gives rise to different structures; extramatrical hyphae, intracellular hyphae, globose or oval termial swellings called "vesicles", intracellular structures with a form like small tree or "arbuscules" as Gallaud (1905) named them and irregular clumps (when the fine branches of the arbuscules are no longer visible) termed "sporangioles" by Janse (1896-97). These different structures have different positions in the root.

(3) In nature the same root can be infected by two different endophytes at the same time. This hypothesis of a double infection was put forward by Peyronel (1924), who on the basis of the endophyte's morphological characteristics, recognised a septate mycelium (one that could be classified as a **Rhizoctonia** type fungus, also responsible for orhid mycorrhizae) as well as another endophyte, similar to a phycomycete and responsible for VA infections. Following this clear distinction, VAM have since been identified as a type in themselves, distinct from other endomycorrhizae. Moreover, for the first time, Peyronel (1924) claimed that the endophyte responsible for VAM formation belonged to the Endogonaceae family.

(4) All the above-listed authors recognised mycorrhiza as a mutualistic, symbiotic association, since the host cell seemed to digest the fungus, thus obtaining nutrients.

Many of these definitions have survived the passage of time, others have been modified and updated with the introduction both of new equipment that have made it possible to understand the physiological characteristics of a VAM infection, and of new techniques, that have allowed more sophisticated morphological analysis.

A. VAM TAXONOMY & MORPHOLOGY

Dangeard (1900) was the first to describe a vesicular-arbuscular mycorrhizal fungus "**Rhizophagus populinus**". Peyronel (1924; 1937) accommodated VAM fungi under **Endogone** Link belonging to family Endogonaceae, (an ancient group which has Devonian fossil representatives) in Mucorales of Zygomycetes. VA Mycorrhizal fungi being obligate biotrophs, do not grow on synthetic media and hence are classified according to morphological characteristics of the spores formed in the soil. One Hundred and Twenty species have been described by Schenck and Perez (1987) and they are grouped in a single family, the Endogonaceae. Until 1990 VA mycorrhizal fungi continued their position under family Endogonaceae in the order Endogonales of the class Zygomycetes and the division Zygomycotina (Benjamin, 1979; Trappe, 1987; Morton, 1988). They were accommodated within six genera of the family *viz., Acaulospora* Gerd. and Trappe em. Berch, *Entrophospora* Ames and Sehn, *Gigaspora* Gerd. and Trappe, *Glomus* Tul. and Tul., *Sclerocystis* Berk. and Br. and *Scutellospora* Walk. and Sand. The genus *Endogone* was retained in the family *Endogonaceae* but it included saprophytic members occasionally forming ectomycorrhiza. (Chu-Chou and Grace, 1979). Pirozynski and Dalpe (1989) separated *Glomus* and *Sclerocystis* from others and accommodated them in a new family Glomaceae under Endogonales. On the basis of morphological, phylogenetic and physiological considerations, Morton and Benny (1990) proposed that VAM fungi should be termed as AM fungi since many mycorrhizal fungi do not produce intramatrical vesicles. They created a new order Glomales to include all soil fungi, which produce arbuscules and form obligate mutualistic associations with the roots of terrestrial plants. The order Glomales was divided into two suborders. The Suborder Glomineae included members forming chlamydospores, arbuscules as well as vesicles. It accommodated Glomaceae (chlamydospores on a fertile hypha) with *Glomus* (chlamydospores singly) and *Sclerocystis* (chlamydospores in sporocarps) and Acaulosporaceae (chlamydospores from a sporiferous saccule) with *Acaulospora* (lateral chlamydospores) and *Entrophospora* (intercalary chlamydospores). The suborder Gigasporineae included members forming azygospores and only arbuscules. It accommodated only Gigasporaceae (azygospores on sporogenous cells) with *Gigaspora* (lateral azygospores) and *Scutellospora* (terminal azygospores). The order Endogonales was retained but it included only *Endogone* under the family Endogonaceae.

The main features of VA Mycorrhizal fungi are summarized by Trappe and Schenck (1982) and Hall (1984). Hall and Abbott (1983) compiled photographic transparencies illustrating diagnostic features of the Endgonaceae. Walker (1983) has suggested a standardized terminology for use in description of endogonaceous spore walls involving a stylized graphic representation of the wall structure ("murograph"). This concept is recommended for future descriptions of new species. Berch (1986) reviewed the taxonomy, specificity, fossil record and phylogeny of Endogonaceous fungi. Schenck and Perez (1987) published a manual for the identification of VA mycorrhizal fungi.

Recently, based on ribosomal phylogeny Schubler *et al.* (2001) placed AM fungi under phylum Glomeromycotina and this phylum contains one class Glomeromycetes. The class Glomeromycetes is divided into four orders Glomerales, Diversisporales, Archaeosporales and Paraglomales. The order Glomerales has a single family Glomeraceae with two genera *Glomus* and *Sclerocystis*. The order Diversisporales includes three families, Diversisporaceae having *Diversispora*, Gigasporaceae having *Gigaspora* and *Scutellospora,* and Acaulosporaceae having *Aculospora* and *Entrophospora*. The third order Archaeosporales including two families, Archaeosporaceae and Geosiphonaceae each with single genus *Archaeospora* and *Geosiphon* respectively. The fourth order Paraglomales includes a single family Paraglomaceae with *Paraglomus.*

More recently based on molecular, morphological and cytological evidences Walker *et al.* (2004) showed that the AM fungus *Glomus scintillans* is more closely related to the genera *Gigaspora* and *Scutellospora* (Gigasporaceae) than to *Glomus* and consequently it contributes to the non morphology of genus *Glomus*. They also separated *Gerdemannia* gen. nov., from *Glomus* and erected new family Gerdemaniaceae fam. nov. in Glomeromycota. Small subunit (ssu) rDNA analysis places *Gerdemannia* in a clade sharing common ancestry with the Gigasporaceae, and therefore within the Diversisporales (Glomeromycota). The AM fungus *Glomus dominikii* was considered to be conspecific with *G. scitillans* and these species are synonymised by Walker *et al.* (2004). Helgason *et al.* (2003) carried out joint analysis of the action and elongation factor (EF) I-alpha sequences suggested that the sister group to the AM fungi was a zygomycete order Mortierellales. Redecker *et al.* (1999) based on phylogenetic analysis of a dataset of fungal 5.8 S rDNA sequences reported that highly divergent copies of internal transcribed spacers (ITS) from *Scutellospora castanea* are of ascomycete origin. Further Hijri *et al.* (1999) based on polymorphism of ITS of rDNA reported that the spores of *Scutellospora castanea* are multinucleate.

IDENTIFICATION OF VAM FUNGI

Vesicular-arbuscular mycorrhizal (VAM) fungi are difficult to identify in the absence of knowledge on their complete life cycle. These fungi have also not been cultured on artificial media. The identification of VAM fungi is therefore done on the basis of morphology of azygospores/chlamydospores/sporocarps and hyphae attached to these structures.

Precise identification of spores from the field is difficult as morphology may vary. This is major limitation for ecological studies since sporulation depends on environmental conditions and on host and fungal genotype. These studies therefore require the assumption that the number of spore types extracted from a soil core are directly related to the colonization of fungi in host roots (Morton *et al.*, 1994) an assumption that is being questioned by the use of molecular methods of identification (Simon *et al.*, 1992) and is likely to give a very unreliable indication of root colonization in the field (Clapp *et al.*, 1994).

Methods that can directly reveal the colonizing fungal taxa within or associated with a root will be of great importance in the elucidation of mycorrhizal biodiversity and ecology. A variety of techniques have therefore been developed to assist in this process. These include the anatomy of infections (Abbott, 1982), selective enzyme staining following gel electrophoresis (Sen and Hepper, 1986), isozyme techniques, serological techniques, immunological probes, DNA library and fingerprinting and PCR techniques combined with

RAPID/RFLP. Out of these, serological and molecular techniques are most potential and have been discussed here.

SEROLOGICAL TECHNIQUES

Serological techniques are dependent on the production of highly specific antibodies in animals when they are inoculated with the chemicals contained in or produced by micro-organisms (antigens). If the antibody and the antigen that elicited its formation are incubated together they will react. A number of tests have been developed that use this to detect the presence of either the antibody or the antigen. Of these tests, fluorescent antibody technique (FAT), enzyme linked immunosorbent assay (ELISA) and the radioimmunoassay (RIA), are the most sensitive.

Fluorescent antibody techniques (FAT)

The FAT has been applied to VAM fungi by Warner and Mosse (1979), Aldwell (1983), Kough *et al.* (1983) and Wilson *et al.* (1983). However, antisera of Warner and Mosse had poor specificity probably because the antigens were prepared from crushed whole spores containing much non-specific chitinaceous spore wall material and were administered by a single intra muscular injection without adjuvant and without boosters (Aldwell, 1983). In contrast Wilson *et al.* (1983) ground, centrifuged and sonicated *Gigaspora decipiens* Hall and Abbott germ tube hyphae and auxiliary cells gathered from spores germinated between Milipore filters and then produced antisera by subcutaneous injection with Freunds's Complete Adjuvant (FCA) and with intravenous boosters. Using these techniques they were able to distinguish between *G. decipiens* and *Acaulospora laevis* Gerd. and Trappe and other non-mycorrhizal mucoraceous fungi.

Aldwell (1983) also developed antisera to mycorrhizal hyphae washed from the surface of roots, later these hyphae also formed the target antigens. Using these techniques he was able to use the FAT to distinguish amongst the genera *Glomus, Gigaspora, Acaulospora* and *Sclerocystis* and to separate these from the non-VAM fungi but was unable to distinguish amongst VAM fungal species within the same genus.

Enzyme-linked immunosorbent assay (ELISA)

The ELISA is now well established as a practical method for the qualitative and quantitative detection of hormones, drugs (Pal, 1978) and plant pathogens (Clark, 1981 a,b). The most significant characteristic of ELISA is their ability to detect antigens occurring in concentrations much lower than those that can be detected by the FAT and have sensitivities comparable with RIA. However, unlike the RIA, the ELISA employs stable reagents and requires less expensive equipment. Aldwell (1983) and Aldwell *et al.* (1983, 1985) therefore chose to apply the ELISA to VAM rather than RIA following Aldwell's failure to distinguish between species within a genus using the FAT.

With the ELISA and antisera prepared using techniques essentially similar to those used to produce antisera for the FAT, Aldwell *et al.* (1983,1985) obtained these results:

1. VAM fungal species within a genus could be distinguished.
2. There was little difference in the antigenic composition of hyphae washed from surface of roots when the host plant was changed.
3. Although hyphae washed from the surface of roots were sometimes contaminated with small quantities of root material, *Pseudomonas* sp. and *Bacillus* sp., these did not elicit the production of detectable quantities of antibodies.
4. Although the antigenic composition of hyphae was affected by the way the hyphae were raised, i.e. normal pot culture, hydroponic sand pot cultures, nutrient film cultures, or germ tube hyphae, these differences were not as great as the differences between species within a genus.
5. There was little difference in the antigenic composition of hyphae from cultures of a species originally from the same source but held and subcultured in different geographical locations for ten years.

Practical uses of serological techniques for identification of mycorrhizal fungi

Serological techniques, in particular the ELISA show considerable promise for use as taxonomic tools for the VAM fungi. Unfortunately Aldwell *et al* (1983, 1985) only prepared antisera to six VAM fungal species and this would be insufficient to identify all VAM fungi. Consequently, considerably more antisera to a wider range of species would need to be raised before the routine use of the ELISA for identifying VAM fungi could be envisaged.

MOLECULAR TECHNIQUES

As mycorrhizal fungi are usually identified on the basis of morphological characters shown by the fruit bodies, spores, vegetative mycelia or symbiotic structures, it is proposed by Lanfranco *et al.*(1995), that the development of molecular techniques provides a valuable alternative approach to identify mycorrhizal fungi, particularly when it is difficult to gather a sufficient amount of data on morphological features.

A prerequisite for these techniques is the recovery of DNA from biological samples. DNA can be extracted from both mycorrhizal spores and mycorrhizal roots. The number of spores necessary to obtain amplifiable DNA has reduced from several thousand to single spores. Justin and Simon have been frequently successful in amplifying from single spores using a Chelax resin DNA extraction method based on Walsh *et al.* (1991).

Various buffers are used to extract DNA from mycorrhizal roots. They tend to contain a combination of hexadecyltrimethylammonium bromide (CTAB), sodium dodecyl sulphate (SDS), and B-mercaptoethanol or Proteinase K and involve incubation at 50-65^{0}C for 30-60 minutes (Gardes *et al.* 1991; Henrion *et al.* 1992; Lanfranco *et al.* 1995). Homogenization with liquid nitrogen prior to extraction is only really necessary if very small quantities of root material are used or quantification of DNA is to be performed (as other methods of homogenization are less standardized). Otherwise, homogenization within the extraction buffer will suffice. The root homogenate is usually incubated with phenol to remove protein. Nucleic acids are then retrieved by alcohol precipitation.

Genus-specific primers and PCR

By coupling the sensitivity of the polymerase chain reaction and the specificity afforded taxon-specific primers, a variety of samples can be analysed including small amounts of colonized roots. Family-specific primers as well as generic primers are described and can be used to amplify small sub-unit rRNA fragments from endomycorrhizal fungi by polymerase chain reaction. The amplified products are then subjected to single-strand conformation polymorphism analysis to detect sequence differences. Among the advantages of this approach are the possibilities of directly identifying the fungi inside field-collected roots without having to rely on the fortuitous presence of spores. Primers specific to the order Glomales (VANS 1) and to the three genera *Gigaspoa/Scutellospora* (VAGIGA), *Acaulospora* (VAACAU), and *Glomus* (VAGLO) have been used in conjunction with PCR (Simon *et al.* 1992).

RAPD-DNA analysis

Random Amplified Polymorphic (RAPD)-DNA analysis is a variation of standard PCR. It uses single, short, arbitrary primers, which anneal more or less frequently throughout the genome. Where the primers anneal in inverted orientation on opposite stands within approximately 4 kb of each other, amplifiable products can be produced. Usually, several products of varying size are formed, which can be visualized by agarose gel electrophoresis and ethidium bromide staining. DNA differences that disrupt or create primer-annealing sites will determine the pattern of products visualized. In this manner DNA polymorphism can be detected without specific nucleotide sequence information. Wyss and Bonfante (1993) have used RAPD analysis to show polymorphism between and within VAM fungal species. Lanfranco *et al.* (1995) isolated species-specific primers for use in standard PCR. In this case a RAPD band, specific for *Glomus mosseae* was identified and sequenced. This information was then used to design primers which amplified the *G. mosseae* RAPD fragment alone. It was found that, using these primers, a fragment was only amplified from *G. mosseae*.

PCR-RFLP technique

A possibility of identifying individual VAMF spores from the field on the basis of a molecular marker, namely the nuclear ribosomal DNA encoding the highly conserved 5.8 S rRNA with the two flanking internal transcribed spacers (ITS region) known to vary between species was explored by Sanders *et al.* (1995). A technique involving polymerase chain reaction followed by restriction fragment length polymorphism analysis (PCR-RFLP) was developed to amplify and characterize the ITS region from single VAMF spore. PCR reactions with extracts from single spores of three VAMF species raised under glasshouse conditions yielded reproducibly a single amplification product of the ITS region in sufficient amounts to allow cleavage with several restriction enzymes. The size of the ITS region C.600 base pairs varied only slightly between species. Digestion of the PCR products with the restriction enzymes HinfI and TaqI resulted in banding patterns that were reproducible for different individual spores of a given species but showed clear differences between the three species tested. The sum of the fragment sizes was sometimes greater than the size of the original PCR product e.g. in *Glomus mosseae*. Clones of the amplification product from a single spore of this fungus were obtained and sequenced. This yielded two closely related but different sequences indicating that two different ITS regions co-existed in the spore. The RFLP pattern of the amplification product of the spore was a result of an amalgamation of these two sequences. Sanders *et al.* (1995) applied this technique to the AMF spores collected from a species rich grassland. They sorted out the spores on the basis of morphological characters e.g. their colour, size and shape and then subjected to PCR-RFLP analysis. A group of *Glomus* spores yielded amplification products in the majority of cases was further investigated. PCR-RFLP analysis on 10 individual spores from the field produced 10 different patterns. Similar results were obtained with other groups of spores. The results suggest that the diversity in natural VAM communities and the genetic diversity within individual spores might be much greater than previously thought.

Practical uses of molecular techniques for mycorrhizal fungi

Molecular techniques are a major impetus in mycorrhizal research. They are tremendously powerful and offer huge advantages over conventional methods. The new information obtained by molecular techniques, especially in mycorrhizal biodiversity, must be founded in current taxonomy. However, it is likely that as the weight of molecular data rises, the current classification of many species and genera will have to be reviewed.

Abbott and Robson (1984) claimed that morphology and anatomy of VAM fungi infecting different hosts remain same and suggested that such features can be used as criteria for identification. Morphological, anatomical, biochemical and molecular characteristics of VAM fungi have been investigated in recent past through most sophisticated techniques like Fluorescent Antibody Technique (FAT), Radio Immuno Assay (RIA) and Enzyme Linked Immunosorbent Assay (ELISA) (Aldwell, 1983; Aldwell *et al.*, 1985), Gel electrophoresis (Walker, 1983). Being the most important criteria for identification and taxonomy, the spores of VAM fungi have been extensively studied. The light influences their colour, however, it varies from hyaline to black or yellow to brown. (Morton, 1988). There exists a wide variation in the size of spores. This is evident even within the same taxon e.g. spores of *Glomus macrocarpum* are 50 µm or less in diameter while the diameter of the spores of *G. mosseae* range between 100-200 µm (Morton, 1988). Wide variations exist also in shape, which ranges from globose to sub-globose, ellipsoid, oblong or clavate. The structure of spore wall is major criteria of great significance especially in taxonomy. Numbers of layers in spore wall, their lamination, ornamentation and other characteristic features have been extensively used for taxonomic considerations (Schenck *et al.*, 1984; Hall, 1984).

Glomus, the most common genus of VAM fungi, has over 50 species which form globose, ellipsoid or rather irregularly shaped spores that range from 20-400 µm. These spores are thick-walled (up to 30 µm), and hyaline, yellow, red-brown, brown or black. Spores are attached to a single subtending hypha (or to two, three or more in exceptional species), and produced in the soil near plant roots, at the soil surface, or occasionally in roots and they may form singly, in groups of a few to many, or in large aggregates called sporocarps.

Sporocarps are most commonly found in undisturbed forest communities with perrennial plants and accumulated organic material. Sporocarpic species, when grown on plants in pot culture, usually produce spores singly or in small aggregates. Some *Glomus* species (e.g. *Glomus lacteum*) are not known to form spores in sporocarps, while others form spores outside the plant root only infrequently (e.g. *Glomus intraradices*). The spores of most species germinate by emergence of a germ tube directly through the subtending hypha.

Sclerocystis species produce chlamydospores that are very similar to those of *Glomus*, except that they tend to clavate rather than globose. All *Sclerocystis* species form spores only in sporocarps up to 700 µm in diameter. The spores, which can be up to 125 µm long developing in a single layer, radially arranged around a central plexus of sterile and sporogenous hyphae. There is no published information on the germination of *Sclerocystis* spores.

Spores of the genus *Acaulospora* appear to form laterally on the neck of a small thin-walled saccule, called variously as mother spore, a vesicle, a hyphal terminus, a sporiferous saccule, or a sporogenous saccule (Walker, 1987). When mature, the spore does not display subtending hyphae because the vesicular structure that gives rise to it loses its cytoplasm and collapses during development of the spore. The spores are globose or ellipsoid from <100 to >400 µm in diameter, and hyaline, yellow or reddish-brown. The surface of the spore wall may be ornamented with pits, projections of various shapes, folds, spines or reticulations and the wall is up to 12 µm thick. One species of *Acaulospora* is now known to produce sporocarps but in other species spores are found in the soil of the rhizosphere.

Entrophosphora is more or less similar to *Acaulospora* except that the spore is formed inside the parent hypha just below the vesicle.

Scutellospora spore can be greater than 600 µm in diameter, and the smallest ones are just under 200 µm in diameter. Spores develop terminally on a bulbous hyphal suspensor, which remains attached at maturity and may bear short lateral projections. Spores are usually globose but can be ovoid, obvoid, pyriform or irregular. Spores are hyaline, white, yellow, pinkish, grey-green or brown. Spore wall structure is composed of at least two wall groups, with one or more flexible membranous or coriaceous walls in the inner group or groups. The genus is characterized by its mode of spore germination. Germination is by means of one or more germ tubes produced near the spore base from a germination shield formed within or upon a flexible inner wall. Thin walled, knobby or broadly papillate auxillary cells (vesicles) are borne in soil on straight or coiled hyphae formed singly or in clusters.

Gigaspora produces spores, which are superficially similar to those of *Scutellospora* spores. The striking difference is the germination of spores, which is by direct growth of one or more germ tubes through the spores wall.

B. VAM: DISTRIBUTION

Majority of land plants including cereals, legumes, millets, plantation and horticultural crops, ornamentals and medicinal plants and forest trees are reported to be hosts of VAM fungi (Chapin, 1980; Barea and Azcon-Aguilar, 1983). Plants of nearly all the important families of Angiosperms including Poaceae, Leguminaceae, Chenopodiaceae, Rosaceae,

Rutaceae, Solanaceae, Plantaginaceae and Liliaceae have been shown to be mycorrhizal. Their wide distribution suggests that they have played a critical role in the evolution of land plants (Malloch *et al*., 1980). In addition to their distribution throughout the plant kingdom i.e. Angiosperms, Gymnosperms, Pteridophytes and Bryophytes (Kendrick and Berch, 1985) they are also geographically universal, occurring in plants of arctic, temperate and tropical regions (Mosse, 1981). They have been reported from natural ecosystems (Chang, 1994), woody plant communities at low altitudes (Read, 1991), frozen soil (Addy *et al*., 1994) and sand dunes (Logan *et al.,* 1989). VAM fungi have been recovered from different soil sites (Land and Schonbeck, 1991) and soil types (Antibus and Lesica, 1985) and soil depths (Abbott and Robson, 1991), even upto a depth of 120 cm (Zajicek *et al*., 1987). They have been observed at unfavourable location like acid lateritic belt with high amount of iron and aluminium (Bagyaraj and Varma, 1987) andosol soil (Ahiabor and Hirata, 1994), sand stone soil (Bellgard, 1993), mine soils (Diaz and Honrubia, 1993), effluent irrigated fields (Kothari, 1992) wastes of zinc refinery (Griffioen, 1994), saline soils (Juniper and Abbott, 1993), alkaline and calcarious soils (Singh, 1993) and coal wastes (Daft and Hacskaylo, 1976). Occurrence of VAM fungi has been reported also in marshy (Rozema *et al.,* 1986) and forest ecosystems (Sharma *et al*., 1987; Fahey, 1992; Raman and Gopinathan, 1992). The abundance and biodiversity of VAM fungi varies with climatic and edaphic factors as well as land use (Giovannetti and Pearson, 1994) and seasons (Lopez and Honrubia, 1992). The seasonal variation in abundance of VAM fungi is reported to be influenced by soil temperature (Raju *et al.,* 1990), soil pH (Medeiros et al., 1994), soil moisture (Safir *et al.,* 1990), soil nutrients (Howeler *et al.,* 1987), phosphorus levels (Li *et al.,* 1991) and nitrogen availability (Davis and Young, 1985, Mosse, 1990). Presence of micronutrients in soil also influences the occurrence and association of VAM fungi (Sreenivasa and Bagyaraj, 1988c). Use of pesticides, herbicides, insecticides and other chemicals and their presence in soil also affects the distribution of VAM fungi (Trappe *et al.,* 1984; Nemec, 1987). Cultural practices like tillage (Schenck and Sequeira, 1987; Jasper *et al.,* 1989), crop rotation (Thompson, 1987), fallow period (Thompson, 1987) and monoculture (Kruckelmann, 1975; Sieverding and Leihner, 1984) also have an important role in the distribution of VAM fungi. Type of vegetation has been shown to exert a selective effect on the abundance and distribution of VAM fungi (Schenck and Kinloch, 1980).

Presence and absence of host plant is of great importance in colonization and sporulation of VAM fungi, possibly because of toxic non-host root exudates (Hayman *et al.,* 1975; Morley and Mosse, 1976). Although VAM fungi have extremely wide host ranges and do not show host specificity, existence of host preference has been reported by many workers (Mosse, 1975; Sanders *et al*., 1975; Schenck and Kinlock 1980). The preferential association between host and symbiont provides highest root colonization, multiplication and growth stimulation.

C. VAM: BENEFITS TO HOST

Plant growth is directly influenced by the availability of nutrients in soil and their absorption through the root system which in turn depends on the concentration of nutrients in the soil and their transport to absorbing sites, either by mass flow or by diffusion (Tinker, 1980;

Chapin, 1980). Because of low concentration and slow diffusion of certain nutrients a depletion zone develops around roots. Extramatrical hyphae of vesicular arbuscular mycorrhizae with their capacity of absorption and transport nutrients to cortical cells improve the nutrient uptake of the root system (Barea *et al.*, 1983; Harley and Smith, 1983). As the extramatrical hyphae extend several centimeters in the soil, they bypass the depletion zone surrounding the roots. They overcome the limitations on the absorption of slow diffusing ions, therefore, they have a great functional importance in nutrient acquisition by the roots (Smith and Gianinazzi-Pearson, 1988). It has been emphasized that VAM fungi complement root system especially when it is less developed or when the environment is stressed, nutrient-poor or competitive (Mosse *et al.,* 1981; Harley and Smith, 1983; Bolan and Robson, 1987).

Improved supply of nutrients

It is well established by Barea and Azcon-Aguilar (1983), Smith and Gianinazzi-Pearson (1988) Gianinazzi *et al.* (1995) and others that VAM fungi function mainly in regulating phosphate absorption, translocation and transfer to host plants. Phosphorus plays a key role in all biological system and is required in large amount by plants. Under normal conditions, concentration of available phosphorus in the soil is very low i.e. only 1.5% of total phosphorus, which is in the form of organic or inorganic compounds (Hayman, 1975). Bolan and Robson (1987) suggested that increased uptake of phosphorus by mycorrhizal plants is due to the production of organic acids by VAM resulting in better availability of available phosphorus.

Improved uptake of phosphorus by mycorrhizal plants is attributed to the exploration of hyphae beyond "P" depletion zone (Cooper, 1984; Koide and Li, 1990; Li *et al.* 1991; Beamer, 1992; CuiMuyi and Nobel Park, 1992). It has been suggested that VAM fungi are able to take up phosphate more efficiently than the roots as well as they take it faster (Harley and Smith, 1983; Gildon and Tinker, 1983; Bolan *et al.,* 1984). Changes in the root-exudation pattern resulting in alteration in the pH of rhizosphere seem to be indirectly associated with the increase in phosphate uptake (Bolan *et al.,* 1984). Proton ATPase system located on plasma membrane of VAM fungi has also been associated with increased uptake of phosphorus (Clarkson, 1985; Smith and Smith, 1986).

Phosphate absorbed by extramatrical hyphae is accumulated in vacuoles in the form of polyphosphate granules (Callow *et al.*, 1978; Cox *et al.*, 1980). It is transported to arbuscules by cytoplasmic streaming (Smith and Gianinazzi-Pearson, 1988). Exchange of carbohydrates and nutrients along with "P" between symbiont and host is considered to be through arbuscular "interface" (Smith, 1993); phosphorus is effluxed from the arbuscular interface and absorbed by the plasma membrane. Carbohydrate from host is transported in opposite direction across the same interface (Gianinazzi-Pearson *et al.*, 1991).

VAM fungi have been reported to increase nitrogen content in plants through an improvement of the biological nitrogen fixation (Hayman, 1984; Barea *et al.*, 1988). This may be due to the fact that nitrogen fixation depends on steady and adequate supply of phosphate to the roots and nodules (Danso, 1988; Barea *et al.*, 1987, 1988). It has been suggested that the uptake of nitrogen compounds from soil and their transfer to the plants takes place with the help of appropriate enzymes (Smith *et al.*, 1985; 1986; Barea *et al.*,

1987). Using ^{15}N labelled fertilizer it has been confirmed that VAM fungi take up nitrogen from soil and transport it to roots (Johansen *et al.*, 1992; Tobar *et al.*, 1994). VAM fungi have been found to indirectly increase nitrate reductase, which is important for nitrate assimilation in plants (Smith and Gianinazzi-Pearson, 1988). Their ability to accumulate ammonium in alkaline soil and derive nitrogen from sources that are less available to non-mycorrhizal plants has been reported (Stevenson, 1986). In neutral and acid soils, where ammonium nitrogen predominates, VAM fungi have been shown to help in the uptake and transport of nitrogen (Harley and Smith, 1983). It is hypothesized that VAM may improve nitrogen economy of non-legumes through the hyphae linking legumes and non-legumes growing together (Newman, 1985; Barea, 1988).

VAM fungi have been reported to play a key role in improvement of uptake of nutrients other than phosphorus and nitrogen by altering acquisition mode of the absorbing system (Rhodes and Gerdemann, 1980; Gildon and Tinker, 1983; Harley and Smith, 1983). They have been shown to be involved in the uptake of Zn, Cu and other trace elements having low mobility in soil. They have also been shown to increase Fe (Rai, 1988) and sulphate uptake (Rhodes and Gerdemann, 1980; Harley and Smith, 1983). Uptake of Br, Cl, and others has been shown to be directly influenced by VAM fungi (Buwalda *et al.*, 1982).

Improvement of Water Supply

VAM plant interaction has been reported to increase absorption of water (Safir *et al.*, 1972), root hydraulic conductivity (Levy *et al.*, 1983) and plant water dynamics (Hardie, 1985). It has been proposed that in phosphate deficient situations, the mycorrhizal plants exhibit lower resistance to the flow of water and improve its absorption (Cooper, 1984). It has been suggested that extramatrical hyphae bypass the dry zones surrounding the root during the drought period and maintain a water continuity across the soil-root interface (Hardie, 1985). However, the suggestion is not in conformity with the reports of Sanders and Tinker (1973); Safir and Nelson (1981); Allen (1982) and Cooper (1984). Absorption of water under the influence of VAM fungi continues to be a matter of controversy, however, it is most likely that by maintaining the uptake of slowly diffusing nutrients, the VAM fungi help the host plant to cope with drought stress (Azcon *et al.*, 1988).

Improvement in Hormone Level

VAM fungi in symbiotic association with plants affect the hormone level; increase the level of cytokinin (Allen *et al.*, 1980) and alter level of gibberellins and abscisic acid (Allen *et al.*, 1982). They also affect the growth regulating substances, which are useful not only in root development and many basic processes of plant growth but also in mobilization of nutrients and their translocation (Letham *et al.*, 1978; Dixon, 1990). The possible mechanism for the production of growth regulators by VAM symbiont or its host plant has not been investigated in detail, however, the production is possibly related to the physico-chemical changes in the host plant brought about by symbiont (Smith and Gianinazzi-Pearson, 1988).

Improvement in Disease Resistance

VAM association with plants has been shown to provide them protection from pathogens (Zambolim, 1986). Many evidences, which show that plants inoculated with VAM symbiont exhibit an increased resistance to fungal root diseases, are available in literature (Schenck. and Kellam, 1978 Jalali and Thareja, 1981; Roseldahl, 1985). Protection from *Olpidium brassicae* in lettuce and tobacco (Schoenbeck and Dehne, 1979), from *Thielaviopsis basicola* in citrus (Davis *et al*., 1978), from *Pyrenochaeta terrestris* in onion (Becker, 1976), from *Fusarium oxysporum* f. sp. *radicis lycopersici* in tomato (Caron *et al*., 1986), from *Macrophomina phaseolina* in many crops (Ramaraj *et al*., 1988; Jalali *et al*., 1990; Kehri and Chandra, 1989a) has been reported. Activities of other pathogens including nematodes have been shown to be suppressed by VAM fungi (Oliveira and Zambolim, 1986; Cooper and Grandison, 1987). Various morphological, physiological and biochemical mechanisms have been proposed to explain the resistance of mycorrhizal plants against pathogens, however, it is generally believed that the protection of host plant is due to accummulation of some metabolites, lignification of root cells and high chitinolytic activities in them or by siderophore production (Dehne and Schoenbeck, 1979; Spanu *et al*., 1989; Hasselwandter, 1995).

D. VAM: STRESSED ENVIRONMENT/ WASTELANDS

Mycorrhizal association in plants has been reported to cope with various stresses posed by physical and chemical environment of soil like nutrient deficiency (Chapin, 1980). In infertile habitats like eroded and degraded soils, the mycorrhizae confer a greater root longevity by absorbing nutrients beyond depletion zone. In some soils, they play a direct role in soil aggregation (Thomas *et al*., 1986). The role of VA Mycorrhizae in land rehabilitation is well established through the report of Gardner (1986), Skujins and Allen (1986), Allen *et al*., (1987), Allen and Allen (1988), Habte *et al*., (1988), Sylvia and Will (1988), White *et al*., (1989), Graf and Gerber (1997).

Salinity problem in arid and semiarid conditions is caused due to nutritional imbalances e.g. excess of Cl^- can interfere with NO_3^- and PO_4^- uptake and that of Na^+ can inhibit Ca^{++} and Mg^{++} acquisition (Plaut and Grieve, 1988). VAM fungi have been reported to increase plant tolerance to salinity (Pfeiffer and Bloss, 1988; Dixon *et al*., 1993). It has been suggested that the tolerance of salinity in mycorrhizal plants is due to improvement of "P" uptake and increase of K^+ content in plants (Allen and Cunningham, 1983; Poss *et al*., 1985). Decrease in Cl^- concentratioin in mycorrhizal plants has also been associated with the tolerance of plants to the salinity (Hartmond *et al*., 1987), however, the mechanism is not clear.

Imbalances and disturbances in the micorbial ecosystem developed by edaphic environment in soils polluted with heavy metals, industrial effluents, acid rain and biocides affect the growth and establishment of plants. VAM fungi play an important role and help the plants to tolerate such situations (Dakessian *et al*., 1986; Mosse, 1986). They have been reported to develop strains capable of facing polluted soils (Mosse *et al*., 1981). They are reported to protect the plants from toxicity as they accumulate the toxic substances and check their translocation to plants (Schuepp *et al*., 1987). They are reported to increase tolerance of plants to zinc and other metals (Dueck *et al.,* 1986). Their cell wall with large number of sites for binding metals saves the host plants from toxicity of metals.

Role of mycorrhizal association in the management of sand dunes and desert through an establishment of a plant cover is well recognized. Many reports regarding distribution and occurrence of VAM fungi in sand dunes and deserts are available (Ernst *et al.*, 1984; Bloss, 1985; Puppi and Riess, 1987; Logan *et al.,* 1989). In these locations, VAM fungi not only enhance the absorption of nutrients but also improve water relations of plants and physical structure of soil (Logan *et al.*, 1989).

Difficulty in developing vegetation over mining wastes may be reduced with the help of VAM fungi. Their presence has been recorded from a variety of wastelands developed due to mining (Daft and Hacskaylo, 1976; Jasper *et al.*, 1987; Lambert and Cole Jr., 1980; Mishra, 1994; Diaz and Honrubia, 1993; Helm and Carling, 1993). VAM fungi have been reported to help in the establishment of plant cover on mining wastes through their favourable "P" support to plants in photosynthesis, hormone regulation and water economy (Allen, 1984). VAM fungi enable their hosts to face other stresses like unfavourable pH and temperature of soil. They have been reported to be helpful in reclamation of alkaline soil through the production of organic acids (Medeiros *et al.*, 1994). Their role in moderating the adverse effect of higher temperature of soil has also been reported (Borges and Chaney, 1989; Baon *et al.*, 1994). It has been postulated that increased plant survival at high temperature is due to VAM modified pattern of root exudation (Bansal and Mukerji, 1994).

It is now well documented that vesicular arbuscular mycorrhizal fungi function and interact in symbiotic association for the benefit of the host in order to have long company. They transform and modify environment in harsh conditions. They are reported to play key roles in mineral cycling, energy flow and plant succession in disturbed ecosystems (Stahl *et al.*, 1988; Chang, 1994).

VAM fungi have been reported to play significant role in improving physical property of soil by binding soil into semistable aggregates making soil more porous (Thomas *et al.,* 1986; Logan *et al.,* 1989). Their role in alleviating the effect of soil compaction has been observed in sweet gum (*Liquidamaer styraciflua* L.) inoculated with *Glomus fasciculatum* (Simmons and Pope, 1988). Chemical property of soil is also found to be improved by influence of VAM fungi by production of metabolites particularly organic acids, which help in the process of solubilizing insoluble forms of phosphorus. There exists a positive correlation between VAM induced changes in root exudation and mycorrhizosphere mycoflora which is beneficial for the root development and chemical enrichment of soil (Linderman, 1988; Bansal and Mukerji, 1994). VAM fungi help the plants in the uptake of various nutrients (Rovira, 1983; Bagyaraj and Varma, 1987). They modify the pH of the soil to favourable scale for the plant growth (Medeiros *et al.,* 1994).

Role of VAM fungi in establishment of plant community in problematic soils i.e. alkaline, acidic alkali-saline, calarious, usar, wastelands of brick-mines and other degraded sites has been recognised by many workers (Daft and Hacskaylo, 1977; Chandra and Kehri, 1994).

Daft and Hacskaylo (1977) achieved success in planting trees inoculated with VAM in critical and degraded sites in United States. In degraded lands, presence of various types of undesirable and toxic substances restricts the development of vegetational cover. Presence of VAM propagules in such lands helps the re-establishment of plant cover (Jagpal and Mukerji, 1988; Bellgard, 1993). VAM inoculation has given positive results in "*Andosol soils*" of different fertilities (Ahiabor and Hirata, 1994). Bellgard *et al*., (1994) have reported potential of VAM in the re-establishment of positive plant community. Revegetation of wasteland developed due to mine spoils in Spain with the help of VAM has also been achieved (Diaz and Honrubia, 1993). The importance of VAM fungi in raising *Populus* in wasteland developed due to effluents (Kothari, 1992) has been reported. Likewise, raising of onion and bell pepper (Hirrel and Gerdeman 1980) *Distichlis spicata* (Allen and Cunningham, 1983) and *Parthenium argentatum* (Pfeiffer and Bloss, 1988) in saline soils has been emphasized by Dixon *et al*., (1993). Advantage of VAM fungi in reclamation of saline soils has been attributed to better transport of phosphorus than sodium ions to host plants (Pfeiffer and Bloss, 1988; Smith and Smith, 1990). It has also been attributed to increased tolerance of the plants to fluctuations in temperature and pH (Borges and Chaney, 1989; Medeiros *et al.,* 1994).

Mycorrhizal association has been shown to play important role in the establishment of green cover over desert, semi-desert as well as sandy and silica sites developed due to mining activities (Bloss, 1985; Puppi and Reiss, 1987; Koske, 1988; Logan, *et al.,* 1989). Williams and Allen (1984) have emphasized the importance of VA Mycorrhiza in reclamation of arid and semi-arid lands. Inoculation of VAM fungi has been reported to improve establishment of desert succulents by improving their nutrient and water uptake (Cui-Muyi and Nobel Part, 1992). Attempt to recover desertified mediterranean ecosystem by plantation of woody legumes with selected arbuscular mycorrhizal fungi and rhizobia has been successful (Herrera *et al*., 1993). Role of VAM in such areas has been attributed to improved flow through internal and external hyphae as well as changes in the biochemistry of host plants (Hirrel and Gerdemann, 1980). It has also been attributed to a secondary response due to favourable nutrition or physiological alteration in the host (Allen *et al*., 1981; Powell and Bagyaraj, 1984).

Numerous attempts have been made to explore the possibility of using VAM fungi in reclaiming polluted sites created by human activities. Inoculation of VAM fungi helped the seedlings of red maple and leucerne to establish satisfactorily on coal wastes (Daft ad Hacskaylo, 1976, 1977). Application of VAM fungi has been shown to help the revegetation of brick-kiln abandoned lands (Vilarino and Arines, 1992). It has also been shown to help the reclamation of mining sites (Jasper *et al*., 1987; Diaz and Honrubia, 1993). Strategies for the reclamation of wastelands created due to iron mining through the application of selected microbes including VAM fungi has been developed (Mishra, 1994). Use of VAM fungi in handling the problem of metallic water has been highlighted by many investigators including Dueck *et al*., 1986; Schuepp *et al*., 1987; Dehne and Schuepp, 1990; Galli *et al.,* 1993 and Griffioen, 1994.

E. VAM FUNGI: INTERACTIONS WITH OTHER MICROORGANISMS

Now it is well documented that a key determinant for the ability of a root system to acquire nutrients from the soil is the extent to which it is symbiotically colonized by appropriate VAM fungi (Mosse, 1973; Harley and Smith, 1983). The formation of symbiosis in turn is affected by the levels of soil fertility (Hayman, 1982; Gerschefske *et al.*, 1988) and microbial as well as biochemical aspects of the rhizosphere (Mosse *et al.*, 1981). Soil microorganisms play a vital role in the rhizosphere region of the plant. Their presence is stimulated by organic substances and exudates released by the roots and decaying organic matter (Rovira, 1979; Tinker, 1980). In turn they increase availability of plant nutrients, improve their uptake, improve the production of growth regulators and enable the plants to protect them against root pathogens (Brockwell, 1980; Moore and Cooksey, 1981).

VAM fungi have been reported to play a vital role along with other soil microbes for an over-all development of the host plants (Tinker, 1982). A positive correlation between the changes in the pattern of root exudation and mycoflora of rhizosphere caused by them have been reported (Bansal and Mukerji, 1994).

Interaction with rhizosphere microflora

Concept of **"Mycorrhizosphere"** is the recent one. It hypothesised that mycorrhizae exert a strong influence on the microflora in the rhizosphere. Mycorrhizal fungi might alter the physiology and exudation of the root resulting in a change in composition of microbial populations that depend on roots for a source of energy. External hyphae of mycorrhizal fungi might provide a physical or nutritional substrate for bacteria. Many of the papers on interaction with specific microbial groups, such as free-living nitrogen fixers, provide evidence for these hypotheses.

In rhizosphere analysis of mycorrhizal and non-mycorrhizal tomato plants Bagyaraj and Menge (1978) showed that greater populations of bacteria and actinomycetes occurred in the mycorrhizosphere compared to non-inoculated control. A number of other authors suggested that the species composition of the rhizosphere microflora may be altered by the formation of mycorrhizae (Newman *et al.*, 1981; Ames *et al.*, 1984). This hypothesis was supported by Meyer and Linderman (1986) and they have demonstrated the differences in populations of taxonomic and functional groups of bacteria associated with VAM and non-VAM plants. Population of total bacteria was higher in the rhizoplane of mycorrhizal sweet corn and clover, compared to non-mycorrhizal plants. Chitinase-producers, *Streptomyces* spp. and fluorescent pseudomonads were more numerous in the rhizosphere of non-mycorrhizal plants. The authors concluded that the microbial equilibrium was altered by the formation of VA mycorrhizae. The realization that VAM fungi can alter the microbial makeup of the rhizosphere has led to the increased scrutiny of control treatments in mycorrhizal experiments. Studies by Ames *et al.*, (1987); Ferrera-Cerrato and Perez-Moreno (1994); Secilia and Bagyaraj, (1987); Schisler and Linderman, (1987); Forster and Nicolson, (1981) etc. showed clearly that the microbial polulations in the rhizosphere are influenced by mycorrhizal fungi. The effects vary with time, the speices of mycorrhizal fungus, the host and other environmental parameters. Paulitz and Linderman (1991) emphasized on the need for the measurement of specific functional groups that break down or produce a certain substance which might give a clearer picture, rather than measuring total numbers of bacteria or actinomycetes. The active and dynamic population of rhizosphere microflora affects the spore germination, growth of mycelia through soil, stimulation and attachment of infective hyphae to the root, penetration of the root etc., thereby affecting the formation of the symbiosis either by stimulating or inhibiting the formation of mycorrhizae.

All these effects must be studied in detail in order to increase the predictability of inoculations with mycorrhizal fungi to increase their consistency of performance and benefit to plant growth and development.

Interaction with symbiotic nitrogen fixers

Rhizobium - Most leguminous plants are simultaneously symbiotic with nodule-forming *Rhizobium* which fixes nitrogen, and benefits plant growth, and the mycorrhizal fungi which benefit the host by increasing the efficiency of uptake of mineral elements and water from the soil as well as altering host metabolism and other physiological parameters. However, the tripartite relationship of host-*Rhizobium*-VAM represents a new dimension unlike dipartite symbiosis. A wide range of leguminous host species including those used for forage, as human food sources and some woody perennial tree species have been extensively studied to determine the nature of the relationship. The subject has been reviewed by Tilak (1985a), Barea and Azcon-Aguilar (1985) and Cluett and Boucher (1983).

When legumes are symbiotic with both *Rhizobium* and VA mycorrhizal fungi, plant growth is generally much greater than control plants symbiotic with either alone. Many reports (Manjunath and Bagyaraj, 1984; Purcino *et al.*, 1986; Bethlenfalvay and Yoder, 1981) indicate that the VAM increased P uptake, which under P stress conditions would intself increase plant growth. Improved P nutrition in turn would favour the nitrogen fixation process by *Rhizobium*. The combination of effects resulted in further growth enhancement (Manjunath *et al.*, 1984; Subba Rao *et al.*, 1986; Azcon-Aguillar and Barea, 1981; Redente & Reeves, 1981; Kawani and Yamamoto, 1986; Ganny *et al.*, 1982). However, factors that influence each symbiont separately may also affect the tripartite relationship. All strains of *Rhizobium* or strains or species of VA mycorrhizal fungi do not affect their host plant in the same manner or to the same degree. Different edaphic factors influence the behaviour and effectiveness of the symbionts, so would affect the combination as well (Hicks & Loynachan, 1987, Bethenfalvay *et al.*, 1987).

Their interaction with *Rhizobium* has been shown to increase nodulation and nitrogen fixing ability of legumes by providing phosphate, trace elements and plant hormones required for nodulation and N_2 fixation (Hayman and Mosse, 1978, 1979; Bagyaraj and Munjunath, 1980; Munns and Mosse, 1980).

Frankia - Actinomycete endophytes in root nodules of certain shrubs and trees can fix nitrogen *in situ*. These actinomycetes, which are in the genus *Frankia*, form symbiotic associations with 15 genera, such as *Alnus, Cerocarpus, Purshia, Ceanothus* etc. (Rose, 1980). These host plants usually colonize nutrient-poor habitats such as forests, sand dunes, and arid soils. Relatively little is known about the ecology and survival of *Frankia* in soil and even less is known about mycorrhizal-antinorrhizal interactions.

The tripartite association between actinomycetes and VAM fungi was examined by Rose and Youngberg (1981) in *Ceanothus*, Gardner *et al.*, (1984) in *Hippophae rhamnoides*. They observed that plants with both actinomycete nodules and VA mycorrhizae had greater plant dry weights, number of nodules, nodule weight and more nitrogenase activity than plants that were inoculated with the actinomycete only *(Glomus gerdemanni)* for *Ceanothus* and *G. fasciculatum* for *H. rhamnoides*).

Interaction of VAM fungi with beneficial actinomycetes has also been reported (Bagyaraj and Menge, 1978; Krishna *et al*., 1982). *Frankia* and *Streptomyces* in association with VAM fungi have been reported to improve nitrogen supply and growth in certain non-leguminous plants (Bond and Wheeler, 1980; Rose and Youngberg, 1981).

Interaction with free-living nitrogen fixers

Although symbiotic nitrogen-fixing bacteria account for a majority of the biologically fixed nitrogen, some free-living bacteria also can fix nitrogen. Free-living diazotrophic species are found in the genera *Azotobacter, Beijerinckia, Clostridium, Pseudomonas* and *Azospirillum*. The ecology of non-symbiotic dinitrogen is reviewed by Subba Rao *et al.*, (1981).

The work done by Tilak (1985b), Subba Rao *et al.*, (1985), Tilak and Singh (1988), Barea *et al*., 1983), Pacovsky (1988), Mohandas (1987), Li and Castellano (1984, 1987), etc. suggest a close spatial association between mycorrhizal fungi and free-living diazotrophic bacteria. These bacteria might utilize P from mycorrizal fungi, or mycorrhizal fungi might transport fixed nitrogen from the bacteria to the plant. Both possibilities should be investigated. *Azospirillum brasiliense* has been isolated not only from the rhizosphere, but also from the root cortex (endorhizosphere). This opens up the possibility of a direct interaction between a VAM fungus and *Azospirillum* within the plant. Interactions might also occur via the production of plant hormones by free-living nitrogen fixers. Tien *et al.*, (1979) suggested that plant hormone production by *A. brasiliense* might account for the increased growth of inoculated pearl millet. *Azotobacter vinelandii* and *A. beijerinckii* stimulated plant growth and VAM infection of lavender, tomato and alfalfa inoculated with *G. mosseae* produced auxins, gibberellins and cytokinins in culture and cell-free supernatants and whole cells of *A. vinelandii* and *A. beijerinckii* (Azcon and Barea, 1975; Azcon *et al*., 1978). These same effects were mimicked by the addition of a combination of gibberellic acid, kinetin and IAA. The involvement of plant hormones, especially cytokinins, in mycorrhizae has been suggested by many workers (Allen *et al*., 1980; Ho 1987; Mudge, 1987). Production of cytokinins by VA mycorrhizal fungi has also been documented (Ng *et al.*, 1982; Barea and Azcon-Aguilar, 1982). The stimulation of VAM infection and plant growth by some free-living nitrogen fixers could be due to hormonal interaction, but more work is needed to definitively prove this hypothesis.

VAM fungi have been reported to interact with free-living nitrogen fixing bacteria *viz*., *Azotobacter, Beijerinkia, Derxia, Clostridium* and *Azospirillum* etc. in a synergistic manner in relation to the growth of plants (Ho and Trappe, 1979; Brown and Carr, 1980). They have been reported to show a positive relationship with free-living nitrogen fixing bacterium.

Interaction with plant growth promoting bacteria (PGPR)

Plant growth promoting rhizobacteria (PGPR) are rhizosphere-inhabiting bacteria that have a beneficial effect on plant growth. This subject has been reviewed by Burr and Caesar (1984). This discovery, that specific strains of the *Pseudomonas fluorescens-putida* group could rapidly colonize potato roots and produced significant yield increases has stimulated research in this area. Exactly how beneficial rhizobacteria promote plant growth is not well understood, but they may produce photohormones or other growth-enhancing compounds that directly affect the plant. These rhizobacteria may also interact with mycorrhizal fungi to produce the growth-enhancing effects by as yet unknown mechanism.

Interaction with phosphate-solubilizing bacteria

Many bacteria produce organic acids and solubilize inorganic and organic forms of phosphorus that are unavailable to plants. VAM fungi, on the other hand, do not take up insoluble P, but more efficiently utilize labile forms. Some studies question this **"Central dogma"** of VAM fungi and show that differences in the availability of soil P to mycorrhizal and non-mycorrhizal plants are not reflected by differences in 32 P labelling. Neverthless, VAM fungi do not seem to take up much P from insoluble sources.

Barea *et al.*, (1975) suggested the possibility of an interaction between mycorrhizal fungi and phosphate-solubilizing bacteria (PSB). They observed a synergistic increase in dry weight and P uptake in maize grown in a low P soil and dually inoculated with *Endogone* and P solubilizing *Pseudomonas* sp. and *Agrobacterium.* Similar results were reported by Delorenzini *et al.,* (1979), Krone *et al.*, (1987), Raj *et al.* (1981), Azcon-Aguilar *et al.,* (1986), Lee and Bagyaraj (1986).

According to some scientists the interactions observed might be due to the production of plant hormones or vitamins by PSB, rather than to P solubization. Barea *et al.,* (1976) demonstrated that PSB produced IAA, gibberellins and cytokinins. Baya *et al.,* (1981) have shown the production of vitamins.

A positive relationship between VA mycorrhizal fungi, *Azotobacter* and phosphate solubilizing fungus, *Aspergillus niger* in triple inoculation was reported by Manjunath *et al.*, (1981) and Kehri *et al.*, (1995). It has been speculated that in such interactions hormones produced by the microbes also exert a positive effect on plant growth (Azcon-Aguilar and Barea, 1978; Mosse *et al.*, 1981).

F. VAM: INOCULUM PRODUCTION

Vesicular-arbuscular mycorrhizal (VAM) fungi behave like obligate symbionts and must be grown in the presence of a living host plant. The potential utilization of VA mycorrhiza in plant production depends on a suitable inoculum, which can be produced in a simple way and be spread on agricultural land with traditional equipment.

Soil-based inoculum production

The most common medium used for this purpose is sterilized soil inoculated with pure stock culture of a particular VAM isolate. The inoculum so produced consists of a mixture of soil, spores, pieces of hyphae and infected root pieces. It generally takes 12-16 weeks for its production and the final product is crude, bulky and heavy and hence, cumbersome to maintain as well as expensive to transport to sites of application. Anywhere from 2,000 to 8,000 kg of the crude inoculum may be needed per hactare, depending on the rate of application.

Gradually, soil was replaced by inert substrates such as vermiculite, perlite, sand or a mixture of these with only a small proportion of soil.

Inoculum production in soil free substrate

A new phase in inoculum production began with the use of calcined montmorillonite clay as a supporting medium (Plenchette *et al.*, 1982). The material marked under the name *Turface* used in conjugation with complete Long Ashton mineral solution (Hewitt, 1966), proved a good substrate for plant growth and abundant mycorrhizal infection eliminating the need for soil in the growth medium.

This was followed by a report of the use of light, expanded clay aggregates called as *Blahton* (Dehne and Backhaus, 1986). Plants were fed with a complete nutrient solution containing 800 mg/l N, 800 mg/l P and 600 mg/l K_2O applied twice weekly. After three months growth of host plants and root inoculum the Blahton particles were strongly colonized by VAM fungi. After air drying and storage at room temperature for one year, they retained 75-100% infectivity. Within three months a production area of 1 m^2, produced enough inoculum for 15-20 m^3 of horticultural compost.

Dehne and Backhaus (1986) have described the production of VAM inoculum in a hydroponic culture system using light expanded clay as a carrier material to solve the problem of inoculum production.

Vestburg and Sukainen (1992) of Finland have described a new method for producing VA Mycorrhizal inoculum in a soil free substrate. They have experimented on the potential use of the polymeric hydrogel product "water works" manufactured in the USA as a soil free substrate for producing inoculum of VA Mycorrhizae for use in micropropagated strawberry. The authors used vermiculite as a substrate together with the dehydrated polymer crystals. A *Glomus* sp. from Finland (probably *G. intraradix*) was used as inoculant. They concluded that properties of polymeric hydrogel greatly stimulate sporulation of *G. intraradix* inside the gelmatrix.

Inoculum production in nutrient flow culture

In nutrient film technique (NFT) roots grow in a shallow layer of a weak nutrient solution that is constantly and rapidly recycled. By careful manipulation of nutrient concentrations and use of rhizobium nodules to supply nitrogen requirements, Mosse and Thompson (1984) obtained root mats of beans (*Phaseolus vulgaris*) containing up to 60% mycorrhizal roots with attached floating films of mycelium, spores and sporocarps.

Root based inoculum production

Efforts to develop efficient procedures for the production, preservation and application of infected root inocula are also going on.

Inocula grown axenically with root organ culture

Typical VAM infections were produced in root organ cultures of clover roots growing in agar in a modified White's (1954) tissue culture medium containing 20-g/l sucrose and potassium phosphate or calcium phosphate instead of sodium phosphate (Mosse and Hepper, 1975). Because the medium had to be adjusted to the tolerance limit for fungal infection, root growth remained relatively restricted. There was no suggestion that this development might lead to large-scale inoculum production. Such a possibility came nearer when RiT-DNA transformed roots of *Convovulus sepium* were used (Mugnier and Mosse, 1987). Such roots, containing root inducing plasmid genes transferred from *Agrobacterium rhizogenes* have a greatly increased growth potential. The culture system consisted of two compartments, one containing the roots growing in a modified Murashige and Skoog (1962) nutrient medium and spreading into an adjacent compartment containing water agar and neutralized peat particles. In this compartment VA infections developed from pre-germinated spores were placed near the roots. Five phases of the infection process were identified and found to be controlled by different nutrient requirements. Although infection spread throughout the water agar compartment and new spores were formed, the need for two separate compartments made a practical exploitation for inoculum production difficult to envisage. This has become a reality with the formulation of a minimal medium (Becard and Fortin, 1988) containing (in mg/l) $MgSO_4.7H_2O$ 731, KNO_3 80, KH_2PO_4 4.8, $Ca(NO_3)_2$ $4H_2O$ 288, sucrose 10,000, NaFeDTA 8, plus trace elements and vitamins as in White's medium. In this medium RiT-DNA transformed carrot roots and VAM fungi can grow together and form abundant mycorrhizal infection. From 103 surface sterilized spores of *Gigaspora margarita* infection occurred within 5 days, and after 7 months up to 500 new, viable spores of *G. margarita* and up to 1000 spores of *G. intraradices* could be harvested from one Petri dish. Generally *G. margarita* is a relatively sparsely sporulating species.

Inoculum production using RiT-DNA transformed roots and expanded clay mineral

The last decade has seen two important new developments, namely the use of some type of expanded clay mineral as a supporting medium in whole plant inoculum production, and the use of RiT-DNA transformed roots grown in a minimal nutrient medium that allows abundant mycorrhizal infection while also supporting reasonable root growth. It remains to combine these two developments. If sterile expanded clay granules could be used instead of agar in the axenic cultures of RiT-DNA transformed roots, then it should be possible to produce a pure, concentrated inoculum with a high spore content and good storage potential after air drying. As the cultures would be aseptic there would probably be no need to remove the dry roots before storage. Such inocula could then be used directly in high value horticultural production. For large scale low-cost operations a two-phase system might be used in which the users themselves would undertake multiplication from a pure starter inoculum, thus minimizing transport costs and other disadvantages of a bulky inoculum.

2

VA MYCORRHIZA : DISTRIBUTION

VAM fungi are of worldwide distribution. They are present in the soil in form of chlamydospores, zygospores and azygospores and have been recovered from the soils of a variety of habitats e.g. agricultural fields, nutrient deficient soils, sand dunes and deserts, industrial wastes, sodic soils, polluted sites, sewage, eroded sites and others like forests, open woodlands, scrub, savana, heaths, grasslands, anthracite and bituminous coal wastes. There are several reports of VAM fungi from diverse habitats from India.

VA-mycorrhizal associations occur widely throughout the plant kingdom. They have been reported to be present in Bryophytes, Pteridophytes, Gymnosperms and Angiosperms in nearly all the geographical regions of the world. Most of the agricultural and horticultural crops belonging to diverse families have been shown to form VA-mycorrhizae. Many species of plants of economically important families have been reported to be mycorrhizal. Likewise, many of the plantation crops like coffee, tea, citrus, timber trees, many of the fruit trees and medicinal plants have been shown to form VA-mycorrhiza.

The purpose of the following description is to provide a current account of the knowledge of the distribution of VAM fungi in the soils and their association with diverse types of plants in different geographical and climatic regions of the country.

Agricultural Crops/ Fields

Godse *et al.* (1976) screened a number of standing crops of Karnataka e.g. capscicum, cowpea, *Eleusine coracana*, maize, *Pennisetum typhoides*, soybean and *P. purpureum* for the occurrence of VA mycorrhizas and indicated that *Endogone* is widespread in cultivated soils in the state.

Parvathi *et al.* (1984) examined 25 legumes from the University Campus of Nagarjuna University, Nagarjuna Nagar, out which 18 showed mycorrhizas. *Glomus mosseae* was present in all of them. The most heavily infected species were groundnut, pigeonpea, *Crotalaria juncea, Dolichos* species, *Phaseolus* species and *Rhynchosia minima.*

Ammani *et al.* (1985) described the occurrence and development of mycorrhizas in the upland rice cv. MTU-17 under field conditions. Infection patterns within the root tissue indicated that a *Glomus* sp. and *Gigaspora calospora* were involved.

Ammani *et al.* (1986) studied the VAM fungal sporocarps associated with *Pennisetum pedicillatum* and described and illustrated the characteristic features of *Sclerocystis sinuosa, S. claviceps, G. fasciculatum, G. macrocarpus* and *Glomus* sp.

Tilak and Murthy (1987) compared the association of VAM fungi with the roots of different cultivars of barley (*Hordeum vulgare*). VAM infection in 4 hulled cvs. was 40.2-

52% compared with 75.5-87.5% in 3 hulless barleys. The later had a high protein content but it was not known whether this was related to the greater mycorrhizal infection.

Vijayalakshmi and Rao (1988b) observed extensive mycorrhizal infection in sesamum growing in the fields. Plants were mycorrhizal even in the early stages. A progressive increase in infection was noted for 6 weeks followed by a slight decline. A gradual increase in vesicle count/1 cm segment occurred as plants aged. *Glomus, Gigaspora* and *Sclerocystis* spp. were found in the rhizosphere soil. Among the spore types recorded, chlamydospores of *Glomus* especially *G. mosseae* were predominant.

Ammani *et al.* (1988) studied the incidence of vesicular-arbuscular mycorrhizae in sugarcane. A survey of 4 sugarcane varieties, Co-7219, Co-6971, Co-6907 and Co-8001 from the fields of Krishna and Khammam districts showed variability in infection except Co-7219 in which the fungus was *Acaulospora scrobiculata*. There was a correlation between infection percentage and clay content of soil. Other edaphic factors including soil phosphorus did not have any influence on mycorrhizal colonization of roots.

Sulochana and Manoharachary (1989) studied the VAM associations in castor and safflower samples collected from Rajendranagar, Hyderabad, 30, 60 and 90 days after sowing and the percentage of VAM infection was calculated. The roots were shown to be mycorrhizal and the number of spores was directly correlated to host age. Castor was associated with 8 VAM fungi and safflower with 4. Both were associated with *Glomus fasciculatum, G. constrictum, G. mosseae* and *Acaulospora* sp. and castor also with *G. epigaeum, G. aggregatum* and 2 *Gigaspora* spp. This study confirms the formation of VAM in castor and safflower in semi-arid tropical soils.

Rao *et al.* (1990) made a survey of the VA-mycorrhizal association in non-cultivated (agro-forestry) and cultivated plants collected from different sites at and around Wagholi, Pune. The spore population of vesicular-arbuscular fungi varied with the plant species and soil profile. The plants belonging to the same genus or family differed in the intensity of mycorrhizal association in particular soil. Differences were also recorded in plants collected from different soils and studied in detail.

Rai (1990a) assessed the status of VA mycorrhizas in potato roots collected from Shimla, Himachal Pradesh and Shillong, Meghalaya. Dark septate hyphae of a dematiaceous nature were recorded but efforts to isolate and culture the hyphae were unsuccessful.

Vyas (1990) studied the occurrence and distribution of endomycorrhizal fungi in soybean and chickpea in a black soil. Results of the study indicated that VAM species such as *Glomus monosporum, G. fasciculatum, G. epigaeum, G. constrictum* and *Aculospora morrowae* were associated with soybean and chickpea. *Glomus monosporum* and *G. fasciculatum* occurred in 40% and 25% of samples and 50% and 25% of samples in soybean and chickpea, respectively. *Glomus epigaeum* occurred in 25% of samples in soybean and 30% of samples in chickpea. *Glomus mosseae* and *G. constrictum* was found to be associated with all cultivars of soybean except with those of JS-72-44 and MAC-58. *Acaulospora morrowae* was recorded in JS-72-44, PK-262 and PK-472 cultivars of soybean and in chickpea, it was recorded only in UJJAIN-21 and ICCC-4. All species of VAM were associated with soybean cultivar JS-72-44 and with UJJAIN-21 of chickpea.

Veeraswamy *et al.* (1990) studied the occurrence of endomycorrhizal fungi in rhizosphere soils of sorghum and foxtail millet. Both the crops developed extensive mycorrhizae. The rhizosphere soil contained many species of *Gigaspora* besides other species of VAM fungi. In all, nine species of *Gigaspora viz., G. aurigloba, G. calospora, G. coralloidea, G. decipiens, G. gilmorei, G. gigantea, G. heterogama, G. margarita, G. pellucida* were recovered. Also *Sclerocystis rubiformis* and few species of *Glomus* were found to occur consistently in these soils.

Padmavathi *et al.* (1990b) investigated the VAM fungi associated with foxtail millet-*Setaria indica*, one of the major rainfed crops being cultivated in the semi arid soils of Anantpur district in Andhra Pradesh. They reported varied population of VAM fungi from the root systems as well as the rhizosphere soil samples. Three species of *Sclerocystis viz., S. sinuosa, S. clavispora* and *S. pakistanica* were found to occur consistently at all the places. The species of *Glomus, G. deserticola, G. etunicatum, G. fecundisporum, G. geosporum, G. invermaium, G. intraradices* and *G. tortuosum* besides several unidentified isolates were most common in heavy black soils. *Gigaspora* and *Scutellospora* were observed along with *Glomus* in alluvial soils.

Janakirani and Manoharachary (1990) studied the occurrence and distribution of VAM fungi in safflower. Out of the five safflower cultivars tested CO 1 showed maximum response to VAM fungi followed by Sagarmutyal in pot culture experiments. The extent of VAM colonization was 96% on the 90th day after planting in CO 1 variety followed by Sagarmutyalu (75%). In view of the promising performance of CO 1 and Sagarmutyalu cultivars in relation to VAM colonization further experiments were conducted with these two cultivars under glass house conditions choosing the most dominant and commonly occurring VAM fungi namely *Glomus caledonicum* and *Glomus mosseae*. Of the two VAM fungi tested *G. caledonicum* has shown better colonization than *G. mosseae* under pot experiments. The dry weight of plant samples suggested that these VAM fungi were beneficial in improving plant growth. The addition of rock phosphate enhanced the VAM effectiveness by increasing plant height, number of leaves and dry weight. Increased levels of macro- and micro- nutrients were also observed in VAM inoculated plants over control.

Veeraswami *et al.* (1992) noted the widespread occurrence of VAM fungi with a predominance of *Glomus* species in black and red laterite soils, in millet fields in Andhra Pradesh. Species of *Glomus* identified included *G. aggregatum, G. deserticola, G. etunicatum, G. fecundisporum, G. geosporum, G. intraradices, G. invermatum, G. leptotichum* and *G. tortuosum*. Red soil contained more fungi than black soil.

Rani and Manoharachary (1994) surveyed 10 safflower fields for VAM fungi in Hyderabad. A total of 14 VAM fungi were identified and most of the propagules were found in soils with low P and neutral pH. They observed a significant correlation between VAM fungi and soil pH, moisture and P content. The most frequently identified VAM fungi were *Glomus* spp. (7 species) followed by *Scutellospora* and *Acaulospora*.

Edathil *et al.* (1994a) studied the VAM status of tomato, brinjal and chilli during the initial establishment period under natural field conditions and in pot culture using nonsterile soil. The soil had a low nutrient status. VAM colonization in roots was maximum at 45, 50 and 60 days after the germination of brinjal, tomato and chilli seeds respectively under field conditions and on the 60th day in pot culture.

Kehri and Chandra (1994d) studied the VA-mycorrhiza formation in crops in drier tracts of Bundelkhand region. A variety of crops *viz.*, paddy, wheat, mustard, linseed, pea, gram, fenugreek, onion, coriander, garlic, dill seed, chilli, tomato and potato were included in the survey. A number of fields raising these crops were marked out at different sites spread over the region. The data presented are based on the observations from 15 fields distributed at 5 sites (3 fields/site). The findings of the study showed that the soils of Bundelkhand region have no scarcity of VAM inoculum, however, the inoculum pool appears to be inappropriate for the crops raised in the region. Poor performance of the crops in the soils of this region is a clear indication of a lack of the efficient VA-mycorrhizal system. They concluded that there is a need to develop efficient mycorrhizal systems for the crops through introduction of effective strains of VAM fungi in the soils of Bundelkhand region.

Singh and Pandya (1995) studied the occurrence of VA mycorrhiza in pearl millet and other hosts. They collected the soil samples from three different soil types from different fields in the vicinity of J.N. Agriculture University, Gwalior. All the plants studied *viz.*, pearl millet, maize, wheat, pigeonpea, chickpea etc. showed the presence of VAM. However, the extent of colonization varied. A variety of spores were recovered from the rhizosphere soils. They mainly belonged to the genus *Glomus*. However, spores of *Sclerocystis* sp. and *Endogone* sp. were also recovered though very rarely.

Dalal and Hippalgaonkar (1995) studied the occurrence of VAM fungi in arable soils of Konkan and Solapur. The soils of Konkan region are red laterite (alfisols) with pH ranging from 5.01 to 6.4. The soils of Solapur region are black soils (vertisols), with pH ranging from 5.62 to 8.72. Both have low phosphorus and nitrogen content. The VAM spore density was also found to be less in vertisols than in alfisols. The predominant VAM species found in both regions were *Glomus fasciculatum, Glomus mosseae* and *Glomus macrocarpum*. Species of *Acaulospora, Gigaspora* and *Sclerocystis* were rare. They concluded that *Acaulospora* was more tolerant to acidic soils, while *Glomus* spp. were favoured by neutral to alkaline soils. They also reported the variation in spore density. It was more in June than October. Soil samples collected in summer season under water stress had high spore density.

Harikumar (1995) studied the VAM association in *Kaempferia galanga* L. in relation to soil type. Root and root region soils of *Kaempferia* (var. Vellanikkara) were coastal sandy, greyish *onattukara*, black *Kori* and Laterite. In general, the plants grown in sandy soils had a higher level of VAM infection as compared to laterite. The spore load in the soil samples from root region of plants grown in black *Kori*, greyish *onattukara*, coastal sandy and laterite soils were 140, 78, 192 and 118/50 cc soil respectively. These spores belonged to genera *Gigaspora* and *Glomus*, however, *Glomus* was reported to be abundantly found in all the 4 soil types studied.

Iqbal and Gautam (1996) reported status of VAM fungi of Aligarh District. Rhizosphere samples of soil and roots collected from agricultural crops in 10 localities showed *Glomus mosseae* and *G. constrictum* as the most abundant of the 6 VAM species recovered.

Wani and Konde (1996) studied the VAM association in garlic. All of 67 garlic root and rhizosphere soil samples collected from different areas in Mahatma Phule Krishi Vidyapeeth, Rahuri, Maharashtra, contained VAM fungi. Colonization ranged from 39 to 62% and chlamydospore counts ranged from 62 to 242/ 50 ml of rhizosphere soil. The VAM fungi observed were *Acaulospora, Gigaspora* and *Glomus*, with *Glomus* predominating. A positive significant correlation occurred between VAM colonization and chlamydospore count, and electrical conductivity of soil. A negative correlation occurred between organic C and available P_2O_5 in soil.

Bhat and Kaveriappa (1997) reported the occurrence of VAM fungi in the tubers and roots of *Colocasia esculenta* (L.) Schott. Infection of VAM was confined to areas of tubers from where roots had originated. Colonization was observed in the outermost parenchymatous tissues. The base of the roots also showed infection in the epidermal and cortical regions. Spores of *Acaulospora foveata, A. scrobiculata, Gigaspora giantea, Glomus caledonium, G. fasciculatum* and *G. monosporium* were isolated from rhizosphere soil.

Kunwar *et al.* (1999) reported occurrence and distribution of VAMF associated with garlic rhizosphere soil. The qualitative and quantitative composition of vesicular arbuscular mycorrhizal fungi (VAMF) associated with garlic in Andhra Pradesh and Uttar Pradesh was estimated. A total of 35 species of VAMF were isolated and identified, with species belonging to all 6 genera of VAMF observed. The species identified consisted of 14 species of *Glomus*, 11 species of *Acaulospora*, 4 species of *Gigaspora*, 3 species of *Scutellospora*, 2 species of *Sclerocystis* and 1 species of *Entrophospora*. The physico-chemical properties of garlic rhizosphere soils was also studied.

Reddy *et al.* (1999) studied the VAM association in different varieties of turmeric *(Curcuma longa*). The roots and rhizosphere soil samples of 12 varieties of turmeric (*Curcuma longa* L.) were screened for vesicular arbuscular mycorrhizal (VAM) fungi. Mycorrhizal infection was recorded in all the varieties. Variation in the percentage of infection, number of arbuscules, vesicles and spores was noted in different varieties. Spores of *Glomus, Gigaspora* and *Sclerocystis* were detected from the rhizosphere soil samples.

Vijayakumar *et al.* (2000) studied the VAM fungal association in *Lycopersicon esculentum* L. grown in semi-arid tropical soils of Puttaparthy, A.P. Soil samples were taken from sites growing tomatoes in Karnatakapalli, Kammavarapalli and Enumanepalli (Andhra Pradesh) in 1998 and 1999. Vesicular arbuscular mycorrhiza found in these red sandy loam soils were identified to species level. A total of 12 species belonging to the genera *Acaulospora* and *Glomus* were found. The percentage VAM root colonization of the tomatoes ranged from 75 to 90%.

Hasan (2002) conducted a study to examine the degree of mycorrhization of chickpea (*Cicer arietinum*), lentil (*Lens culinaris*), mung bean (*Phaseolus aureus* [*Vigna radiata*]), pea (*Pisum sativum*), pigeon pea (*Cajanus cajan*) and urd bean (*Phaseolus mungo* [*Vigna mungo*]) grown on marginal lands of seven districts of eastern Uttar Pradesh. (Bahraich, Faizabad, Gonda, Jaunpur, Mirzapur, Sultanpur and Varansi). All six pulses were colonized by arbuscular mycorrhizas to varying degrees (18-98%) in the districts surveyed. The average percentage of colonization of these crops was lowest (44.6-52.2%) in Sultanpur compared to the other districts (51.8-78.9%). The average percentage of mycorrhization in the pulses of Bahraich district was the highest (65.3-78.9%).

Singh and Adholeya (2002) studied the diversity of arbuscular mycorrhizal fungi in 114 samples collected from 11 different wheat-growing regions of India. The spore count, root colonization, species diversity and dominant species varied with the region as well as the soil nutrient conditions. Thirty-three species scattered over 5 genera were recovered. The genus *Glomus* was ubiquitous. The distribution of other genera *i.e. Entrophospora, Gigaspora, Sclerocystis* and *Scutellospora* was limited to only a few regions, indicating the adaptability of *Glomus* to varied soil conditions. Other genera showed a narrow range of host-environment adaptation. Spores of *Glomus albidum* and *Glomus fasciculatum* were the most frequent ones (33% frequency) among the spore communities. Other species recorded include: *Entrophospora* sp., *Gigaspora gigantea, Gigaspora margarita, Glomus aggregatum, Glomus ambisporum, Glomus botryoides, Glomus caledonium, Glomus clarum, Glomus constrictum, Glomus dimorphicum, Glomus etunicatum, Glomus fulvum, Glomus intraradices, Glomus macrocarpum, Glomus microcarpum, Glomus monosporum, Glomus mosseae, Glomus multisubstensum, Glomus pubescens, Glomus reticulatum Sclerocystis coremioides, Scutellospora calospora, Scutellospora coralloidea* and many other unidentified species. Although 33 species was recovered from the fields, the number of species per field ranged only between 1 and 5.

Fruit Crops/ Orchards

Girija and Nair (1988) recorded the incidence of VAM in banana varieties. Root samples from 8 popular banana varieties grown in Kerala, were examined for VAM fungal infection incidence, which was highest in Palayanthodan (60%) and lowest in red banana (57.6%). Infection levels for Montas, Padatti, Rasthali, Nendran, Pacha Kappa and Chingan were 38.5, 26.6, 24.4, 22.7, 19.0 and 18.1% respectively. Variation in the extent of VAM association were thought to be determined by the host as these varieties were grown in similar soil and environmental conditions.

Katiyar *et al.* (1989) reported VA mycorrhizal association in mulberry. Roots of 6 mulberry cultivars were examined for VAM infection and the mode of initial mycorrhizal infection was observed. There were no significant differences in the total percentage of colonization among the different cultivars. However, root colonization did vary, with Mysore Local having the high percentage, followed by S_{36}, Kanva-2, S_{54}, S_{30} and S_{41}. Most of the VAM fungi were *Glomus* spp. but there were also *Gigaspora, Acaulospora* and *Sclerocystis* species.

Rajagopal *et al.* (1989) studied the VAM fungi in the roots of mulberry at Hyderabad. They isolated spores from the infected roots of mulberry and inoculated to Sorghum. When VAM developed, the Sorghum roots were cut off and mulberry cuttings planted in the same pots. The roots become heavily infested with VAM. The infection patterns indicated that the VAM is a *Glomus* sp. Sprouting and growth rates of plants were high in cuttings planted in pots with VAM inoculum.

Pandey and Singh (1990b) investigated endophytic mycorrhizal status in different varieties of *Litchi chinensis* growing in Bihar.

Mohandas (1990b) studied the occurrence of vesicular arbuscular mycorrhizal fungi in root stocks of mango. Rhizosphere soil samples collected from rootstocks of mango namely, Olour, Bappakai, Chandrakaran, Necker, Peach and Vellakulamban were wet-seived to determine spore density. Mycorrhizal colonization of the root was also determined. Response of Vellakulamban rootstock to mycorrhizal colonization was studied in a pot culture study by inoculating soil with 280 g of *Glomus mosseae* inoculum. Root colonization by the fungus, plant height and dry matter were determined after 5 months of growth. The level of VAM colonization was highest in cultivar Chandrakaran. Density of mycorrhizal spores was higher in surface soil compared to sub-surface soil. The spores belonged predominantly to *Glomus, Acaulospora* and *Endogone* species.

Padma and Sullia (1990) studied the development of vesicular arbuscular mycorrhizal fungi in cultivars of mulberry. The root samples of all cultivars of mulberry namely, Berhampore, Berhampore local, Channapatna local, Jakkur local, Kosen, Mysore local, Popua, S-41 and Tsugasaguwa showed the presence of arbuscules and vesicles. The colonization of roots by VAM fungi was observed on the 30th day after planting in Berhampore local, Channapatna local, Jakkur local, Mysore local, S-41 and Tsugasaguwa where as in Berhampore, Kosen and Popua in was noticed on the 60th day after planting. The number of arbuscules varied with plant age and maximum number of arbuscules were noticed between 60 and 150 days. Among the cultivars of mulberry studied, maximum number of arbuscules were noticed in the cultivar Berhampore at 120 days. The number of arbuscules decreased with plant age and at 240 days degeneration of arbuscules was noticed. The number of vesicles also increased with age and they appeared after 90 days of planting. In the early phase of plant growth (30 days) no vesicles were seen. At 60 days, a few vesicles were noticed in some cultivars, and in others they were totally absent. The highest number of vesicles was observed in the cultivar Popua at 210 days.

Raverkar *et al.* (1990) surveyed 4 fruit and 5 ornamental (flowering) plants growing in the orchards of Department of Fruit culture, Y.S. Parmar University - Solan, Himachal Pradesh, for their association with VA-mycorrhiza. According to them all fruit and ornamental plants examined showed VA mycorrhizal association, however, the levels of natural infection were low (5.72 - 16.4%) and they suggested that artificial inoculum may prove beneficial for improved crop management.

Thakur and Lakhanpal (1990) studied the mycorrhiza of apple (*Malus domestica* Borkh.) growing in the orchards of Himachal Pradesh. According to them 6 VAM fungal spore ypes were associated with the roots of apple cultivar *Royle Delicious viz., Glomus mosseae, G. candidus, G. fasciculatum, G. macrocarpum, Endogone increseta* and *Gigaspora* sp.

Gupta and Lakhanpal (1993) identified and characterized mycorrhizal types, mycorrhizosphere mycoflora and vesicular-arbuscular mycorrhizal (VAM) fungi in apple. Anatomically, the roots of younger apple plants (1 to 3 years old) showed ectendomycorrhizas whereas the mature plants showed only intracellular infection i.e. VAM fungi. Mycorrhizosphere studies on apple revealed the presence of 30 fungi and 1 form of unidentified basidiomycetous mycelium. *Aspergillus niger, Penicillium chrysogenum, P. citrinum, Rhizopus nigricans* and *Trichoderma viride* were found in all the samples analyzed in a 3 month study. Six different types of fungal spores of VAM were recovered from soils of apple orchards. *Glomus macrocarpum* and *G. mosseae* were observed in all the soil samples.

Gowda and Rao (1994) studied the rhizosphere ecology of *Morus alba* var. Mysore local, inoculated with VA mycorrhizal fungus *Glomus intraradices*. A total of 37 and 33 fungi from amended and unamemded soils, respectively, is reported. The restricted appearance of fungi within the rhizosphere and non-rhizosphere of inoculated and uninoculated plants grown on both types of soils indicated that the VAM fungus influenced the presence of fungi. *Trichoderma* sp. was unaffected by the presence of VAM fungi but *Aspergillus flavus, A. fumigatus, A. japonicus, A. nidulans, A. terreus, Eurotium* sp., *Paecilomyces* sp., *Penicillium chrysogenum, P. citrinum, P. digitatum, P. expansum* and white sterile mycelium were commonly found on inoculated and uninoculated plants grown on both types of soil.

Awasti *et al.* (1999) surveyed ten well-managed peach orchards in the Rajgarh area of Himachal Pradesh, to determine the relationship between soil microflora (vesicular arbuscular mycorrhizas and *Azotobacter*) and fruit yield. A simple linear relationship worked out between all possible combinations indicated that % root colonization and spore number showed a significant positive correlation with the fruit yield. However, the *Azotobacter* population exhibited a very weak, non-significant positive correlation with fruit yield, except in one orchard, where it showed a negative correlation.

Setua *et al.* (2001) had undertaken a study in West Bengal during 1996-98 to assess the rate of inherent colonization of arbuscular mycorrhizal fungi, percent of arbuscules and vesicles in roots and the endo-mycorrhizal spore population in rhizosphere soil of six high yielding mulberry cultivars at various soil moisture levels in the field. The isolation and identification of predominant strains of AM fungi were also primary objectives. A wide range of variations in all the parameters was observed irrespective of the seasons, soil moisture levels, cultivars and cultivar × season interaction. Maximum root colonization (72%), arbuscules (17%), vesicle (19%) and arbuscules with vesicle (50%) formation were observed. The highest population of spores (13 spores/g rhizosphere soil) was found at low moisture levels during the summer followed by winter season. *Glomus* was the predominant genus in all the mulberry cultivars. The study, thus provides a scope for increasing spore population in filed soil through artificial inoculation of an efficient strain of local endophyte such as *Glomus* sp., to exploit, commercially, the arbuscular mycorrhizas for their beneficial effects in terms of improvement in quantity and quality leaf production in mulberry at reduced rates of phosphorus especially under tropical conditions.

Bharat and Bhardwaj (2002) surveyed various apple growing orchards and nurseries in Himachal Prades, falling under the agroclinatic zones II, III and IV for the colonization of vesicular arbuscular mycorrhizal (VAM). VAM colonization was observed in all orchards and nurseries surveyed, with nurseries recording higher VAM colonization and spore population compared to the orchards. Among the zones, zone II recorded the highest VAM colonization. Spores of *Glomus mosseae, G. fisciculatum* and *G. heterosporum* were the most abundant in the soil samples collected. Mycorrhizal structures such as vesicles and mycelium were observed in most of the root samples collected.

Medicinal and Aromatic Plants

Manjunath and Bagyaraj (1982) reported mycorrhizas in *Elettaria cardamomum, Piper betel* and black pepper. The number of endomycorrhizal spores in root zone soil as influenced by cultivars of *Lablab purpureus* was recorded.

Rao *et al.* (1987) investigated the VAM in aromatic plants. All aromatic plant species examined, growing in P deficient soil at the University Horticulture Farm, Bangalore, had VA mycorrhizal fungi in their root systems. Root colonization was highest in *Mentha piperata* and *M. spicata*, but was not correlated with chlamydospores counts in the rhizosphere soil, which was highest in *Artemisia pallens*.

Rao *et al.* (1989) studied the VA mycorrhizae in medicinal plants. They collected root and root zone soil samples from 25 medicinal plants growing in red sandy loam soil in Bangalore, Karnataka. All the plants examined harboured VAM fungi in their root systems. *Mentha arvensis* had maximum colonization (100%) and *Glycyrrhiza glabra* had minimum colonization (38%). Mycorrhizal spore counts in the root zone soil were highest in *Digitalis lanata* (113 in 25 g soil). There was no correlation between percentage mycorrhizal root colonization and the chlamydospores number in the root zone soil.

Selvaraj and Subramanian (1990a) studied the incidence of vesicular arbuscular mycorrhizal fungi in medicinal plants growing in four different soil types of Thanjavur district. All plant species were colonized with VAM fungi and the species of VAM fungi colonizing these plant species varied. 25 VAM fungal species belonging to genus *Glomus, Acaulospora, Scutellospora, Sclerocystis, Entrophospora* and *Gigaspora* were recovered. A certain degree of site specificity and more particularly host specificity was noticed. *Glomus fasciculatum, G. geosporum* and *Sclerocystis sinuosa* were common at all the four sites whereas the remaining occurred only with specific host and in particular soil types. Of the twenty five VAM species recorded, only seven species were found to have colonized the roots of host plants. Among them, five were species of *Glomus* and the remaining two were *Acaulospora appendicula* and *A. laevis*.

Sullia and Sampath (1990) studied the occurrence of vesicular arbuscular mycorrhizal fungi in medicinal plants. In this study root samples from 114 plants belonging to 50 families were used. The results indicated the presence of VAM fungi in 110 plants. The plant species such as *Euphorbia tirucalli* (Euphorbiaceae), *Crataeva nurvala* (Capparidaceae), *Hibiscus rosa-sinensis* (Malvaceae) and *Sweitenia mahogony* (Meliaceae) were not colonized with

VAM fungi. Out of 110 medicinal plants found to be colonized with VAM fungi, arbuscules were found in 107 plants. Vesicles were found in all plants species used with the exception of two plant species. With the exception of *Syzygium cuminii* mycelium was seen along with vesicles and arbuscules in all plants species screened. Along with the VAM association in the root cortex of *Azadiracta indica,* spores of *Thielaviopsis basicola* were also present. In *Merremia tridentata* spores of *Stemphylium* sp. were present as an endophyte along with VAM fungi in the root cortex.

Lakshman and Raghavendra (1990a) studied the occurrence of vesicular arbuscular mycorrhizal fungi in medicinal plants. Forty species of medicinal plants were examined for the presence of VAM colonization. All plant species examined in the study were susceptible for colonization by VAM fungi. However, the extent of VAM colonization varied. The response of few species of medicinal plants to inoculation of soil with VAM fungi was determined in a pot culture experiment. The results of the study indicated that inoculation of soil with VAM fungi improved growth and dry matter yield of at least three species of medicinal plants.

Janardhanan *et al.* (1990) studied whether the infection by fungal endophyte (*Balansia sclerotica* enhance the VAM association in Lemongrass (*Cymbopogon flexuosus* (Steud.) Watts) or not. Lemongrass is an important essential oil bearing plant commercially cultivated in India. These plants are generally infected by *B. sclerotica* which causes systemic disease, but the infected plants produce significantly higher tillering and biomass than the healthy plants. When healthy and *B. sclerotica* infected plants were examined for VAM association, the level of infection by *Glomus* spp. was 89.6% in infected plants (with *B. sclerotica*) and 44.6% in healthy plants. The VAM association supported the growth of the endophyte infected grasses.

Barthakur and Bordoloi (1990) made a survey of seven aromatic *Cymbopogon* species cultivated in Jorhat, Assam, for the VA mycorrhizal associations. *C. citratus* showed maximum colonization (82.2%). Arbuscules were observed only in 4 of the 7 *Cymbopogon* spp. tested.

Sharma and Roy (1991) observed some hypoglycaemic plants *viz., Alpinia galanga, Argyreia nervosa, Catharathus roseus, Coccinia grandis* and *Ichnocarpus frutescens* for their mycorrhizal associations. They have observed the maximum colonization in *A. galanga* (100%). Two VAM fungal species *Glomus dimorphicum* and *G. australe* were also identified.

Khaliq and Janardhanan (1994) studied the variation of native VA mycorrhizal association on cultivated species of mint (*Mentha arvensis* subspecies *haplocalyx, M. citrata, M. piperata, M. spicata, M. cardiaca* and *M. viridis*). All species of mint had abundant VAM association and it varied from 37.2% to 56.0%. The highest level of VAM colonization (56%) was observed in the roots of *M. spicata* and *M. citrata.* The rhizosphere soil of these plants had a VAM population ranging from 416 to 707, 100/g soil. The highest VAM spore population was observed in the rhizosphere of *M. spicata.* The results showed that cultivated species of mint have significant variation in their responses to native VAM fungi. This is the first report of the occurrence of *Glomus aggregatum, G. fasciculatum, G. mosseae, G. geosporum, Entrophospora* sp. and *Sclerocystis* sp. on mints.

Lekha *et al.* (1995) studied the VAM fungi associated in the rhizosphere soil samples of black pepper *Karimunda* from forest soil of kerala especially the Idukki district. They recovered *Glomus fasciculatum, G. macrocarpum, Sclerocystis coremioides, Sclerocystis clavispora* and *S. rubiformis* from the samples. They reported that *G. fasciculatum* was consistently present in all locations with high spore counts.

Ratti and Janardhanan (1995) studied the VA mycorrhizal association in some alkaloid bearing plants. They collected the roots with the rhizosphere soils of *Datura metel, D. innoxia, Duboisia myoporoides, Rauvolfia serpentina, R. tetraphylla* and *Catharanthus roseus* growing at CIMAP, Lucknow. All the plants were found to be mycorrhizal. The range of infection was between 22.6 (in *R. tetraphylla*) to 72.4 (in *D. innoxia*). The VAM fungi present in the rhizosphere were identified as *Glomus aggregatum, G. fasciculatum, G. geosporum, G. mosseae* and one unidentified species of *Glomus*.

Srivastava and Basu (1995) studied the occurrence of vesicular-arbuscular mycorrhizal fungi in samples of medicinal plants collected from Azamgarh, Uttar Pradesh. Twenty medicinal plants belonging to 14 families were screened. Most of the plants were mycorrhizal but few e.g.*Cassia obutsifolia, Asphodelus tenuifalius* and *Scoparia dulicis* were non-mycorrhizal.

Homkar *et al.* (1998) investigate the development of vesicular arbuscular mycorrhizas on 30 species of herbaceous and woody medicinal plants growing in the gene bank at the State Forest Research Institute at Jabalpur, Madhya Pradesh (India), and the plants categorized in to 4 groups based on their root infection (high, 9 species; moderate, 5 species; low, 13 species; and zero, 3 species).

Gupta *et al.* (2000) studied the VAM fungi associated with *Ocimum* species. (*O basilicum, O. canum [O. americanum]. O. gratissimum. O. sanctum [O. tenuiflorum]* and *O. kilimand-scharicum.* Vesicular-arbuscular mycorrhizal (VAM) fungi, *Glomus aggregatum, Glomus fasciculatum, Glomus geosporum, Glomus mosseae, Gigaspora* spp., *Sclerocystis sinuosa,* and *Acaulospora* spp, were identified and quantified. *Glomus aggregatum* and *Glomus fasciculatum* were predominant and associated with all the *Ocimum* species. Root colonization studies revealed the highest colonization (62.8%)in *O. basilicum* and the lowest colonization (35.2%) in *O. gratissimum*

Rani and Bhaduria (2001) studied vesicular arbuscular mycorrhizas (VAM) association in the roots of some medicinal plants growing on alkali soil in the Mainpuri district, Uttar Pradesh. Fifteen plant species were screened as positive for VAM fungi infection. The highest VAM colonization percentages were in *Cassia fistula, Cassia alata, Leucaena leucocephala* and *Convolvulus orvensis* (94, 69, 100 and 100% respectively). Results indicate that all medicinal plants growing on alkali soil show mycorrhizal infection, and that VAM infection possibly increases the capacity of plants to accumulate phosphate as it is released by other microorganisms.

Muthukumar and Udaiyan (2001) examined the vesicular arbuscular mycorrhizal (VAM) association in medicinal plants from Maruthamalai hills of Western Ghats. Of the 60 species of medicinal plants examined, VAM fungi colonized 58, with colonization levels ranging from 12 to 90% of the root length. The spore numbers in the soils ranged between 12 and 72 spores/ 10 g soil. Spores of 10 VAM fungal species belonging to *Acaulospora, Gigaspora, Glumus* and *Scutellospora* were isolated. The results indicate that soil phosphorus influences root length colonization.

Babu and Manoharachary (2003) studied the occurrence of arbuscular mycorrhizal fungi in rhizosphere soils of some medicinal plants. Nineteen species of arbuscular mycorrhizal fungi (AMF) were isolated and identified from the rhizosphere soils of 15 medicinal plants in Osmania University, Hyderabad, Andhra Pradesh, India. The isolated species belong to the genera *Glomus* (9 species), *Acaulospora* (5 species), *Entrophospora* (1 species), and *Gigaspora* (1 species). The physicochemical characteristics of the rhizosphere soils are presented. The data revealed that the AMF endophytes were widespread in all the soils investigated but varied both in number and the type of spores and sporocarps. No correlation was found between soil pH and spore numbers.

Weeds and Grasses

Venkataramanan *et al.* (1986) recorded and described the vesicular-arbuscular mycorrhizal association in some common weeds. *Ageratum conyzoides* revealed maximum infection followed by *Solanum khasianum, Hydrocotyle asiatica* and *Eclipta alba*. Arbuscles were present in *Ageratum conyzoides* and *Alteranthera sessilis*.

Kehri and Chandra (1989b) studied the mycorrhizal association in certain leguminous weeds. Most of the leguminous weeds growing at Allahabad showed a heavy infection of *Glomus* spp. in their roots, maximum being in *Cassia tora* L. The spore population varied with the weeds, however, highest population was noted in *C. occidentalis* L.

Rathi and Singh (1990) made a preliminary survey of mycorrhizal fungi in weeds growing at Meerut. The range of VAM infection in roots was recorded. Maximum infection was found in *Peperomia pellucide* while few weeds showed no infection.

Muthukumar and Udaiyan (1995) studied the vesicular-arbuscular mycorrhizae in dicots of a nutrient deficient semi-arid grassland. The soil was low in available nutrients. The two widespread grasses, *Eragrostiella bifaria* (Vahl.) Bor. and *Heteropogon contortus* (L.) Beauv ex. Roem and Schut, had a significantly ($P < 0.05$) higher root colonization ($73 \pm 4.0\%$) and $72 \pm 5\%$) compared to the dominant *Cymbopogon caesius* ($27 \pm 5\%$). Out of the 60 dicot species (representing 25 families) examined, 47 were infected by VAM fungi. Mycorrhizal infection was more frequent (100%) in eight families. Thirteen species belonging to Amaranthaceae, Apocynaceae, Caryophyllaceae, Cleomaceae, Molluginaceae, and Verbenaceae lacked VAM infection. VAM fungal spores belonging to 11 species (of three genera) were recovered from the rhizosphere soils. In a grassland, low in nutrients as in the present case, the competition for nutrients and water is a decisive factor in plant establishment and survival. The abundance of VAM fungi suggests the importance of VAM in this infertile semi-arid ecosystem. This study indicates that the ability to form associations with VAM fungi would have enabled the dicots in these grasslands to compete successfully with the dominant and widespread grasses.

Muthukumar *et al.* (1996) examined 24 species of sedges (representing six genera) from different vegetation types in the Western Ghats for VAM association. All the sedges had VAM fungal infections with hyphae and vesicles, but arbuscules were observed only in 42% of the total species. The VAM fungal colonization varied considerably between species ranging from 9 to 62%. Root colonization was positively and negatively correlated with root diameter and root hair length, respectively. The number of VAM fungal spores in the rhizosphere varied from 5 to 86/g soil.

Kapoor *et al.* (1996) screened two wild crucifers *Coronopus didymus* and *Sisymbrium irio* for VAM fungi colonization. The rates of colonization at different plant growth stages were also assessed. VAM was associated with the roots of both plants. *Glomus macrocarpum* and *G. fasciculatum* were the dominant species. Plant development was divided into 5 stages and VAM colonization was maximum for both species at the third stage.

Gupta and Mukerjee (1996) studied the association of VAM fungi in the roots of 14 grasses. Most of the grasses were colonized by the VAM fungi showing 20-100% infection. In *Brachiaria reptans* and *Setaria tomentosa* VAM colonization was seemingly absent, in most other grasses arbuscules and vesicles were present.

Jain *et al.* (1997) quantified mycorrhizal fungi in roots and rhizosphere soil of 15 weed species growing with 3 forage legumes, lucerne, *Trifolium alexandrium* and *Leucaena leucocephala* in Jhansi, Uttar Pradesh. All weeds were found to have mycorrhizal associations, but this varied both quantitatively and qualitatively with weed species. *Trianthema portulacastrum* was the most heavily mycorrhizal infected weed. Mycorrhizal fungi isolated included 4 *Glomus* spp. (*G. fasciculatum, G. aggregatum, G. intraradices* and *G. mosseae*) and an unidentified *Gigaspora* sp.

Chaurasia and Khare (1999) surveyed VAM fungi associations in a protected grassland of a tropical deciduous forest region in Sagar, Madhya Pradesh. Information on the occurrence and association of VAM with different dominant and common grassland plants was collected Root colonization ranged from 46-100% and was higher in grasses than in herbs. *Glomus* spp. were the most dominant forms.

Muthukumar and Udaiyan (1999a) studied the vesicular arbuscular mycorrhizal association in medicinal plants of Western Ghats of Southern India. Fifty two medicinal plants belonging to 28 families growing in Western Ghats in Southern India were examined for their vesicular arbuscular mycorrhizal (VAM) status. Forty eight of the 52 species of medicinal plants examined were colonized by VAM fungi ranging from 12% to 90% of the root length. Three angiosperms (*Portulaca oleracea, Phyllanthus amarus, Cyanodon dactylon*) and two ferns (*Drynaria quericifolia, Osmunda regalis*) lacked VAM association. VAM fungal spores of 14 species (belonging to *Acaulospora, Glomus, Sclerocystis* and *Scutellospora*) were isolated from rhizosphere soils. Of these, species of *Glomus* constituted 50% of the total species recorded.

Aquatic Plants

Bagyaraj *et al.* (1979b) reported VA mycorrhizas in 3 of the 12 species of Tropical aquatic plants examined *viz., Cyanotis cristata, Eichhornia crassipes* and *Salvinia cucullata.*

Salvinia cucullata, a rootless aquatic fern showed mycorrhizal association in its modified submerged leaves acting as roots.

Chaubal *et al*. (1982) studied the VA mycorrhizal association in subtropical aquatic and marshy plant communities growing in the ponds. Only fewer endophytes were found in samples from terrestrial habitats and the spores were absent from water samples. Plants without mycorrhizal associations grew poorly. All the endophytes could establish mycorrhizal associations in pot cultures.

Firdaus-E-Bareen (1990) studied the VA-mycorrhizal association in aquatic plants. A number of common aquatic and marshy plants of Punjab were shown to be mycorrhizal. Colonization was also observed in non-root plant parts. Greater colonization was found in plants of marshy than aquatic environment, possibly owing to low availability of spores in water.

Sengupta and Chaudhury (1990) reported VA mycorrhiza in marshy plants in West Bengal. Mycorrhizal colonization in four species including two of Chenopodiaceae has been reported. Five common species of VAM fungi were recorded from the rhizosphere soils of these plant species. Population of spores of VAM fungi and effective inoculum potential of these fungi in rhizosphere soils as determined by the MPN method were low.

Sengupta and Chaudhuri (2002) examined thirty-one species of mangrove and associates and 23 species of transported flora, belonging to 25 families at fopur physiological stages of successions of the mangrove plant community at the terminal part of Ganges river rstuary for arbuscular mycorrhizal (AM) root association. Dominant members of the mangrove plant community were all AM, mostly with 'Paris 'type structures many of the known mycotrophic plant families, except the Cyperaceae, also showed AM association, with intra cellular hyphae and vesicles as the most discernible endophyte structures. Intensity of AM colonization varied both with the species and situations of their occurrence, being more intense and also more extensive in less saline dry ridge mangroves than in more saline formative and developed swamp mangroves. Introduced exotic tress on the ridges and embankments were infected by AM, but less than the declining mangroves in the same location. Seven species of AM fungi in common with those of the upstream mesophytic plants were isolated from root-free rhizosphere soils of the mangroves, three of which predominated in root association. These species, individually an as mixture infected roots of salinity herbs and trees in both location silt and upstream alluvial soil with obvious improvement in their biomass yield and phosphorous nutrition. AM infective potential of root free rhizosphere soils of the dominant members of the mangrove community were negatively related to salinity level of the sediment soil of the successional stage. The evidences of AM association of mangroves and other salt marsh plants obtained and those reported elsewhere are discussed.

Jha *et al.* (2002) observed a new arbuscular mycorrhizal association in *Typha angustata* under waterlogged condtion. *Typha* roots were collected from two sites, one was water logged throughout the year and the other was waterlogged for a brief period during the rainy season in December 1999. The root samples were collected from plant rhizosphere zone.

Rhizosphere soil was sieved to obtain whole root biomass. The procured root samples were homogenized thoroughly. Extensive infection with arbuscular mycorrhizal was recorded in *T. angustata.* Oil droplets were more frequently observed in samples collected from permanently waterlogged than temporarily waterlogged conditions. Results showed that plants collected from permanently waterlogged had a higher percentage (74.5%) of root length infection with mycorrhizas than plants collected from temporarily waterlogged conditions (63.25%).

Jaiswal *et al.* (2003) carried out studies on the arbuscular mycorrhizal status of seven (*Rhizophora, Exocarya, Avicennia, Derris, Acanthus, Candelia* and *sanneratia*) dominating mangrove plant species from Goa. Only two of the seven plant species selected for the study, exhibited the presence of arbuscular mycorrhizal fungal colonization. The spore density of arbuscular mycorrhizal fungi varied from 26-spores/100 g rhizosphere soil to 88-spores/100 g rhizosphere soil. Ten species of arbuscular mycorrhizal fungi belonging to three *viz., Acaulospora, Glomus* and *Scutellospora* were recovered during the study. *Glomus macrocarpum* Tulasne and Tulasne was the most frequently occurring species of AM fungi, associated with the rhizosphere soil of the mangrove species.

Forest

Thapar and Khan (1973) carried out studies on endomycorrhizae in some forest species. They reported endomycorrhizae in several conifer and broad-leaved species. Vesicular, arbuscular, vesicular-arbuscular type of infections were recorded, among which the vesicular-arbuscular type was common. The hyphal characters of the endophyte in the extramatrical and internal phase were of diagnostic value in establishing the identity of lower symbionts which were identified as different species of *Endogone*. Sporocarps and spores of *Endogone* species were recorded in forest soils underneath different tree species and were quantitatively assessed and described. The occurrence of spores was shown to be influenced by land usage, soil moisture and texture.

Thapar *et al.* (1985) described the occurrence of six species of Endogonaceae *i.e., Gigaspora aurigloba, G. margarita, G. heterogama, G. gilmorei, Glomus fasciculatus* and *Glomus monosporous* in forest soils.

Thapar and Khan (1985a) made assays of forest soils for occurrence and frequency of VA mycorrhizal fungi at 21 locations in 8 states. 15 species belonging to 4 genera of *Endogonaceae*, i.e., *Acaulospora, Gigaspora, Glomus* and *Sclerocystis* were isolated. Spores of *Glomus macrocarpus* were widely distributed followed by *Sclerocystis* spp., *Gigaspora gigantea* and *Acaulospora* sp. New forest soils represent maximum number of species of *Endogonaceae*.

Venkataramanan *et al.* (1985) studied the occurrence of endotrophic mycorrhizal fungus in agarwood plant *Aquillaria agallocha* (Roxb). Vesicles were abundant in the entire infected roots, about 3-4 per field (16x10). Spherical to elliptical 10 um diameter to 13x6 um in size, deeply stained. Some vesicles contained young spore and hyphae running parallel to the long axis of the roots, inter- and intracellular measuring 3.0 um. No difference was noted in the width of the external and internal hyphae. The root pieces were infected to an extent of 75" of their length. Only the fine lateral branches of the roots, showed the endophyte. In older segments, it was absent or sparsely distributed.

Mohankumar and Mahadevan (1986) reported the vesicular-arbuscular mycorrhizae in mangrove vegetation of the Pitchavaram forest, Tamil Nadu.

Mukerji and Kapoor (1986) made a study of 16 species of trees and shrubs in a natural forest stand in old Delhi. All the plants showed mycorrhizal association. *Glomus, Gigaspora* and *Acaulospora* were present in the rhizosphere soils and their numbers varied from 200 to 300/100 g of soil. Infection in field recovered roots ranged from 20 t0 50%.

Sharma *et al.* (1986b) investigated the status of mycorrhizae in sub-tropical ecosystem of Meghalaya. All the tree species investigated in an evergreen mountain forest were mycorrhizal. VA mycorrhizae were dominant. Eighteen tree spp. were endomycorrhizal, 5 ectomycorrhizal, 2 ericoid and 1 with ectendo-mycorrhizal association. The soil was sandy, acidic with high organic matter content.

Thapar and Khan (1988) studied the seasonal variation in the population of *Endogone* spores all round the year under hoop pine (*Araucaria cunninghamii* Sweet) and teak (*Tectona grandis* L.f.) in forest soils. The spore numbers increased after rains during summer due to high soil moisture.

Mohankumar and Mahadevan (1988c) studied the viability of VAM spores in a tropical forest soil. The efficiency of VAM spores to infect plant roots was examined from soil samples collected from high altitude evergreen and low altitude reed forests (from Kalakad Reserve Forests, a tropical forest). Soil samples were collected in July 1983 and stored in dark at room temperature. To check the viability of spores, soil samples were withdrawn at intervals of 6, 12, 16, 24 and 30 months and tested on the infectivity of onion roots. A spurt in spore population and per cent infection were recorded after 6 months in evergreen forest soils, but declined thereafter. In reed forest soils, the number of spores and quantum of infection declined sharply after 6 months and no VAM infection was recorded after 18 months.

Parameswaran and Augustine (1988) studied the distribution and ecology of VAM in a scrub jungle. Mycorrhizal association in 44 plant species collected from Madras Christian College Campus, Tambaram, Tamil Nadu was investigated. Mycorrhizal infection was recorded in 31 species. Generally, number of spores was more in 2 cm depth than in 10 cm depth soils. Spores of *Glomus aggregatum, G. citricolum, G. fasciculatum, G. geosporum, G. pansihalos* and *Sclerocystis pachycauliis* were collected from the soil samples. Variation in per cent mycorrhizal infection was noted during different period of collection within the same plant species. Soil physical and chemical parameters (pH, temperature, moisture, N,P,K,Zn, Cu, Fe) showed positive relationship with both the number of spores and per cent infection.

Kshattriya *et al.* (1990) studied the development of mycorrhizae in more disturbed and less disturbed forest communities. They have shown that plant species growing in a more degraded site (slash-burn agriculture on a 5-year cycle) had higher mycorrhizal infection than those in a less disturbed site. In general, there was heavy infection in the spring and summer months. Fewer endogenous spores were present in soil of the less disturbed site. Maximum spores were counted in the winter months.

Kanakdurga *et al.* (1990) studied occurrence of endomycorrhizal fungi in teak. Several VAM fungal species namely *Acaulospora laevies, Glomus caledonicum, G. fasciculatum, G. mosseae, G. multicaule, G. tortuosum, Scutellospora gregaria* and *Sclerocystis rubiformis* were present in these soil samples. Many of the VAM species were found associated with Telangana forest soils supporting teak and these soils were deficient in phosphorus and moisture. The forest soils from Ryalaseema and Coastal Andhra did not support many VAM fungi. A survey of lateral roots collected from naturally existing forest regions showed the abundant root colonization by VAM fungi. *Glomus fasciculatum* and *G. caledonicum* which were isolated from these soils in large numbers are being mass produced using sorghum and *C. ciliaris* as hosts using sterilized soil as growth medium.

Maitra and Gautam (1994) carried out a preliminary survey of *D. strictus* associated VAM fungi for different rhizosphere soils in bamboo growing regions of Central Madhya Pradesh. Five VAM forms were associated with *Dendrocalamus strictus* of which *Glomus* emerged as the most prevalent genus. Of the *Glomus* spp. observed during the study, *G. macrocarpum* and *G. fasciculatum* were the most common. The relationship of VAM and *D. strictus* forest was evaluated based on spore density and percent root infection. It was observed that mature, undisturbed, bamboo forests had the highest VAM spore density and percent root infection compared with mature disturbed, young undisturbed and young disturbed *D. strictus* forests.

Bhat *et al.* (1994b) studied the mycorrhizal association in the roots of forest trees of Nilgiris. They have presented data for 11 forest tree species, all of which were mycorrhizal except *Pinus putida.*

Vijaya *et al.* (1995) studied the occurrence of endomycorrhiza in some forest soils of Royalaseema region of Andhra Pradesh. According to them the spore density in the soil varied from one species to another. The *Glomus* species was found predominantly in the rhizosphere soil collected from *Albizia amara* and *A. odoratissima* where the soil pH was slightly alkaline (7.3 -7.4), *Acaulospora species* was found occurring more in soil where pH was 6.6.

Santhaguru *et al.* (1995) studied the VA mycorrhizal spores in the rhizospheric soils of 20 species of tree legumes and a liana in the reserved forest of Alagar Hills in the Eastern Ghats of Tamil Nadu and the extent of root mycorrhizal infection of the legumes was investigated. Nearly 75% of species showed VAM infection, but with varying frequency. Altogether, 21 species of VAM fungi belonging to 6 genera, *Acaulospora, Entrophospora, Gigaspora, Sclerocystis* and *Scutellospora* were recorded.

Raman and Nagarajan (1995) studied the incidence of mycorrhizal association in forest tree fire site in Servarayans hills, Tamil Nadu. In all, they studied VAM association in 12 plant species belonging to different families of Angiosperms and reported 18 to 52% root colonization in fire-prone areas. Fifteen VAM fungal species were reported from fire-prone regions: *Gigaspora* (1 sp.), *Glomus* (10 spp.), *Sclerocystis* (1 sp.) and *Scutellospora* (3 spp.). They included *Gigaspora gigantea, Glomus aggregatum, G. albidum, G. constrictum, G. deserticola, G. etunicatum, G. fasciculatum, G. geosporum, G. intraradix, G. macrocarpum, G. microcarpum, Sclerocystis pakistanica, Scutellospora erythropa, S. nigra* and *S. percica.*

Pradhan *et al.* (1996) isolated and characterized the VAM symbionts of an actinorrhizal Himalayan alder *Alnus nepalensis* from forest stand. Altogether 7 species of VAM fungal symbionts of *A. nepalensis* were found - 4 *Glomus* spp. (*G. intraradices, G. macrocarpus* var. *geosporus {Glomus macrocarpum* var. *geosporus}, G. mosseae, G. tortuosum*), 2 *Gigaspora* spp. (*G. gigantea, G. gregaria*) and one species of *Acaulospora (A. scrobilata)*. They showed variable seasonal and temporal distribution patterns. The forest soil also exhibited distinct seasonal and temporal variations in physico-chemical properties (such as temperature, moisture and pH). Diversity in the VAM fungal species in a single forest stand showed wide and flexible *Alnus* VAM endophyte compatibility-characteristic of mycorrhizal actinorrhizal species. The tendency of fungal symbionts to grow luxuriantly under moderate acidic conditions coincided with soil pH range of the actinorrhizal host *A. nepalensis*.

Singh and Tiwari (1999) investigated the impact of soil degradation on distribution of vesicular-arbuscular mycorrhizal (VAM) fungi population in forest soils. Degraded, least degraded and undegraded forest sites were selected and distribution of VAM spore population was assessed. VAM population was inversely correlated to the degree of soil degradation irrespective of the soil depth in all the sites. Maximum VAM population was recorded in the undegraded forest soil followed by least degraded and lowest population in case of the highly degraded site. The higher and lower VAM population in undegraded and degraded sites were directly related to the higher value of soil moisture content, WHC and pH except in case of soil bulk density. The bulk density was negatively correlated to the distribution of VAM population in all the sites. It suggested that degree of soil degradation exerted a major impact on the distribution of VAM fungi in forest soil.

Rahangdale and Gupta (1999) evaluated the occurrence of VA-mycorrhizal symbiosis in forest trees in 46 (multipurpose) species from 19 families selected for biomass energy plantation of Raipur, Madhya Pradesh. Seedlings were raised in the nursery using both nursery bed and polypot methods, and the local (Vertisol) soil. Roots of the seedlings were examined for VAM colonization after 150-180 days of growth. Out of the 46 species tested all but *Schrebera swietenioides* had VAM colonization of their roots. VAM colonization of the remaining species varied from 10% in *Schleichera oleosa* to 94% in *Cassia fistula*. The results indicate that the widespread occurrence of VAM symbiosis in plants belonging to many families in the tropical forests of Madhya Pradesh. The VAM fungi associated with the tree species mainly belonged to species of *Glomus, Acaulospora* and *Gigaspora.*

Srinivas *et al.* (1999) conducted an extensive survey on Godavari belt forest soils (Andhra Pradesh) and revealed a rich vesicular arbuscular mycorrhizal (VAM) fungal flora. The fungal population both qualitatively and quantitatively varied with forest plant community and place of collection. A total of 54 fungal species representing 5 genera were recorded. *Glomus* species dominated in all types of soils, *Scutellospora* was least in its occurrence with regard to the number of species recorded in different places. Among different places, Bhadrachalam soils were rich in VAM fungi. VAM fungal composition also varied with plant species. Among the different host plant species evaluated, i.e. *Albizia lebeck, Acacia melanoxylon, A. nilotica* and *Enterolobium saman* [*Samanea saman*], no VAM fungus was associated with a specific host plant collected from the different places.

Talukdar and Thakuria (2001) collected soils and root samples from teak and gomar plantations at 18 locations under 11 reverse forests of Assam and were analysed for spore numbers, levels of vesivular arbuscular mycorrhizal fungi (VAMF) colonization and VAMF species. Effect of two VAMF on teak (*Tectona grandis*) and gomar (*Gmelina arborea*) stumps was also investigated in pot experiment. The number of VAMF spores detected in plantation soils ranged from 14-1718 per 100 g soil. VAMF colonization in roots also varied widely. The VAMF spores isolated were representative of one species of each of *Acaulospora, Entrophospora, Gigaspora* genera and three species of *Glomus*. The *Gigaspora* sp. and *Glomus fasciculatum* were detected in all the sites. The spore inocula of these two species improved growth of teak and gomar grown in a sterilized forest soil. *Gigaspora* sp. was more effective in improving plant growth and *Glomus fasciculatum* showed an antagonistic effect on its performance. This result emphasized investigation on whether antagonistic effect of one VAMF species on another species can occur under natural soil in field, where VAMF affect plant as a community.

Khade and Rodrigues (2003) surveyed the prevalence of arbuscular mycorrhizal (AM) fungi in Mollem forest area of Western Ghats of Goa. A total of 25 tree species belonging to 18 families were screened. AM colonization in the tree species varied with family and host genera. The highest mean root colonization (100%) was recorded in *Macaranga paltata, Xylia xylocarpa, Zanthoxylum rhetsa* and *Randia ruglosa.* Maximum mean spore density of 745 spores 100 g^{-1} rhizosphere soil was recorded in *Leea indica.* A total of 18 AM fungi belonging to five genera, namely, *Acaulospora, Gigaspora, Glomus, Sclerocystis* and *Scutellospora* were found to be associated with three tree species studied.

Trees

Reddy and Goud (1988) studied the occurrence of VAM association in *Cassia occidentalis*. Occurrence of VAM in the root system of *Cassia occidentalis* was observed. The morphology of the endophyte and its mycotropic habit and the effect of soil characteristics on mycorrhiza formation were studied. The roots with VAM fungus were completely devoid of secondary thickening as compared to normal roots which were otherwise invariably with secondary growth. The light textured soil (loamy sand) which was slightly alkaline was more favourable for mycorrhizae formation.Rhizosphere and rhizoplane mycoflora of mycorrhizae were also assessed and compared with non-mycorrhizal rhizosphere and non-rhizosphere soil fungi.

Sidhu *et al.* (1990b) reported the clonal selectivity in *Populus deltoides* for vesicular arbuscular mycorrhizal association. Nine clones of *Populus deltoides* (59-101, ST-148, 15-29, 55-264, D-212, S7C15, S7C20, G-48 and AST-242) growing at Banthra Research Station were screened for association of VAM fungi. Only three clones (15-29, ST-148 and 59-101) out of the nine clones investigated were found to form endomycorrhizae. Two VAM species were found to colonize the roots of these three clones. These fungi were identified as *Glomus intraradices* and *Scutellospora gigantea*. Though all the soil samples showed the presence of spores of *Glomus intraradices* and *Scutellospora gigantea* only these three clones formed mycorrhizae.

Thapar and Vijayan (1990) presented a quantitative evaluation of the intensity of VAM infection in nursery seedlings of hardwood tree species belonging to 19 genera in 14 families at Dehra Dun. Low frequencies indicated the need for evaluation of nursery stocks with efficient VAM endophytes for improved performance.

Kaushik *et al.* (1992) carried out an investigation on 2 important semi-arid plants, *Prosopis juliflora* and *Zizyphus jujuba* from the Mehrauli ridge, Delhi to study VAM association in them. Several VAM fungi (notably *Glomus fasciculatum, Glomus macrocarpum, Glomus mosseae, Glomus multisubstensum, Glomus constrictum, Gigaspora* sp., *Sclerocystis* sp., *Acaulospora elegans, Endogone aggregatum*) were recovered.

Thapar *et al.* (1992) studied the vesicular-arbuscular mycorrhizal associations and root colonization in some important tree species. The plant species belonging to 14 families sampled in the study differed in their mycorrhizal status. Four species were occasionally mycorrhizal and the remaining species were consistently mycorrhizal. Five species showed vesicular type of infection and 9 were of vesicular-arbuscular type. Four showed arbuscular infection. Highest level of root colonization (92%) was recorded in *Michelia champaca* and *Toona ciliata* whereas very low (1 to 6%) levels were recorded in *Albizia lebbek, Grewia robusta* and *Madhuca longifolia*. The percentage infection was almost the same in *Terminalia arjuna, Dendrocalamus strictus, Dalbergia sissoo, Acacia auriculiformis, Emblica officinallis, Olea glandulifera, Acacia nilotica* and *Seraca asoca* (upto 40%). Spores of *Glomus macrocarpus* and *Sclerocystis coremioides* were found to occur in rhizosphere soil. Heavy arbuscular infection was found in *Madhuca longifolia* extramatrical spores in *Acacia auriculiformis* and heavy vesicular infection in *Acacia auriculiformis, Dalbergia sissoo* and *Delonix regia*. The highest frequency of VAM infection was recorded in *Toona ciliata* (83.8%), followed by *Michelia champaca* (79.54%), *Jacaranda mimosifolia* (67.25%) and *Dendrocalamus strictus* (65.5%) while it was lowest in *Albizia lebbek* (8.33%).

Sagar *et al.* (1993) isolated vesicular arbuscular mycorrhizal spores from soils in 10 localities in association with infected roots of *Celtis australia* - an agroforestry tree. The predominant species were *Glomus fasciculatum* and *G. macrocarpum*, followed by *G. constrictum, G. fulvum, G. mosseae* and *Sclerocystic coremioides*.

Vasanthakrishna *et al.* (1994a) conducted a preliminary survey of VAM fungi in the root zone of 2-4 year old *Casuarina equisetifolia* plantations at Bangalore. Root zone soil samples upto 15 cm. depth were collected and VA mycorrhizal spores were isolated. Five dominant spore types were identified. The spores were identified up to the species level. All of them belonged to the genus *Glomus*.

Manoharachary and Reddy (1995) studied the VA-mycorrhizal association in *Tectonia grandis, Terminalia alata* and *Dalbergia sissoo*. All the three were found heavily colonized both quantitatively and qualitatively with VAM fungi during mansoon and winter. Altogether, 20 VAM species were collected and identified up to the species level. Of these, *Glomus*, followed by *Sclerocystis* and *Acaulospora* were dominant.

Bhattacharya *et al.* (1995) studied the mycorrhizal association in different species of bamboo in different waste land soils of West Bengal. The three most widely available and economically important species of bamboo *viz., Bambusa balcoca, B. vulgaris* and *B. tulda* were included in the study. All the three species were naturally infected by VAM to all the soil situations, but *B. tulda* showed the low infection (below 40%). The number of extramatrical spores in the rhizosphere varied with the species, highest being in *B. balcooa* (617/100g soil) and the lowest in *B. tulda* (250/100g soil). Only single genus *Glomus* was revealed in different soil situations. They indicated that low inherent infection intensity and low extramatrical spore development may open a possible scope for introduction of an efficient VAM inoculant for bamboo plantation in wasteland situation.

Reddy *et al.* (1995a) assessed the endomycorrhizae (VAM) of two species of Eucalyptus (*E. camaldulensis* and *E. tereticornis*) in Godavari belt of Andhra Pradesh by wet sieving and decanting technique. In all, 28 VAM fungal species representing 6 genera were associated with two species of *Eucalyptus*. The degree of root infection varied both with the season and soil. Interestingly, species of *Gigaspora* could not be recorded. A correlation was sought between physico-chemical conditions and incidence of VAM fungi and degree of infection and prevailing environmental conditions.

Kulkarni *et al.* (1995) studied the VA mycorrhizae of *Casuarina* from Konkan. The rhizosphere soil and root samples of *C. equisetifolia* (about 10 year old) growing along the coastal region of Konkan, were taken for study. They were collected from Ganapatiphule, Dapoli and Ratnagiri. The maximum number (/25 g rhizosphere soil) of spores were isolated from Ratnagiri (112) then from Ganapatiphule (58) and minimum (38) spores from Dapoli. Same trend was observed in percent root infection(60%, 50% and 40%) respectively. They isolated *Gigaspora gigantea, G. decipens, Glomus botryoides* and *Sclerocystis pakistanica* from Ratnagiri and Ganapatiphule soils and *Gigaspora gigantia, G. margarita, G. decipens* and *Sclerocystis pakistanica* from Dapoli soils. They concluded that VAM fungi help in the luxuriant growth of *Casuarina* plants on seashores where less moisture is retained (5-6%).

Rani *et al.* (1995) studied the occurrence of VAM fungi in forest nursery seedlings in Assam. They surveyed different forest nurseries of Udali- Tinali, Lanka, Nagaon district (Assam) and collected root and soil samples of *Tectona grandis, Gmelina arborea* and *Michelia champaca*. All the samples showed VAM association in the roots and spores in the rhizosphere soil but no definite correlation could be established between spore count and VAM root colonization.

Singh and Kapoor (1995) studied the root colonization of *Casuarina equisetifolia* in different soil samples of the country. They collected the soil samples from different parts of the country, Bangalore, Baroda, Barrackpore, Coimbatore, Delhi, Hyderabad, Jabalpur,

Jaipur, Kanpur, Parbhani, Ranchi, Ratangarh, West Godavari and *Casuarina* plants were raised under potted conditions in greenhouse. Mycorrhizal infection was observed in 120 days old seedlings/ plants. In general, root colonization of *Casuarina* by native VAM fungi was observed in almost all the soil samples except four *viz.*, Ratangarh, Kanpur, Jaipur and Delhi. Highest root colonization was observed in plant roots grown in Jabalpur soils and lowest in Barrackpore soil.

Kaushal and Srivastava (1995) studied the biology of VAM fungi associated with *Prosopis cineraria* and *Acacia nilotica* growing in and around Jodhpur, Osian, Sumerpur and Jhalawar. Their observations revealed rich VAM fungal flora belonging to *Glomus, Gigaspora* and *Acaulospora. G. fasciculatum* dominated over *Glomus aggregatum, G. macrocarpum* and *G. mosseae*. The colonization of young feeder roots was observed in plants of varying ages, ranging from four-days-old seedlings to 20-30 year-old plants. 70% root colonization in *A. nilotica* and 82% in *Prosopis cineraria* was reported. The spore counts/were low in summer, maximum in August and decreased slowly in winter.

Ravi *et al.* (1995) conducted an intensive survey to assess the occurrence of VAM in locally cultivated agroforestry tree species at National pulse Research Centre, Vamban, Tamil Nadu. All the 45 tree species screened showed VAM infection. Most of the VAM were represented by species of *Glomus* and *Gigaspora*. They grouped the tree species into effective (root infection 62-78% all Acacias, *Jambolina, Azadirachta*, teak etc., 20 spp.), moderate (root infection 56-58%, in *Cassia siamea* and *Mangifera indica* etc 15 spp.) and less infective group e.g. *Cardia*, 22% infection, 10 spp.)

Mohan *et al.* (1995) made an extensive survey of the roots and rhizosphere soils of neem (*Azadirachta indica*) from various Forest Department nurseries and plantations in Rajasthan for the distribution of VAM fungi. The mean number of VAM fungal propagules and percentage root infection of plants were greater in plantation samples than in nursery samples. Various VAM fungi were isolated. *Glomus* spp. were dominant in nurseries (5 species) and plantations (9 species), but *Sclerocystis* sp. was also common in plantations. Among the *Glomus* spp. *G. fasciculatum* was predominant in both nurseries and plantations.

Jamaluddin and Chaturvedi (1996) studied the colonization of VAM in *Dendrocalamus asper*. They recorded the extent of VAM infection in tissue culture plants of this edible bamboo. Many VAM species were detected and the extent of infection increased with age of the plants. Rhizome and tiller number increased with age and the extent of infection.

Ginwal *et al.* (1997) studied the natural occurrence of arbuscular mycorrhizae in different provenances of *Acacia nilotica* Willd. ex Del. seedlings. Seedlings of 32 provenances of *A. nilotica* (including 6 subspecies /varieties) from Sudan, Yemen, India, Pakistan and Senegal were examined for vesicular arbuscular mycorrhiza root colonization in nursery experiments (at the Tropical Forest Reseach Institute, Jabalpur). *Glomus etunicatum, G. mosseae, G. aggregatum*, 2 unidentified *Glomus* spp., a *Gigaspora* sp., a *Scutellospora* sp. and 2 *Acaulospora* spp. were recorded. The percentage root colonization varied significantly (from 1.7 to 50.0%) in different provenances. Mycorrhizal infection was significanly correlated with seedling height, collar diameter and seedling biomass.

Mohan and Singh (1997) studied the status of VAM fungal association in Accacia (*A. nilotica*) in various forest nurseries and plantations in the arid zone of Rajasthan. The extent of VAM colonization in roots and spore populations in rhizosphere soils was quantified. Variation was observed in percentage root colonization and number of VAM propagules in the rhizosphere of both nursery and plantation samples. The mean number of VAM propagules and percentage root colonization were greater in plantation samples than in nursery samples. Spores from the rhizosphere soils were isolated and tentatively identified as different species of *Glomus, Gigaspora* and *Sclerocystis*. Amongst these *Glomus* was dominant in the rhizosphere of both the tree species screened.

Abhijeet *et al.* (1997) showed the VA mycorrhizas associated with *Casuarina equisetifolia* roots and the rhizosphere soil of plants growing on the sea shore of Konkan area in Maharashtra. They isolated and identified the spores of *Glomus, Gigaspora* and *Sclerocystis* spp. (*Glomus mosseae, G. fasciculatum, Gigaspora gigantea, G. rosea, G. margarita, G. decipens* and *Scleroscystis pakistanica*).

Kaur *et al.* (1997) observed VA mycorrhizal association in *Grewia optiva* - an important agroforestry species. Soil samples collected from (10 districts of Himachal Pradesh) rhizosphere soil of *G. optiva* showed the occurrence of *Acaulospora elegans, Gigaspora candida, Glomus fasciculatum, G. fulvum, G. intraradices, G. macrocarpum, G. mosseae* and *G. reticulatum*.

Mohan and Mishra (1998) studied VAM association in Marwar Teak (*Tecomella undulata* (Smith) Seeman) in desert soils of arid zone of Rajasthan. Almost all the soil samples contained spores of different species of *Glomus, Gigaspora* and *Sclerocystis. Glomus fasciculatum* was the most commonly isolated fungus.

Srinivas *et al.* (1998) studied the dynamics of vesicular-arbuscular mycorhizae in association with *Albizia lebbeck*. Association of VAM fungi with *Albizzia lebbeck* under different edaphic conditions of Godavari belt was assessed. In all 13 species representing 6 genera could be recorded. *Glomus mosseae* and *Sclerocystis pakistanica* were most predominant. *Albizzia lebbeck* was found to be VAM dependent as there was considerable increase in growth rate, biomass production. chlorophyll content and phosphorus content in plants grown in unsterilized soil than in sterilized soil. The degree of VAM infection varied with the place as well as the time of sampling.

Prasad (1999) carried out a preliminary survey of VAMF in the root zone of one-year-old field grown *Prosopis juliflora* plants at the Biomass Research Center experimental plot at Bettiah [in Bihar]. Rhizospheric soil samples were collected randomly and VAM spores were isolated by the sieving and decantation technique. A higher spore count supported good growth. All the VAM fungi identified were *Glomus* spp. (7). *G. fasciculatum* was the commonest species (55.45% occurrence). The other species were *G. claroideum* (8.33%), *G. constrictum* (10.45%), *G. geosporum* (34.32%), *G. macrocarpum* (33.35%), *G. mosseae* (33.33%) and *G. vermiforme* (31.45%).

Nagabhushanam *et al.* (1999) studied the association of VAM (vesicular arbuscular mycorrhizal) fungi with 40 leguminous (multipurpose) trees growing in the Godavary belt of Andhra Pradesh , and the number and type of mycorrhizal spores in the soil were investigated. All the trees showed the VAM infection, but with varying incidence. Altogether 37 species of mycorrhizas were recorded - *Acaulospora* (12), *Entrophospora* (2), *Gigaspora* (3), *Glomus* (16), *Sclerocystis* (1) and *Scutellospora* (3).

Khan and Uniyal (1999b) studied arbuscular mycorrhizal associations in 11 bamboo species (6 *Bambusa* spp. and 5 *Dendrocalamus* spp.) in a species trial at a reforested site at Kurukshetra in Haryana. A total of 36 AM fungi belonging to 6 genera were isolated. The genus *Glomus* was dominate, being represented by 18 species, followed by *Acaulospora* and *Scutellospora* with 7 and 6 species, respectively. *Scutellospora, Sclerocystis, Entrophospora* and *Gigaspora* were represented by 2 and 1 species. Amongst the host species *Dendrocalamus strictus* and *Bambusa nutans* had the highest spore population of AM fungi, and *B. baccifera*. The results showed a significant variation in abundance and distribution of AM fungi between the bamboo species.

Rao and Gupta (1999) collected the soil samples from the rhizosphere of *Prosopis cineraria* from different localities of Rajasthan state. Eight different VAM fungi belonging to the genera *Acaulospora, Gigaspora* and *Glomus* were isolated and used to study mycorrhizal preference. Among VAM fungi tested *G. fasciculatum* showed preference for *Prosopis cineraria.*

Gurumurthy and Sreenivasa (2000) conducted a survey to isolate the predominant native VAM fungi in Northern Karnataka, at different locations (Dharward, Prabhunagar and Sirsi) in the rhizosphere of five tree species, namely teak silver oak, Casuarina, shisham and tamarind. Occurrence of native endophytes of VAM fungi in the rhizosphere soil of teak and silver oak differed in their spore count at different locations. They were found to be higher in the rhizosphere soil samples collected in Dharwad followed by Sirsi and Prabhunagar. Two predominant VAM spore types were observed in the rhizosphere soils of teak and silver oak. Highest spore counts were observed in the rhizosphere soil samples of tamarind collected in Dharwad followed by Sirsi and Prabhunagar, while in Casuarina and shisham the spore counts were higher in soil samples collected from Dharwad followed by Prabhunagar and Sirsi. Two predominant VAM spore types were observed in the rhizosphere soils of tamarind, shisham and Casuarina.

Uniyal and Uniyal (2000) studied the population dynamics of arbuscular mycorrhizal fungi in *Dalbergia sissoo* Roxb. Arbuscular mycorrhizal (AM) fungi associated with *Dalbergia sissoo* in natural and planted stands were assessed for their qualitative and quantitative distribution in soil samples (from the surface to a depth of 25 cm around the lateral roots) from 9 localities in Uttar Pradesh (India). A total of 17 species of *Glomus*, 5 species of *Scutellospora*, 6 species of *Sclerocystis*, 4 species of Acaulospora and 2 species of *Gigaspora* were recorded. The frequency of species varied from 2 to 20. Among the genera, *Glomus* was the most common followed by *Acaulospora, Sclerocystis* and *Scutellospora.*

Rajeswari *et al.* (2000) conducted a survey to assess the natural arbuscular mycorrhizal (AM) colonization in the rhizosphere of 2-3 year old *Casuarina equisetifolia* seedlings in various places in Tamil Nadu. High levels of AM fungal spores were present in Cuddalore, with lower levels in Marakkanam and Kanyakumari, although AM fungal infection percentage was greater of Kanyakumari than Cuddalore and Marakkanam. AM spore presence and infection were not affected by soil pH (4.5 - 7.5). *Glomus* was the most common genus of AM found, being present at all ten sites surveyed.

Devagiri *et al.* (2001) screened twenty nine seed sources representing natural distribution range of *Dalbergia sissoo* in India and in Nepal for the variations in mycorrhizal colonization and nutrient uptake and to identify good mycorrhizas forming seed sources. VAM [vesicular arbuscular mycorrhizas] colonization rate varied significantly between 24 to 54% and 17 to 46% while mycorrhizal root length ranged from 6-32 and 5-22 cm among seed sources bring the first and second year, respectively. This showed influence of seed source on mycorrhizal colonization. Variations in number of nodules and nodule dry weight were significant and were positively correlated with mycorrhizal colonization and seed source. Seedling growth, dry matter accumulation and nutrient uptake varied significantly and improved in seed sources with good mycorrhizal formation which indicated mycorrhizal effect and this response was likely to be seed source dependent. Among the seed sources tested, Likhimpurkheri, Tanakpur, Allahabad, Badaun and Kalpi appeared to be the best germplasms with regard to mycorrhizal fomation, seedling growth and biomass production. Therefore, these seed sources could be considered while making genotypes with good symbiotic response and for making nursery stock for afforestation of degraded lands. The utility of seed source/fungus genotype- dependent differences in VAM symbiosis in tree breeding is discussed.

Tamuli and Boruah (2002) studied the vesicular-arbuscular mycorrhizal (VAM) association of agarwood tree in Jorhat District of the Brahmaputra Valley. A survey for collection of root and rhizosphere soil samples of agarwood (*Aquilaria malaccensis*) plant from various plantations of Jorhat district of the Brahmaputra Valley, Assam, India, revealed variation in percent root colonization and number of VAM propagules in the rhizosphere soil samples. Different VAM fungi were isolated, of which the genus *Glomus* was dominant. Among the *Glomus* spp., *G. fsciculatum* was the most dominant followed by *G. aggregatum.*

Vijaykumar and Abraham (2002) undertaken a survey of mycorrhizal and rhizosphere fungi of some endemic trees (*Gluta travancorica, Bentinckia condapanna* and *Myristica malabarica*) of Western Ghats, Kerala was conducted in 1997. Significant variations were observed in the species composition of the rhizospheric fungi and arbuscular mycorrhizal (AM) fungi associated with the different tree species selected. Seasonal variation of the spore population in the rhizosphere zone was observed. The lowest spore population in the rhizosphere zone was observed during June and July and the spore population was highest during December to February. *Glomus* was the most commonly occurring AM fungus of the plants selected.

Raj and Chandrashekar (2003) conducted a study in Karnataka to determine the assocition of endomycorrhizal fungi with four endemic dipterocarps such as *Dipterocarpus indicus, Hopea parviflora, H. ponga* and *Vateria indica.* Soil samples from the rhizospheric regions and mixed to form a composite sample. The spores isolated from the soil were identified and feeder roots collected from the trees were treated with 10% KOH and stained by trypan blue. A total of 36 species of mycorrhizal fungi belonging to five genera were associated with the dipterocarps of the Gersoppa region. Of these, *Glomus* was represented by 20 species, *Acaulospora* by 8 species, *Scutellospora* by 4 species, and *Gigaspora* and *Entrophospora* by only 2 species each. *Acaulospora* sp., *Glomus ambisporum, Glomus fasciculatum, Glomus macrocarpum, Glomus* sp., and *Scutellospora* sp. were associated with all the four species of dipterocarps. *A scrobiculata, E. colombiana, Glomus fragilistratum, Glomus leptotichum, Glomus pustulatum* and *Glomus tenerum, Glomus leptotichum, Glomus pustulatum* and *Glomus tenerum* were associated only with *V. indica. V. indica* was associated with a maximum of 31 mycorrhizal species, followed by *D. inducus* with 22 species, *H. ponga* with 14 species and *H. parviflora* with 16 species. The infection percentage varied from 84% in *H. parviflora* to 92% in *V. indica.* The degree of infection was almost the same in the four species of Dipterocarpaceae.

Specific Habitat/ Regions

Kehri *et al.* (1987) studied the occurrence and intensity of VAM in weeds, ornamentals and cultivated plants at Allahabad and areas adjoining it. A variety of spores were recovered from the soils mainly belonging to the genus *Glomus*, however, azygospores of *Acaulospora* or *Gigaspora* and sporocarps of *Sclerocystis* were also recovered but very rarely. The spore population of *Glomus* spp. varied with the plant species and soils. Even the plants of same family differed in the intensity of mycorrhizal association in a particular soil. Similar differences were recorded in plants of same species collected from different soils.

Mohan and Natrajan (1988) studied the VAM association in sand dune plants in the coromandel coast. The occurrence of VAM fungi in the roots of 26 plant species belonging to 16 families of Angiosperms growing in the beach between Ennore and Mahabalipuram was studied. Twenty plant species belonging to 13 families had VAM fungal infection in form of vesicles. VAM fungal spores were also observed in the sand samples collected from the root zones of 26 plant species. Based on the spore characters, 8 species of the genus *Glomus* were identified: *Glomus clarus, G. claroideum, G. fulvus, G. intraradices, G. microcarpus, G. occultrum* and *G. pubescens*.

Mohankumar *et al.* (1988) made a survey of 53 plant species from 30 families growing along the Madras (Tamil Nadu) coast (Sandy beach soils). They revealed that most plants harboured VAM fungi. The VAM species were *Entrophospora schenckii, Glomus claroideum, G. clarum, G. intraradices, G. microcarpum, G. monosporum, G. occultum, G. pubescens* and *G. pustulatum*. Soil temperature and moisture status of the soil influenced the infection of VAM fungi.

Vijaylakshmi and Rao (1988a) tested the vesicular-arbuscular mycorrhizal association 10 members of Asteraceae and 7 of Amaranthaceae growing in red laterite soil. All the plant species examined exhibited mycorrhizal infection but the incidence varied with the species, the highest being found in *Blumea* and the lowest in *Zinnia*. Chlamydospores of *Glomus mosseae* and *G. macrocarpum* were common in rhizosphere of almost all plants. Azygospores of *Gigaspora margarita* were abundant in the rhizosphere of *Tridax*. Sporocarps of *Glomus ambisporum, G. macrocarpum, Sclerocystis clavispora, S. pakistanica* and *S. sinuosa* were also found in the rhizosphere of several plants. There was a positive correlation between the relative abundance of spores of VAM fungi in the rhizosphere soil samples and percent infection of the host species concerned.

Nair and Girija (1988) recorded occurrence of VA mycorrhiza in 20 randomly selected tree crops of Kerala. VAM fungi were present in all the trees but the percentage infection varied. The maximum was 71.9% in rubber, followed by 70.6% in bread fruit, 62% in coconut, 59.4% in *Eucalyptus grandis*. The incidence of VAM fungi was much lower in silk cotton (31.1%), rambuttan (29.3%), arecanut (22.1%) and sugar apple (19.9%).

Ragupathy *et al.* (1988) studied the distribution of VAM in Thanjavur district flora. Mycorrhizal association in 98 plant species collected from Thanjavur district, in Tamil Nadu was studied. In 48 species, mycorrhizal infection was recorded. Of these, 7 plants were aquatic. *Spirodela polyrhiza,* an aquatic plant displayed heavy VAM infection. Generally, the quantum of infection ranged from 20-57%. Spores of *Entrophospora schenkii, Glomus aggregatum, G. amvisporum, G. citricolum, G. etunicatum, G. fasciculatum, G. inraradices, G. microaggregatum, G. occultum, G. pubscens, G. pustulatum, Sclerocystis clavispora* and *S. pachycaulis* were collected from the soil samples.

Srinivas *et al.* (1988a) made a survey for the occurrence of native VAM fungi. A survey was conducted on the occurrence of VAM fungi in Coimbatore, Periyar and Nilgiri districts of Tamil Nadu. Four *Acaulospora* species, i.e., *A. bireticulata, A. elegans, A. laevis,* and *A. mellea,* 6 *Glomus* species, i.e., *G. claroideum, G. constrictum, G. geosporum, G. fasciculatum, G. mosseae* and *G. tortuosum* and one *Gigaspora* sp. were isolated and identified. The spore characters and spore density of these Endogonaceae have been described.

Kannan and Lakshminarasimhan (1988) made a survey of VAM of maritime strand plants. Forty-eight plants species from 32 families were screened for mycorrhizal association in samples collected from 4 different sites in 1985 and 1986, in Point Calimere, a village on the east coast of India bordered by Bay of Bengal on the east and Palk Straight on the south. The mycorrhizal plants were of vesicular type and vesicular-arbuscular type. The extent of mycorrhizal colonization in each species and the spore abundance in the rhizosphere soil of each species were studied. Of the 6 VAM fungal species distinguished, spore types of *Glomus fasciculatum, G. macrocarpum* and *Sclerocystis rubiformis* were the most abundant; also present were *Gigaspora* sp., *S. sinuosa* and *Acaulospora bireticulata.*

Rachel *et al.* (1989) investigated the VA mycorrhizal infection in different angiospermic plant species growing in semi-arid soil of Andhra Pradesh. With the exception of *Solanum surattense* all the plant species under investigation representing 20 angiospermic families revealed mycorrhizal association. Percent colonization varied with the plant species even within the family.

Kiran Bala *et al.* (1989) screened different plant species of Indian Desert for their mycorrhizal association. All the 17 spp. screened, were mycorrhizal however, the intensity of infection varied. Root infection >50% was recorded for *Acacia tortilis, Azadirachta indica, Bauhinia racemosa, Crecentia alata, Prosopis juliflora, Gliricidia maculata* etc.

Glomus and *Gigaspora* were common VAM genera found associated with roots. In *Opuntia* spp. and *Euphorbia* spp. infection rate was comparatively higher (30-100%) than in trees. In *Azadirachta indica* (neem tree) the infection was observed in roots up to 250 cm depth. Intensity of VAM infection varied with availability of water. Deep-rooted growth habit along with VAM infections of desert vegetation may be survival mechanism in competition for water and nutrients with shallow-rooted and fast growing plant species.

Kannan and Lakshminarasimhan (1990) studied the development of endomycorrhizae in *Phyla nodiflora* growing in saline sandy loam soil. They have collected root samples and rhizosphere soil samples at monthly intervals from February to June 1986 (post-monsoon to summer). Percent root colonization, the presence of diagnostic structures of VAM fungi and spore density were determined. The soils were found to be saline-alkaline and nutrient deficient. In the samples examined, the phenology of mycorrhiza formation namely, formation of mycorrhizal structures such as infection threads, arbuscules, vesicles, intraradical spore and extramatrical spore production were recorded. Abundant mycorrhizal colonization was observed in summer. The occurrence of mycorrhizal spores in rhizosphere soil and high level of root colonization indicated that the VAM association made the plants salinity tolerant.

Tarafdar and Rao (1990) surveyed Indian arid-zone species for the occurrence of VAM infection. According to them VAM infections were present in 11 of the 12 plants from 9 environmentally harsh sites in Rajasthan. The percentage and density of infection varied with the plant and the location. Most of the VAM fungi belonged to genera *Gigaspora, Glomus, Acaulospora* and *Endogone*.

Mathew *et al.* (1990) reported the VAM fungi from the rhizosphere of desert cacti. Most of the samples of root and rhizosphere soil from *Opuntia* spp. collected in Jodhpur, Rajasthan contained VAM fungi with *Glomus macrocarpum* predominating. Percentage root infection ranged from 12 to 35.6% and greater number of spores were isolated in spring than in winter. The fungi were able to survive very harsh conditions with soil moisture contents of 3.5% in winter and 2.0% in summer and pH values from 8.3 to 8.9.

Thapar and Uniyal (1990) surveyed VAM fungi of saline soils of Haryana State. A total of 10 species of Endogonaceae were isolated from the alkaline soils deficient in organic matter and able to support sparse vegetation. The greatest number of VAM spores were detected under *Acacia nilotica, Prosopis juliflora, Dalbergia sissoo*, some of the few plants able to survive in the soils.

Chaudhury and Sengupta (1990) studied the VA mycorrhizal association in successional stages of mangrove vegetation at the Ganges river estuary. Their study revealed the presence of VA mycorrhizal association in 24 mangroves, 10 mangrove associates and 18 nonlittoral species. In general, root infection intensities of mangroves at the early successional stages were less than at later stages. In the saline areas non-mangrove showed less root infection than true mangroves. VA mycorrhizal inoculum and spore densities of the successional stage rhizosphere soils were inversely related to levels of soil water salinity. Spores of 9 different Endogonaceous spore types were isolated from the soils of the ecosystem.

Venkataramanan *et al.* (1990) gave the details of qualitative and quantitative distribution of VA mycorrhizal spores in soil samples from Assam, Arunachal Pradesh, Manipur, Mizoram and Nagaland. The abundant spore counts were present in the plants of Assam, while hilly regions contained fewer spores and those from Mizoram none. It has been suggested that in hilly areas inoculation trials with VAM fungi may prove beneficial to crops such as *Capsicum*, soyabean etc.

Gupta and Ali (1990) reported the incidence of VAM fungi in Raipur soils, Chhattisgarh region of Madhya Pradesh. Soil samples showed the spores of 8 *Glomus* spp. and a *Sclerocystis* sp. Root association with VAM was reported in 19 plants belonging to 10 families including *Amaranthus spinosus* and *Eclipta alba* previously regarded as non-mycorrhizal.

Thapar *et al.* (1991) surveyed the native VAM fungi of sodic soils of Haryana State. They reported data from 3 districts, Kurukshetra, Hisar and Rohtak, on isolations of species of Endogonaceae - which form the normal component of soil microflora in the sodic soils of this state under barren and vegetated conditions. Land use included fallow, grasslands and planted *Prosopis juliflora, Acacia nilotica, Dalbergia sissoo,* and *Eucalyptus*. They reported VAM species, with *Glomus* most abundant, 13 spp.; *Scutellospora* 1 sp., *S. nigra* and *Sclerocystis* spp.

Ragupathy and Mahadevan (1993) examined 737 plant species belonging to 121 families of angiosperms and 4 species of Pteridophytes for mycorrhizal association in the Thanjavur District of Tamil Nadu. Only 372 species showed mycorrhizal infection. The percentage of mycorrhizal colonization varied from 10 to 90% with 35 species showing >75% colonization. Of 4 Pteridophytes examined, *Isoetes coromandelina* showed mycorrhizal association by *Entrophospora schenkii* and *Glomus aggregatum* in its rhizosphere. A total of 40 VAM fungal species belonging to 6 genera (*Acaulospora, Entrophosphora, Gigaspora, Glomus, Sclerocystis* and *Scutellospora*) were isolated from the rhizosphere soils of different ecosystems, with *Glomus fasciculatum* (13.8%) and *G. aggregatum* 9.11%) prevalent and *Acaulospora foveata* (0.2%), *A. longula* (0.5%) and *Glomus hoi* (0.9%) the least represented as mycorrhizal spores.

Rachel *et al.* (1993a) studied the distribution of VA-mycorrhizal fungi in four different soils of Andhra Pradesh. Studies on distribution of vesicular-arbuscular mycorrhizal fungi in four different soils (sandy loam, black loam, red loam and clay) with reference to 21 angiospermic hosts revealed that majority of plants exhibited more infection in sandy loam soil followed by red loam, black and clay soil. The percentage of infection varied from 2.0-83.4; 21.7-72.6; 24.5-64.2 and 20.8 - 59.8 in sandy loam soil, red loam, black and clay soils. All the four important genera of VAM fungi namely *Glomus, Sclerocystis, Gigaspora* and *Acaulospora* were found to be commonly distributed in the present four soils.

Lakshminarasimham and Vijaykumar (1994) gave details of VAM fungi associated with 4 different soil types in semi-arid regions of Andhra Pradesh used for groundnut cultivation.

Mathur and Vyas (1994) studied the VA mycorrhizal relationship of *Simmondsia chinensis* collected from 8 localities in Jodhpur. All the plants showed well developed VA mycorrhizal associations with 6 fungi *Glomus fasciculatum, G. mosseae, G. macrocarpum, Gigaspora margarita, Scutellospora calospora* and *Acaulospora* spp. of which *G. fasciculatum, G. mosseae* and *S. calospora* were most common.

Sengupta and Chaudhuri (1994) showed associations of a typical fungal endophytes in addition to typical VA mycorrhizas in the roots of mangrove plants and associated vegetation inhabiting different successional stages at the Ganges river estury. Two dark septate mycelial endophytes, widely distributed and forming intracellular hyphal coils and microsclerotia like structures in both mangrove and non-mangrove roots, were isolated from infected roots and cultured. One resembled a *Rhizoctonia* sp., another formed pycnidia in stromatic masses (unidentified). Both dark septate endophytes infected wide variety of non-mangrove plants roots following inoculation, forming intracellular structures similar to those in mangrove roots and comparable to VAM. When the isolates were used for root inoculation of *Cajanus* seedlings significant increase in growth occurred under conditions of nutrient stress, partly through mobilization of insoluble phosphate.

Muthukumar *et al.* (1994) examined 38 wild legumes, including woody species from the 3 sites in the Western Ghats region of southern India (Tamil Nadu) for their symbiotic association with vesicular-arbusculr mycorrhizal fungi. This study reports the extent of root colonization and the density and distribution of VAMF spores in the rhizosphere soils. All the plant species examined invariably showed a high percentage (66.7-100%) of root colonization. Species of *Acaulospora, Gigaspora, Glomus, Sclerocystis* and *Scutellospora* were isolated and *Glomus* spp. constituted 74% of the total number of species recorded. The spore density ranged from 11 to 126 spores g/litre soil. the intensity of root colonization and spore density varied not only with plant species and genera within a family but also with sites.

Hariram and Santhaguru (1995) studied the occurrence of VAM spores in the rhizosphere of plant species belonging to 23 families in Madurai. Altogether 38 species of VAM fungi belonging to 6 genera *viz., Acaulospora, Entrophospora, Gigaspora, Glomus, Sclerocystis* and *Scutellospora* were recorded. The pH of the soil samples ranged from 5.7 to 8.0 and the EC value ranged from 0.1 to 1.5 d sm-1. The K and P content were high in comparison with N content. The richness of the number of spores of VAM fungi correlated with low soil N content.

Bisen *et al.* (1995) studied the VAM fungus *Glomus mosseae* in soil samples of Bhopal. Soil samples were collected from natural plantations, Artificial plantations, Agricultural lands and Grass fields. They reported the maximum spore counts in winter and minimum in summer, irrespective of their sites.

Raja *et al.* (1995) studied the mycorrhizal association of pteridophytes of Nilgiris and Kodaikanal Hills, South India. Vesicular-arbuscular mycorrhizae were present in 24 of the 27 species in Nilgiris and 13 of the 16 species in Kodaikanal pteridophytes. *Angiopteris arecta, Dicranopteris linearis, Loxogramma paraleta, Pteris arguriea, P. confusa* and

Adiantum uneipinnulum showed more than 75% colonization. The highest mycorrhizal frequency occurred in Angiopteridaceae, Dicranopteridaceae, Drynariaceae, Pteridaceae, Adiantaceae and Lindsaeaceae. No mycorrhizal infection was observed in Athyriaceae and Thelypteridaceae. In terricolous species growing in both hills, mycorrhizal infection was 100%. In epiphytic species growing in Nilgiris and Kodaikanal hills, it was 88 and 63% respectively. Thirteen VAM spores were isolated from the rhizosphere soils of 10 terrestrial ferns growing in Nilgiris; 2 species from *Gigaspora*, 10 species from *Glomus* and one from *Sclerocystis*. Of the 6 terrestrial ferns, 9 VAM species were isolated from Kodaikanal rhizosphere soils, one each from *Acualospora* and *Gigaspora*, 5 from *Glomus* and 2 from *Sclerocystis*.

Mohan and Verma (1995) studied the VA mycorrhizae association in seedlings of forest tree species in arid zones of Rajasthan. Their investigation revealed that VAM fungi are most common, occurring in all the tree seedlings examined from various nurseries in the arid zones of Rajasthan. The major tree species were *Acacia (A. nilotica, A. senegal, A. tortilis), Prosopis (P. cineraria, P. juliflora)*,neem (*Azadirachta indica*), rohida (*Tecomella undulata)*, eucalyptus (*Eucalyptus camaldulensis, E. tereticornis*) etc. All the tree species had mycorrhizal associations at the nursery seedling stage. Rhizosphere samples of all three seedlings contained VAM fungal spores. They have identified *Glomus aggregatum, G. fasciculatum, G. intraradices, G. microcarpum, G. monosporum* and *Gigaspora* species. They concluded that prevalence of VAM fungi may be of great importance for plant growth in tropical soils, especially in arid and semi-arid deserts which have extremely low soils moisture and are relatively poor in mineral nutrients.

Mehrotra (1995) studied the VAM associations in plants colonizing overburdened soil at an opencast coal mine site situated at Durgapur, Maharashtra. According to him the plants growing over the mine spoil showed mycorrhizal infection which ranged from 20 to 95%, the lowest being in *Eucalyptus* hybrid, and highest in *Pithecellobium dulce*. He identified six VAM fungal species *viz., Entrophospora colombiana, Glomus ambisporum, Glomus aggregatum, Acaulospora scrobiculata, Scutellospora calospora* and *Glomus* sp. unidentified.

Rodrigues (1995) studied the mycorrhizal status of naturally occurring and cultivated legumes of iron ore mine wastelands in Goa. He showed that all the legumes growing on the iron ore mine rejects were infected by VAM fungi. Percent root infection varied from 19 to 98% lowest being in *Atylosia scarabaeoides* and highest in *Desmodium polycarpum* and *Smithia sensitiva*.

Jamaluddin and Chandra (1997) studied the distribution of VAM fungi in bauxite mine over burden plantation of Amarkantak, Madhya Pradesh. In all, 11 plantation tree species, *viz., Pinus caribaca, P. roxburghii, P. kesiya, Shorea robusta, Eucalyptus* hybrid (*E. tereticornis*), *E. camaldulensis, Grevillea pteridifolia (G. banksii), G. robusta, Acacia auriculiformis, Pongamia pinnata, Tectona grandis* established in overburden in a bauxite mined area in 1980 and 1984 were screened. Measurements of tree growth and soil VAM fungi and colonization were made in plantations of the same tree species in a nearby degraded forest area, and of VAM fungi in adjacent barren bauxite overburden. The data collected showed that barren bauxite mine overburden soil was deficient in VAM fungi but the forest plantations enhanced the VAM population. VAM colonization and spore density varied with species and with age of the species. The population of VAM fungi was more in the plantation in the degraded forest area than in plantations on the overburden, and the tree growth was also greater in the plantation in the degraded forest area than in the overburden site.

Tarafdar and Rao (1997a) screened 37 plant species including legumes, grasses and sedges growing naturally on gypsum-mined soils at Jodhpur. All these species carried vesicular arbuscular mycorrhizal (VAM) fungal infections, which varied from plant to plant. The percentage root infection by VAM fungi was higher among legumes, grasses, and sedges growing on mined spoils than on plants growing on normal soil. Plants on mined spoil had significantly higher K, Ca, Mg and Fe concentrations and lower concentrations of N, P, Zn and Cu. The possibility of employing VAM fungi for rehabilitating gypsum mined spoils has been discussed.

Srivastava *et al.* (1997) studied the occurrence of VAM fungi in vegetated sand dunes of Indian Thar desert. The vegetation in sand dunes of Thar Desert is sparse due to poor rainfall, high wind velocity, sandy texture of soil with poor water holding capacity and nutrient deficiencies. Examination of periodically collected root and soil samples showed that most of the species growing on and around the sand dunes develop mycorrhizas of VAM type under natural ecosystems. Perrenating spores of VAM fungi belonging to genera *Glomus, Gigaspora* and *Acaulospora* were recovered from the mycorrhizosphere soil/root samples of *Acacia jacquemontii, A. nilotica, A. senegal, Aerva persica, Aristida adcensionis, Calotropis procera, Cappans decidua, Cenchrus biflorus, C. ciliaris, C. prieuri, C. segiterus, Crotalaria bushia, Lasiurus scindicus, Maytenus, panicum turgidum, Prosopis cinerasia, Tephrosia purpurea* and *Zizyphus nummularia*. The isolates of VAMF symbionts were cultured in pots on the living roots of *Cenchrus ciliaris* and identified on the basis of developmental structures of the spores. *Glomus fasciculatum* invariably dominated over *G. aggregatum, G. macrocarpum, Gigaspora gigantea* and *Acaulospora elegans*.

Bhaskaran and Selvaraj (1997) studied the seasonal incidence and distribution of VA mycorrhizal fungi in native saline soils of Tamil Nadu. They identified 15 species of VAM fungi from colonized roots and the spore population in root-zone soils. Sporocarps of *Glomus aggregatum* and *Sclerocystis pakistanica* were recorded as dominant. The VAM fungal spore density was relatively high during the summer season at all the sites. Exclusive colonization of *Sclerocystis* sp. in the roots of *Spinifex littoreus* was observed.

Rodrigues and Bukhari (1997) assessed the mycorrhizal association in 29 naturally occurring herbaceous plant species belonging to 26 genera and distributed among 16 families collected from a 12 year old iron ore mine reject dump at Goa. All the plant species showed mycorrhizal infection which varied from 10 to 98% in the different species. Plant exhibiting mycorrhizal infection had typical vesicular or arbuscular and inter- and intracellular hyphal infection.

Raman and Sambandan (1998) investigated the mycorrhizal profile of tannery effluent polluted wastelands in North Arcot of Tamil Nadu. A total of 15 mycorrhial species were isolated and identified.

Ragupathy *et al.* (1998) examined twenty one plant species of coastal sand binder for vesicular-arbuscular mycorrhizal (VAM) association. Only 14 species showed mycorrhizal colonization. VAM colonization ranged from 20 to 85%. The VAM species were *Acaulospora longula, Gigaspora albida, G. gigantea, Glomus clarum, G. claroideum, G. intraradices, G. occultum, G. pustulatum, Sclerocystis pachycaulis* and *S. microcarpus*.

Mehrotra (1998) carried out an extensive survey of the soil and root samples collected beneath some pioneering plants colonizing reclaimed mine spoil at an opencast coal mine site in Maharashtra state, to examine the host and edaphic influence on the occurrence of VAM fungi. 13 plant species were studied to determine the mycorrhizal root colonization and the number of spores of individual VAM fungal species in the rhizosphere. Typical VAM fungal infection was observed in roots of all the plant species examined. Mycorrhizal infection ranged from 20 to 95%. Soil samples contained six spore-forming species of VAM fungi *i.e. Acaulospora scrobiculata, Entrophospora colombiana, Glomus aggregatum, Glomus ambisporum Scutellospora calospora* and a previously undescribed species of *Glomus*. *E. colombiana* and *G. ambisporum* were the leading species in terms of spore densities and frequencies. Phenotypic variations in spores of VAM fungi were possibly related to the physical and chemical properties of the mine spoil. He concluded that the pioneering plant species on mine spoils can cause the development of different populations of VAM fungi. The study also indicated that certain species of VAM fungi have broad environmental requirements.

Sastry and Johri (1999) studied the diversity of vesicular arbuscular mycorrhizal fungi of stressed soils of Bailadila iron ore sites in Bastar region of Madhya Pradesh. Eighteen soil and 10 plant samples were collected. Soil pH varied between 6.5 and 7.5; N content between 0.05 and 1.25% and P content between 2 and 13 ppm. Eighty-nine species of VAM fungi scattered over 6 genera were recovered. *Glomus* species were the most dominant, consisting 56.82% of the total isolates followed by *Acaulospora* (21.35%), *Scutellospora* (15.73%) and *Gigaspora* (3.37%). *Entrophospora* and *Sclerocystis* were represented by a single species each. Natural VAM infection in plants collected from various sites varied between 25 and 90%. Spore population was strikingly low in the soils (0-2000 kg^{-1} soil). There was no direct relationship between soil nutrient status and percentage of infection or spore density. All the eight species of angiosperms and a single species of pteridophytes showed an average to high level of mycorrhizal infection (25-90%). The VAM dependency of the host species in the iron stressed habitats seems to be quite high and may play a significant role in establishment under metallotoxic conditions. The capacity of the native VAM isolates to survive under iron stress may be instrumental in reclamation of disturbed sites.

Pande *et al.* (1999) examined the vesicular arbuscular mycorrhizal infection of agroforestry systems in arid zone. Vesicular arbuscular mycorrhizal (VAM) infections were examined in six tree species of the Indian arid zone, aged 12-15 years, both under rainfed and irrigated conditions. The intensity of infection varied among the species. Maximum infection of root was recorded for *Azadirachta indica* whereas maximum VAM propagules were found in *Acacia nilotica. Glomus* (80%). *Gigaspora* (16%) and *Acaulospora* (4%) were the only VAM genera found associated with the roots. Irrigation reduced 14% of root infection and 16% of VAM spore population, although it enhanced dehydrogenase activity by 9-53%. The intensity of VAM infection varied with the availability of water. The deep rooted growth habit along with VAM infection of desert trees may be a survival mechanism.

Rattan *et al.* (1999) studied the variation in the propagule density of vesicular arbuscular mycorrhizal fungi at rehabilitated water-logged sites in Haryana, where waterlogging was a usual phenomenon that extended for more than nine months annually. Vesicular arbuscular mycorrhizal (VAM) fungi distribution in all the plots represented by various tree species, *viz., Terminalia arjuna, Syzygium cumini, Populus euphratica* and naturally growing *Typha elephantina* was found to be significantly different. Distribution profile showed the dominance of two genera, *viz., Glomus* and *Gigaspora. Glomus* was dominant in the *T. arjuna* plot and showed positive correlation to Olsen P (r=0.74, p=0.05). *Gigaspora* showed positive correlation to organic matter content (r=0.74, p=0.05). Both the genera also predominated at the naturally growing *T. elephantina* plot.

Rodrigues and Jaiswal (1999) carried out the investigations to find out the diversity of vesicular arbuscular mycorrhizal (VAM) fungi in vegetation of coastal sand dunes. Nine species belonging to 4 genera of vesicular arbuscular mycorrhizal (VAM) fungi were recorded. The spores were obtained as azygospores and chlamydospores. The identified species included *Acaulospora scrobiculata, A. spinosa, Gigaspora coralloidea, G. gregaria, G. margarita, Glomus monosporum, G. multisubtensum, Sclerocystis coremoides* and *S. microcarpus*. Average spore count was 300 spores + 17 sporocarps per 100g rhizospheric soil.

Srinivas *et al.* (1999) investigated the mycorrhizal dependency of two multipurpose tree (MPT) species *viz., Acacia melanoxylon* and *A. nilotica* in eleven coal mining affected soils. Mycorrhizal colonization and resting spore population varied from place to place. Both the species, in general, exhibited a high amount of dependency in coal mine disturbed soils than in relatively undisturbed soils. A positive correlation was observed between the per cent of VAM colonization and mycorrhizal dependency in *A. melanoxylon* but no such correlation was recorded in *A. nilotica*.

Wath and Deotare (1999) examined seven species of *Alysicarpus* in Amravati region, Vidharbha, dominant in natural grassland ecosystem for their vesicular arbuscular status. This study reports the percent root colonization, number and distribution of VAM spores in undisturbed ecosystem. All the plant species of *Alysicarpus* examined showed the high degree of root colonization. The response of VAM colonization with soil characteristics was analyzed. The spores of *Glomus, Acaulospora, Scutellospora* and *Gigaspora* were isolated from the rhizosphere soil, of which *Glomus* showed the highest spore number in per 100 grams of soil. Soil pH near to neutral and high potassium content played a positive role in VAM colonization and spore number.

Mohanan and Sebastian (1999) studied the mycorrhizal status of nineteen species of bamboos in Kerala. An investigation on ecto- and endomycorrhizal association in 19 species of bamboo belonging to 8 genera, *viz.*, *Bamboo, Dendrocalamus, melocanna, Oxytenanthera, Phyllostachys, Thyrsostachus, Arundinaria* and *Ochlandra* in natural forests and plantations in the State was carried out. Detailed investigations on relationship between high incidence of emerging culm disease, the most potential disease of *Bambusa bambos* caused by *Fusarium* spp. and mycorrhizal association was made in experimental plots laid out at different localities in Wynad and Nilambur Forest Division. Bamboo root and rhizosphere soils samples collected from 44 localities in the State were assayed by standard procedures. Very high percent infection of vesicular-arbuscular mycorrhizal fungi (VAM) was recorded in *Bamboo* roots collected from the experimental plots in Wynad (81-90%) and Nilambur (88-99%) Forest Divisions. VAM spore count in rhizosphere soil ranged from 65 to 205/10g soil. Spores of *Glomus* spp., *Gigaspora* spp., *Acaulospora* spp., and *Scutellospora* spp., were identified from the bamboo rhizosphere soil. *Glomus fasciculatum, G. mosseae, G. australe* and *G. geosporum* were the most frequently encountered VAM species in bamboo rhizosphere soil. No correlation could be made on the incidence of rot of emerging culm in the plants and level of mycorrhizal infection of the roots of the disease affected clumps. Study on the mycorrhizal status of bamboos in Kerala revealed a very high per cent VAM infection in all the 19 species of bamboo collected from 44 localities of the State.

Rao *et al.* (1999b) studied the VAM status of some plants growing naturally in open cast coal mine soils. They investigated the VA mycorrhizal status of 23 angiospermic plant species, representing 14 families growing on the dumps of open cast coal mine soil. All the plant species investigated exhibited the mycorrhizal colonization. The percentage of colonization varied between 26.2-75.2. Mycelial and arbuscular colonization was observed in all the species. Resting spore population in the rhizosphere soil was found to be very less. *Glomus* and *Sclerocystis* species were found to colonize the plants.

Beena *et al.* (1999) conducted a survey along ten sand dunes across the Karnataka coast to assess the diversity of vesicular arbuscular mycorrhizal (VAM) fungi associated with a dominant sand dune plant *Ipomoea pes-caprae*. The samplings were carried out during wet and dry season during 1996-97. Altogether 25 species of VAM fungi were recorded. The edaphic features (moisture, pH, sodium, potassium, calcium, salinity, nitrogen, phosphate and organic carbon) of rhizosphere sand were analyzed in relation to the status of VAM fungal association. This study also addressed the VAM fungal status at disturbed and undisturbed locations. It was evident that the VAM fungal population was considerably reduced at disturbed locations. The VAM fungal population at different locations and seasons was assessed based on diversity indices.

Selvaraj and Bhaskaran (1999) investigated some sporocarpic VA-mycorrhizal fungi associated with *Spinifex littoreus* and *Aeluropus lagopoides* in saline soils. Altogether ten sporocarps of VAM fungi were collected and identified. *Glomus ambisporum* was recorded as dominant. It was interesting to observe that the colonization of VAM fungi in root cortical cells of *Spinifex littoreus* showed exclusively *Sclerocystis*, though sporocarps and spores of *Glomus* were found in its rhizosphere. The VAM fungal colonization and sporocarp density were relatively very high in summer season in all the sites.

Chandra and Jamaluddin (1999b) studied the distribution of VAM fungi in coalmine overburden dumps. The development of vesicular arbuscular mycorrhizal (VAM) fungi in different tree species in coal mine soils was studied at Kusmunda in Bilaspur district of Madhya Pradesh. All the species (*Acacia catechus, Dalbergia sissoo, Peltophorum pterocarpum, Ailanthus excelsa, Eucalyptus hybrida [E. tereticornis], Acacia auriculiformis, Pithecellobium dulce, Gmelina arborea, Acacia nilotica, Albizia procera* and *Cassia siamea [Senna siamea]*) planted in coal mine soils were infected with VAM fungi. The maximum % root colonization was observed in *A. catechu*, followed by *D. sissoo* in 5-year-old plants, whilst 2- and 3-year-old plants exhibited comparatively lower VAM infection. *Acaulospora* was the most dominant fungus in the coalmine soils and *Glomus* and *Gigaspora* species were also common. *Scutellospora* and *Sclerocystis* were scarce in overburden soils.

Beena *et al.* (2000) investigated the diversity of vesicular arbuscular mycorrhizas (VAM) associated with *Ipomoea pes-caprae* on the coastal sand dunes of the West coast of India. The impact of rhizosphere edaphic features and disturbance on the species richness and diversity was investigated in 10 locations consisting of moderately disturbed dunes (MDD) and severely disturbed dunes (SDD) during wet and dry seasons. Vegetation cover, VAM colonization and species richness and diversity were greater in MDD than in SDD, irrespective of season. Nitrogen and phosphate content were significantly different between MDD and SDD. In both dune types *Glomus mosseae* was the dominant VAM followed by *Glomus dimorphicun, Gigaspora gigantea, Acaulospora taiwania, Glomus fasciculatum* and *Glomus* sp. 27SS.

Muthukumar and Udaiyan (2000b) undertaken a survey of the arbuscular mycorrhizal (VA) status of plants growing in the Western Ghats region of Southern India. Root and soil samples of plants growing in the four-vegetation types forest, grassland, scrub, and cultivated land or plantation were examined. Of the 329 spores (representing 61 families) examined, 174 were mycorrhizal. AM association was recorded in 81 species for the first time, including species from several families assumed to be non-mycorrhizal, e.g. Amaranthaceae, Capparaceae,Commenlinaceae, Cyperaceae and Portulacaceae. AM fungal spores of 35 species belonging to *Acaulospora, Gigaspora, Glomus, Sclerocystis* and *Scutellospora* were recorded. AM fungal species richness was found to be highest in scrub and lowest in agricultural and plantation soils. Mean colonization levels were dependent on plant life form, life cycle pattern and vegetation type.

Bhaduria and Yadava (2000) conducted field surveys to determine the status of VAM fungal associations in *Acacia nilotica* of various forest nurseries and plantations in alkaline (Usar) soil sites of Mainpuri, Uttar Pradesh. The extent of VAM colonization in roots and spore populations in rhizosphere soil were quantified and variations in percentage root colonization and number of VAM propagules in the rhizosphere of both nursery and plantation samples were observed. Mean number of VAM propagules and percentage root colonization were greater in plantation samples than in nursery samples. It was observed that *Glomus fasciculatum* showed the highest frequency (100%) and density (90.5) in the rhizosphere of *A. nilotica* at the plantation site. Similar density and frequency distributions of VAM fungi in the rhizosphere of *A. nilotica* in sampling sites were screened out. *Glomus fasciculatum* showed the maximum percentage of frequency (75.0%) and density (98.5).

Gupta *et al.* (2001) collected soil and root samples from the root zone of different species growing on native saline soils in polybags with non-saline garden soils and non-saline soil under open field conditions. Mangroves growing under different situations in both the saline and non-saline soils showed distinct patterns of AM infection and colonization. Out of the 25 species studied, 13 showed positive mycorrhization in their roots, which ranged from 10 to 60%. The roots collected from plants growing in non-saline open field conditions showed vesicles and arbuscular association in *Acanthus ilicifolius* (10%), *Ipomoea biloba [I. Pescaprae]* (10%), *Bruguiera parviflora* (20%) and *Brownlowin tersa* (20%). *Thespesia populnea* exhibited AM fungi infection, which ranged from 30 to 40% in both saline and non-saline soils. About 60% colonization was observed in *Algaia cucullata* grown on non-saline open field. The saline soil samples from Paradeep port area in the mangrove ecosystem of the Mahandi delta showed the presence of AM fungi, such as *Glomus geosporum, G. macrocarpum, G. magnicaule, F. fasciculatum, G.reticulatum, Entrophospora* spp. and a single species of *Gigaspora.* The plants grown in polybags containing non-saline soil showed dominant association of species of *Glomus, Enterophospora* and *Acaulospora, glomus constrictum, Glomus* spp. *Acaulospora* sp. And *Sclerosystis* spp. was associated with plants grown in non-saline open field conditions.

Harikumar (2001) studied arbuscular mycorrhizal colonization in wetland plants from Kerala. Of the 46 plants screened, 29 had mycorrhizal associations attributed to *Gigaspora* and *Glomus*. The highest percentage of root colonization and intensity of infection was exhibited by *Impatiens chinensis*, followed by *Lymnophylla gratisima* and *Hydrocotyle asiatica.*

Jaiswal and Rodrigues (2001) studied the occurrence and distribution of arbuscular mycorrhizal fungi in costal sand dune vegetation of Goa. The qualitative and quantitative distribution of vesicular arbuscular mycorrhizal (VAM) fungi in coastal sand dune vegetation of Goa was determined. Rhizosphere soil samples comprising three root zones from 9 host plants (and *Achyranthes aspera, Calotropis gigantea, Cusuarina equisetifolia, Cocos nucifera, Flueggea sp., Hyptis suaveolens, Ipomoea biloba [I. pescaprae], Lantana carnara and Vitex trifolia*) were sampled in March 1999, at a depth of 0-25 cm. Average mean spore count was 725 spores + 15 sporocarps/100 g rhizosphere soil. The highest mean spore density was recorded in *Cocos nucifera* 11 156 spores + 60 sporocarps/100 g soil), followed by *Casuarina equisetifolia* (864 spores + 4 sporocarps/100 g soil). The lowest mean spore density was observed in *Lantana camara* (dd8 spores + 16 sporecams/100 e soil). The AM fungal spores obtained belonged to five genera, i.e. *Acaulospora, Gigaspora, Glomus, Sclerocystis* and *Scutellospora*. Amongst the identified species, *Gigaspora coralloidea* was the most frequently occurring fungus(89%),followed by *Scutellospora verrucosa* (67%), *Gigaspora margarita* and *Sclerocystis sinuosa* (44%). The highest number of AM fungal species was observed in root zone soil of *Calotropis gigantean* (7 spp.) and the least was found in *Casuarina equisetifolia* (2 spp.).

Singh and Singh (2001) studied the root and soil samples from various chickpea-growing areas in different districts of Punjab (Ropar, Hoshiarpur, Gurdaspur, Bathinda) to determine the incidence of vesicular arbuscular mycorrhiza (VAM) associated with chickpea. Maximum

(72.5%) and minimum (55.0%) colonization were recorded from Ludhiana and Hoshiarpur samples, respectively. VAM spore count was highest (485 spores/ 50 g soil) in rhizosphere soil samples from Ludhiana and lowest (360 spores/ 50 g soil) in samples from Hoshiarpur. *Glomus* species were the most prominent in all the samples. Mycelium was abundantly present in most of the samples.

Selvaraj *et al.* (2001) conducted a survey in 10 localities in the Vellore district of Tamil Nadu to study the arbuscular mycorrhizal (AM) associations of *Cichorium intybus* in relation to soil physicochemical characters. Maximum root colonization percentage and number of AM spores were found in soil samples from Vellore (92% and 382 propagules/100 g soil, respectively) and Tirupathur (84% and 260 propagules/100g soil respectively), the soil being P-deficient sandy loam with a neutral to alkaline pH (7.2 to 7.4). The lowest colonization percentage and number of AM spores were in samples from Ambur (42% and 73 propagules/ 100 g soil, respectively), where the soil was sandy clay loam, high in phosphorus and had an acidic pH (6.1). A total of 11 different species of AM fungi were collected and identified as belonging to the genera *Glomus, Acaulospora, Gigaspora, Sclerocystis* and *Scutellospora* spp. *Glomus aggregatum* was the most frequent fungus (100% site frequency), followed by *Sclerocystis pakistanika* (80%) and *Gigaspora margarita* (70%). Some *Glomus* species were very common in neutral or alkaline soil but not in acidic soils, while *Acaulospora* and *Scutellospora* species were only found in acidic soils.

Hasan *et al.* (2001) cónducted a survey in 6 salt-affected districts of Uttar Pradesh, to study the degree of colonization, occurrence and spore density of arbuscular mycorrhizas (AM) on weeds. Soil and root samples were collected from 11 commonly occurring weeds. Root colonization in all samples varied from 0 to 80%. The highest degree of colonization occurred on Opuntia spp. (50-61%), followed by *Saccharum spontaneum* (35-59%), *Cynodon dactylon* (30-46%), and *Ipomoea carnea* (30-49%). The least colonized was *Croton sparsiflorus* (17-33%). In general, the spore densities around *Opuntia* spp., *Cynodon dactylon* and *S. spontaneum* were higher than those around the other weeds. *Glomus fasciculatum, Acaulospora spinosa* and *Gigaspora margarita* were the most frequent AM species in the survey sites.

Beena *et. al.* (2001) performed an inventory of arbuscular mycorrhizal (AM) fungal status of 28 plant species belonging to 14 families established on the coastal sand dunes of Southwest coast of India (Karnataka). Roots of 23 plant species were colonized by AM fungi, whereas the rhizosphere of only 20 plant species possessed AM fungal spores. *Canavalia cathartica* had the highest root colonization (83%) by AM, while the rhizosphere of *Borreria articularis* had the highest number of AM spores ($1.6g^{-1}$). Among 30 AM fungi recovered, *Scutellospora erythropa* showed a wide host range (colonized 13 plant species), while *S. gregaria* had high spore abundance per plant species (12.75). The mean species richness on the west coast dunes was 4.4, highest being in *Ipomoea pes-caprae* (11). Among the mat forming creepers, the AM fungal diversity was highest in *Alysicarpus rugosus.*

Guatam and Mahmood (2002) surveyed some cultivated legume crops of Aligarh district to determine the occurrence of AM fungi. Seven root samples of six commonly cultivated leguminous crops (pea, chickpea, pigeonpea, cowpea, mungbean, and clover) were collected to measure root colonization of AM fungi. The AM fungal population was affected by pollutants released for the Kasimpur thermal power station. Pollutant free sites always yielded higher populations of AM spores compared to polluted sites. Among the crops, clover supported the biggest production of spores and cowpea recorded the lowest yield in all 4 sites investigated. At the polluted sites, mungbean shown the second highest number of AM spores followed by chickpea, pea, and pigeonpea. At pollution free sites, the highest and lowest spore numbers were recorded in clover and cowpea respectively. The second highest number of spore was recorded in mungbean, followed by chickpea and pea. The percentage colonization of AM fungi was affected by pollutants. The percent colonization in clover was highest, 60-64% and 70-74%, at the polluted and non-polluted sites respectively. The colonization percentage in mungbean was the second highest in all the 4 sites, with an average of 44%. In pea, pigeonpea, and chickpea the average percentage of root colonization was 54% and 56.5%, respectively.

Muthukumar and Udaiyan (2002a) characterize the population of arbuscular mycorrhizas (AM) in a semiarid grassland and scrubland in the Maruthamalai Hills in Combatoire. Soil samples collected from the rhizosphere of *Acacia* sp. and *Cymbopogon caesius* between October 2000 and January 2001 yielded total of 10 and 8 species from grasslands and scrublands, respectively. These species consisted of *Acaulospora scrobiculata, Scutellospora calospora, S. heterogama, Glomus aggregatum, Glomus geosporum, Glomus intraradices, Glomus microaggregatum, Glomus mosseae, Glomus sinuosum* and *Glomus taiwanense.* The successive trap-culture technique used for the isolation of AM fungi successfully yielded three additional species (*Acaulospora* sp., *Glomus* sp. and *Gigaspora gigantean).* Seven species (*A. scrobiculata, Glomus aggregatum, Glomus geosporum, Glomus mosseae, Glomus sinousum, Glomus viscosum* and *S. calospora)* were detected in both vegetation types. Among these the spor population of *Glomus geosporum, Glomus sinousum* and *S. calospora* were significantly affected the spore populations of *Glomus geosporum, Glomus sinuosum and S. calospora*. The average spore numbers of Gigasporaceae species across both sites were low (3.68/100 g of soil compared to the average spore numbers of many members of Aculosporaceae (9.66/100 g of soil) and Glomaceae (10.32/100 g of soil).

Ammaiyappan *et. al.* (2002) surveyed the presence of arbuscular mycorrhizal (AM) species in an abandoned lignite fly ash pond, overburden dumps and reclaimed overburden dumps. In ash pond, 15 AM (*Acaulospora gerdemannii, Gigaspora decipiens, Gigaspora gigantean, Gigaspora margarita, Glomus citricola, Glomus fasciculatum, Glomus formosanum, Glomus fulvum, Glomus maculosum, Glomus magnicaule, Glomus mosseae, Glomus tenebrosum, Sclerocystis pachycaulis, Scutellospsora erythropa* and *Scutellospora fulgida)* were isolated. From overburden and reclaimed overburden dumps, 4 (*Glomus fasciculatum, glomus mosseae, Sclerocystis microcarpus* and *Scutellospora verrucosa*) and 13 AM fungal species (*Acaulospora gerdemannii, Entrophospora colombiana, Gigaspora gigantean, gigaspora margarita, glomus fasciculatum, glomus macrocarpum, glomus mosseae, glomus vesiculiferum, Sclerocystis pachycaulis, Sclerocystis sinuosa, Scutellospora persica)* was isolated, respectively. In all the sites, *Glomus mosseae* was the dominant AM fungus. From *Acacia auriculiformis* plantation in the overburden dumps, fruit bodies of ectomycorrhizal fungus *Pisolithus tinctorius* were recorded. Soil pH, electrical conductivity, organic carbon and N, P, K content of rhizosphere soils were analyzed to find their influence on mycorrhizal colonization.

Sampathkumar and Ganeshkumar (2003) conducted a study to screen spores and sporocarps of vesicular arbuscular mycorrhizal (VAM) fungi from rhizospheric soil supporting cowpea (*Vigna unguiculata*) and black gram (*V. mungo*) from four localities (Pattukottai, Orathanadu, Papaasam and Vallam) in Tamil Nadu. Soil physicochemical characteristics, VAM colonization, host specificity and fungal plants association were also investigated to determine the effect of VAM fungi and rhizobium applied separately or in combination on cowpea and black gram. The soils were phosphorus-deficient and were of four types: sandy loam, clay sandy loam, black sandy loam and red sandy loam. Root colonization percentage of cowpea and black gram was highest in Vallam (94.8 and 92.5%, respectively). The VAM spore number was 520-970 per 100 g soil. Sixteen VAM fungal species associated with both plants were observed in the rhizosperic soil of the four sites. The genus *Glomus* dominated. Among the VAM species identified *Glomus fasciculatum, Glomus aggregatum, Gigaspora margarita* and *Glomus mosseae* had the broadest host ranges in the sites. *Glomus fasciculatum* and *Glomus aggregatum* were the predominant symbionts. Dual-inoculated plants had significantly increased shoot and root length, total nodule number and nodules dry weight compared to plants inoculated with individual endophytes. Fewer nodules with increased biomass were formed in dual inoculated plants compared to rhizobium-inoculated plants. It is concluded that dual inoculation with mycorrhizal fungi and rhizobium species is more effective in increasing legume growth and biomass production.

Miscellaneous

Zaidi and Mukerji (1983) studied the incidence of VA mycorrhizas in diseased and healthy plants. They studied the rhizosphere VAM spore population of 10 plant species at the seedling stage before the onset of various fungal diseases. When VAM spore population was again observed at the time of maximum disease development there were higher concentrations of extramatrical spores in healthy than in diseased plants. However, more active endomycorrhiza in terms of arbuscules, vesicles and hyphal proliferation were observed in roots of diseased plants.

Mukerji and Ardey (1986) studied the vesicular-arbuscular mycorrhiza in *Kalanchoe spicata* and *Eclipta alba*. The formation of VAM in these 2 plant spp. described. The extent of infection and size and shape of vesicles varied.

Ardey and Mukerji (1986) have given a short note on the association between *Olpidium brassicae, Glomus fasciculatum* and *Glomus fuegianum* in *Sorghum bicolor* and *Eclipta alba*.

Selvaraj *et al*. (1986) reported the occurrence of VA mycorrhizal fungi in roots and scale-like leaves of *Canna indica* (Cannaceae).

Singh *et al.* (1987) investigated the VA mycorrhizal fungi associated with diseased and healthy plants of *Vicia faba*. Roots of diseased faba bean plants in experimental plots were reported to be infected by *Fusarium oxysporum* and the senescent and dried aerial parts were invaded by the saprophyte *Aspergillus sydowii*. Diseased plants showed reduced growth, yield and root nodulation but had more VA mycorrhizal infection than healthy plants. Soil samples from the vicinity of diseased roots also contained more endomycorrhizal spores than did those from healthy roots.

Rekha Rani *et al.* (1987) studied the mycorrhizal status of some ornamental plants at their reproductive and vegetative stage. They reported maximum VAM infections at the flowering stage and least during the vegetative stage in 6 of the 8 plants (7 families), the exception being *Antirrhinum* and *Dianthus*.

Harinikumar *et al.* (1988) surveyed air-dried earthworm cast collected over a period of one year from culture troughs and examined for the presence of VAM propagules by inoculating the cast to onion (*Allium cepa*) plants grown in sterilized soil. Mycorrhizal propagules were present in the earthworm cast. The most probable number of infective propagules varied from 2 to 54 per gram sample. Maximum mycorrhizal propagules were observed in the worm cast collected during winter. The mycorrhizal propagules survived in earthworm casts for up to 11 months, suggesting the role of earthworms in the dissemination of VAM fungi.

Joseph *et al.* (1988) studied the VAM infection in different leguminous cover crops of rubber. With a view to study the mycorrhizal formation in 4 common cover crops of rubber, *Puereria phaseoloides, Mucuna bracteata, Centrosema pubescens* and *Calopogonium mucunoides,* a field study was conduced. The cover crops differed significantly in mycorrhizal infection and count of spores in the soil. *M. bracteata* recorded highest percentage of infection followed by *C. mucunoides, P. phaseoloides* and *C. pubescens*. However the number of spores of mycorrhizal fungi in soils was highest in *C. pubescens* followed by *M. bracteata, C. mucunoides* and . *P. phaseoloides*. There was no direct relation between the percentage of infection and number of spores in soil. This study clearly shows the mycorrhiza formation in leguminous cover crops of rubber under Indian conditions.

Harinikumar and Bagyaraj (1990) examined 4 different species of wasps namely, *Sceliphron madraspatanum, S. spinoli, Eumenes conica* and *Rhynchium* sp. Out of 174 samples tested, onion plants did not grow in 101 samples, which constituted 58% of the samples tested. It is possible that nest mud used by wasps was phytotoxic and did not encourage the growth of onion seedlings. Out of 73 samples in which onion seedlings grew, only 27 samples (constituting 37%) were positive for VAM colonization. Considering the individual species of wasps VAM colonization was positive in 47% of the *Sceliphron madraspatanum* samples, 40% of the *S. spinoli* samples and 29% of the *Eumenes conica* samples, in which plants established. None of the samples of *Rhynchium* sp. showed mycorrhizal propagules. *Rhynchium* spp. are known to secrete a gummy substance from the body which is used in the construction of nests. It is likely that this gummy substance may inhibit the development of VAM fungi, which warrants further investigation. The extent of VAM colonization was maximum in *Sceliphron madraspatanum* nest samples (15%) followed by *S. spinoli* (12%) and *Eumenes conica* nest samples (9%). Similar trend was seen in the number of infective propagules. *Sceliphron madraspatanum* nests had 310 propagules while those of *S. spinoli* and *E. conica* had 190 and 92 propagules respectively.

Sceliphron and *Eumenes* which are the common nest building wasps in India thus may act as potential vectors in disseminating VAM fungi.

Vasanthakrishna *et al.* (1990) studied occurrence of vesicular arbuscular mycorrhizal fungi in tropical ornamental plants. belonging to 6 families namely, Bromeliaceae, Piperaceae, Acanthaceae, Uriticaceae, Gesneriaceae and Araceae. All the 16 plant species harboured VAM fungi in their root system. The mycorrhizal spore numbers in the root zone ranged from 128 to 136/50 g soil. Maximum spore numbers (136) were observed in the root zone of plants belonging to the family Acanthaceae and the minimum (128) in family Araceae. Data showed that these beneficial fungi may be utilized to produce healthy, vigorously growing ornamental plants with more market value.

Ghazala Nasim (1990) investigated the VAM association in portions other than root. A list of mycorrhizal plants consumed as food or used as medicines is extended and the mycorrhizal status of nonroot portions of many of these reported for the first time. Non root portions had heavy vesicular infections and lacked arbuscular infections, the VAM mycelium became septate and frequently filled the cell lumen.

Rangeshwaran *et al.* (1990) studied the VAM association in coffee. According to them VAM infection of arabica coffee in 3 different growing areas in Karnataka, was in the range 28-38, 32-64 and 21-30% respectively, while in robusta it was 14-38, 35-68 and 10.34%. *Glomus macrocarpum* and *Sclerocystis rubifermis* spores were isolated from soils of 2 areas while the third yielded *Gigaspora* spp.

Lakshman and Raghavendra (1990b) found the spores in worm casts, mud nests of wasps, balls of dung and fowl droppings and they concluded that earthworms are an important means of distribution of these spores in the soil.

Mago *et al.* (1993) investigated the colonization by VAM fungi of scale leaves of *Zingiber officinale* rhizomes, *Colocasia esculenta* tubers and *Amorphophallus campanulatus* corms. The highest degree of colonization occurred in *Z. officinale*, with abundant hyphae present. No infection was observed in scale leaves of *A. campanulatus*.

Mehrotra and Baijal (1993) studied the vesicular-arbuscular mycorrhizal formation in an ornamental plant, *Codiaeum pictum* L. Intensive VA mycorrhizal colonization was observed in roots of *Codiaeum pictum*. Sixty per cent of the root segments were found to be colonized by VAM species. *Glomus aggregatum* found associated with the host plant.

Harinikumar and Bagyaraj (1994) investigated the effects of earthworms, termites, ants, and millipedes on the dissemination of propagules of vesicular-arbuscular mycorrhizal (VAM) fungi. Earthworm (*Lumbricus terrestris*) casts were air dried and stored; the presence of VAM fungi in the cast material was demonstrated by the successful inoculation of sterile grown onion (*Allium cepa*) plants with soil subsamples. Soil from ant nests (*Camponotus compressus*) contained 790 infective propagules per g of nest material. Faecal pellets of the millipedes (*Phyllogonostreptus nigrolabiatus*) also contained VAM propagules, but their viability was lost after 4 days of storage. Samples of the termitaria of both mound-building and subterranean termites contained non-viable VAM propagules.

Iyer and Bhat (1995) carried out a survey to identify various VAM fungi prevalent in arecanut, coconut and oil palm farming systems from different regions of India so that cultures of predominant VAM species could be established for future studies and field use. They presented data on infection grading crop-wise and location-wise. Among the three palm studied *viz.,* Arecanut, coconut and oil palm, oil palm was mycorrhizal to a greater extent when compared to coconut or arecanut. When grown under similar conditions, loose soils with more of sand supported VAM colonization better than soils containing clay fractions, irrespective of the crops grown.

Kumaran and Santhanakrishnan (1995) studied the VAM fungi in tea (*Camellia sinensis*) soil of Nilgiris, Tamil Nadu. They observed that VAM fungi propagules, in different soil samples screened, ranged from 40 to 450 / 100 ml of soil. Different types of VAM fungal spores belonging to *Glomus, Gigaspora* and *Sclerocystis* were encountered, however, the *Glomus* species were found to be abundant in most of the soils. Infected roots showed well developed coenocytic mycelium and oval or spherical vesicles but the arbuscules were not clearly discernible. Percent VAM colonization varied from 0-24%, highest being in Ooty region. They recorded a negative correlation between the available P level and infective propagules of tea soils.

Joseph *et al.* (1995a) studied the vesicular-arbuscular mycorrhizal colonization in ginger and its influence on the natural incidence of rhizome rot. Idukki and Wynad districts of Kerala were selected for the present study as these two districts are the main ginger-growing tracts of Kerala. Altogether, 22 locations were marked in the Idukki district and 25 in the Wynad district. Samples of healthy and diseased ginger plants were collected from these locations. The data indicated that ginger was a heavily mycorrhizal plant and the percentage of VAM colonization and intensity varied with locality. The percentage of colonization and intensity was found to be more in the Idukki district than in the Wynad district. Generally, the percentage of colonization and intensity were higher in the roots of healthy plants. Similarly, in those samples where the disease intensity was higher, the percentage of VAM colonization and intensity was comparatively less. The study indicates that ginger is largely colonized by VAM fungi and the high degree of VAM colonization tends to reduce the intensity of rhizome rot. A higher percentage of VAM colonization and intensity was observed in healthy plants when compared to that in the diseased plants.

Lee *et al.* (1996) studied the vesicular-arbuscular mycorrhizal fungi in earthworm casts and surrounding soil in relation to soil management of semi-arid tropical Alfisol. Spores and infective propagules of vesicular-arbuscular (VA) mycorrhiza were examined in earthworm casts and field soils collected from three different soil management treatments (zero tillage with straw amendment, deep tillage without amendment (bare), and zero tillage with previous perennial cropping of *Stylosanthes hamata*) on an Alfisol in the semi-arid tropics (SAT) of India. The average mean of spore counts and most probable number (MPN) of infective propagules of VA mycorrhiza (VAM) were significantly ($P<0.05$) higher in earthworm casts than in field soil across the three soil management treatments. There was no significant difference in the number of VAM spores or propagules among field soils from the three different soil management treatments, but the number of VAM spores and propagules in the earthworm casts from the deep tillage bare treatment was significantly higher ($P<0.05$) than in the earthworm casts from the other two treatments. In the deep tillage bare treatment, the number of spores and MPN of infective propagules were significantly ($P<0.05$) higher in earthworm casts than in field soil. Therefore, it was concluded that earthworms can concentrate VA mycorrhizal spores and propagules in their casts.

Vijaykumar and Abraham (1998) studied the arbuscular-mycorrhizal association of *Bentinckia condapanna*, Berry ex Roxb., an endemic palm of Western Ghats. *Glomus intraradices* was the most frequent fungus in rhizosphere. Inoculation studies were carried out with this fungus alone and with the addition of 'P' as KH_2PO_4. VAM inoculation enhanced seedling growth of *B. condapann*. The addition of 'P' did not reduce the dependency of the plants on the VAM fungus.

Srivastava and Chandra (1998) made an extensive survey of mycorrhizal association in some floricultural crops of importance in national/international trade viz. chrysanthemum, gladiolus, carnation, dahlia, lily, rose, tagetes, cosmos, hollyhock, phlox and verbena. Samples of roots and rhizosphere soils were collected from nurseries, gardens, parks and fields in and around Allahabad and analyzed for intensity of mycorrhizal infection in the roots and population of mycorrhizal spores in the rhizosphere soils. All the plant species included in the survey exhibited mycorrhizal infection in their roots, however, the intensity varied with the sites, plant species and their stages of growth. Irrespective of sites or stages of growth, some of the crops e.g. chrysanthemum, tagetes, rose etc. showed very high range of root infection (upto 70%). The population of mycorrhizal spores in the rhizosphere soils also varied with the sites, plant species and stages of growth. Comparatively higher population (up to 125 spores/ 20 g oven dry soil) was recorded in the rhizosphere soils of some crops e.g. chrysanthemum, dahlia, phlox etc. A critical analysis indicated that the intensity of root infection and population of spores in the rhizosphere soils were not directly correlated in majority of plant species. The VA mycorrhizal fungi associated with the crops included in the survey were identified and it was found that forms belonging to *Glomus (G. fasciculatum, G. macrocarpum, G. mosseae, G. multicaule*) were the most common. Forms belonging to *Gigaspora, Acaulospora, Sclerocystis* were also recovered occasionally.

Sreeramulu (1998) conducted a survey of different cardamom plantations of Karnataka and Kerala and revealed the existence of mycorrhizal association in cardamom. However, preferential association of native mycorrhiza varied with different cultivars of cardamom, soil type and agro climatic conditions. *Glomus fasciculatum* was predominant in Karnataka and *Glomus caledonicum* was predominant in Kerala.

Shinde and Nair (1999) studied the VAM fungi from earthworm casts of grape vine orchards. Earthworm casts rhizosphere and non-rhizosphere soil samples from grape vine orchards were analyzed for physical and chemical parameters. VAM spore population was assessed by wet sieving and decanting method. One species of *Acaulospora*, three species of *Glomus* and two species of *Scutellospora* were reported from earthworm casts.

Korikanthimath and Gayatri (1999) carried out studies on the association of vesicular arbuscular mycorrhiza (VAM) and other beneficial microbes in the robusta coffee, pepper and cardamom mixed cropping system. The population of VAM spores, phosphate solubilizing bacteria and fungi were compared in the rhizosphere of robusta coffee (*Coffee robusta*), pepper (*Piper nigrum*) and cardamom (*Elettaria cardamomum*) in mono- and mixed cropping systems. The results indicated that in the mixed cropping systems, rhizosphere of robusta coffee harboured significantly more (34 vs 32 per 100 g soil) number of VAM spores and phosphate solubilizing bacteria (4.32 vs 2.52 x 10^4) than cardamom. Mixed cropping brought about significant increase in the number of VAM spores (32 vs 21) and decrease in the phosphate solubilizing bacteria (2.52 vs 3.24 x 10^4) in the rhizosphere of cardamom. The overall microbial association had a better beneficial impact in the mixed cropping systems.

Baby and Premkumar (2001) conducted an experiment to study the incidence of vesicular arbuscular mycorrhizas in tea soils. *Glomus* was present in all plots in various planting districts in India; *Gigaspora* was present only in Anamallais and *Sclerocystis* in Central Travan core. *Scutellospora* was present in Anamallais and Nilgris, while *Enterophospora* was present in High Range. Nilgiri-Wynaad and Wynaad, *Acaulospora* was present in all locations except in Anamalais and Central Travancore. Spore density was highest in High Range followed by Nilgris and least in tea areas of Chickmagalur.

Muthukumar and Udaiyan (2002c) collected root and soil samples of three potted or ground-grown cycads (*Cycas circinalis, C. revolute, Zamia* sp.) between November 1999 and June 2000 and surveyed for arbuscular mycorrhizal (AM) colonization and spore populations. AM fungi were associated with all root systems and rhizosphere samples examined. Root colonization was of a typical Arum type and AM colonization levels differed significantly between species and between potted and ground-grown cycads. Mycorrhizal colonization levels were inversely related to root hair number and length. Spores of nine morphotypes belonging to three genera (*Acaulospora, Glomus, Scutellospora*) were extracted from soil. The percentage root length colonized by AM fungi was not related to soil factors, but total AM fungal spore numbers in the rhizosphere soil were inversely related to soil nitrogen and phosphorus levels. AM fungal spore numbers in the soil were linearly related to root length colonized. The co-occurrence of septate non-mycorrhizal fungi was recorded for the first time in cycads. These observations and the relationship between plant mycorrhizal status and soil nutrients are discussed.

Singh *et al.* (2002a) conducted a study at the Meerut College Campus, in Uttar Pradesh to determine the occurrence of mycorrhizal associations in 62 species of wild (including weeds) and cultivated plants. Of the 62 species evaluated, 54 showed mycorrhizal associations. The vesicular arbuscular mycorrhizal associations in species of *Alternanthera, Cassia, Calotropis, Cyperus, Eichhornia, Hymenocallis, Melia, Phoenix, Polygonum, Ranunculuc, Ruellia, Rumex, Tridex* [*Tridax*] and *Tephrosia* constitute the first reports from Meerut. *Glomus* and *Gigaspora* were the dominant VAM fungi in the area under evaluation.

Khade and Rodrigues (2002) studied the arbuscular mycorrhizal fungi associated with some pteridophytes from Western Ghat region of Goa. Commonly occurring pteridophytes from two sites namely Mollem and Chorlem located in the Western Ghats region of Goa, were selected for the study, including: *Selaginella* sp., *Lygodium flexuosum, Lindsaea heterophylla, Pteris vittata, Adiantum lunulatum, Athyrium hohenackerianum, Blechnum orientale, Gleichenia dichotoma, Pityrogramma calomelanos* and *Chriistella dentate.* All the pteridophyte species examined during the study, exhibited the presence of arbuscular mycorrhizal (AM) association. The highest arbuscular mycorrhizal root colonization (75%) was recorded in *Pityrogramma calomelanos* whereas the highest spore density (155 spores 100 g^{-1} rhizosphere soil) was recorded in *Adiantum lunulatum.* A fairly good diversity of AM fungi was observed in the rhizosphere of pteridophytes of this region. A total of 17 AM fungal species belonging to five genera namely *Acaulospora, Glomus, Gigaspora, Sclerocystis* and *Scutellospora* were recorded.

Dwivedi *et al.* (2003) studied the qualitative and quantitative distribution of arbuscular mycorrhiza (AM) fungal spores and rhizosphere soils of betel (*Piper betle*) vine in 4 villages (Tilli, Purabyau, Kakaganj and Deori) of Sagar, Madhya Pradesh in October 2001. *Glomus australe, Glomus clarum, Glomus fasiculatum* [*Glomus fasciculatum*], *Glomus geosporum, Glomus gerdemanii, Glomus hoi, Glomus mosseae, Acaulospora bireticulata, A. lacunose, A. mellea, A. scrobiculata*, and *Gigaspora albida* were identified. *Glomus australe* and *A. scrobiculata* were found in all the 4 sites with 100% frequency. *Glomus gerdemanii, Glomus mosseae* and *A. bireticulata* were differentially distributed in the 4 sites with 75% frequency. *Glomus gerdemanii, Glomus mosseae* and *A. bireticulata* were found at Tilli and Deori; *Glomus mosseae* and *A. bireticulata* at Purbayau; and *Glomus gerdemanii* at Kakaganj. The other AM fungi were poorly distributed and showed 25-50% frequency. The maximum number of AM fungi was recorded at Tilli (9 species) followed by Kakaganj (7 species), Deori (7 species) and Purbayau (5 species). The physicochemical properties of the rhizosphere soils from the 4 sites are mentioned.

Gupta *et al.* (2003) studied the distribution of AM fungi in the rhizosphere of different tea clones. The distribution of vesicular arbuscular mycorrhiza (VAM) in the rhizosphere of 15- to 17-year-old tea cultivars (Betzan, Nandadevi, tenali, TV1, TV18) was determined in a field experiment conducted in Orissa, India. Thirty-four VAM species belonging to the genera *Acaulospora, Glomus* and *Sclerocystis* were recorded. The rhizosphere of both Nandadevi and Betzan recorded VAM species belonging to all 3 genera, with the rhizosphere of Nandadevi recording the highest VAM distribution. The rhizosphere of Tenali was devoid of *Acaulospora,* whereas those of TV1 and TV18 were devoid of *Acaulospora* and *Sclerocystis. Glomus* species was distributed in the rhizospheres of all tea cultivars and highest in the rhizosphere of Nandadevi.

Mishra *et al.* (2005) discussed the occurrence of endophytic fungi in the Rhynie Chert fossils, especially the morphology and anatomy of the endophytes. They have reported the inter and intracellular hyphae of nonseptate and septate type endophyte. However, nonseptate hyphae were prevelant, which form the dense cluster or coils inside the lumina of cells and produced characteristics arbuscules, thick or thin walled resting spores *i.e.* chlamydospores or vesicle of varying size and shape. They have also observed the dual influence of endophytes consisting as *Glomus* and *Rhizoctonia* in the section of axis.

CONCLUDING REMARKS

In all 137 crops including 5 cereals, 13 millets, 10 pulses, 16 oilseed crops, 5 fibre crops, 2 sugar/ starch crops, 32 fodder crops, 54 vegetable crops, 9 fruit crops, 10 condiments and spices, 11 dyes and narcotics and 12 plantation crops have been mentioned in *Handbook of Agriculture* (ICAR Publications, New Delhi). Indian mycorrhizasts have attempted to gather information about the occurrence of VAM fungi in fields located at different geographical regions of the country raising these crops. However, the information appears to be still incomplete in view of the vastness of the arable land in the country, diversity of crops being raised there and climatic variations. Since all these factors cause a marked effect on the distribution of VAM fungi, more extensive survey seems to be a prerequisite to secure a complete knowledge of the quantitative and qualitative spectrum of VAM fungi in the arable land of the country.

The Indian mycorrhizasts have attempted to gather information also about the VA mycorrhizal associations in different cereals, millets, pulse crops, oilseed crops, fibre crops, sugar crops, fodder crops and vegetable crops being raised in different geographical regions of the country. However, a perusal of published reports clearly shows that major thrust of survey has been on only few crops including barley, paddy, pearl millet, sorghum, Indian millet, chickpea, pigeonpea, sunflower, sesame, sugarcane, soyabean, cowpea, tomato, brinjal and potato. Considering the vast majority of crops of the country and the importance of VAM as an established biotool in improving the growth and productivity of crops, the information on the VA mycorrhizal association in the crops appears to be far from complete.

Association of VAM fungi with the fruit crops and their occurrence in the orchards has been surveyed by a number of Indian mycorrhizasts. However, published account shows that out of more than three dozen fruit crops work has been restricted to only few crops including banana, mulberry, litchi, mango, apple and guava. In view of the fact that VA mycorrhizal association has been shown to be beneficial to the performance of fruit crops, more information about the association of VA mycorrhizal fungi with other fruit crops will go a long way to boost the fruit production in the country.

The country has the proud privilege of raising a variety of plantation crops (12). In addition, a vast strech of its land is utilized to cultivate plants yielding condiments and spices as well as dyes and narcotics. It is a sad state of affairs that only few of these have been surveyed for the VA mycorrhizal association. Variety of grasses, weeds and aquatic plants has been identified to be growing in the country. Many of them are of economic importance, therefore raised on a commercial scale. Only fragmentary information is available on the VA mycorrhizal association also of these economically important plants. It is felt that there is need to explore the association of VA mycorrhizal fungi in them.

VA mycorrhizal association has been shown to be beneficial not only to annuals or herbs and shrubs but also to perennials including trees of various types. Even the forest trees which have been shown to be benefited by ectomycorrhizal associations are reported to have better performance when associated with VA mycorrhizal fungi. Variety of trees including forest trees of the country have been surveyed for their association with VA mycorrhizal fungi. Because of a variety of reasons many tree plantation programmes are being actively executed in the country. However, these programmes are facing more failures than success. In view of its versatility as biotool, VAM fungi may be helpful in ensuring more success in the tree plantation programmes. Although commendable work has been done to gather information about VA mycorrhizal associations in trees including forest trees, however, in view of their importance more concerted effort is needed in this direction.

Qualitative and quantitative characteristics of VA mycorrhizal association in plants depends not only on the factors related to physical, chemical and microbiological quality of soil but also on the climate and environment. In view of this, many Indian mycorrhizasts had been engaged in exploring the VA mycorrhizal association in the plants growing in specific habitats under different climatic / environmental conditions. However, a perusal of the published reports shows that many specific habitats and regions have not been worked out with this reference. There appears to be a need to explore such habitats and regions to acquire a complete information regarding this matter.

3

EFFECT OF VA MYCORRHIZA ON GROWTH

In recent years the importance of VAM fungi for plant growth in most soils has been accepted and acknowledged by soil chemists, agronomists, horticulturists and farmers. Ever since it was shown that VAM could increase nutrient uptake and water uptake from soil in plants, researchers have been trying to extend and manipulate this effect of mycorrhiza in improving the growth and productivity in various groups of plants. Mycorrhizasts of the country have also explored the possibility of utilizing this tool for improving the performance of various crops including cereals, millets, pulses, vegetables, oilseeds and fruit crops. In addition, they have tried to explore the possibility of utilizing this biotool also for improving the growth of various types of trees including forest trees. Following account is related to the work done by the mycorrhizasts of the country on the effect of VA mycorrhiza on the growth and productivity of various groups of plants including cultivated crops.

CEREALS AND MILLETS

Swaminathan and Verma (1979) studied the response of maize and wheat plants raised in pots from seeds inoculated with *Glomus macrocarpum*. In general, inoculated plants exhibited root infection and assimilated more Zn from soil than non-mycorrhizal plants. For maize, grown in Zn deficient soil, Zn application appeared to stimulate endomycorrhizal efficiency in collecting Zn for the host plant; for other crops and in other soils, Zn fertilizer did not improve uptake of Zn, and it was concluded that mycorrhizal infection, by itself, could take care of the total Zn requirements of the plants. The mycorrhizal effect appears to be restricted to the more easily available Zn fractions in deficient soils. They have discussed the value of mycorrhizal inoculation under field conditions to upgrade nutrient deficient soils.

Champawat *et al.* (1987) studied the influence of VAM fungi on growth and nutrition uptake in barley (*Hordeum vulgare*). Inoculation with *Glomus constrictum*, *G. fasciculatum* and *Gigaspora calospora* significantly enhanced shoot and root dry weight. Total P in the shoot and root was significantly greater in plants inoculated with *G. constrictum*. The effect of mycorrhizal inoculation on plant growth and P uptake was greater in inoculated sterilized soil than in uninoculated sterile soil. Unsterile soil led to better plant growth and P uptake than sterilized (uninoculated) soil, possibly due to the presence of indigenous mycorrhizal fungi.

Tilak and Singh (1988) reported that inoculation with *Acaulospora* sp. *Gigaspora margarita* or *Glomus fasciculatum* in the presence of either single superphosphate or rock phosphate resulted in greater dry matter production by *Pennisetum americanum* and greater

P uptake than in the uninoculated control. Among the 3 VAM species, *G. fasciculatum* either singly or in combination with *A. brasilense* gave better results than the other two species.

Singh and Tilak (1990c) studied the response of different cultivars of sorghum (*Sorghum vulgare*) to inoculation with *Glomus versiforme*. A pot culture experiment was conducted using steam sterilized sandy-loam soil to evaluate the mycorrhizal dependency of five different sorghum genotypes *viz.,* CSH-9, CSH-5, PC-9, PC-6 and PC-23. Response of different genotypes of sorghum to inoculation with *G. versiforme* was highly variable. Among the 5 genotypes, the growth promoting ability of *G. versiforme*, measured in terms of mycorrhizal efficacy varied from 24-43 per cent. The maximum increase was observed in variety PC-23, followed by CSH-5, CSH-9, PC-9 and PC-6 varieties of sorghum. Total phosphorus uptake by the mycorrhizal plants of different genotypes was significantly higher than the control plants. The phosphorus uptake by plants due to inoculation with *G. versiforme* among the various cultivars varied from 45-136 per cent, the maximum increase being observed in cultivar PC-9. Growth and phosphorus uptake of sorghum (*Sorghum bicolor*) on steamsterilized, phosphorus deficient soil was improved by the soil inoculation with VAM fungus, *Glomus versiforme*. Among the 5 sorghum genotypes *viz.* CSH-9, CSH-5, PC-9, PC-6 and PC-23, tested for their mycorrhizal dependency, PC-23 produced maximum mycorrhizal efficiency and phosphorus uptake with *G. versiforme*.

Secilia and Bagyaraj (1990) studied the response of upland rice to vesicular arbuscular mycorrhizal fungi. Maximum grain yield was recorded in plants inoculated with *Glomus fasciculatum*. Regarding the grain phosphorus content, again *Glomus fasciculatum* inoculated plants recorded the maximum value followed by *Acaulospora* sp. inoculated plants. Phosphorus content of the grain in plants treated with twelve more fungi also showed significant increases and were at par with the above two fungi. Straw P content in plants inoculated with *Acaulospora* sp. *Glomus fascicu!atum* (Riv.), *A. laevis* and *Glomus albidum* showed significantly higher values and the increases by these four fungi were statistically at par with each other.

Singh and Tilak (1990b) studied the response of sorghum (*Sorghum bicolor* L.) to inoculation with *Glomus versiforme* in field. Inoculation with *G. versiforme* increased the shoot dry weight of sorghum by 10 and 20 per cent in plots grown to wheat and chickpea respectively. In the plots left fallow also an increase of 28 per cent in dry weight was observed. Phosphorus addition, however gave maximum increase of 33 to 39 per cent in all the rotations. Phosphorus uptake also increased significantly, though its concentrations remained similar in different treatments. Results of the study indicated the effectiveness of VAM fungal inoculation even in the presence of varying levels of indigenous VAM fungi.

Tewari *et al.* (1993) studied the host genotype dependency and growth enhancing ability of VA-mycorrhizal fungi for *Eleusine coracana* (finger millet). Inoculation of plants with one of the six different vesicular, arbuscular mycorrhizal (VAM) fungi increased plant biomass, height, leaf area and absolute growth rate: however, effectiveness of the various VAM fungi varied significantly. Maximum root colonization and mycorrhizal efficacy was observed with plants inoculated with *Glomus caledonicum*. Among five host genotypes tested for mycorrhizal dependency against *G. caledonicum*, genotype HR-374 gave the highest plant biomass, mycorrhizal efficacy and root colonization, the inoculation resulting in increased mineral (phosphate, nitrogen, Zn^{2+} and Cu^{2+}) content and uptake in shoots.

Champawat and Pathak (1993) studied the effect of VAM fungi on growth and nutrition uptake of pearl millet. They observed greater shoot and root dry weight, P and N uptake in plants inoculated with *Gigaspora calospora* than in the non-mycorrhizal control. Their results obtained in P deficient soil suggested that inoculation of pearl millet with VAM fungi could be extremely beneficial for plant growth and nutrition uptake.

Gaonker *et al.* (1993) studied the rhizosphere microflora, growth and yield of wheat as influenced by inoculation of *Glomus fasciculatum* in conjunction with organic amendments. In a pot experiment wheat DWR-39, DWR-163 and DWR-187 were grown in soil inoculated with *G. fasciculatum* and were given soil amendments biogas spent slurry, FYM, *Glyricidia*, groundnut cake, jowar straw and rice bran. Soil bacteria, fungi, actinomycetes and free-living N2 fixers and root colonization by VA mycorrhiza were increased by mycorrhizal inoculation and organic amendments. Seed yield and shoot P concentration were highest in DWR-39 and DWR-187 with inoculation + FYM and in DWR-163 with biogas spent slurry + inoculation.

Secilia and Bagyaraj (1994) screened a total of 18 VAM fungi for their symbiotic responses with wetland rice *cv. Prakash* (IET 2254) in pot culture. Out of these *Acaulospora* sp., *Glomus fasciculatum* and *G. mosseae* were most effective in improving plant height, tiller number, total biomass, panicle number, grain weight and plant P and Zn concentration. Increase in grain yield resulting from inoculation with *Acaulospora* sp., *G. fasciculatum* and *G. mosseae* was 62, 59 and 35% respectively.

Gaonker and Sreenivasa (1994) studied the response of 3 genotypes of wheat, DWR-39, DWR-163 and DWR-187 to inoculation with *G. fasciculatum* in conjunction with locally available organic amendments in lab experiments. In general, all organic amendments had a positive influence on the proliferation of *G. fasciculatum* . Farm-yard manure stimulated the greatest root colonization in artificially inoculated DWR-39 and DWR-187, and biogas spend slurry caused greatest root colonization in DWR-163; a corresponding trend was observed in plant height, shoot biomass, seed yield, shoot phosphorus and seed protein concentration in these treatments. DWR-39 was the superior genotype, compared with DWR-163 and DWR-187 and organic amendments with a narrow C:N ratio had a greater influence on VA mycorrhizal proliferation compared with amendments with a wider C:N ratio.

Kehri and Chandra (1994c) attempted to select VA-mycorrhizal inoculants for some upland varieties of rice. Response of five upland varieties of rice (Saket-4, Pant-4, Ratna, T-3 and Jaya) to the inoculation of six different VAM fungi (*Glomus fasciculatum, G. mosseae, G. macrocarpum, G. aggregatum, G. multicaule* and *G. claroideum*) in terms of growth and yield was evaluated under pot conditions. Variety Jaya was most responsive to mycorrhizal inoculation. *G. macrocarpum* and *G. claroideum* were most versatile in promoting yield since four varieties were benefited by their inoculation.

Singh and Mishra (1995) studied the effect of VA-mycorrhiza - (*Glomus etunicatum*) on growth and phosphorus - uptake of P-deficient tolerant and susceptible varieties of paddy (RCPL-101 and RCPL-104) under different soil phosphorus levels *i.e.* 0 kg P/ha, 20 Kg P/ha, 40 Kg P/ha, 80 Kg P/ha and 160 Kg P/ha. The two varieties responded to mycorrhizal infection differently at different soil phosphate levels. The phosphorus-deficient tolerant variety KCPL 101 responded to mycorrhiza upto soil phosphate level of 20 Kg/ha whereas RCPL 104, susceptible to P deficiency, responded up to 80 Kg P/ha. The variety requiring high phosphorus depended on VAM even at high soil phosphate level. An increase in growth due to mycorrhizae can be achieved in a variety having high phosphorus requirement even at high soil phosphate levels.

Prabakaran *et al.* (1995a) studied the performance of sorghum cultivars to VA-mycorrhizae inoculation in alfisol. They have conducted experiment with Co 26 sorghum and different VAM fungi namely, *Glomus fasciculatum, G. mosseae, Gigaspora calospora* and *Acaulospora laevis*. VAM inoculation induced significant increase in vigour index, plant growth, root biomass and possible level of VAM infectivity in Co 26 sorghum in nitrogen and phosphorus depleted acid soils over untreated control. Among the endomycorrhizae *Glomus fasciculatum* recorded a significant increase in plant growth and yield attributes due to its 72% endophytic root infection as compared to other VAM fungi and control (12%). According to them, the increased biogrowth and yield attributes in sorghum might be due to the improved uptake of nutrients, enhancing other soil biological systems, regulation of water potential, control of root pathogens, and increased accumulation of polyphosphates.

Rathore and Singh (1995a) studied the influence of vesicular-arbuscular mycorrhizal fungi and phosphate on maize. Response of maize to VAM fungal strains inoculation and graded doses of phosphorus was studied in terms of root colonization, fertilizer substitution, plant growth and nutrient uptake in unsterilized soil. *Gigaspora margarita, Glomus fasciculatum* and VAM mixed endophyte individually caused 54, 50 and 58% root colonization, respectively. Mycorrhizal root infection significantly decreased with increasing levels of P. Phosphorus application and VAM inoculation increased dry matter accumulation and uptake of N, P and K by maize shoots at 30 and 45 days of growth. The maximum response of inoculation was recorded at 60 kg P/ha. Inoculation increased available soil phosphorus. Mycorrhizal inoculum may substitute phosphatic fertilizer equivalent to = 30 kg P/ha.

Ammani and Rao (1996) studied the effect of two VAM fungi, *Acaulospora spinosa* and *A. scrobiculata* on eight varieties of upland rice. All the eight varieties showed a positive response to inoculation with the VAM fungi in terms of levels of colonization of roots, formation of vesicles/spores, plant biomass and grain yield. Out of 8 varieties, 3 showed greater response than the others. *A. spinosa* was more effective than *A. scrobiculata* on all the varieties.

Chhabra and Jalali (1997) investigated the influence of inoculation with the VAM fungus *Glomus mosseae* on the utilization of rock-phosphate (50 and 100 ppm) by wheat (*cv.* WH-147) grown in nutrient deficient soil. Results showed enhanced plant growth at both the levels of rock-phosphate in combination with *G. mosseae*. An increase in plant height, total dry weight of root and shoot, number of grains per ear, density of mycorrhizal colonization in roots and nutrient content of host plants was recorded.

Khare *et al.* (1998) conducted an experiment to study the role of native VAM fungi in wheat in Zn-deficient black soil. Wheat was grown in rotation with fallow, soyabeans or sorghum at 3 sites on soil amended with 0-15 kg Zn/ha. At Surai and Khanmond the survival and multiplication of VAM fungal propagules was greater when wheat was rotated with sorghum or soyabeans than in the fallow-wheat system. Zn concentration in grain was also lower at 2 sites in wheat grown after fallow, and P uptake was lower in the fallow system at all the sites. Wheat yield was significantly lower after fallow at Surai and Khamond.

Setua *et al.* (1999a) conducted an experiment at Berhampore, West Bengal during 1994-97 to study the effect of direct inoculation of VAM through sowing of maize seeds (40 kg/ha) coated with VAM-slurry (75kg/ha) in between the rows of 2-year old mulberry or applying AM-slurry (200 kg/ha) in the root zone of 2-year-old mulberry through holes to 20 cm depth. Plant height (119.5 cm), number of leaves/plant (233.3), leaf yield (25.636 kg/ha per year) and leaf moisture (74.4%) were highest when *Glomus fasciculatum* as applied on maize seed and 30 kg P_2O_5/ha was applied annually, though the effects on leaf moisture and plant height were not significant. Phosphorus uptake in leaves was not significantly affected by treatment. The combined use of VAM and 30 kg P_2O_5 have a higher yield than 180 kg P_2O_5 /ha without VAM. Economic analysis showed the application of VAM on maize seed plus 30 kg P_2O_5 was more profitable than application of 180 kg P_2O_5.

Saha *et al.* (1999) compared the two rice varieties, a primitive landrace and an improved high yielding variety, for their mycorrhiza responsiveness against varying levels of available soil phosphorus (P) through challenge inoculation in a nutrient deficient laterite soil. The phosphorus responsive improved variety showed high relative dependency to vesicular arbuscular mycorrhiza (VAM) at low soil P levels but the landrace responded negatively to inoculation. The landrace having higher growth potential as compared to the improved variety had a very high root-shoot ratio which did not change with VAM inoculation. The lower root-shoot ratio of the improved variety increased with increases in P - supply and also VAM inoculation. Phosphorus acquisition efficiency which was higher in the landrace increased with VAM inoculation in both varieties, but the magnitude of such increase was more in the improved variety than the landrace. Improvements in phosphorus acquisition efficiency was high at low P levels. The root colonization intensity persisted in the landrace even at high P level. The results, in the light of previous evidences, have been interpreted to conclude that primitive, wild genotypes of rice well adapted to nutrient and other stress factors are more efficient for nutrient, particularly phosphorus acquisition than the improved high yielding genotypes through higher development of roots. This and their possible lower phosphorus demand may allow them to perform independent of mycorrhiza. Changes with regard to either of both in the modern improved varieties might have made them mycorrhiza dependent.

Srivastava *et al.* (1999) studied the effect of VAM fungi on the increased mineral nutrient uptake in maize. In pot experiments, maize plants were inoculated with *Gigaspora margarita* or not inoculated. Inoculation increased N and P uptake and plant growth.

Singha *et al.* (2000) studied the occurrence and performance of VAM fungi on the growth of rice plant under rain fed lowland condition of Assam. Out of 15-tested transplanted rice cultivars, only 7 were found to be colonized by vesicular-arbuscular mycorrhizal (VAM) fungi. Three VAM fungi were found; *Acaulospora* sp., *Glomus* sp. and *Scuttelospora* [*Scutellospora*] sp. Studied with inoculation of rice *cv. Ranjit* with 20-200 g/pot of *Acaulospora* sp. showed that a level of 100 g/pot was best for plant growth and root colonization. Dry weight of plants and root colonization were significantly higher with than without VAM fungi, but root proliferation was lower. Uptake of P was significantly higher with than without VAM fungi. Plant growth was better with VAM fungi.

Abdel-Fatiah and Mohamedin (2000) studied the effect of the interaction between a vesicular-arbuscular (VA) mycorrhiza *(Glomus intraradices* no. LAP8) and *Streptomyces coelicolor* strain no. 2389 on the growth response, nutrition and metabolic activities of sorghum (*Sorghum bicolor*) plants grown in non-sterilized soil amended with chitin waste in a greenhouse over 8 weeks. Chitin amendment resulted in an increase in the microbial population and chitinase activity in soils. Growth of mycorrhizal (*G. intraràdices* no. LAP8) and non-mycorrhizal sorghum plants increased as compared with other treatments either in the presence or absence of *S. coelicolor* strain 2389. VA-mycorrhizal inoculation significantly increased the growth, photosynthetic pigments, total soluble protein and nutrient contents of sorghum compared to non-mycorrhizal sorghum. Such increases were related increased mycorrhizal colonization. Inoculation with *S. coelicolor* 2389 significantly increased the intensity of mycorrhizal root colonization and arbuscular formation, but the levels of mycorrhizal infection and their beneficial effects were significantly reduced with the addition of chitin waste to the soil. Analysis of the content of total amino acids and ammonia in leaves on the basis of dry matter production showed that, in most instances, total amino acids of mycorrhizal plants were significantly higher than those of non-inoculated plants. The microflora of the rhizosphere was highly affected by mycorrhizal inoculation. Quantitative changes in àcid and alkaline phosphatase activities of the roots in response to the mycorrhizal inoculation are discussed.

Singh *et al.* (2002a) conducted a greenhouse study to characterize variation among 26 maize lines (inbred, hybrid and composite) for their relative ability to respond to arbuscular mycorrhizal (AM) symbiosis. A mycorrhizal inoculum consisting of spores, hyphae, and infected roots from plants grown in the same soil was mixed with autoclaved soil at 1:3 (inoculum: autoclaved soil mixture). Uninoculated plants served as the control. Maize lines varied considerably with respect to mycorrhizal colonization. Mycorrhizal colonization varied from 22.4 (CM 129) to 45.32% (CM 126×CM 127) for the 26 lines. The highest mycorrhizal colonization was recorded in CM 126, FH 3054, FH 3048, FH 3077, FH 3084, FH 3085, FH 3086, HIM 129, VL 89 and CM 126 × CM 127. Significant variation within composite and hybrid lines was also observed. However, composites generally had a lower level of infection than the hybrids. Root colonization by mycorrhizal fungi varied considerably depending on the genotype. The correlation between mycorrhizal responsiveness based on shoot dry weight and colonization was non-significant. A low correlation between arbuscular frequency and mycorrhizal responsiveness was observed. FH 3084 and FH 3077 showed a relatively high mycorrhizal responsiveness despite having a low degree of arbuscular colonization.

Singh *et al.* (2002c) studied growth responses of maize genotypes, i.e., VL90 and VL16 of low and VL89 and HIM129 of relatively high P requirements, to inoculation with a native consortium of arbuscular mycorrhizal (AM) fungi in a sandy loam soil fertilized with single superphosphate (SSP) to provide 0 (P_0), 60 (P_1), 90 (P_2) and 120 (P_3) mg P kg^{-1} soil. AM colonization significantly increased vegetative plants growth after 45 days at P_0 and P_1 levels for genotype VL90 and VL16, at P_2 and P_3 levels for VL89, and P_0, P_1, P_2 levels in HIM 129.Except for VL89, a growth depression in roots and shoot of AM plants was abserved at either P_2 or P_3 levels. This was most pronounced in VL90 and VL16. In contrast, in non-mycorrhizal plants no growth depression was observed; genotype VL90 at P_3 was an exception. Root and shoot P, and shoot Zn concentration were significantly higher in AM compared with non-mycorrhizal plants, at various P levels. However, the pattern of phosphorus and zinc uptake in mycorrhizal and non-mycorrhiza shoots was genotype specific. Alkaline phosphatase (ALPase) activity of mycorrhizal and non- mycorrhizal roots was variable with respect to genotype and soil P level. In mycorrhizal roots ALPase activity was significantly higher at P_0 and P_1 levels in VL90, P_0, P_1, P_2 levels in VL16, P_2 and P_3 levels in VL89 and at all P levels in genotype HIM129.

PULSES

Saxena *et al.* (1977) studied the differential effect of various endomycorrhizal fungi on nodulating ability of greengram by *Bradyrhizobium* sp. (*Vigna)* strain S 24. *Bradyrhizobium* sp. interacted differentially with 8 VAM fungi and caused significant variations in nodulation and growth parameters of greengram. Co-inoculation with *Scutellospora calospora* resulted in the highest nitrogenase activity and dry biomass. The nodulation competitiveness of S 24 was significantly higher (60-63%) in presence of *G. mosseae*, *G. fasciculatum* and *Scutellospora calospora* when compared to treatment with single inoculation of S 24 (51%). Percent VAM colonization was higher in treatments having higher nodule occupancy of introduced strain (S 24).

Jalali and Thareja (1985) studied the plant growth responses of *Cicer arietinum* to VA mycorrhizal inoculation in soils (P-deficient) incorporated with rock phosphate. The plants responded significantly to rock phosphate + VA mycorrhizal treatments in terms of total dry matter production and P uptake. In nutrient rich soils, the root and shoot dry weights remained unaffected with rock phosphate treatment alone, but when such plants were inoculated with VAM fungus (*Glomus* sp.) significant increase in these growth parameters was recorded.

Manjunath and Bagyaraj (1986) studied the response of blackgram (*Vigna mungo*), chickpea (*Cicer arietinum*) and mungbean (*Vigna radiata*) to inoculation with *Glomus fasciculatum*, with and without added P (22 kg P/ha) in P-deficient unsterilized soil. Inoculation increased the dry weight and P content of the shoot and root over the uninoculated control. Application of P did not reduce the percentage root colonization by VAM fungi, but increased the number of extramatrical chlamydospores in the soil. P application also increased the dry weight and P content of the inoculated plants, suggesting a need for application of P to obtain maximum benefit from VAM in P-fixing soils.

Ramaraj and Shanmugam (1986) studied the effect of VA mycorrhizae on pulses. Seed inoculation with three mycorrhizal species in pot trials increased the growth of black gram (*Vigna mungo*) greengram (*V. radiata*) and cowpea (*Vigna unguiculata*). Inoculation with *Glomus etunicatum* was more effective than inoculation with *G. fasciculatum* or *G. mosseae*. Placement of the inoculum below the seed was more effective than when mixed with the topsoil or applied as seed treatment.

Raverkar and Tilak (1988) studied the relative efficiency of different VAM on soyabean (*Glycine max*) under varying levels of phosphorus. A pot experiment was conducted with sterilized sandy-loam soil (pH 7.3) deficient in P to study the relative efficiency of different VAM fungi (*Glomus fasciculatum, G. mosseae, Gigaspora gilmorei* and *Acaulospora* sp.) Phosphorus was supplied at the rates of 25, 50, 75 and 100 kg P2O5ha-1 as a basal dressing at the time of sowing as single superphosphate. There was a wide variation in the growth promoting efficiency of different VAM fungi on soybean *cv.* Lee at varying levels of P. The linear response of different VAM fungi in respect of shoot and root biomass and acid phosphatase activity was observed up to certain level of P (50 kg P2O5 ha-1) and at higher levels of P, it was inhibited. Maximum increase in shoot and root biomass was recorded with *Gigaspora gilmorei* followed by *G. fasciculatum* , *G. mosseae* and *Acaulospora* sp. Maximum colonization of roots by the endophyte was noticed with *G. gilmorei*. There was a reduction in the rate of mycorrhizal colonization by different VAM fungi in higher level of P.

Singh and Tilak (1989) investigated the response of *Cicer arietinum cv.* C-235, already inoculated with *Rhizobium leguminosorum* to inoculation with *Glomus versiforme* under field conditions in a P-deficient sandy loam soil at New Delhi. Inoculation with *G. versiforme* increased the rate of VAM development. Nodule weight and nodule number per plant were higher in inoculated than uninoculated plants. P content of the shoots and total P uptake were increased by either the application of 40 kg P2O5 as super phosphate/ha or by inoculation with *G. versiforme*. Inoculation increased shoot dry weight and seed yields by 12 and 25%, while P application increased them by 33 and 60% respectively.

Champawat (1989) worked on chickpea *cv*. RS-10. He raised the plants in 30 cm pots containing 5 kg sterilized P deficient soil (3 mg available P/kg soil). The treatments included the following: of sterilized soil with inoculum of *Glomus fasciculatum* (a) *G. constrictum* (b) or *Gigaspora calospora* (c), uninoculated unsterilized soil (d) uninoculated sterilized soil (e). Plants were harvested after 60 days. The shoot and root dry weight and P uptake were higher with VAM inoculation than without inoculation. They were higher also in unsterilized soil than in sterilized uninoculated soil. Dry weight and P uptake were highest in (a) and lowest in (e) Root colonization by mycorrhiza was 85.7, 82.0, 76.3, 77.3 and 0% in (a), (b), (c), (d) and (e) respectively.

Mathew and Johri (1989) studied the effect of indigenous and introduced VAM fungi on growth of mungbean. The indigenous vesicular-arbuscular mycorrhizal fungi (VAM) recovered from soil at the Crop Research Centre of G.B.Pant University were unable to promote growth of mungbean. Introduced *Glomus caledonicum* was effective in improving plant growth. Dual inoculation with *G. caledonicum* and indigenous VAM fungi depicted a growth response that was much greater than that obtained with single inoculation. The tripartite system, indigenous VAM, *G. caledonicum* and *Rhizobium phaseoli* improved plant growth and also resulted in increased nodulation and nitrogenase activity.

Reddy and Bagyaraj (1990) studied the response of pigeonpea to vesicular arbuscular mycorrhizal colonization in an alfisol. The eleven fungi used in the experiment were collected locally as well as procured from different places, were screened against the pigeonpea cultivar Hyd-3C. The plants were harvested after 120 days. Pigeonpea varied in its response to inoculation with different VAM fungi. Generally, inoculated plants had higher mycorrhizal root colonization, shoot and root dry weight and shoot P content. *Glomus leptotichum, G. macrocarpum, Acaulospora laevis, Gigaspora margarita, Glomus fasciculatum* (Riverside) and *G. fasciculatum* (Rothamsted) were equally good in improving shoot and root dry weights. Plants colonized with these 6 fungi also had significantly higher shoot P content and the increases were at par with other. Of these 6 fungi maximum shoot weight, root weight and shoot P content was recorded in plants inoculated with *G. leptotichum* followed by *G. macrocarpum* and *A. laevis*. Plants inoculated with *Gigaspora margarita, G. fasciculatum* (Riverside) and *G. fasciculatum* (Rothamsted) ranked 4, 5 and 6. Significantly higher mycorrhizal colonization was seen in plants inoculated with all the VAM fungi except *G. mosseae* and *G. fasciculatum* (Rothamsted).

Ramraj and Shanmugam (1990) studied the effect of vesicular arbuscular mycorrhizal inoculation on cowpea. For placement method of inoculation the soil inoculum was applied in the seed furrows and for broadcasting method the soil inoculum was uniformly broadcasted in the plot and stirred. The quantity of soil inoculum used in both the cases was 250 g/plot. The experimental design adopted was randomized block design with four replications. The available P content of the soil was 10 ppm. The spore density of the inoculum used was 1,200 spores/100 g of inoculum. The data indicated that VA mycorrhizal (*Glomus etunicatum*) inoculation increased shoot weight of cowpea significantly compared to the untreated control.But carbendazim seed treatment along with *G. etunicatum* did not significantly increase the plant biomass. This might have been due to its fungicidal effect on *G. etunicatum*.. There was not much difference in the shoot weight and root weight between the placement method and broadcasting method of inoculation even though the per cent colonization was more in placement method. Besides its growth promoting effect, *G. etunicatum* was also to some extent effective as a biocontrol agent against root rot.

Devi and Sitaramaiah (1990) studied the effect of superphosphate, rock phosphate as sources of phosphorus in combination with *Glomus fasciculatum* on root colonization, growth and chemical composition of blackgram. Inoculation with *G. fasciculatum* in combination with P increased percentage mycorrhizal root colonization, dry weight of root and shoot, total chlorophyll content and chemical contents of *Vigna mungo* over uninoculated plants with P alone. Increased application of P decreased percentage root colonization. Superphosphate was a better P source than rock phosphate for both mycorrhizal and non-mycorrhizal plants.

Champawat (1990a) studied the effect of different VAM fungi under varying levels of phosphorus on growth and nutrition uptake of pigeonpea. *Glomus constrictum* was better than the other 2 VAM fungi tested in enhancing plant dry weight at all levels of P except at 75% kg/ha when *Gigaspora calospora* was superior. Significant increases in shoot and root P and N uptake was observed with all the 3 VAM fungi at all levels of P application.

Kapoor and Mukerji (1990) carried out experiment to determine growth responses of soybean inoculated with selected VAM fungi. Scanning Electron Microscope (SEM) studies were carried out to taxonomically characterize the VAM fungi (*G. fasciculatum* , *G. macrocarpum*, *G. multisubstensum* for their proper application and selection as inoculant fungi for soybean (*Glycine max*). The effect of five VAM fungi (*G. fasciculatum* , *G. macrocarpum, G. fuegianum, G. multisubstensum* and *G. mosseae*) was observed on plant growth in terms of dry weight and total phosphorus. *G. macrocarpum* significantly influenced growth of soybean as compared to other inoculated VAM fungi. The effect of different phosphorus levels and mycorrhizal inoculation was observed on growth of soybean in terms of dry weight and total P with the intention to find out the most suitable phosphorus level of this association. Plants showed greater response to mycorrhizal inoculation at intermediate levels of applied phosphorus. Mycorrhizal colonization decreased approximately three folds at higher levels of phosphorus. Total P in the tissues was maximum as compared to other inoculated treatments. The total P in the tissues of mycorrhizal plants was much more than that of control plants. *G. macrocarpum* was efficient VAM fungus for soybean over other VAM species.

Chhabra *et al.* (1990) investigated the impact of VA mycorrhiza on growth and development of cowpea (*Vigna unguiculata* L. Walp.). According to them VA mycorrhizal inoculation increased plant height and total dry matter of root and shoot compared with the uninoculated controls. Extensive mycorrhizal development (53.78%) took place in the roots of inoculated cowpea plants. Inoculation also increased the N, P and K contents significantly.

Reddy and Bagyaraj (1991) evaluated the symbiotic efficiency of pigeonpea to VA mycorrhizal inoculation in an alfisol and vertisol. In pot trials, pigeonpea *cv*. Hyd.-3C were grown in an alfisol and vertisol without inoculation or inoculated with 1 of 11 mycorrhizal fungi. Plants were harvested after 120 days. Root colonization varied from 40% without inoculation to 56% with *Gigaspora margarita* inoculation in the alfisol and from 44% without inoculation to 67% with *Glomus leptotichum* inoculation in the vertisol. Nodule dry weight, root dry weight and P content/plant were higher with inoculation by *Glomus leptotichum* inoculation in the alfisol and with *G. margarita* in the vertisol.

Tilak and Dwivedi (1991) studied the effect of *Glomus versiforme* inoculation on the yield of chickpea at varying N levels. According to them yields per plant were higher in microplots inoculated with *G. versiforme* than in non-inoculated ones at all N levels. A non-inoculated microplot at 50 kg N/ha yielded almost as much as a VAM inoculated one receiving only 25 kg N/ ha in the presence of *Rhizobium*.

Champawat (1992) studied the effect of vesicular-arbuscular mycorrhizal fungi on growth and nutrition of chickpea. Chickpea *cv*. RS-10 plants grown in pots in sterilized sandy loam soil were inoculated with *Glomus fasciculatum*, *G. constrictum* or *Gigaspora calospora* or were not inoculated. Inoculation with each species increased plant DW and P uptake compared with uninoculated plants in sterile soil, while uninoculated plants in nonsterile soil gave intermediate results. Root colonization by mycorrhiza was 76-86% in inoculated plants and 77% in nonsterile soil.

Meenakumari and Nair (1992) studied the effect of phosphate fertilizer application on VA mycorrhizal infection in cowpea. Cowpea *cv*. C-152 was grown in pots in soil inoculated with *Glomus macrocarpum* and *Rhizobium* and given rock phosphate (30 kg P2O5/ha) produced the greatest plant dry weight after 45 days growth than various combinations of *G. macrocarpum* or local *Glomus* sp. and rhizobia, rock or superphosphates.

Chhabra and Jalali (1995) studied the role of VAM fungus, *Glomus mosseae* inoculation in the utilization of less or unavailable sources of phosphate into an available form for the growth and development of pigeonpea plants grown in nutrient - deficient soil. The two levels of rock phosphate 50 and 100 ppm were used in experimentation. VAM + rock phosphate - 100 ppm was most effective in promoting plant growth and followed by VAM+rock phosphate-50 ppm as compared to controls. The nutrient content also significantly increased in rock phosphate amended and mycorrhiza inoculated plants as compared to control sets. Their study indicated that mycorrhizal inoculation helps in the effective utilization of rock phosphate by changing it into available form, which later is taken up by the plants for their better growth and development.

Mathur and Vyas (1995f) reported production of highly nutritive pods of *Cyamopsis tetragonoloba* by VA mycorrhizae. They have conducted pot experiments to assess the effects of VAM fungi, *Acaulospora morrowae*, *Glomus constrictum*, *G. fasciculatum* , *G. mosseae*, *Sclerocystis rubiformis* and *Scutellospora calospora*, collected from rhizosphere soils of *C. tetragonoloba*. They have shown that mycorrhizal plants had increased levels of protein and sugar compared with non-mycorrhizal plants. Of the 6 VAM species *G. fasciculatum* caused a > 2 fold increase in protein content and 750% increase in total sugar content in pods compared with controls.

Santhi and Kothandaraman (1995) studied the effect of endomycorrhizal colonization and phosphorus sources on the phosphorus recovery, nutrient uptake and growth of greengram. In a pot experiment at Coimbatore, Tamil Nadu, green gram {*Vigna radiata*} *cv*. Co.3 was grown in clay loam soil inoculated with or not inoculated with *Glomus fasciculatum* and given equivalent to 0, 25 or 50 kg P2O5/ha as single superphosphate (SSP), diammonium phosphate (DAP) or Mussoorie rock phosphate (MRP). Inoculation with *G. fasciculatum* increased dry matter (DM) yield. The DM yield/pot was highest in soil inoculated pots with 50 kg P2O5 as DP. The yield increases with the soil inoculation were highest where 25 kg P2O5 was applied. P use efficiency was highest with MRP. Soil inoculation increased the uptake of N,P,K, Ca and Mg.

Tiwari *et al.* (1995) studied the response of VA mycorrhizal colonization in the rhizosphere soil of *Cajanus cajan* at different growth stages. Percentage of root colonization, average number of vesicles/cm of root bit, and number of spores/100g of rhizosphere soil significantly increased with increasing age of the plant (upto 120 days). The increase was much more in unsterilized soil than in sterilized soil.

Bhandari and Rawat (1995) studied the establishment of association of VAM and nutrient utilization by soybean. They inoculated the seedlings with *Glomus fasciculatum*. On the basis of their results they have shown that colonization of root by VAM increased root proliferation, resulting in increased uptake of nutrients which was also reflected in increased biomass of VAM colonized plants over uninoculated plants.

Kehri and Chandra (1995a) studied the interaction between vesicular-arbuscular mycorrhiza and phosphorus application in mungbean. The recommended dose of P fertilizer for mungbean is 70-100 kg/ha for optimum yield. Considering that inoculation of mungbean with an efficient strain of VAM fungi may ensure a better fertilizer use efficiency and curtail its dose, the study was undertaken. A number of VAM fungi *viz., Glomus claroideum, G. multicaule, G. aggregatum, G. fasciculatum* , *G. mosseae* and *G. macrocarpum* were evaluated for their efficacy in improving the performance of the crop with four different doses of P fertilizer (superphosphate: 100 kg, 50 kg, 25 kg and O kg/ha). Inoculation of mungbean with some of them at quarter dose of fertilization enabled it to give the same performance which it gave without them at full dose of fertilizer. The results indicated that an appreciable amount of P fertilizer recommended for mungbean can be saved by inoculating it by these fungi.

Kehri and Chandra (1995b) studied the relative efficacy of single and multiple inocula of VAM fungi for chickpea. The VAM inoculants evaluated for chickpea individually as well as in different combinations caused an improvement not only in mycorrhizal colonization in root, spore count in rhizosphere soil, and nodulation but also in shoot biomass, yield, and P content in shoots. However, as expected the magnitude of improvement varied with the VAM inoculants and their combinations. While maximum improvement in spore count was caused by an inoculum of *G. fasciculatum* , maximum improvement in nodulation was caused by a combination of *G. mosseae* and *G. aggregatum*. On the other hand, combined inoculum of all the three VAM fungi was most effective in improving mycorrhizal colonization, shoot biomass, and P content in shoots. It was also most effective in improving the yield. Considering the yield to be an important parameter of crop performance, comparatively higher yield in chickpea due to triple inoculants is noteworthy and deserves due attention and further exploration. Repeated observations of this kind with other crops will confirm the superiority of multiple inocula and open a new avenue to achieve better productivity in crops.

Rao and Rao (1996) studied the response of blackgram (*Vigna mungo*) and green gram (*V. radiata*) to VAM inoculation under different sources of phosphorus fertilizer. In pot experiment the soil was inoculated with *Acaulospora spinosa, A. morrowae* or *Glomus epigaeum* (*G. versiforme*). P was applied as superphosphate, rock phosphate or tricalcium phosphate. In the inoculated treatment, root colonization was 68-89% in blackgram and 69-

93% in greengram. The colonization in both species was lowest when pots inoculated with *A. morrowae* were given superphosphate, while the best colonization was given by *G. epigaeum* without added P. The shoot dry weight was increased by both mycorrhizas and applied P, and was highest with superphosphate + *G. epigaeum*.

Tarafdar and Rao (1997b) studied the response of arid legumes to VAM fungal inoculation. They conducted a field survey in Rajasthan to determine the effect of VAM fungi on growth and nutrient uptake of legumes including mungbean (*Vigna radiata* L. Winlczek). Nodulation, nitrogenase activity, percent root infection by VAM fungi and the number of VAM spores in the soil were increased significantly upon inoculation. Phosphatase activity, dry matter production grain yield and N,P,Cu and Zn was enhanced significantly in plants inoculated with *Glomus mosseae* and *Glomus fasciculatum*.

Pandey and Tarafdar (1999) examined the critical level of vesicular-arbuscular (VA) mycorrhizal fungi for moth bean (*Vigna aconitifolia*), mung bean (*Vigna radiata)* and pearl millet (*Pennisetum glaucum). Glomus mosseae* was used at different spore levels (0,25,50,75,100, 200 and 300 spores/ 100 g soil). Mycorrhizal inoculation increased plant height; dry matter yield, root length and percent root infection. Seventy per cent infection was sufficient for optimum response by legumes (moth bean and mung bean) whereas 80% infection was required for effective VAM for cereal (pearl millet) the critical level for moth bean, mung bean and pearl millet was 75, 100 and 200 spores/100 g soil respectively. However the effect on plant growth was observed upto 225 spores/100g soil for moth bean, 240 spores/100 g soil for mung bean and >300 spores/100g soil for pearl millet. The results demonstrate that saturated amounts of effective VAM fungi are required for optimum crop response.

Gupta *et al.* (1999) studied the response of different vesicular arbuscular mycorrhizal fungi on cowpea. Cowpeas were inoculated with *Glomus mosseae, Glomus constrictum, Glomus fasciculatum, Gigaspora gilmorei* and *Acaulospora levis* in nutrient deficient sterilized soil using earthen pots. Plants inoculated with these mycorrhizal fungi showed increased plant height, dry weight, nutrient content and mycorrhizal colonization. *G. fasciculatum* gave the best response in cowpea, thereby indicating the need of mycorrhizal inoculation to derive maximum benefit for growth in nutrient deficient soils.

Hazarika *et al. (*1999) studied the effect of VA-mycorrhizal fungi inoculation on growth and nutrient uptake of blackgram. Blackgram (*Vigna mungo*) inoculated with two species of vesicular arbuscular mycorrhizal (VAM) fungi, *Glomus fasciculatum* and *G. mosseae*, grown in sterilized and unsterilized soil with or without phosphorus had significantly greater plant growth, plant dry matter, yield attributes, nutrient uptake, spore density and % root colonization than non mycorrhizal plants. The plants grown in unsterilized soil inoculated with VAM fungi alone and in association with phosphorus had significantly increased plant growth, dry matter production, yield attributes, nutrient uptake, root colonization and spore density than non mycorrhizal plants, however, these parameters were comparatively less than the mycorrhizal plants grown in sterilized soil. There were no significant differences between the two mycorrhizal symbionts in any treatment combination. These results suggest that inoculation of blackgram with an efficient VAM fungus in phosphorus deficient acid soils of Assam would greatly improve plant growth, nutrient uptake and yield.

Meshram *et al.* (2000) studied the effect of VAM fungi and P-sources on nodulation, dry matter and yield of chickpea. During the rabi season of 1991-92 in Maharashtra, the effects of 4 P sources (single superphosphate, rock phosphate (RP) bone meal and press mud) and 6 microbial treatments (*Glomus mosseae*, Gl; *Gigaspora calospora*, Gg; *Acaulospora* sp., Acal; a Gl+Gg+Acal mixture; the P-solubilizing culture Microphos; and an uninoculated control) on the nodulation, dry matter and yield of chickpea were investigated. Cultivar Phule G-12 was firstly treated with 2 g captan per kg seeds, then with the 4 vesicular arbuscular mycorrhizal (VAM) fungal treatments at 25g/kg seed or with 50g Microphos per kg seeds. Seeds were subsequently sown in fields treated with VAM chlamydospores (250 spores/50 ml soil suspension) and infected root segments of guinea grass, and supplied with nitrogen (urea) and the 4 P sources. Data were collected on available N and P, nodulation, dry weight of nodules, shoots and roots per plant at 50% flowering, grain yield and the total dry matter at harvesting. Seeds treated with Gl+Gg+Acal had the highest available N and P (128 and 26.87 kg/ha, respectively) and the highest values for number of nodules per plant (29.50), dry weight of nodules (157.42 mg per plant), shoots (6.91 per plant), root (500 mg per plant), total dry matter (44.08 q/ha) and grain yield (37.83 q/ha). Among the P treatments, RP recorded the highest values for available N and P (118 and 25.25 kg/ha), shoot dry weight (6.97 g per plant), total dry matter (42.50 q/ha) and grain yield (35.21 q/ha).

Kehri and Chandra (2001) studied the performance of black gram, *Phaseolus mungo*, with VAM inoculation and phosphate fertilization. They have conducted an experiment to evaluate six VAM fungi for their effectiveness in improving the performance of blackgram at four levels of phosphorus. Some of the VAM fungi were more effective in improving the root, shoot and nodule biomass, mycorrhizal infection, N, P,K uptake and grain yield. Grain yields of var. T-9 inoculated with *G. macrocarpum* at 1/4th P or *G. fasciculatum* at ½ P dose were as high as those receiving full phosphate fertilization. The grain yields of var. PU-19 inoculated with *G. macrocarpum, G. mosseae* and *G. multicaule* at 1/4th dose were similar wioth those treated with the full P doses. It is therefore possible to reduce the phosphate recommendation for blackgram with inoculation of effective VAM fungi.

Tholkappian *et al.* (2001) conducted a study to determine the effect of AM-fungi, *Glomus fasciculatum* inoculation (at 50 g per pot) on soybean (*Glycine max cv.* CO 1) on proline accumulation, nitrogen content and pod number under water stress conditions. Water stress was imposed on the 10-days-old potted plants. Soil moisture was maintained at three levels: 25, 50 and 100% (control) throughout the entire crop growth period. Proline accumulation increased with increasing levels of soil moisture stress. The mycorrhizal soyabean plants recorded lower levels of proline than the non-mycorrhizal plants under moisture stress. The nitrogen content and pod number per plant increased in mycorrhizal soyabean than the non-mycorrhizal plants at 25% soil moisture.

Devi and Sitaramaiah (2001) studied the response of endomycorrhizal fungi at different levels of inoculum in blackgram. The suitable inoculum level of 4 endomycorrhizal fungi

(*Glomus fasciculatum, G. constrictum, G. epigaeum,* [*G. versiforme*] and *Acaulospora morrowiae*), at which maximum black gram cv. LBG 20 was induced, was investigated. The endomycorrhizas, except *G. constrictum*, were inoculated in black gram seedlings at 0, 100, 200 and 400 chlamydospores per kg soil. *G. constrictum* was inoculated at 200, 400 and 800 spores per kg soil. Results showed that mycorrhizal colonization of black gram with all endomycorrhizas increased with increasing inoculum level. At all inoculum levels tested, the highest root colonization was with *G. constrictum*, followed by *G. fasciculatum, G. epigaeum* and *A. caulospora*. Plants inoculated with the mycorrhizal fungi species showed more root and shoot dry weight, and chlorophyll content compared to uninoculated plants. Vegetative growth and chemical composition of N, P, K, Ca and Mg increased with increasing inoculum density. *G. epigaeum* was the most efficient in increasing dry matter of root and shoot. *A. morroweae* induced the maximum increase in chlorophyll content among the mycorrhizal fungi regardless of inoculum level.

Muthukumar and Udaiyan (2002d) studied the growth and yield of cowpea as influenced by changes in arbuscular mycorrhiza (AM) in response to organic manuring. They studied spore populations of arbuscular mycorrhiza (AM) fungi, mycorrhiza formation, and the growth and yield of cowpea [*Vigna unguiculata* (L.) Walp] grown in soils amended with sunhemp (*Crotalaria juncea* L.), pongamia [*Pongamia pinnata* (L.) Pierrel], cow dung and sheep manure. Spore populations and colonization of cowpea roots by AM fungi were higher in soils amended with organic manure than in non-amended soils. The effects of organic amendment on cowpea growth, nutrient uptake and yield varied with the type of organic amendment. These responses to organic amendment were strongly related to changes in AM colonization levels and spore populations. Cowpea growth and yield were positively related to spore populations of *Glomus aggregatum, G. geosporum* and *Scutellospora calospora*, but were negatively correlated to spore populations of *Acaulospora scrobiculata*. They suggested that changes in the growth and yield of cowpea in response to organic amendment are a result of the differential effects of the types of amendment on the proliferation of different AM fungi within the community.

Gill *et al.* (2002) compared the root colonization and spore population of *Glomus fasciculatum, Glomus mosseae, Gigaspora gigantea* and *Gigaspora margarita* on chickpea *cv.* GL 769. The populations of the vesicular arbuscular mycorrhizas were investigated after 30, 60, 90 and 120 days. The maximum plant growth was observed in *Glomus fasciculatum* pots. The lowest plant height was observed in *Gigaspora margarita* pots. The fresh and dry weights were higher in *Glomus fasciculatum* pots. The highest number of spores was observed in *Glomus fasciculatum* pots.

Kumar *et al.* (2002a) inoculated chickpea *cv. Gaurav* plants grown in nutrient-deficient soil in 20 cm diameter earthen pots with five vesicular arbuscular-mycorrhizal fungi (*Glomus mosseae, Glomus fasciculatum, Glomus constrictum, Acaulospora laevis* and *Gigaspora gimorei*). Inoculation with all the mycorrhizal fungi resulted in increase in plant height, dry weight, nutrient content, mycorrhizal colonization, sporocarp numbers in soil and number of pods per plants as compared to the control. Among these mycorrhizal fungi *Glomus mosseae* exhibited most pronounced effects. This indicated the need for mycorrhizal inoculation to derive maximum benefit for growth in nutrient deficient soils.

Singh *et al.* (2002d) conducted a study to investigate the effect of *G. aggregatum* inoculation and P amendments (at 0, 40 and 80 mg P_2O_5/kg soil) on colonization and spore population and consequent yield enhancement in chickpea. Colonization and spore population were observed at 45, 75, 105 and 130 days after sowing. *G. aggregatum* inoculation significantly increased the colonization of chickpea roots up to 130 days. The maximum colonization and spore population was observed at 40 mg P_2O_5/kg soil.

VEGETABLES

Swaminathan and Verma (1977) studied the symbiotic effect of vesicular-arbuscular mycorrhizal fungi on the phosphate nutrition of potatoes. According to them, *Glomus macrocarpum*, which occurs widely in Indian soils, increases P uptake from the soil. Inoculating the seed tuber with a slurry of the spores gave better results than direct inoculation of the soil. In an infertile soil mycorrhizal plants took up x8 more P and made x5 more growth than non-mycorrhizal plants. Under natural conditions only 23% of the potato fields surveyed carried spores or root infection by endomycorrhizal fungi.

Mishra and Verma (1983) studied the effect of different mycorrhizal treatments on the growth of onion. Seedlings infected with *Glomus tenuis* and *G. fasciculatum* showed marked differences in growth as compared with controls. *G. tenuis* penetrated the root cortex earlier than *G. fasciculatum* and had greater competitive colonizing ability. Treatment with mixed endophytes was particularly effective in stimulating growth.

Senapati *et al.* (1987) studied the effect of a mixture of *Glomus fasciculatum*, *G. mosseae* and *G. macrocarpum* on 3 okra (*Hibiscus esculentus*) cultivars in laterite soil. They reported significantly higher plant height, number of leaves/plant, number of fruits/plant, pod length and the number of seeds/pods in inoculated plants than in non-mycorrhizal controls.

Indi *et al.* (1990a) studied the response of brinjal genotypes in terms of dry weight and phosphorus uptake as influenced by VAM inoculation. *Solanum writti* and 4 aubergine genotypes (PS-8, Dorli, Pragati and Borgaon-1) were more responsive to inoculation with *Glomus fasciculatum* than other 6 genotypes tested. In general, the results indicated a genotype dependent variation in dry matter and P uptake as influenced by inoculation with a vesicular-arbuscular mycorrhizal fungus.

Singh *et al.* (1990) studied the effect of VA mycorrhizas on growth stimulation of onion *Allium cepa* in Maharashtra region. They raised onion plants in 30 cm diameter earthen pots filled with 8 kg sterile soil. N at 776 mg/pot was applied in two equal split doses at the time of transplanting and one month later. Pots were inoculated with cultures of 3 VA endophytes (*Glomus mosseae*, *G. fasciculatum* and *Glomus* sp.) at the time of transplanting. The plants were kept in wire mesh house and assessed after 60 days. Uninoculated plants were maintained as control. Inoculation with VA endophytes resulted in greater plant height, number of leaves, number of extramatrical chlamydospores/50 ml soil and percentage root infection compared with the control. *Glomus* sp. proved to be the most effective.

Rai (1990b` reported increased yields of potato by inoculation of cultivars *Kufri Lalima* and *Kufri Sindhuri* with *Glomus mosseae* and *G. fasciculatum*.

Neeraj *et al.* (1990) studied the effect of soil inoculation with endomycorrhizal fungi on growth of *Cyamopsis tetragonaloba*. *Cyamopsis tetragonaloba* var. *Pusa Navbahar* was used for the experiments conducted in unsterilized soil. All the plants were colonized by the VAM fungi forming arbuscules and/or vesicles. For shoot production, the best response was obtained with *G. fasciculatum* , followed by *G. mosseae*. The plants treated with other VAM fungi were also better than the control (untreated) plants. The roots also showed a similar growth pattern. In all the cases the total dry weight of leaves increased, *G. fasciculatum* treatment showing the maximum leaf dry weight. Thus *G. fasciculatum* was the most suitable VAM fungal species for *C. tetragonaloba*.

Sreenivasa *et al.* (1993) studied the response of chilli (*Capsicum annum* L.) to the inoculation of an efficient vesicular arbuscular mycorrhizal fungus. Thirty-days-old plants of *C. annum cv. Byadagi* were transplanted into 8 litre pots each containing 5 kg of a P deficient unsterilized black clayey soil (pH 7.0). P fertilizer was supplied as superphosphate (soluble) or Musoorie rock phosphate (soluble) at 0, 25, 50 or 100 % of the recommended rates. The soil was inoculated with *Glomus macrocarpum* or *G. fasciculatum* or not inoculated. The crop was harvested 90 days after transplanting. Inoculation with mycorrhizal fungi increased fruit yield, shoot P concentration, and Zn, Cu, Mn, and Fe uptake compared with uninoculated controls. Inoculation with *G. macrocarpum* had a greater effect than inoculation with *G. fasciculatum*. Mycorrhizal fungus inoculation also increased shoot DW but not when rock phosphate was applied. It is concluded that application rates of soluble P fertilizers could be reduced by 50% if *C. annum* plants were inoculated with VAM fungi.

Edathil *et al.* (1994a) assessed the VAM status of tomato, brinjal (aubergine) and chilli (*Capsicum*) during the initial establishment period in natural field conditions and in pot culture using non-sterile soil. The soil had a low nutrient status and no manurial application was made during the 60-d study. VAM colonization in roots was maximum at 45, 50 and 60 d after the germination of aubergine, tomato and *Capsicum* seeds, respectively, under field conditions and on the 60th day in pot culture. None of the plants showed 100% root colonization. Ascending and descending trends in colonization were noted. In all cases, the original spore count of 3.90 g/dry soil increased markedly in the rhizosphere of plants. Spore number was more pronounced in pot than in field culture.

Edathil *et al.* (1994b) studied the effect of VA mycorrhizal exposure period to their colonization and spore production in tomato seedlings and on host biomass. They have presented a method for the estimation of VA mycorrhizal index for a better understanding of VAM colonization intensity. The VAM fungal colonization by different species is increased with increase in the exposure period to VAM soil prior to transplanting upto Day 20, and then was constant to Day 30. The VAM infection and biomass production by the transplanted tomato plants also increased when the length of the pre-transplanting exposure increased, but the relationship was inexact. Total leaf area and biomass were significantly positively correlated.

Raverkar and Bhandari (1995) tested the efficacy of VAM fungi to produce sturdy tomato seedlings with minimum phosphate fertilizers under nursery conditions. Soil was inoculated with pot culture of *Gigaspora gilmorei* and four levels of phosphorus (0, 50,

100, 150kg, P_2O_5/ha). Uninoculated soil served as control. The data showed that inoculation of nursery soil with *Gigaspora gilmorei* improved the health of seedlings in terms of accumulation of dry biomass in shoot, root and P accumulation in shoots. A saving of 50 Kg P_2O_5 /ha was also shown.

Sundaram and Arangarasan (1995a) studied the effect of inoculation of VAM fungi on the yield and quality attributes in tomato *cv* Co3. Four different VAM fungi namely *Glomus fasciculatum, G. mosseae, Acaulospora laevis* and *Gigaspora margarita* were used in the experiment. Out of these four, *G. fasciculatum* gave the highest fruit yield (35% increase over control followed by *G. mosseae*-24%, *Gigaspora margarita*-21% and *Acaulospora laevis*-19%). Inoculation of these fungi also improved the quality attributes such as Vitamin C content and total soluble sugars.

Kumar and Mangal (1995b) made an attempt to find out the optimum dose of P to which VAM can be combined to get maximum growth and yield of onion. They conducted a pot experiment with recommended doses of N and K (120 kg/ha & 25 kg/ha respectively) and four levels of phosphorus *i.e.* O (control), 50% of recommended level, 75% of recommended level and full dose of recommended level and VAM inoculation - *Glomus mosseae*. VAM plants significantly produced greater bulb yield over their respective non-VAM plants at all levels of P except at 100%. VAM inoculation at 50% recommended P recorded the greatest increases in plant height, shoot dry weight and bulb yield over its respective control. The results demonstrated that efficient utilization of P fertilizers in onions was obtained in presence of VAM fungi and thus a net saving of P input was obtained.

Sreeramulu *et al.* (1996) studied the growth of two important leafy vegetables *Amaranthus viridis* and *Trigonella foenum-graecum* in an unsterilized sandy loam soil in response to P fertilizer application (Zero, or 25, 50 or 100% of the recommended rate i.e. 50 kg P_2O_5/ha) was observed in a pot experiment in a greenhouse. Height and leaf number of plants of both species at 40 days after sowing were significantly greater when soil was inoculated with *Glomus fasciculatum* than when not inoculated. Inoculated plant also had significantly higher shoot and root biomass than corresponding uninoculated control plants.

Edathil *et al.* (1996) studied the interaction of multiple VAM fungal species on root colonization, plant growth and nutrient status of tomato seedlings. They grown the tomato seedlings in sterile, phosphorus deficient soil and inoculated with 4 species (*Glomus aggregatum, Glomus fasciculatum, G. geosporum* and *G. sinuosum*) of VAM fungi in 15 possible combinations. Mycorrhizal plants exhibited a significantly higher shoot length and biomass, than non-mycorrhizal plants ($P > 0.05$). VAM also increased host tissue N and P concentration. Among the VAM combinations, inocula containing all 4 endophytes promoted markedly better shoot length and biomass than other combinations, although spore yields were lower in mixed inocula. Mycorrhizal dependency (MD), % increase of biomass, compared with controls, was also increased host tissue N and P concentration. Among the VAM combinations, inocula containing all 4 endophytes promoted markedly better shoot length and biomass than other combinations, although spore yields were lower in mixed inocula. Mycorrhizal dependency, % increase of biomass compared with controls, was also more in quadruple combination than in either triple or double combinations. A negative interaction on plant growth and biomass was observed between *G. aggregatum* and *G. fasciculatum* . This effect was nullified when more species were added to this combination.

Dhinakaran and Savithri (1997) investigated the effects of different P treatments and VAM fungi inoculation in tomato *cv.* CO3 grown in pots in red non-calcareous soil (Paralythic Ustochrept). P was applied as single superphosphate (SSP) or Mussoorie rock phosphate (MRP) at 75 or 100 kg P_2O_5/ha. P applied as superphosphate at 100 kg P_2O_5/ha, significantly increased the yield of tomatoes, but the application of VAM fungi had no effect on yields. Studies using ^{32}P confirmed increased uptake of P at 100 kg P_2O_5/ha. P content and P uptake of tomato fruits increased with increasing applied P and VAM fungal inoculation. VAM fungal inoculation increased fertilizer use efficiency, and the percentage P derived from fertilizers.

Gaur *et al.* (1998a) compared AM fungi inoculants using *Capsicum* and *Polianthes* in marginal soil amended with organic matter. Different types of nursery inocula formations, namely mixed indigenous cultures and *Glomus intraradices*, were compared with commercially available inoculants of AM fungi in a pot experiment using two horticultural crops, *Capsicum* and *Polianthes*. Soil based inocula and soil beads produced the highest response in both crops. *Glomus intraradices* resulted in the highest yield in both *Polianthes* (45% increase in spike length) and *Capsicum* (112% increase fruit yield). Among the commercial inocula tested, only Mycorise enhanced spike length (33%) and fruit yield (11%) in the two hosts. Overall AM colonization was higher in *Polianthes* than in *Capsicum*. Sheared root inocula of *G. intraradices* resulted in high colonization (upto 68%) but the yield enhancement was lower than with soil-based formulations. The mixed indigenous culture produced the highest number of spores and propagules and commercial inocula the lowest.

Iqbal and Mahmood (1998) studied the effect of single and multiple VAM inoculants on the growth parameters of tomato. All the three VAM fungi *Glomus constrictum* Trappe, *Glomus mosseae* (Nicolson, Gerdemann) Gerdmann and Trappe, and *Glomus fasciculatum* (Thaxter Sensu Gerd.) Gerdemann and Trappe were tested for their efficacy on tomato singly and in various combinations. All the treatments showed improvement in the mycorrhizal colonization, number of chlamydospores recovered from the rhizosphere soil, plant biomass, and P contents of the shoots and roots. However, all the three VAM fungi tested behaved differently in different combinations and differed from one species to another species. Maximum growth in plants and spore count was observed by the inoculation of *G. mosseae* followed by *G. constrictum* and *G. fasciculatum* respectively. In combination, the maximum increase was observed in the plants which received *G. mosseae* and *G. constrictum* followed by *G. constrictum* and *G. fasciculatum* and *G. mosseae* and *G. fasciculatum* respectively. However, highest increase was observed in the shoot biomass, mycorrhizal colonization and P content in the shoots and roots when the plants were inoculated with all the three VAM fungi. These enhancements in the plant growth parameters, mycorrhizal colonization, and biomass production may be reflected in higher yield of tomato due to the multiple inoculation of VAM and this deserves due attention and further emphasis to confirm the suitability of multiple inoculum as compared to single inoculum for better crop production.

Wani and Konde (1998a) studied the response of garlic to inoculation with VAM. Vesicular arbuscular mycorrhizas (VAM) (*Glomus mosseae*, *Gigaspora margarita* and *Acaulospora calospora*) significantly increased the plant height, number of functional leaves, dry matter, fresh and dry weight of bulbs, and N and P uptake by garlic plants and available P_2O_5 in soil. The percentage root colonization was 85.83, 78.42 and 76.17% by *G. mosseae*, *G. margarita* and *A. calospora*, respectively. Native VAM root colonization was 34.83%.

Wani and Konde (1998b) studied the effect of *Glomus mosseae* inoculation using different P sources on garlic. The results of greenhouse experiment on garlic conducted in stone soils revealed that *Glomus mosseae* used the insoluble P sources (press mud, rock phosphate and bone meal) rather than soluble P source (superphosphate) for enhancing growth and yield. Inoculation with *G. mosseae* significantly increased plant height, number of functional leaves/plant, plant dry weight, bulb diameter, bulb weight and N and P uptake over the non-mycorrhizal treatment.

Mangal and Kumar (1998) conducted a two year pot trial on onion (*Allium cepa cv. Hisar* 2) using sand with a pH of 8.5 and ECE of 0.15 d/Sm. Ten grams of VAM inoculum was placed below the onion seedlings at the time of planting in each polyethylene bag of 30 x 40 cm size containing 15 kg steam sterilized soil with recommended dose of N and K. Five salinity levels, *i.e.* 0.75 (control), 4, 6, 8 and 10 d/Sm, two phosphorus levels, *i.e.* 0 and 25 ppm and two VAM treatments *i.e.* with and without VAM. At each salinity level, VAM inoculation in presence of P increased plant growth; fresh and dry weight of shoot and bulb but efficiency of VAM in absence of P was significantly better for most of the parameters. A 50% reduction in the bulb yield was recorded at 9.00 to 9.20 EC when VAM was inoculated in absence of P, but similar reduction (50%) was obtained at 8.40 to 8.50 EC under non-inoculated media in presence of P. In absence of both P and VAM, a similar reduction in yield was noticed only when soil salinity ranged between 6.30 and 6.35 d/Sm EC. VAM inoculation without P showed significantly higher sulfur contents in the bulbs whereas total soluble solid content was higher under high salinity with P application. Respiration pattern of plants was lowered at each salinity level when VAM was inoculated under P deficient media.

Harikumar and Potty (1998) compared the relative performance of 2 geographical isolates of the vesicular arbuscular mycorrhizal fungi *Glomus microcarpum* var. *microcarpum* on infectivity, yield and nutrient uptake in sweet potato through a pot culture study under different P regimes. Colonization by both M1 (from southern Kerala) and M2 (from northern Kerala) strains of *G. microcarpum* var. *microcarpum* was greatest at P_0 levels (no added P), followed by P_1 level (half the recommended P rate). The highest colonization percentage was observed in roots of both M1 - and M2 treated pots at 40-80 days after inoculation (DAI). Root and tuber yield of sweet potato inoculated with M1and M2 strains showed almost the same trend at 20 DaI irrespective of the P levels. However, at 40-80 DAI, a general increase in yield trend caused by mycorrhizal inoculation was observed. Strain M1 was more efficient in enhancing yield and P uptake of the crop than strain M2. The beneficial effect of this strain was better under poor/low P level.

Harikumar and Potty (1999) compared the relative performance of 2 geographical isolates of the vesicular arbuscular mycorrhizal fungi *Glomus microcarpum* var. *microcarpum* on infectivity, yield and nutrient uptake in sweet potato through a pot culture study under different P regimes. Colonization by both M1 (from southern Kerala) and M2 (from northern Kerala) strains of *G. microcarpum* var. *microcarpum* was greatest at P_0 levels (no added P), followed by P_1 level (half the recommended P rate). The highest colonization percentage was observed in roots of both M1 and M2 treated pots at 40-80 days after inoculation (DAI). Root and tuber yield of sweet potato inoculated with M1and M2 strains showed almost the same trend at 20 DaI irrespective of the P levels. However, at 40-80 DAI, a general increase in yield trend caused by mycorrhizal inoculation was observed. Strain M1 was more efficient in enhancing yield and P uptake of the crop than strain M2. The beneficial effect of this strain was better under poor/low P level.

Narayana and Gudige (1999) studied the growth response of carrot (*Daucus carota* L.) to a selective VA mycorrhizal inoculant in an unsterile soil. After preliminary trials, *Glomus mosseae* was selected and tested on 2 varieties of carrot (Local and SKG) to study its efficiency on growth and yield in unsterile soil. The dry weight and fresh weight of both root and shoot system were found an increase, and the yield increased by 27% in Local and 42% in SKG over non-inoculated control plants.

Shrihari and Sreenivasa (1999b) conducted a field trial to study the possibility of P savings in tomato (*Lycopresicon esculentum* Mill.) through inoculation of *Glomus fasciculatum* (Thaxt.) Gerd. and Trappe, under black clay and red sandy loam soils. Six levels of P2O5 {0, 20, 40, 60, 80 and 100 per cent recommended dose (PRD)} were tried with inoculation. An uninoculated control (UIC) with 100 PRD of P_2O_5 was maintained. Mycorrhizal development, P and N concentrations in plant growth, yield and fruit quality parameters were recorded. Percent root colonization and extramatrical spore counts increased significantly with increase in P_2O_5 levels up to 60 PRD beyond which both these mycorrhizal parameters decreased significantly. All other parameters studied increased with increase in P_2O_5 levels. The performance of inoculated plants supplied with 60 PRD and 80 PRD of P2O5 was at par with UIC given 100 PRD of P_2O_5 under black clayey and red sandy loam respectively. This suggested savings of 40 PRD of P_2O_5 (40 kg ha-1) under black clayey soil whereas; the savings were 20 PRD of P_2O_5 (20 KG ha-1) under red sandy loam.

Ramana and Babu (1999) studied the response of onion (*Allium cepa*) to inoculation of VAM fungi at different levels of phosphorus. A field trial was conducted to study the response of onion to the inoculation of three VA mycorrhizal fungi in the presence of three levels of phosphorus on an unsterilized alfisol. Inoculation of onion in the nursery with *Gigaspora margarita* at a P level of 17.5 kg per ha enhanced the number of leaves, dry matter accumulation, uptake of P and yield of bulbs. *G. calospora* recorded higher percentage of root infection at lower levels of P. Among the three inoculants, *G. margarita* , proved to be the most effective inoculant in influencing the growth and performance of onion. Both uptake of P and yield of onion bulbs, inoculated with *G. margarita* at half the recommended P level, were significantly higher than that achieved by the uninoculated plants with full levels of P fertilization. This suggested that the field performance of onion plants can be improved by inoculation in the nursery with VA mycorrhizal fungi with reduced quantity of P.

Devi and Phookan (1999) studied the relative mycorrhizal dependency of french bean (*cv. contender*) to different *Glomus* species *viz. G. mosseae, G. clarum, G. constrictum* and *G. fasciculatum* . All the four mycorrhizal species had significant effect on the growth of the crop. Amongst them, *G. mosseae* showed significantly the best effect on plant height, days to first flowering, fruit set and fresh and dry weight of the crop.

Nagaraju *et al.* (1999) conducted an experiment to screen the efficient VAM fungi for aggregatum onion. *Allium cepa* var. *aggregatum* was inoculated with *Glomus mosseae, G. fasciculatum* or *Acaulospora laevis*, at Annamalainagar. *G. mosseae* was found to be the best root colonizer with a percentage increase of 13.86 and 22.14 over *G. fasciculatum* and *A. laevis,* respectively. *G. mosseae*-treated plants had the highest number of bulbs (7.60), bulb diameter (3.55 cm), bulb weight (168.26 g) and plant height (49.53 cm). Similarly, percentage increase for leaf number in *G. mosseae* was 59.42, followed by *G. fasciculatum* (34.04) and *A. laevis (21.8).* The percentage increase for the number of tillers was also highest in *G. mosseae*, followed by *G. fasciculatum* and *A. laevis. G. mosseae*-inoculated plants had higher chlorophyll, ascorbic acid and pyruvic acid contents compared to *G. fasciculatum*-and *A. laevis*-inoculated plants.

Shrihari and Sreenivasa (1999a) conducted nursery bed trials at the Main Research Station, University of Agricultural Sciences, Dharwad, Karnataka to select an efficient arbuscular mycorrhizal (AM) fungus for tomato var. L-15 under black clay and red sandy loam soils. Five different AM species (*Glomus fasciculatum, G. macrocarpum, Gigaspora margarita, Acaulospora laevis* and *Sclerocystis dussii*) were screened for their efficiency for root colonization, extramatrical spore count, population of free living N_2 fixers and phosphate solubilizers, plant P concentration, and growth and yield parameters of tomato. Plants inoculated with *G. margarita* showed significantly the highest population of free-living N_2 fixers in their rhizosphere as compared with all other treatments. Inoculation of *G. fasciculatum* significantly increased all the other parameters studied in comparison with the rest of the treatments. *Gigaspora margarita* was second to *G. fasciculatum*. The performance of the different AM fungi screened did not change with the soil type.

Ramananda *et al.* (2000) observed that onion cultivars N-53 and Bellary Red responded well to inoculation with different AM fungi. Percentage root colonization, spore count and population of free living N_2-fixers and P-solubilizers were significantly higher in N-53 inoculated with *Gigaspora margarita* compared with the other AM fungi screened. Bellary Red responded to inoculation with *Acaulospora laevis.* Plant growth and yield were significantly higher in N-53 and Bellary Red onion seedlings inoculated with *G.margarita* and *A. laevis,* respectively. All the investigated parameters were lowest in the uninoculated control.

Sharma and Adholeya (2000b) conducted a field experiment to evaluate the benefit to growth of *Alium cepa* inoculated with a mixed culture of indigenous arbuscular mycorrhizal (AM) fungi. Four locally adapted onion cultivars Pusa White Flat (PWF), Pusa white round (PWR), Early Grano (EG) and Madhvi (PM) woere grown at 2 P rates (25 and 50 kg P ha^{-1}) in a P deficient alfisol. Inoculation significantly increased mycorrhizal formation over that caused by native AM fungi present at the site. At harvest, all inoculated onion cultivars showed higher values for bulb diameter, fresh weight, shoot dry matter, shoot P content and bulb yield than uninoculated plants. However, the magnitude of AM response for yield in a given onion cultivar was found to be different at different rates of P. This holds true in all the cultivars tested. Inoculated plants of cultivars PM and PWR gave greater bulb yields at 25 kg P ha^{-1} whereas PWF and EG plants showed a similar response at 50 kg P ha^{-1}. The percentage root length colonized by AM fungi of inoculated plants did not differ significantly between the P rates. However, the extent of colonization varied among the cultivars.

Gaur and Adholeya (2000a) multiplied a mixed Culture of indigenous endomycorrhizal fungi in a nursery and tested for its ability to promote growth and yield of three agricultural crops in Haryana. Onion, potato and garlic were successfully inoculated with VAM fungi in nutrient deficient soil amended with composted leaves of *Albizzia lebbek* (albizzia), *Populus deltoides* (poplar) and *Leucaena leucocephala* (leucaena). Colonization in onion reached approx. 85% of root length, followed by garlic and potato (65%). Inoculation response in terms of yield increase was maximum in onion (70%) whereas garlic and potato showed 30% and 48% increases respectively. Growth response to VAM fungus inoculation at the higher level of organic amendment as well as production of infectious VAM population was demonstrated.

Ramananda and Sreenivasa (2000) studied the influence of different AM fungi on growth of onion. During the kharif season of 1997, onion cultivars Bellary red and N-53 were grown in potted Vertisols (10 kg per pot) containing arbuscular mycorrhizal fungi: *Gigaspora margarita* and *Glomus fasciculatum* (both at 50 g per pot, having 0.19 x 10^4 infective propagules g^{-1}) and *Glomus macrocarpum, Acaulospora laevis*and Sclerocystis dussii(all at 52.89 per pot, having 0.18 x 10^4 infective propagules g^{-1}). Bellary red inoculated with *A. laevis* and N-53 inoculated with *Gigaspora margarita* gave the highest values for root colonization and shoot P content concentration percentages, chlamydospore count, dehydrogenase [oxidoreductase] and alkaline phosphatase activities, and fresh bulb weight.

Nagaraju *et al.* (2000) conducted an experiment to study the effect of VAM on growth and yield of onion (*Allium cepa* L. var. *aggregatum* Don). Along with the application of graded rates of single super phosphate (SSP) and rock phosphate at 25,50, 75 and 100%. The inoculated plants receiving 50% treatment showed significant increase in plant height, number of leaves number of tillers, bulb diameter and weight of bulbs compared to plants that were supplemented with 100% phosphorous and with no VAM inoculation. However the treatment receiving100% SSP with *G. mosseae* recorded the highest plant growth response with respect to all biometric and biochemical characters. Nutrient uptake was significant under inoculated conditions compared to the control. Availability of nutrients in growing media varied according to the fertilizer application in which VAM-inoculated treatments

,showed lesser availability compared to uninoculated treatments. Fifty percent SSP with *G. mosseae* inoculation excellence 100% SSP without inoculation of VAM for yield attributing characters and is therefore suggested as the best treatment.

Kashappanavar and Sreenivasa (2000) conducted a pot experiment to study the effects of inoculation of *Gigaspora margarita* (50 g/pot) compared with *Acaulaspora laevis* (52.8 g/pot) at different P-levels (0, 50, 75 and 100%) in onion *cv*. N-53. Percent root colonization (82%) and spore counts (161/50 g soil) were highest in *G. margarita* inoculated plants and 75% P level. The alkaline phosphatase activity, shoot P concentration, total soluble solids, ascorbic acid content, plant dry biomass and fresh bulb weight were also highest in *G. margarita* inoculated plants. However, the difference between 75 and 100% P level was non-significant.

Subramanian *et al.* (2001) studied the role of mycorrhizas in tomato production under drought stress conditions. Tomato plants were inoculated or not inoculated with *Glomus intraradices* at 50 g/m^2 in a field in Tamil Nadu under 4 moisture regimes (0.50, 0.75, 1.00 and 1.25 IW: CPE ratio). Drought induced reduction in crop growth and yield were minimized by mycorrhizal symbiosis. Fruit quality of tomato was also improved by mycorrhizal inter action under drought stress. Crop yield data showed that Mycorrhizae-plant interactions were 50, 33, 23 and 27% under severe (0.50 IW: CPE), moderate (0.75 IW: CPE). mild (1.00 IW:SPE) and no drought (1.25 IW:CPE) conditions. Results indicate that mycorrhizal association was more beneficial under severe drought stress than under mild or no drought stress.

Nagaraju *et al.* (2001) studied the effect of vesicular arbuscular mycorrhizas (VAM, *Glomus mosseae*) inoculation at 25 g/plant under varying phosphorus levels (25, 50, 75 and 100% of recommended dose) on onion at Agricultural College, Annamalai University, Annamalainagar, Tamil Nadu. P was supplied in the form of single superphosphate (SSP) and rock phosphate. Results revealed that VAM inoculated onion in combination with the application of P fertilizers in the form of SSP at 100 per cent recommended dose increased the yields. Single super phosphate was comparatively better than rock phosphate, through the percentage of infection was higher in the rock phosphate- applied plants. There was no significant difference between 100 per cent rock phosphate and 50 per cent superphosphate in combination with VAM for growth and yield of onion.

Harikumar and Potty (2002a) screened two hundred and fifty seven genetic stocks of sweet potato for mycorrhizal colonization and their growth response. Among the accessions screened for percentage root colonization, 20% had 25% natural infection while another 20% had 50% infection in their root system and the remaining 60% had a very high colonization level. When the intensity of infection was taken, the trend was reversed indicating that a higher number of accessions had very low infection intensity. With regard to spore load, majority of accessions had a spore count between 100 and 300 per 50 ml rhizosphere soil with the predominance of *Glomus microcarpum* followed by *Acaulospora laevis, Gigaspora aurigloba, Gigaspora margarita, Glomus fasciculatum,* and *Glomus mosseae.* Correlation studies between fungal characters showed a positive correlation in majority of interactions. When plant characters were compared with mycorrhizal association, underground (root and tuber) characters had positive correlation with mycorrhizal colonization and the aboveground portion had no direct bearing on the interaction between different characters.

Haripriya and Sekharan (2002) conducted a pot experiment to evaluate the effect of vesicular arbuscular mycorrhizas (VAM) on the growth and yield of onion. The treatments comprised different species of VAM, *i.e. Glomus mosseae, G. fasciculatum* and *Acaulospora laevis,* as well as an uninoculated control. *G. mosseae* was the most efficient in colonizing the roots and in improving different characters such as plant height, number of tillers, number of bulbs, bulb diameter and bulb weight. Plants treatment with *G. mosseae* showed high contents of chlorophyll, ascorbic acid and pyruvic acid. In general, VAM inoculation gave higher root infection and improved the growth and yield of onion compared with the uninoculated control.

Singh *et al.* (2002b) conducted a field experiment in Varansi, Uttar Pradesh, during 1999/2000 and 2000/2001 to study the effect of vesicular arbuscular mycorrhiza (VAM) inoculation, along with nitrogen, phosphorus, and potassium fertilizer (100:50:100 kg/ha recommended rate) application, on onion growth and yield. At 90 days after transplanting (DAT), VAM+N+P+K was superior to the other treatments except N+P+K during 1999/2000 in increasing plant height. Leaf number per plant was highest with VAM+0.5 N+P+K, VAM+N+P+K, and N+P+K. The greatest shoot girth was obtained with VAM+0.5 N or P+K, VAM+N+P+K, and N+P+K. Neck thickness, bulb diameter, bulb weight, and horizontal bulb diameter were most pronounced under VAM+0.5 N+0.5 P+K and VAM+N+P+K. Bulb fresh and dry weight were highest with N+P+K singly or in combination with VAM. The inoculation of VAM along with the application of N+P+K or 0.5 N+0.5 P+K gave the highest yields (15.98 and 15.54 t/ha, respectively).

Hulkoti and Sreenivasa (2002) studied the least number of AM infective propagules required to initiate root colonization in onion. The lowest number of infective propagules of arbuscular mycorrhizal fungi (AM) required to initiate root colonization in onion cultivars *N 53* and *Bellary Red* was determined by most probable number technique. The original inoculum of *Gigaspora margarita* 10.19×10^4 propagules per g) and *Acaulospora laevis* (0.18×10^4 infective propagules per g) were diluted up to 19×10^{-7} and 18×10^{-7} infective propagules per g) were diluted up to 19×10^{-7} and 18×10^{-7} infective propagules per g. respectively, using sand:soil mixture (1:1). Five tubes were used in each dilution. The number of positive tubes for mycorrhizal colonization was counted six weeks after seed germination. In *G. margarita,* infective propagules of more than 19 per g gave 3-5 positive tubes. In *A. laevis,* 4-5 positive tubes were recorded for more than 180 infective propagules per g. These results indicate that the greatest root colonization in N-53 required 19 infective propagules per g of *G. margarita* inoculum; the corresponding value for Bellary Red cultivar was 180 infective propagules per g of *A. laevis* inoculum. The infective propagules of *G. margarita* at 0.19/g and *A. laevis* at 0.18/g did not initiate root colonization. When inoculated with 0.19 infective propagules per g of *G. margarita* and 1.8 infective propagules per g of *A. laevis* were used one of the five tubes was positive for mycorrhizal colonization by both species. Thus a minimum of 1.8 infective propagules of *A. laevis* and 0.19 infective propagules of *G. margarita* per g was required to initiate mycorrhizal colonization in onion.

OILSEED CROPS

Krishna and Bagyaraj (1982) studied the influence of vesicular-arbuscular mycorrhiza on growth and nutrition of *Arachis hypogaea*. They have discussed the possibility of the responses of mycorrhizal infection being mediated by plant growth hormones.

Krishna *et al.* (1982) studied the response of groundnut to VA mycorrhizal inoculation in black clayey soil. They used the chlamydospores and root segments of *Panicum* infected by *Glomus fasciculatum* as mycorrhizal inoculum for pot experiment. They recorded increase in dry matter and yield due to inoculation, which was 3 fold in sterilized soil, and 1-2 fold in unsterilized soil. Mycorrhizal groundnut plants contained larger amounts of K, Mg, Zn, Cu and Fe.

Champawat (1988) studied the influence of three VAM fungi on nutrient uptake and growth in groundnut (*Arachis hypogaea*). Plants treated with the 3 mycorrhizal fungi *Glomus fasciculatum, G. constrictum* and *Gigaspora calospora* enhanced shoot and root dry weight and uptake significantly over non-mycorrhizal plants. Mean root infection of *G. fasciculatum* , *G. constrictum* and *G.calospora* treated plants in sterilized soil was 75%, 91% and 80% respectively. Mycorrhizal treatment resulted in increased spore number in root zone soil. *G. constrictum* was significantly superior over *G. fasciculatum* and *G. calospora* in enhancing growth, nutrient uptake and root colonization.

Sulochana *et al.* (1989) studied the growth response and root colonization in cultivars of sesame to 7 VAM fungi isolated from field-grown sesame plant. *G. margarita, G. monosporum* and *G. fasciculatum* enhanced growth of all the cultivars to a great extent.

Gupta *et al.* (1990b) studied the growth response of *Brassica juncea* to vesicular arbuscular mycorrhizal colonization. *Glomus fasciculatum* inoculation induced significant increase in plant height, average branch length, total dry matter production of root and shoot, total pod weight and yield, at all the stages of plant growth as compared to control. VAM endophyte colonized extensively the root system of mustard. The results also indicated that mycorrhizal inoculation increased nitrogen, phosphorous and potassium concentration of the shoot and root significantly compared to uninoculated plants. Of these nutrients, the pronounced increase was observed in the uptake of phosphorus. These studies indicated that mycorrhizal inoculation either resulted in higher survival and growth of mustard plant or influenced growth by modifying the physiology of the plants.

Harinikumar *et al.* (1990b) screened eight different VAM fungi for their symbiotic efficiency against sunflower (*Helianthus annuus* L.) var. Modern. The plants were harvested 90 days after sowing. The uninoculated plants had least plant height, shoot dry weight and P content. Plants inoculated with *Acaulospora laevis* exhibited maximum plant height and shoot dry weight, closely followed by *Glomus fasciculatum*. The per cent mycorrhizal root colonization was also the highest in plants inoculated with *A. laevis.* The shoot P concentration was maximum in plants inoculated with A. laevis. Least symbiotic response was observed in plants inoculated with *Glomus versiforme* and *Gigaspora margarita*. It was concluded that of the fungi screened, *Acaulospora laevis* was the most efficient fungus for inoculating sunflower. The next best fungus was *Glomus fasciculatum*.

Prabhakar *et al.* (1992) studied the effect of VA mycorrhiza on growth, phosphorus nutrition and kernel yield of groundnut (*Arachis hypogaea* L.). In tests using pots of sterilized sandy loam soil, inoculation with VAM significantly increased plant growth, P concentration, number of flowers/plant, number of pods/plant and grain weight/plant. Higher levels of P significantly decreased root infection by VAM suggesting that VAM fungi would be more effective in soils with low P. Out of three VAM fungi studied (*Acaulospora* sp., *Glomus* sp. C_5 and *G. fasciculatum*) *Glomus* sp. C_5 was superior in increasing plant growth and number of pods/plant.

Mane *et al.* (1993) studied the comparative performance of vesicular arbuscular mycorrhizal fungus and *Rhizobium* inoculation in groundnut. In field experiment during the monsoon season of 1989/90 at Parbhani, Maharashtra, groundnuts cv. LG-19 plants inoculated either with vesicular arbuscular mycorrhiza *Glomus fasciculatum* or *Rhizobium* increased root length, root weight, biomass production, nodulation, N_2 fixation, P uptake and seed yield (by 235-330 kg/ha).

Rachel *et al.* (1993b) worked on sunflower with reference to the performance of VAM fungi in comparison to superphosphate application in a sandy loam soil of Andhra Pradesh. In pot experiments inoculation of sunflower plants with *Glomus constrictum* gave yield responses similar to those obtained with superphosphate at 50 kg/ha, while *G. mosseae* gave results equivalent to 40 kg/ha fertilizer and *Acaulospora morrowae* up to 70 kg/ha. Mixed inoculation showed no significant advantage over individual inoculations. Plant biomass was increased in VAM inoculated soils and in those receiving phosphorus amendments.

Mehrotra and Baijal (1994) studied the effect of VAM fungi on the growth and nutrition of sunflower plant at different levels of added N and P fertilizers. The effect varied with the levels of N and P.

Chandrashekara *et al.* (1995) studied the VA-mycorrhiza mediated P effect on growth and yield of sunflower (*Helianthus annuus* L.) at different P levels in a field experiment on a vertisol at Dharwad, Karnataka. Plots inoculated with *Glomus fasciculatum* were given 16, 24 or 32 kg P/ha. At the vegetative stage, percent mycorrhizal root colonization, spore count, dry weight and P uptake did not differ significantly between inoculated and non-inoculated control plants. However, at flowering and maturity all the parameters were significantly higher in inoculated plants than non-inoculated plants. The total dry weight, P content and seed yield increased with increasing P level in non-inoculated plants, whereas no significant difference was observed between 16 and 32 kg P/ha in inoculated plants. The positive effect of mycorrhizal inoculation decreased with increasing P level above 16 kg P/ha, due to decreased percent root colonization and spore count at higher P levels.

Joshi (1995) studied the prospects of groundnut improvement by VAM fungi in medium-black calcareous soils of Gujrat. Thirteen groundnut cultivars were screened for VAM root infection by native fungi and observations were made 30 and 60 days after sowing on nodule weight, dry plant biomass, VAM root infection and plant nutrients (N and P), nitrogenase activity and micronutrient (Fe, Mn, Zn and Cu) uptake by plants. Thirteen groundnut cultivars belonging to different botanical groups were found to have wide range of root infection (30 to 92%) by native VAM fungi. Similarly, they differed greatly in plant biomass. In general, the Virginia runner and bunch cultivars were superior to the spanish bunches and Balencia cultivars in terms of VAM root infection, nodulation, plant biomass, uptake of nutrients, nitrogen fixation and pod yield.

Jadhav and Patil (1996) studied the effect of VAM inoculation on growth and N, P, K uptake of groundnut (*Arachis hypogaea* L. *cv.* JL-24). The seeds were sown in a medium black clay loam soil in boxes in greenhouse experiment. Spores of naturally occurring indigenous fungi were collected from the rhizosphere of locally growing crop plants for root inoculation of groundnut. The vesicular-arbuscular mycorrhizal (VAM) inocula, which were multiplied and maintained on maize plants, consisted of a mixture of spores of endophytes belonging to *Acaulospora* spp., *Entrophosphora* spp., *Glomus fasciculatum*, *Glomus mosseae* and *Scutellospora pellucida*. Plants were harvested every 30 days after sowing. VAM increased shoot and root dry weight at 30, 60 and 90 days of crop growth and increased the uptake of N, P and K. It was concluded that VAM fungi can be established in previously undisturbed and unsterilized soil to the benefit of groundnut crops.

Chandrashekara and Patil (1997) studied the influence of VAM inoculation and phosphorus on dry matter production and seed yield of two sunflower (*Helianthus annuus* L.) genotypes. In a field trial at Dharwad, Karnataka in summer 1993, sunflowers cv. Morden and MSFH-8 were inoculated with *Glomus fasciculatum* at three P levels (38, 56 and 75 kg P_2O_5/ha, equal to 50, 75 and 100 per cent of the recommended dose of phosphorus, respectively) on a Vertisol. MSFH-8 performed better than Morden. Application of P fertilizer increased the uptake of phosphorus, dry matter production, at all the growth stages except at 30 days after sowing (DAS), and seed yield, with the responses generally significant up to 56 kg P_2O_5/ha. Inoculation with *Glomus fasciculatum* increased seed yield, oil yield, total dry matter (16%) and stalk yield over uninoculated plants. The total dry matter production in leaves and stem (at 60 DAS) and in stem and head (at harvest) of mycorrhizal plants at 38 kg P_2O_5/ha were more than the non-mycorrhizal plants receiving 75·kg P_2O_5/ha, The positive effect of mycorrhizal inoculation decreased at P levels higher than 38 kg P_2O_5/ha.

Vijaykumar and Bhiravamurthy (1999) conducted experiments for selection of efficient indigenous local VAM fungi for improving the groundnut crop in semi-arid tropical soils of Anantpur district, Andhra Pradesh. Out of eight local indigenous VAM species tested, *Glomus fasciculatum* was significantly superior in producing the highest root-shoot length, dry weight, phosphorus content, % root colonization, compared to other VAM species inoculated in soil grown to groundnut crop (potculture experiment).

Gupta and Mukerji (1999) studied the effect of VAM fungus on the growth and flowering of *Helianthus annus* L. var. *Peredovik* EC-68414. Soil inoculation of sunflowers *cv. Peredovik* EC-68414 with *Glomus macrocarpum* significantly increased root and shoot length 20, 40 and 60 days after inoculation compared with untreated controls.

Sharma *et al.* (2000) treated the seeds of red and white stem castor [*Ricimus communis*] with: (A) farmyard manure (FYM) + soil (1:1)- (B) arbuscular mycorrhizal fungi (AMF,

100 2 inoculum at 200 spores/100 g soil) +soil, (C) vermicompost +soil (l:l), (D)FYM+soil +AMF, (E) vermicompost + soil + AMF and (F) soil only (control) to study the efficacy of AMF in improving the biomass production of castor plants. Eri silkworms, *Philosamia ricini* [*Samia cynthia ricini*] (race Borduar), were reared on leaves of AMF-inoculated rod and white stem castor (RSMC and WSMC, respectively) and AMF uninoculated red and white stem castor (RSNMC and WSNMC, respectively) and secondary food plants (*Carica papaya, Ziziphus mauritiana, Ailanthus excelsa* and *Plumeria rubra*) to observe the effect of mycorrhizas and food plants on the growth of eri silkworms and cocoons during October-November 1997 and March April 1998. The highest values of germination (10%). Survival (95-98%), plant height (94-95%), leaf number (20-24 leaves per plant), fungal root infection (90-100%) and concentration of mycorrhizal spores (95-100 spores/IOC v soil) for both red and white stein castor were recorded from treatment E. The spore count and percent mycorrhizal infection was higher in red stem castor. Also, the percent infection increased up to 5-10% in E. All treatments (except FYM) increased seed germination percentage in red and white stem castor. Percent survival of red stem castor increased in treatments B, D and E, while in white stem castor, all treatments (except FYM) increased this parameter. Higher larval weight was recorded from red stem castor. The growth of the larvae was better during March-April than October-November. Larval growth ceased in third instar in *C. papaya* and *A. excelsa* during both periods. In *Z. mauritana*, all the eri silkworms died in the second instar and with *2*, 100% larval mortality was recorded from second instars during October-November and in third instars during March-April. There was no cocoon formation on secondary food plants. Cocoon yield was 98-99% with RSMC treatments during both seasons. Average wet silk ratio for RSMC increased to 15 during March-April. The dry silk ratio was generally higher during March-April. Castor was concluded to be the sole food plant for eri silkworms.

Srinivasulu and Lahsman (2002) conducted a study to determine the response of niger (*Guizotia abyssinica*) and linseed (*Linum usitatissimum*) to mycorrizal inoculation. Although the native vesicular arbuscular (VA) mycorrhiza in unsterile soil caused 47.30 and 54.12 colonization in niger and linseed plants respectively, inoculation with *Glomus fasciculatum* increased both mycorrizal colonization and the number of spores in the rhizosphere soil. Inoculation with *G. fasciculatum* increased the dry weight of shoot significantly in the two oil-yielding plants. The increase in root weight was significant with linseed plants, although some increased root weight was not observed in niger. Inoculation significantly increased the phosphorus and zinc content in shoots, but had no significant effect as manganese uptake in the plant. A positive response to VA Mycorrhizal inoculation occurred in both niger and linseed plants in unsterile soil, although the natural infectivity was significantly lower when compared to inoculated plants. The observations of dry matter produced, and phosphorus and zinc content, indicate that native VA- mycorrizal fungi are less efficient in stimulating plant growth the inoculant strain of *G. fasciculatum.* Inoculation plants did not differ significantly from non- mycorrizal plants in their manganese content.

Spices

Bagyaraj and Sreeramulu (1982) raised chilli in inoculated nursery beds with *Glomus fasciculatum* and 3 local isolates. Mycorrhizal chilli seedlings were transplanted to the field, two levels of P fertilizer were applied. Isolate I_4 increased growth, P and Zn nutrition, flowering and yield of the plants and ascorbic acid content of green fruits. Yield of I_4 inoculated plants given half the recommended P level was slightly more than that of uninoculated plants with the full level of P.

Babu *et al.* (1988) studied the response of chilli (*Capsicum annuum* L.) plants of early inoculation with mycorrhizal fungi at different levels of phosphorus. Plants of the *cv.* Pusa Jwala were inoculated in the nursery bed with *Gigaspora calospora*, *G. margarita* or *Glomus fasciculatum* and transplanted to plots supplied with 0, 25 and 50% of the recommended P level (65 kg/ha). Uninoculated plants were transplanted at 0 and 100% of the recommended P. Plants height, shoot DW, P uptake, root colonization were measured. The plants responded to inoculation and P. Inoculation plus 25 and 50% P had beneficial effects similar to uninoculated plants.

Suvercha and Mukerji (1988) studied the influence of VAM fungi on the growth of two cultivars of *Capsicum annnum*. In one set of experiment, the seedlings were inoculated (I, II) with *G. fasciculatum* (infected roots & spores) while the other set was left uninoculated (I^0, II^0). Plants were harvested after 15, 30 and 60 days of transplantation, their heights and dry weights were recorded. Roots were cleared and stained to estimate mycorrhizal infection. A difference in percentage of infection between the cultivars of *C. annuum* was recorded. The infection ranged from 41% to 83% in *cv*. X-235 while it varied from 33% to 70% in the *cv. Pusa Jwala*. An increase in mean plant height was observed in the inoculated cultivars. In the inoculated and uninoculated sets of *cv*. X-235, it ranged from 14.12 cm to 22.25 cm and 9.7 cm to 15.75 cm, respectively. Similarly, in *Pusa Jwala*, it varied from 12.87 cm to 20.16 cm and 10.32 to 19.62 cm in inoculated and uninoculated sets, respectively. The mean dry weight also increased from 0.208 g to 0.902 g and 0.102 g to 0.65 g in the inoculated and uninoculated sets of *cv*. X-235. While in Pusa it ranged from 0.192 g to 0.76 g, in the inoculated set and 0.188 g to 0.34 g in the uninoculated set.

Sreenivasa (1992) studied the crop response to inoculation with VAM fungi *Glomus fasciculatum*, *Gigaspora margarita*, *Acaulospora laevis*, *Sclerocystis dussi* and a local isolate of *Glomus macrocarpum* in *Capsicum annuum cv. Byadagi* growing in pots of unsterilized soil. The local isolate of *G. macrocarpum* was the best for improving growth and P, Zn, Cu, Mn and Fe uptake.

Sreenivasa and Gaddagimath (1993) studied the mycorrhizal dependency of 6 chilli (*Capsicum annuum*) cultivars. In glasshouse trials, all the cultivars responded well to soil inoculation with *Glomus macrocarpum*, with the highest percentage root colonization and spore count in *Byadagi*, and the lowest in G-3. Similar trends were observed for shoot P, plant dry mass and yield.

Sreenivasa *et al.* (1993) transplanted 30 day-old plants of *Capsicum annuum cv. Byadgi* into 8 liter pots containing 5 kg each of a P-deficient unsterilized black clayey soil (pH 7.0). P fertilizer was supplied as superphosphate (soluble) or Musoorie rock phosphate (insoluble) at 0, 25, 50 or 100% of the recommended rate. The soil was inoculated with *Glomus macrocarpum* or *G. fasciculatum* or not inoculated. They harvested the crop 90 days after transplanting. Inoculation with mycorrhizal fungi increased fruit yield, shoot P concentration and Zn, Cu, Mn and Fe uptake compared with uninoculated controls, inoculation with *G. macrocarpum* having a greater effect than inoculation with *G. fasciculatum* . Mycorrhizal fungus inoculation also increased shoot dry weight but not when rock phosphate was applied. It was concluded that application rates of soluble P fertilizers could be reduced by 50% if *C. annuum* plants were inoculated with VAM fungi.

Anandraj and Sharma (1994) investigated the effect of the vesicular-arbuscular mycorrhiza (VAM) fungus, *Glomus fasciculatum* on the rooting of *Piper nigrum cv. karimunda* cuttings, consisting of a single node and leaf taken from either lateral (intended for fruit production) or runner (stolon) roots. Percentage rooting of all cuttings was higher in media containing VAM compared with that observed in sand. The addition of the commercial rooting powder, ceradix B, decreased the percentage rooting of runner shoots in the presence of VAM, although no differences were observed for lateral cuttings.

Sreenivasa (1994b) studied the response of chilli (*Capsicum annuum*) to inoculation with VA mycorrhizal fungi and organic amendments. All the organic amendments increased the proliferation of VAM fungi. Inoculation of *Glomus macrocarpum* in conjunction with worm casts with a narrow C:N ratio was the optimum treatment for improving the growth and yield of chilli.

Sreenivasa (1994a) studied the response of chilli (*Capsicum annuum*) to VA mycorrhiza at different phosphorus levels in the yield. Root colonization and sporulation by both VA mycorrhizal fungi tested (*Glomus macrocarpum* and *G. fasciculatum*) increased with the addition of P up to 56.2 kg/ha but decreased with further increase in P. At 56.2 kg/ha root colonization and sporulation was higher with *G. macrocarpum* than with *G. fasciculatum* . Rates of application of P fertilizer could be reduced through the efficient use of suitable mycorrhizal fungi.

Hazarika and Phookan (1995) studied the growth response of chilli to VA mycorrhizal infection with different levels of applied phosphorus in acid soil. They used three VAM fungi *viz. Glomus mosseae*, *G. fasciculatum* and *G. aggregatum* with three levels of single superphosphate and recommended doses of urea and muriate of potash. Inoculation of all the three VAM fungi in the nursery bed exhibited significant influence on growth and yield of chilli grown under the recommended package of practice of the state of Assam. Lower levels of phosphorus were observed to be at par or sometimes better than the higher levels in producing vigourous growth and yield of inoculated plants.

Gurumurthy and Sreenivasa (1995) carried out an investigation to study the influence of different levels of an efficient local isolate of VAM fungus, *Glomus macrocarpum* on the growth and yield of chilli in two soil types-black clayey soil and red soil. The results revealed that the percentage of root colonization and the number of extramatrical chlamydospores increased with increase in the level of inoculum in both the soil types. However, the increase in the percentage of root colonization beyond 50g inoculum level was not significant. Plant dry weight, number of chilli fruits, fruit weight and shoot P concentration increased with increased levels of inoculum in both the types of soil, however, they were highest in the red soil. Increase in these parameters beyond 50g inoculum level was not significant.

Gaur *et al.* (1998) working on *Capsicum annuum* compared different types of nursery inocula formulations, namely mixed indigenous cultures and *Glomus intraradices*, with commercially available inoculants of VAM fungi in a pot experiment. Soil-based inocula and soil beds produced the highest response. *Glomus intraradices* resulted in the highest yield (112% increase in fruit yield). Among the commercial inocula tested, only Mycorise enhanced spike length and fruit yield. The root inocula of *G. intraradices* resulted in high colonization (up to 68%) but the yield enhancement was lower than with soil-based formulations. The mixed indigenous culture produced the highest number of spores and propagules and commercial inocula the lowest.

Sreeramulu and Bagyaraj (1998) studied the response of *Cardamom, cv. Mysore* to 6 promising arbuscular mycorrhizal fungi. Selected from a primary screening trail and 5 dominant native mycorrhizal fungi isolated from the local cardamom plantations were tested in a location trial at Myladrapara, Foukki district. Kerala. In general, mycorrhizal inoculation was found beneficial in improving the plant growth compared to controls. Among the 11 different VA mycorrhizal fungi tested seedling inoculated with *Glomus monosporum* exhibited growth, dry matter content and increased uptake of phosphorus.

Anandaraj *et al.* (1999) tried to identify the efficient strains of vesicular arbuscular mycorrhizae for black pepper (*Piper nigrum* L.). Inoculation of VAM in black pepper nursery was found to enhance growth and produce healthy cuttings. Efficient strains of VAM were identified for major varieties of black pepper for use both as a biofertilizer and a biocontrol agent. These strains belonged to two species *viz. Glomus fasciculatum* (Thaxter sensu Gerd.) Gerd. and Trappe, and *Gigaspora gigantea* (Nicol. and Gerd.) Gerd. and Trappe.

Kapoor and Mukerji (1999) investigated the effect of inoculation with two species of VAM fungi (*Glomus macrocarpum* and *G. fasciculatum*) on growth and essential oil of coriander. Mycorrhizal plants showed better growth than control plants. However, the effects of individual VAM fungal species varied with growth parameters. The essential oil content in seeds also increased due to VAM fungal inoculation. GLC characterization of essential oil showed increased concentration of geraniol and linalol on *G. macrocarpum* and *G. fasciculatum* inoculation, respectively over control.

Sreeramulu and Bagyaraj (1999) studied the efficiency of improving seedling growth and nutrient uptake by six promising arbuscular mycorrhizal (AM) fungi (*Acaulospora laevis, Gigaspora margarita, Glomus fasciculatum, G. intraradices, G. monosporum* and *G. mosseae*) for cardamom selected from a primary screening trial conducted at the University of Agricultural Sciences, Bangalore and five dominant native fungi (*G. caledonium* and *Glomus* spp.) isolated from local cardamom plantations of Kerala were tested against Vazukka cultivar of cardamom under nursery conditions at Myladampara, Kerala. Results showed that mycorrhizal inoculation was beneficial in improving the seedling growth, dry weight ad phosphorus uptake compared to uninoculated plants. Amongst the eleven different AM fungi tested, *Glomus monosporum* performed best in improving the seedling growth.

Gaur *et al.* (2000) inoculated *Coriandrum sativum, Trigonella foenum-graecum* and carrots with vesicular-arbuscular mycorrhiza (VAM: *Glomus, Gigaspora, Scutellospora*) fungi and grown in nutrient deficient, sandy-loam soils amended with organic matter (compost made from *Albizzia [Albizia*] and *Populas* leaves). Under field conditions, shoot and root dry weights and total uptake of P and N were significantly increased in all inoculated plants. VAM colonization was 76, 60 and 63% in *C. sativum, T. foenum-graecum* and carrots, respectively. Infection propagules produced in greater numbers on *C. sativum* and *T. foeum graecum*. The extent of green yield increase following VAM inoculation was greater in plots amended with leaf compost at a soil: compost ratio of 1:1 than in plots amended at a ratio of 2:1. High levels of VAM colonization and large numbers of infection propagules demonstrated the potential of these crops as substrates for inoculum production VAM also increased crop yield.

Thanuja *et al.* (2002) carried out an investigation to study the influence of arbuscular mycorrhizae (AM) on induction of rooting and other root characteristics of pepper cultivar Panniyur-I. Among three AM fungi tried, the maximum percentage rooting was observed in cuttings inoculated with *Acaulospora laevis* (69.69%). Orthotropic cuttings showed a higher rooting percentage on inoculation with AM. Mycorrhizal inoculation also improved other root characteristics. Among the different AM fungi tried, *Gigaspora margarita* was efficient in producing the maximum number of primary roots and dry weight of roots, whereas the longest primary root was observed with the inoculation of *Glomus fasciculatum*. The plant P content was found to be higher in the inoculated plants. It was significantly higher in cuttings inoculated with *G. fasciculatum* (0.202%). The percentage root colonization and spore count were found to be higher in cuttings inoculated with *A. laevis*. The present investigation clearly showed the beneficial effects of inoculation with AM fungi which resulted in enhanced rooting and root growth in black pepper cuttings.

FRUIT CROPS

Manjunath *et al.* (1983) studied the response of citrus to inoculation of VAM fungus *Glomus fasciculatum* in 4 Indian soil types under greenhouse conditions. Inoculation increased the dry weight of shoot and root and nutrient content of plants. Plants grown in sandy soil responded more than those grown in clay or laterite soil. The results indicated that mycorrhizal inoculation could be beneficial in certain unsterile soils in spite of the presence of native endophytes.

Kandasamy *et al.* (1986b) studied the effect of VAM fungal infection on the growth and nutrient uptake of the mulberry cultivars S-41, MR-2, S-54 and Kanva-2. All showed enhanced root and shoot length, although dry weight and leaf area was reduced in S-54.

Mohandas (1988) studied the response of papaya (*Carica papaya*) to VAM fungal inoculation. Papaya (*Carica papaya cv. Coorg Honey Dew*) plants inoculated with VAM fungi *Glomus fasciculatum* and *G. mosseae* responded favourably to the treatment. Plant height, dry matter, phosphorus content of roots and leaves of the seedlings increased significantly at all levels of P. The differences were more pronounced at 1/4 dose of P. Zinc and nitrogen content also improved in the inoculated seedlings. In general *G. mosseae* was more beneficial than *G. fasciculatum*.

Vinayak and Bagyaraj (1990a) investigated the response of Trifoliate orange (*Poneirus trifoliata*), an important citrus rootstock, to 18 VAM fungi. *Glomus velum, G. caledonicum, G. merredum, G. macrocarpum* and *Acaulospora* spp. were the most effective in improving growth and nutrition resulting in great plant height, stem diameter, plant biomass, P, Zn, and Cu content. All the 5 fungi selected were statistically equivalent in improving growth and nutrition of *P. trifoliata*. Inoculation of the root stock with these fungi induced a stem girth of pencil thickness *i.e.* ready for budding in 14-15 months in contrast to 22-24 months without mycorrhizal inoculations.

Katiyar *et al.* (1990) studied the interaction between vesicular arbuscular mycorrhizal colonization and phosphorus application in mulberry. The results revealed that *G. fasciculatum* and *G. mosseae* inoculation at lower levels of phosphatic fertilization (30 kg and 0 kg p/ha/year) resulted in better response of mulberry compared to inoculated plants receiving higher doses (90 kg P and 60 kg P/ha/year) and uninoculated control receiving recommended level of 120 kg P/ha/year. Though the highest leaf yield (36,114 kg/ha/year) was obtained in uninoculated treatment receiving recommended level of phosphatic fertilizer (120 kg P/ha) it was statistically at par with the inoculated treatment receiving 30 and 0 kg P/ha/year. *G. fasciculatum* inoculation at 0 kg and 30 kg P yielded 35,431 and 35,836 kg/ha/year leaf respectively, while *G. mosseae* inoculation at the same levels of P yielded 34, 164 and 36,076 kg leaf/ha/year respectively. A similar trend was seen in the total above ground biomass (stem+ leaf yield). This indicated that inoculation of mulberry with both the VAM fungi enhanced plant growth at reduced phosphatic fertilizer application. Root colonization due to *G. fasciculatum* and *G. mosseae* inoculation was also more at lower levels of P fertilizer application (0 and 30 kg P) compared to higher levels of P fertilizer application. The height of plants and number of leaves/plant at 0 and 30 kg P were almost at par with the uninoculated control receiving 120 kg P. The trend was same for leaf N and P content. The results suggested that inoculation of mulberry with efficient VAM fungi may become a great boon to mulberry cultivation through curtailing the application of phosphatic fertilizer.

Padma and Kandasamy (1990) studied the effect of interaction between VA-mycorrhizae and graded levels of phosphorus on the growth of papaya. According to them the optimum dose of P for the crop was 75% of the recommended level when mixed VAM inoculum (*Glomus mosseae* + *G. fasciculatum* + *Gigaspora margarita*) was added with the fertilizer. When large amounts of P were added to the seedlings there was a decrease in the intensity of VAM colonization.

Singh *et al.* (1993) studied the influence of VA mycorrhizal inoculation and phosphorus amendments on growth and nutrient uptake by rough lemon seedlings. Inoculation of soil with *Glomus fasciculatum* prior to planting increased the spore population of the fungus in soil and the mycorrhizal colonization, dry weight and N, P and K content of rough lemon seedlings. Root colonization, mycorrhizal spore density, dry matter production and nutrient uptake of seedlings was not significantly influenced by the level of phosphorus in the soil.

Onkarayya and Mohandas (1993) studied the dependency of citrus rootstocks to VAM inoculation in Alfisol soil. Seedlings of the rootstocks of rough lemon, Rangpur lime (*Citrus limonia*), *Poncirus trifoliata, Troyer citrage, Carrizo citrange, Citrumelo* and *Cleopatra mandarin* with 3-4 leaves were inoculated with *Glomus fasciculatum* and planted in pots containing 5 kg of Alfisol soil (pH 6.8, 5 ppm P, 0.72% C). All pots received 100 ppm N as calcium ammonium nitrate and 50 ppm K as KCl. No P was supplied. Growth was assessed 160 days after planting. In inoculated plants VAM infection ranged from 47% in Troyer citrange to 69% in P. trifoliata. In non-inoculated controls, some natural infection was observed and varied between 18 and 38% shoot height, root length and shoot dry weight all increased in response to inoculation and the mean shoot: root ratio decreased. When mycorrhizal dependency was calculated as the percentage improvement in dry weight due to inoculation, *P. trifoliata* was found to be the most dependent rootstock, followed by rough lemon. *Troyer citrange, Rangpur lime* and *Carrizo citrange, Cleopatra mandarin* was least dependent. The contents of P, N and Zn also increased significantly in inoculated plants.

Ambika *et al.* (1994) studied the roots and rhizosphere soils of 25 mulberry genotypes for local VAM colonization in Mysore, Karnataka. In all, 5 VAM genera, *Glomus, Acaulospora, Sclerocystis, Gigaspora* and *Scutellospora* were present in the rhizosphere soil of different genotypes. Although root samples of all the genotypes showed the presence of vesicles and arbuscules, there was no significant correlation between VAM colonization and growth and yield of mulberry. However, in some mulberry genotypes, the uptake of leaf nitrogen and phosphorus was influenced by natural VAM association.

Sivaprasad *et al.* (1995a) had undertaken *Artocarpus heterophyllus* to test the effects of inoculation with VAM fungi on growth and establishment of tissue culture plantlets of jackfruit. Five species, namely, *Acaulospora morroweae, Glomus fasciculatum, Glomus etunicatum, Glomus constrictum*, and *Glomus mosseae* were used for the experiment. Jackfruit plantlets inoculated with different VAM fungi at the time of planting showed high mycorrhizal colonization and spore count on the 150th day of planting. The control plants, though non-mycorrhizal at the time of transplanting of garden pots filled with non-sterile potting mixture, recorded a colonization percentage of 13.4 and spore count of 21/50 g of soil due to native VAM propagules. The maximum colonization was in *G. fasciculatum* inoculated plants which recorded 88% with spore count of 81/50 g rhizosphere soil. All the VAM fungi used in the present study significantly increased plant survival rate, the maximum being in *G. etunicatum* inoculation (100%) followed by *G. fasciculatum* (87%). Non-mycorrhizal control treatments showed 31.3% survival. The VAM fungi inoculated plantlets recorded significantly higher plant height and shoot and root fresh weight over control plants. *G. mosseae* was the most effective culture in stimulating plant growth, and recorded the maximum plant height and fresh weight (4.38 g) as against 5.85 cm height and 1.81 g fresh weight recorded for control plants. Analysis of the total and orthodihydric phenol content indicated that the inoculation with VAM fungi resulted in significant increase of phenol content in root tissue over control plants. *G. etunicatum* recorded the highest concentration of phenols in the root tissue.

Das *et al.* (1995a) conducted an experiment in the nursery to understand the role of VAM inoculation on the growth and development of mulberry saplings. The results revealed that inoculation of nursery beds with *Glomus mosseae* increased growth, development and survival of mulberry saplings in comparison with the uninoculted control. The inoculated saplings raised in the nursery beds were much healthier and contained more nitrogen, phosphorus and potash in leaves and stems than uninoculated saplings.

Katiyar *et al.* (1995) conducted a field experiment for 4 years (1989-93) on a red loam alfisol soil at Mysore, using mulberry (*Morus alba*) var. *kanva-2* preinoculated with *Glomus fasciculatum* and *Glomus mosseae* and fertilized with 4 doses of single superphosphate (0, 30, 60 and 90 kg P/ha/yr) and examined their effects on plant growth, leaf yield and quality. Pooled data for 4 years showed that inoculation with *Glomus mosseae* in combination with 30 kg P ha/yr gave similar values for plant growth, leaf yield and leaf chemical constituents to the uninoculated control (which received the "full" dose of phosphate fertilizer-120 kg P/ha/yr) being the amount normally applied to mulberry raised for sericulture.

Sharma and Bhutani (1995) studied the response to mycorrhizal inoculation and P-fertilization of apple seedlings. They used 4 VAM fungi *viz. Glomus mosseae*, *G. fasciculatum*, *G. macrocarpum* and *G. versiforme* and two different P-sources, namely single superphosphate and rock phosphate, in concentrations of 0, 25, 50, 100, 150, 200 and 250mg P/kg soil. Among the mycorrhizal fungi tested, *Glomus macrocarpum* showed greatest plant height, stem diameter, leaf area, and dry weights of shoot and root of apple seedlings. It was followed by *G. fasciculatum* , *G. mosseae* and *G. versiforme* in descending order. The apple seedlings fertilized with 50 ppm P were comparable in their height, stem diameter, leaf area, and dry weights of shoot and root with the seedlings inoculated with *G. macrocarpum* but given no P fertilizer.

Varma and Schuepp (1995) investigated the mycorrhization of micropropagated plantlets of strawberry (var. Vanta, early maturing and var. Elsanta, late maturing). On the basis of their observations they concluded that inoculation of micropropagated plantlets with active cultures of VAMF appears to be critical for the survival and growth, which avoids transient transplant shock and stunted growth on transfer to field.

Fathima *et al.* (1995) investigated the role of VAM inoculation on growth and development of three promising mulberry varieties under the influence of different sources and levels of phosphorus. They conducted the experiment in the nursery inoculated with either *Glomus mosseae* or *Glomus fasciculatum* or a mixed inoculum of both. Four sources of phosphorus *viz.* single superphosphate 16% P_2O_5, diammonium phosphate 48% P_2O_5, rock phosphate 20% P_2O_5 and Massorie rock phosphate 20% P_2O_5 were used. Significantly higher rate of survival, height of the plants and fresh and dry weights were observed due to VAM inoculation over uninoculated control. Among inocula, *G. mosseae* inoculum was the best.

Mohandas (1995) studied the potential of VA-mycorrhizal inoculum (*Glomus fasciculatum* and *G. mosseae*) for banana cultivation under field conditions. They noted a significant increase in the colonization of roots upto 15 months. The uninoculated controls in the adjacent plots showed 30% colonization at the end of 15 months. The plant nutrient contents like Zn, copper increased significantly. Yield, which was calculated in terms of total number of fingers and bunch weight increased by 25%. *G. mosseae* and *G. fasciculatum* treatments did not show marked difference in their effectiveness. The yield and nutrient contents in plants of superphosphate added plots did not differ from plots where it was not added.

Reddy *et al.* (1996a) screened 13 species of VA mycorrhizal fungi for their symbiotic effect on acid lime in pot experiments. Inoculation of pot with *Glomus macrocarpum, Glomus mosseae, Acaulospora laevis, Glomus caledonicum* (*G. caledonium*) and *Gigaspora margarita* resulted in greater plant biomass, plant height, plant girth, leaf number, leaf area and leaf P and Zn concentrations. With an emphasis on plant biomass and P nutrition, the best VAM fungi for lime were considered to be *G. macrocarpum* and *G. mosseae.*

Reddy *et al.* (1996b) screened 13 different VAM fungi for their symbiotic efficiency with 50 days-old seedlings of papaya *cv. solo.* Plant growth, percentage mycorrhizal colonization and P and Zn content of shoots and roots were assessed 90 days after inoculation. The highest biomass (9.9 g/plant), height (56 cm) stem girth (1.6) cm and leaf area (607.7 cm2) and high concentrations of P (69 and 30.4 mg/plant in shoots and roots respectively) and Zn (606.8 and 644.9 µg/plant in shoots and roots respectively) were observed in plants inoculated with *Glomus mosseae* (ICRISAT). These plants exhibited a high degree of mycorrhizal colonization (76% compared with 14% for the uninoculated control).

Reddy *et al.* (1998) studied response of S13 mulberry variety to VAM inoculum under semiarid condition. Saplings were raised in nursery beds soil inoculated with the vesicular arbuscular mycorrhizal (VAM) fungus *Glomus mosseae* and given, P fertilizer. VAM inoculated saplings showed better growth and development as well as higher leaf N, P and K contents.

Shirsath *et al.* (1998) studied the development and performance of mycorrhizal seedlings of ber (*Zizyphus mauritiana* L.). Greenhouse studies were conducted to test the possibility of developing mycorrhizal seedlings of ber (*cv. Umran*). Following inoculation with different VA-mycorrhizal fungi (*Acaulospora calospora, Glomus mosseae* and *Gigaspora margarita*), either singly or in combination, ber seedlings exhibited VA-mycorrhizal root colonization ranging from 69.33 to 77.33%. There was good establishment of VA-mycorrhizae in the rhizosphere soil (457.66-538.33 chlamydospores/50 mg at 240 days after sowing). All the VA-mycorrhizae either singly or in combination significantly increased root and shoot dry weights as well as P uptake over non-mycorrhizal treatment. Combined inoculation of *A.calospora + G. mosseae + G. margarita* and single inoculations of *G. mosseae* were superior in increasing the dry weight of ber seedlings as compared with the other inoculation treatments.

Sreeramulu *et al.* (1998) studied the influence of VAM on establishment and growth of Khirni (*Manilkara hexandra* (Roxb.) Dub.) seedlings. The effects of inoculating the growing medium (2:1:1 mixture of soil, sand and FYM) with inoculum from 11 VA mycorrhizal fungi on the growth of *M. hexandra* (used as a rootstock for sapodillas) were studied in pot experiments at Bangalore. Plant height at 6 and 12 months after planting and total plant DW were highest following inoculation with *Glomus intraradices.* All plants inoculated with *G. intraradices* survived for 30 days after planting.

Mathur and Vyas (1999) reported improved biomass production, nutrient uptake and establishment of in vitro raised *Zizyphus mauritiana* due to VA mycorrhiza. *Glomus constrictum*, one of the commonly occurring VAM fungi of arid and semi-arid regions was multiplied on modified MS medium in association with *Zizyphus mauritiana* under *in vitro* conditions. The in vitro multiplied VAM strain established efficient symbiosis with *in vitro* raised plantlets of *Z. mauritiana*. This VAM strain significantly improved the biomass production, nutrient uptake and acclimatization of the in vitro produced plantlets of *Z. mauritiana* into the field.

Setua *et al.* (1999c) conducted experiments to study the effect of vesicular arbuscular mycorrhizal association in 'S-1' mulberry (*Morus* sp.) at nursery stage. An experiment was conducted during 1995-97 to study the effect of vesicular arbuscular mycorrhizal (VAM) inoculation on survival, growth characters, leaf yield and phosphorus uptake in 'S-1' mulberry saplings at the nursery stage. Inoculation of plants with *Glomus mosseae* resulted in an increase in survival (10.4%), plant height (5.7%), leaf area (11.8%), root length (9.8%), number of lateral roots/plant (56.1%) and plant dry biomass (27.4%). Phosphorus uptake was also enhanced by 13.5%

Shashikala *et al.* (1999b) conducted experiments under glass house conditions, to study the effect of *Glomus mosseae* at various levels of phosphatic fertilizer on the growth of hardened micropropagated banana plantlets. The results indicated that high soil P levels inhibit the extent of root colonization of *G. mosseae* and thus curtail the growth and nutritional benefits of the symbiont to the plant. It is concluded that plantlets raised in soil inoculated with *G. mosseae* and amended with 50% of recommended P fertilizer exhibited a significant increase in plant height, number of leaves, stem girth, root volume, total biomass and total P content over those grown with 100% recommended P fertilizer.

Setua *et al.* (1999b) undertaken a field experiment during 1994-97, to study the effect of vesicular arbuscular mycorrhizal association in mulberry saplings S1 with reduced levels of phosphate on growth characters, leaf yield and phosphorus uptake on irrigated, alluvial soil (pH 7.4) in the Gangetic plain of West Bengal, India. Pooled data analysis of 8 seasons revealed that all 3 strains of *Glomus* at reduced phosphate doses were promising in promoting plant growth characters and leaf yield. *Glomus mosseae* at the lowest phosphate dose were promising in promoting plant growth characters and leaf yield. *Glomus mosseae* at the lowest phosphate number of branches/plant (7.7), number of leaves/plant (191.7) and leaf yield (23526.0 kg/ha/year) and was found to be at par (132.8 cm, 7.0, 177.0and 22023.0 kg/ha/year respectively) with the control (180 kg phosphate/ha/year respectively) with the control (180 kg phosphate/ha/year) and other treatments. With regard to leaf quality *G. mosseae* with 30 kg phosphate/ha/year influenced maximum leaf moisture contest (74.61%) and phosphorus uptake in leaves (3.91 mg/g dry leaf) except leaf area and were similar in control and other treatments. The association of *G. mosseae* in roots of mulberry S1 mulberry can reduced the requirement of 80% phosphate fertilizer (30 kg/ha/year) over recommended dose (180 kg/ha/year) for irrigated gardens without any detrimental effects and enabled the saving of Rs 2125/ha per year.

Rupnawar and Navale (2000a) studied the effect of VA-mycorrhizal inoculation on growth of pomegranate layers. A pot culture experiment using sterile and P-deficient soil was conducted to study the symbiotic relationship of pomegranate layers with VAM fungi (*Glomus epigaeum* [*Glomus versiforme*] (GE), *Glomus mosseae* (GM), *Gigaspora calospora* (GC), singly or in mixtures), applied at a rate of 150 g per pot, with respect to growth parameters like height, root length, number of leaves, dry weight of shoot and root and mycorrhizal dependency percentage. It was observed from the results that mycorrhizal treatments were superior to non-mycorrhizal treatments of pomegranate layers. GM alone and the GE+GM+GC mixture recorded the maximum height, root length, number of leaves, dry weight of shoot and root and mycorrhizal dependency percentage.

Singh *et al.* (2000b) conducted a field experiment to show the effect of fertilizers on growth, yield and quality of sweet orange (*Citrus sinensis*) cv. Mosambi. The height, trunk diameter, canopy volume, root growth and biomass production, yield and quality were affected by the different treatments (biofertilizers along with chemical fertilizers). The treatment combination of 3/4 P + vesicular arbuscular mycorrhizas (VAM) + N (T_{10}) was the best treatment for producing better growth and yield of high quality fruits in Mosambi. It indicated the suitability of biofertilizers (*Azospirillum brasilense* culture and VAM) inoculation in combination with chemical fertilizers for better growth, yield and fruit quality. It will help bring down the cost of chemical fertilizers particularly nitrogen and phosphorus and help improve soil fertility by maintaining the physical conditions of the soil.

Reddy *et al.* (2000) studied the effect of vesicular arbuscular mycorrhiza (*Glomus mosseae*) and phosphorus on growth of mulberry S13 under semi-arid conditions. Growth was significantly enhanced by VAM compared to the additional of phosphorus or the control.

Fathima *et al.* (2000) studied the effect of different levels and sources of phosphorus on VA-mycorrhizal root colonization and spore load in mulberry (*Morus alba* L.). Mulberry *cv*. Kanva-2 saplings inoculated with *Glomus mosseae* or *G. fasciculatum* were treated with P (30 or 60 kg ha year) in the form of single superphosphate (SSP), diammonium phosphate (DAP) or Mussoorie rock phosphate (MRP) in a 2-year field trial conducted in Karnataka, during 1993-95. Controls were either inoculated with VAM fungi without P application, given P at 120 kg ha year as SSP without VAM inoculation, or given neither. In saplings inoculated with *G. fasciculatum* root colonization was highest with P at 30 kg ha year in the form of DAP. In the case of *G. mosseae*-inoculated saplings, root colonization was highest with a P application of 30 kg ha year in the form of MRP. Spore load in the mulberry rhizosphere was significantly higher with P applied at 30 kg ha year as MRP, than with P at the recommended rate of 120 kg ha year as SSP.

Setua *et al.* (2000) studied the quality and quantity of mulberry leaves var. *S*1 associated with three strains of vesicular arbuscular mycorrhiza (VAM) separately and with graded levels of phosphorus (180,90,60 and 30 kg P ha^{-1} $year^{-1}$) at CSR&TI, Berhampore, West Bengal. Two-year average data revealed the overall improvement in leaf yield, leaf moisture (%) and P uptake by leaves in all the VAM-treated mulberry plants particularly those associated with *Glomus mosseae* at 30 kg P ha^{-1} $year^{-1}$ which was statistically found to be at par with control (180 kg P ha^{-1} $year^{-1}$ without VAM) and other treatments. The silkworm rearing result with the Multi × Bi ($CB_5 \times P_5$) hybrid indicated the overall improvement in larval weight, single shell weight, S.R.%, absolute silk content in mycorrhizal leaves of any of the strains under reduced rates of P when fed to silkworms, while ERR, filament length, denier and non-breakable filament length were found to be at par with the control. It was further observed that *G. etunicatum* with 50% reduced rate of P during favorable season and *G. fasciculatum* with 67% reduced rate of P during unfavorable season yielded better results in terms of important economic characters of rearing and reeling.

Manjunath *et al.* (2002a) studied the effect of different VAM fungi on growth parameters of papaya *cv. Sunset Solo*. The effects of vesicular arbuscular mycorrhiza (VAM): *Glomus fasciculatum, Sclerocystis dussii* and *Acaulospora laevis)* soil inoculation and P fertilizer (50, 75 and 100%of 625 kg P/ha as single superphosphate) on the growth of papaya *cv. Sunset solo* were studied in Arabhavi, Karnataka, India [date not given]. Observations were recorded every month from 30 to 240 days after transplanting. Generally, *G. fasciculatum* and 75% of the recommended P rate, alone or in combination, were most effective in increasing plant height, stem girth, petiole length, and leaf number.

Phirke *et al.* (2002) conducted a semi-commercial banana plantation trial from July 1998 to October 1999 at the North Maharashtra University, in Maharashtra, to minimize irrigation, eradicate pests, sustainably cultivate banana and demonstrate integrated plant nutrition management (IPNM) practices to farmers. The banana orchard harboured diversified species of vesicular arbuscular mycorrhizal (VAM) fungi belonging to the genera *Glomus, Gigaspora, Sclerocystis* and *Scutellospora*. These fungi infected the roots severely (78.3%) and showed elevated spore density (68.7 spores/g soil) in the mycorrhizosphere of plants for IPNM system compared to chemical fertilizers alone (traditional farming). This resulted in retardation of the establishment of nematodes (*Radopholus similis* and *Helicotylenchus* spp.) in root zone of IPNM treatment (2.1 nematodes/g soil) compared to chemical treatment (5.5 nematodes/g soil). VAM enabled proper nourishment and protection of crop in IPNM treatment compared to the traditional way, which was evident from reduced plant mortality (12%) of planted sword suckers; enhanced greenery (foliar chlorophyll; 778 µg/g); improved nutrient uptake on dry weight basis (3.62% N, 0.31% P and 6.45% K); elevated finger number (137.6 per bunch), size (length 21.6 cm and girth 12.4 cm) and weight (130.8 g); and overall vigour and robustness. Hence, exploitation of VAM through adoption of IPNM system revealed the relatively increased banana productivity (76.5%, 72 t/ha) and recyclable biowaste (36.6 kg); saving of 50% chemical fertilizers; and permitted control over soil fertility in farmer's favour over traditional cultivation practices. These findings related to native VAM fungi in IPNM, traditional farming and typical soil systems for cultivation of bananas are discussed in detail.

Eswarappa *et al.* (2002) studied the effects of vesicular arbuscular mycorrhiza (VAM) on banana *cv. Yalakkibale* cultivation in Malur Taluk, Karnataka. VAM was applied to the soil at 500 g per plant. VAM application increased yields by 9.02% (46.75 t/ha vs. 42.10 t/ha in the control) in 1997/98 and 15.75% (47.75 t/ha vs. 41.25 t/ha in the control) in 1998/99. VAM-treated plants had greener and larger leaves, and greater bunch weight and number of fruits per bunch than the untreated plants. The onset of flowering was also earlier in treated plants than in the control. VAM application resulted in an 18.9% decrease in costs by reducing the rate of inorganic P fertilizer by approximately 50%.

Sharma *et al.* (2002) carried out the study to investigate the effect of vesicular arbuscular mycorrhizal fungi (VAM) and P soil application on leaf nutrient content of apple seedlings and to determine the effect of P application on the threshold levels of nutrients in the soil so that optimum doses can be determined with suitable VAM fungi. Individual soil lots were amended with 25, 59, 100, 150, 200 and 250 mg P/kg soil. The VAM fungi tested were *Glomus fasciculatum, G. mosseae, G. macrocarpum* and *G. versiforme.* About 10 g of the inoculum consisting of soil, spores, hyphae and mycorrhizal chopped roots was placed in the rhizosphere of seedlings. Regardless of P added to the soil, VAM inoculation increased the concentration of P, Zn, Cu, Mn, and Fe in leaves. There was no indication of nutrient interaction between P and Zn with respect to VAM application. Plants grown with supplemental P had greater P tissue concentration than those not receiving P. The maximum nutrient concentration was observed when the plants were supplied with 50 ppm P and inoculated with *G. macrocarpum.* The threshold level of P in soil was 15 ppm above, which there was, declined trend on the availability of soil Zn and Cu to plants.

Manjunatha *et al.* (2002b) studied the effect of different vesicular arbuscular mycorrhizal (VAM) fungi on the yield and yield parameters of papaya *cv. Sunset Solo* under filed conditions (Karnataka) at different phosphorus levels. Three VAM fungi, *Glomus fasciculatum, Sclerocystis dussii* and *Acaulospora laevis* along with uninoculated control were tried at three levels of phosphorus (50,75 and 100% of the recommended dose of phosphorus (RDP; 625 kg/ha)). Among the VAM fungi, *G. fasciculatum* and *S. dussii* were at par in terms of the number of fruits and yield, which were higher than those of the remaining fungi and uninoculated control. The highest mean fruit weight was recorded in *G. fasciculatum*-treated plants. Treatment with 100% RDP recorded significantly higher mean number of fruits (28.24 fruits per plant), mean fruit weight (0.445 kg) and mean yield (38.85 t/ha) compared to those of 75% RDP (25.99 fruits per plant, 0.413 kg and 33.42 t/ha) and 50% RDP (21.78 fruit per plant, 0.372 kg and 25.01 t/ha).

MEDICINAL/AROMATIC/DYES/ NARCOTICS

According to Sirohi and Singh (1993) inoculation with *Glomus fasciculatum* produced better root infection, root length, shoot dry weight, P uptake, organic and inorganic 'P' fractions in leaf tissues and oil yield in peppermint, as compared with the uninoculated control, both at deficient and high levels of native soil P.

Chatterjee *et al.* (1994) evaluated the efficacy of selected combinations of organic amendments (*Calotropis procera*) and VA mycorrhiza (*Glomus fasciculatum*) in improving the alkaloid content of *Catharanthus roseus* under microplot conditions in sandy loam soil. Treatments gave an improvement in the mycorrhizal infection in roots and the spore population in the rhizosphere. They also reported an improvement in the N,P,K and C contents in the soil and plants.

Gupta *et al.* (1994) reported the association of VA-mycorrhiza (*Glomus* unidentified sp.) in palmarosa (*Cymbopogon martini* var. *motia*). Glasshouse trials showed that the mycorrhizal association enhanced growth and productivity in palmarosa significantly. A significant increase in phosphorus and potassium uptake was observed in mycorrhizal plants compared with non-mycorrhizal plants. Their results indicated that VAM inoculation could enhance the productivity in essential oil-bearing plants.

Chandra *et al.* (1995b) screened 12 isolates of VAM fungi recovered from medicinal plants for their colonization potential to the roots of *Ageratum conyzoides, Catharanthus roseus* and *Papaver rhoeas*. Only four of them proved to be most efficacious in colonizing the roots at a faster rate. In the final screening *G. claroideum* proved to be the most versatile in improving the overall performance of *Ageratum conyzoides, G. fasciculatum* proved to be so for *Catharanthus roseus* and *G. multicaule* proved to be so for *Papaver rhoeas*. These endophytes proved their superiority in improving the length, fresh weight as well as dry weight of roots, shoots and leaves and number of pods or capsules and dry weight of seeds in the respective medicinal plants. When the additional inoculum of selected VAM fungi (*viz., Glomus claroideum* for *Ageratum, G. fasciculatum* for *Catharanthus* and *G. multicaule* for *Papaver*) was incorporated in the soil along with the organic amendments, an increase in their efficacy in improving the performance of medicinal plants was recorded. The extent of this increase, however, varied with the VAM-amendment combinations and medicinal plants. Even in the presence of VAM inoculant, *Eichhornia, Calotropis* and *Lantana* retained their superiority in impvoving the mycorrhizal status and performance of *Ageratum, Catharanthus* and *Papaver*, respectively. These findings clearly indicated that treatments involving *G. claroideum* and *Eichhornia, G. fasciculatum* and *Calotropis* and *G. multicaule* and *Lantana* may be used for improving the performance of *Ageratum, Catharanthus* and *Papaver*, respectively.

Prasad and Sailaja (1995) studied the effect of vesicular-arbuscular mycorrhizae inoculation on periwinkle (*Catharanthus roseus* (L.) G. Don.). Rhizosphere soil samples and root pieces of *Catharanthus roseus, C. pusillus, Wrightia tinctoria, Thevetia pernviana, Nerium indicum*, and *Alstonia scholaris* were found to be colonized by a variety of VAM fungi. Percentage of mycorrhizal infection was found to be more in *A. scholaris, Catharanthus roseus, C. pusillus*, and *Wrightia tinctoria*. The hyphal density was related with the infection percentage. Vesicles were abundant in all the plant species while arbuscules were more in periwinkle. Nineteen VAM fungi were collected and identified up to species level. In pot culture experiments, *Glomus aggregatum* and *Glomus mosseae* enhanced the plant growth significantly. Mixed inoculum also improved the plant growth. Hyphal infection was noticed two weeks after the transplantation. Vesicles were observed at six weeks. Enhancement in plant shoot and root dry weights were evident in all the treatments.

Subhashini and Krishnamurty (1995) studied the influence of VA-mycorrhiza on phosphorus economy, yield and quality of tobacco in rain fed alfisols. Among the 10 treatments, 5 doses of P with and without VAM (P0, P20, P40, P60, P80), VAM infection, spore number and hyphal density were highest at low P-levels. VAM inoculation with *Glomus fasciculatum* resulted in a significant increase in the total nitrogen concentration and nicotine content in the leaves. The pot culture study clearly revealed that application of VAM in combination with 20 Kg P2O5/ha could improve the performance of tobacco.

Kothari and Singh (1996) studied the nutrient acquisition and growth of *Cymbopogon winterianus* in deficient pasteurized sandy loam soil to determine the effects of mycorrhizal symbiosis and soil compaction. The soil was inoculated either with rhizosphere microorganisms excluding VAM fungi or with the VAM fungus, *Glomus intraradices*, and supplied with 0, 50 or 100 mg P/kg soil. The soil was compacted to a bulk density of 1.2 or 1.4 mg/m^3 (dry soil basis). *G. intraradices* substantially increased root and shoot biomass, root length, nutrient uptake (K, P, Zn and Cu) per unit root length and nutrient concentration in plant compared with inoculation with rhizosphere microorganisms when the soil was at the low bulk density and not amended with P. Little or no plant response to the VAM fungus was observed when the soil was supplied with 50 or 100 mg P/kg soil and / or compacted to the higher bulk density. At higher soil compaction and P supply the VAM fungus significantly reduced root length. Non-mycorrhizal plants at the higher soil compaction produced relatively thinner roots and had higher concentrations and uptake of P, Zn and Cu than at lower soil compaction, particularly under conditions of 'P' deficiency. The quality of the essential oil (extracted from green foliage), measured in terms of citronellol concentrations, did not vary appreciably between soil treatments.

Khaliq and Janardhanan (1997) evaluated the influence of 3 VAM fungi (*Glomus aggregatum*, *G. fasciculatum* and *G. mosseae*) on the shoot biomass production and essential oil content of 6 cultivated mints, *Mentha arvensis, M. cardiaca,* (*M. gracilis*)*, M. citrata, M. piperita, M. spicata* and *M. viridis* in a glasshouse study. Their results showed significant variation in effectiveness of different VAM fungi on mints. Inoculation of VAM fungi did not affect the essential oil content or composition in any of the mints examined. VAM inoculation, in general, improved the shoot biomass production of all the species. The results suggest the potentiality of VAM fungi in improving the productivity in commercially cultivated mints.

Vineeta and Singh (1997) studied the growth responses of mint plants infected by vesicular arbuscular mycorrhizal (VAM) fungus. Inoculation of *Mentha spicata* plants with *Glomus fasciculatum* increased root length, root surface area and leaf area of the plants which resulted in greater yields. Infected soil and root fragments were found to be good sources of inoculum.

Basu and Srivastava (1998) reported occurrence of VAM endophytes in certain medicinal plants. Fifty-five medicinal plants, collected from four districts of Uttar Pradesh, belonging to 28 families of flowering plants, were screened for mycorrhizal association. But for two plants, all were found to be mycorrhizal. They tested the efficacy of VAM fungi in improving the growth and medicinal value of five medicinal plants, *Allium sativum, Coriandrum sativum, Occimum basilicum* (essential oil); *Sesamum indicum* (essential oil and vitamin) and *Trigonella foenum-graceum* (alkaloid). It was observed that mixed inoculum was better than single inoculum in improving the growth and root shoot biomass for all the test plants.

Arya (1999) studied the effect of application of VAM fungi on the growth of neem seedlings (Margosa, *Azadirachta indica* A. Juss, multipurpose tree medicinally important due to presence of limonoids and azadirachtin). An effort was made to assess the improvement in the biomass of neem seedlings due to inoculation of three VAM isolates. One of the three isolates used caused about 5 times more growth of root and 4 times more growth of shoot system as compared to control.

Sumana and Bagyaraj (1999) conducted a greenhouse investigation to screen and select an efficient VA mycorrhizal fungus for inoculating neem (*Azadirachta indica*) which is an important multipurpose evergreen tree species native to Indian subcontinent. The neem seedlings (45 days old) were inoculated with nine different VAM fungi obtained from various sources. Inoculated seedlings generally had greater plant height, stem girth, biomass, P content, Zn concentration, biovolume index and quality index than uninoculated control plants. They also had more mycorrhizal root colonization and spore numbers in root zone soil. The results revealed that neem seedlings responded best to inoculation with *Glomus mosseae* (local), followed by *Glomus fasciculatum*.

Kapoor *et al.* (2001) investigated the effect of inoculation with two species of VAM fungi (*Glomus macrocarpum* and *G. fasciculatum*) in the presence or absence of fertilizer on the growth and yield of essential oil in the seeds of *Anethum graveolens* was investigated. Significant increase was observed in the vegetative growth of mycorrhizal treated plants. The essential oil concentration increased by 19.18 and 41.3% with *Glomus macrocarpum* and *G. fasciculatum* respectively in plants grown with the recommended level of phosphorus in soil. The potential role of VAM in low-input cultivation of *A. graveolens* is suggested.

Gupta *et al.* (2002) studied the effects of inoculation with vesicular arbuscular mycorrhizal (VAM) fungus *Glomus fasciculatum* on the root colonization, growth, essential oil yield and nutrient acquisition of three cultivars of menthol mint (*Mentha arvensis*): Kalka, Shivalik and Gomati, under field conditions. The VAM inoculation significantly increased the root colonization, plant height, fresh herbage and dry matter yield, oil content and oil yield as compared to non-inoculated cultivars. The effect of VAM inoculation on the root colonization, growth and yield of mint was more pronounced with the *cv. Shivalik*. VAM inoculation significantly increased the uptake of N, P and K by shoot tissues of mint, but most markedly increased the uptake of P. The VAM-inoculated mint plants depleted the available N, P and K in the rhizosphere soil as compared to non-inoculated control plants, however, the extent of nutrient depletion was greater for P than N and K. We conclude that the VAM inoculation could significantly increase the root colonization, growth, essential oil yield and nutrient acquisition of mint for obtaining economic production under field conditions.

Kapoor *et al.* (2002) evaluated the effects of application of two arbuscular mycorrhizal (AM) fungi (*Glomus macrocarpum* and *G. fasciculatum*) on shoot biomass and concentration of essential oil in *Anethum graveolens* and *Trachyspermum ammi*. Results revealed significant variation in effectiveness of the two AM fungal species. AM fungal inoculation in general improved the growth of the plants. On mycorrhizal, the concentration of essential oil increased up to 90% in dill and 72% in carum over their respective controls. *Glomus macrocarpum* was more effective than *G. fasciculatum* in enhancing the oil concentration. The constituents of the essential oils were characterized by gas liquid chromatography. The levels of limonene and cavone were enhanced in essential oil obtained from *G. macrocarpum*- inoculated dill plants, while *G. fasciculatum* inoculation resulted in a higher level of thymol in carum.

Kumar and Murugesh (2002) inoculated ten medicinal plants (*Abrus precatorium, Cynodon dactylon, Euphorbia tirucalli, Gymnema elegans, Hemidesmus indicus, Ocimum basilicum, Plumbago zeylanica, Phyllanthus amarus* [*P. niruri*], *Sida acuta* and *S. rhombifolia*) with 3 vesicular arbuscular mycorrhiza (VAM; *Glomus mosseae, G. fasciculatum* and *G. monosporum*) to determine the effects of VAM on the plants. *Arbus precatorius* recorded the highest VAM infection (86%), whereas *Phyllanthus amarus* recorded the lowest (36%). Plants inoculated with VAM spores were taller compared to those of control, with inoculation with *G. fasciculatum* resulting in the tallest plants. VAM improved the seedling growth, dry weight and phosphorus uptake of the medicinal plants.

Earanna *et al.* (2002) conducted a pot culture experiment to study the effects of vesicular arbuscular mycorrhiza (VAM; *Acaulospora laevis, Gigaspora margarita, Glomus fasciculatum, Sclerocystis dussii,* and *Scutellospora calospora*) inoculation on the growth and P content of six-week-old *Rauvolfia tetraphylla* seedlings. Plant height, number of leaves and branches, stem girth at the collar region, and P content before harvesting at 8 months after planting, were determined. All VAM fungi improved the growth parameters (except *Scutellospora calospora*), but did not significantly increase the P content (except *Glomus caledonium*) of *R. tetraphylla.* Plants inoculated with *Glomus caledonium* were tallest and had the greatest leaf number and stem girth. *Glomus caledonium* and *Glomus fasciculatum* gave the highest number of branches.

Ratti *et al.* (2002) inoculated potted *Vetiveria zizanioides* with soil culture of *Glomus aggregatum, G. fasciculatum , G. intraradices* and *G. mosseae* to determine the effects of the arbuscular mycorrhizal fungi on the growth and productivity of the plant. Additon of *G. fasciculatum* soil culture to the soil of potted *V. zizanioides* resulted in the highest P content in roots (0.15%) and stems (0.11%), acid and alkaline phosphatase activity in roots, fresh (31 g) and dry weight of leaves (11.3 g) and total dry weight (72.3 g). Fresh and dry weight of roots and total fresh weight (283 g) were highest with *G. mosseae* inoculation. Oil content of *V. zizanioides* was highest in the control (1.67%). *G. fasciculatum* recorded the highest colonization of *V. zizanioides* roots (88.3%) and spore number (820/100 g soil), whereas *G. intraradices* recorded the lowest (78.7% and 625/200 g soil, respectively).

TREES

Thapar *et al.* (1984) studied the effect of fertilizers on development of endomycorrhiza in *Agathis robusta*. The application of compost proved favourable for plant growth, development of endomycorrhiza and increased the number of red brown laminate spores. This trend was in conformity with the results obtained with agricultural crops. However, the beneficial effect was disturbed when mineral fertilizers were added, particularly in higher dosages.

Thapar and Khan (1985b) reported the effects of VA mycorrhizal fungi on the growth of hoop pine (*Araucaria cunninghamii* Sweet), in sterilized and unsterilized soil. Significant improvement in growth was recorded in sterilized soil.

Nalini *et al.* (1986) investigated the response of *Leucaena* cultivars to inoculation with 20 isolates of *Glomus, Gigaspora, Acaulospora* and *Sclerocystis* spp. The isolates varied in their ability to improve shoot dry weight and the test varieties varied in their response.

Reddy and Bagyaraj (1988) carried out an investigation to select an efficient VAM fungus for inoculating *Leucaena* in oxisols and vertisols. Seven VAM fungi were screened for symbiotic response with 2 Hawaiian giant cultivars (K8 and K28) of *Leucaena*. *Glomus mosseae*, proved to be the best mycorrhizal fungus in providing healthy, vigorously growing seedlings in both oxisols and vertisols.

Srinivas *et al.* (1988b) studied the effect of VAM fungi on growth and nutrient uptake of forest tree seedlings. The growth response of some forest tree seedlings, *i.e., Eucalyptus tereticornis, Leucaena leucocephala, Acacia nilotica* var. *indica* and *Casuarina equisetifolia* to inoculation with *Glomus fasciculatum, G. epigaeum* and *Acaulospora morrowae* was tested in a P deficient unsterile soil. Inoculation of VAM fungi increased biomass, P, N and K uptake in all the forest tree seedlings.

Bagyaraj *et al.* (1989) studied the selection of an efficient inoculant VA mycorrhizal fungus for *Leucaena.* Seven VA mycorrhizal fungi were screened for symbiotic response with four Hawaiian giant cultivars, K8, K28, K67 and K72 of *Leucaena leucocephala*. Of the seven fungi, three were promising ones from a preliminary screening trial conducted earlier with fungi obtained from different parts of the world, and the other four were local (Indian) isolates obtained from the rhizosphere of *Leucaena*. A local isolate of *Glomus mosseae* was found to be the best mycorrhizal fungus for inoculation of leucaena in order to obtain healthy, vigorously growing seedlings.

Adholeya and Cheema (1990) studied the effect of mycorrhizal inoculation (with a mixture of VAM spores of *Glomus* and *Acaulospora*) at 3 different stages of establishment in 2 clones of *Populus deltoides*(G-3 and D-121) micropropagated from explants taken from the leader shoots of field growing trees. The 3 stages were: (1) Shoots given a rooting stimulus with 5 mg/ litre IBA and rooted plantlets derived from them either. (2) before and (3) after hardening. Marked beneficial effects were noted after inoculation at all the 3 stages in 20 week transplants, but in 60 week transplants, inoculated rooted and hardened plants gave the best results.

Narayanan *et al.* (1990) studied the response of shola plant, *Rhododendron nigricans* to VAM inoculation. *R.. nigricans* responded well to inoculation with all 8 VAM fungi tested, showing increase in height after 6 months (15.85 to 39.02% increase over control). Maximum increase was given by an isolate of *Glomus mosseae* from Bangalore.

Reddy *et al.* (1990) studied the phosphate response curve of *Leucaena* inoculated with *Gigaspora.* An experiment was carried out to determine the response of *Leucaena*, uninoculted and inoculated with *Gigaspora margarita* using superphosphate and Mussorie rock phosphate at different levels (0, 5, 10, 15, 20, 30 and 40 kg per hactare). The plants were harvested 75 days after sowing. Out of two sources of P, superphosphate proved to be a better source producing higher shoot biomass compared to Mussorie rock phosphate. The results also showed that the shoot biomass production with the recommended level of P (when Mussorie rock phosphate was added as the source of P) was 7.22 g/plant. Nearly the same amount of shoot biomass was produced with half the recommended level of P when the plants were inoculated with *Gigaspora margarita*. When superphosphate was used as the source of P, the observations revealed that nearly 75% of recommended P can be saved through inoculation with *Gigaspora margarita*.

Reena and Bagyaraj (1990) studied the growth stimulation of *Tamarindus indica* by selected VAM fungi. Seedlings of this slow-growing forest tree were inoculated with 13 VAM fungi, obtained from various sources round the world. Inoculated plants had greater plant height, leaf number, stem girth, biomass, phosphate and Zn content. Numbers of VAM spores in the soil, percentage root colonization and external hyphae (measured by soil aggregation) were also higher. *T. indica* seedlings responded best to inoculation with *Gigaspora margarita*.

Sidhu *et al*. (1990) studied the occurrence of vesicular arbuscular mycorrhiza of *Cassuarina equisetifolia* L. *Glomus fasciculatum* and *Scutellospora calospora* colonized the roots of this fast growing fuel wood tree. These VAM fungi can be used successfully when this species is planted, especially in degraded soils.

Lakshman (1992) studied the development and response of VAM fungi in *Terminalia bellerica* (Roxb.). According to him root colonization by VAM fungi was higher in *Terminalia* seedlings raised in pots in garden soil than in those raised in sterilized non-inoculated soil. There was a progressive increase in number of spores and rootlets in plants inoculated with chopped mycorrhizal roots of *Chloris gayana.* Growth of inoculated seedlings was also superior.

Beniwal *et al.* (1992) studied the effect of VA mycorrhizal fungi and phosphorus on growth and nodulation of *Acacia nilotica* (L.) Willd ex Del. Dual inoculation with *Glomus mosseae* and *Rhizobium* at all phosphorus concentration tested (30, 60 or 90 ppm) increased growth and nodulation of *A. nilotica* compared with single inoculation with *Rhizobium*, with the effect increasing with increasing P concentration. It is suggested that the increased growth may be attributed to increased nodulation. Nitrogenase activity also increased with increasing P in the presence of *G. mosseae* and/or *Rhizobium*, and total N was higher in dual-inoculated plants than in plants inoculated with *Rhizobium* alone. It was concluded that VA mycorrhizae can assist nodulation and N-fixation in *A. nilotica* inoculated with *Rhizobium* at different P levels, and could improve the efficiency of the system, especially in P-deficient soils.

Pokhriyal *et al.* (1992) reported the effect of VAM inoculation on nodulation, nitrogenase activity in three different potting substrates, *viz*., soil + compost, soil + sand + compost and soil+sand in *Albizia lebbek* under pot culture conditions. No significant differences were observed in VAM inoculated plants in various characters, *i.e.* nodule number, nodule weight, nitrogenase activity, plant height, root length, and total biomass as compared to control under three different soil media.

Bhat *et al.* (1993) studied the response of subabul (*Leucaena leucocephala*) to VAM inoculation in sterile and unsterile soil. 7 VAM fungi, *Acaulospora laevis*, *Gigaspora margarita*, *Glomus etunicatum*, *G. fasciculatum*, *G. mosseae*, *G. versiforme* and *Sclerocystis rubiformis* were used to inoculate the seedlings of *Leucaena* growing in pots in sterile and unsterile soil under greenhouse conditions. *Glomus versiforme* was the best VAMF based on the mycorrhizal dependency value. The response to other VAMF varied. There was a positive correlation between total seedling biomass and percentage colonization.

Vasanthakrishna and Bagyaraj (1993) conducted experiments to screen and select efficient VAM fungi for inoculation of *Casuarina equisetifolia*. They inoculated seedlings with 13 different VAM fungi that had been obtained from various global sources. Inoculated plants exhibited increased plant height, biomass, and P and Zn contents. They also had increased mycorrhizal root colonization and spore numbers in the soil. Seedlings of *C. equisetifolia* responded best to inoculation by *Glomus mosseae* followed by *G. versiforme* both from a local source (Bangalore).

Bhat *et al.* (1994a) investigated the field response of three forest tree seedlings, *Acacia nilotica, Casuarina equisetifolia* and *Leucaena leucocephala* to inoculation with VAM fungi when transplanting to field conditions in a P deficient oxisol at the Tamil Nadu Agricultural University, Coimbatore. VAM fungi for each species were selected after an initial screening test under pot culture. The species used for *A. nilotica* were - *Glomus versiforme, G. mosseae* and *G. etunicatum*; for *C. equisetifolia* - *G. fasciculatum* , *G. mosseae* and *Sclerocystis rubiformis* and *L. leucocephala* - *G. mosseae, G. versiforme* and *G. margarita* . The best VAMF for each tree species yielded marked biomass increase (36-41%) over control.

Mandal and Kaushik (1994) studied the effect of VAM inoculations on growth and development of *Acacia nilotica* (L.) Willd.ex. del. The response of seedlings of *Acacia nilotica* to inoculation with *Glomus constrictum*, *Glomus mosseae*, *Gigaspora gilmorei* (collected from old plantations of *A. nilotica*) and combinations of these was studied in a phosphorus deficient sterile soil in pots in the greenhouse. Inoculation with VAM fungi improved collar diameter, root-shoot dry weight and length in comparison with uninoculated controls. *G. mosseae* combined with *G. gilmorei* was the most effective treatment and *G. constrictum* the least effective.

Mandal *et al.* (1994, 1995) studied the response of *Acacia nilotica* to inoculation with *Glomus mosseae*, *Glomus constrictum*, *Gigaspora gilmorei* alone and in combination in a phosphorus deficient sterile soil. VAM inoculation improved nodulation, root colonization, dry weight, N, P and K content of plants in comparison with the uninoculated control.

Sarwar and Mukerji (1995) studied the response of *Acacia catechu* with appropriate inoculation of VAM fungus (*Glomus fasciculatum*). They raised the plants under stressed and non-stressed soils with and without VAM inoculum and data on root length, shoot length, root/shoot dry weight was collected. The VAM inoculation resulted in an appreciable increase in all the growth parameters. Mycorrhizal plant showed better growth as compared to control plants. The better response of plants to mycorrhizal inoculation was recorded in both stressed and non-stressed soils.

Balasubramanian and Srivasan (1995) studied the response of *Dalbergia sissoo*, *Tectona grandis* and *Ailanthus excelsa* to four VAM fungi namely *Glomus fasciculatum*, *G. mosseae*, *G. monosperma* and *Gigaspora marginata*. On the basis of results they concluded that though a number of VAM fungi are available for inoculation, their compatibility with a particular host is more important for establishing symbiosis. The study revealed that *Gigaspora margarita* was compatible with teak, *G. mosseae* with *Dalbergia* and *G. monosperma* with *Ailanthus*.

Ragupathy and Mahadevan (1995) studied the effect of different mycorrhizal inoculations *viz.*, *Gigaspora albida*, *Glomus aggregatum* and *G. mosseae* on the growth of fast growing multipurpose tree species *Peltophorus pterocarpum*. The results revealed that shoot height and width significantly increased with the inoculation of VAM fungi. The increase was well pronounced in the seedlings inoculated with *G. mosseae*. The total phosphorus content of roots and shoots was high in the mycorrhizae-inoculated plants. VAM fungal species differed in their ability to improve the plant height and width.

Agarwal and Chauhan (1995) studied the effect of *Glomus fasciculatum* inoculation on the seedlings of *Dalbergia sissoo*. The VAM inoculated plants produced greater biomass in comparison to non-inoculated plants. However, the height of the plants was not proportional to extent of colonization.

Lakshman and Raghvendra (1995) studied the effect of leaf litter and soil in different proportions on growth and biomass production in six timber tree species colonized with VA mycorrhizae. Litter was mixed in sandy loam soil in 1:1; 1:0.75; 1:0.50 and 1:0.25 ratio. At the time of seedling transplant soil was inoculated with *G. mosseae*. Six experimental plants namely *Artocarpus heterophyllus, Eugenia bracteata, Lagerstroemia lanceolata, Pterocarpus marsupium* and *Terminalia tomentosa* grown under different combinations of soil and litter exhibited varied responses. However, the growth and biomass production of the host was favoured most by the litter-soil proportion of 1:025. *E. bracteata, A. heterophylla, T. tomentosa* and *T. bellirica* showed significant increase in their heights (from 19.4 to 25.2 cm) in litter-soil proportion of 1.025 ratio.

Durga and Gupta (1995) studied the effect of the VAM fungi *Glomus fasciculatum* and *Glomus mosseae* + rock phosphate (100 mg/plant) on initial establishment and mineral nutrition of 3-month-old teak stumps in polybags. Measurements were made of the uptake of various macronutrients (N,P,K) and micronutrients (Cu, Zn, Fe, Mn), and of various growth parameters. *G. mosseae* + *G. fasciculatum* treated plants registered maximum growth. They also showed an increase in concentrations of phosphate, K and Mn, but a decrease in the concentration of N, Cu, Zn, Fe in the shoots. The plants treated with *G. fasciculatum* + rock phosphate gave the next best growth, followed by those treated with *G. fasciculatum* alone. The *G. mosseae* + *G. fasciculatum* + rock phosphate treatment was deleterious to growth.

Gautam and Maitra (1995) studied the impact of VAM fungi on the growth of *Dendrocalamus strictus*. Varying proportions of different VAM fungi, *G. macrocarpum* 35%, *Glomus* sp. 30%, *Gigaspora* sp. 20%, *Scutellospora* sp., *Sclerocystis* sp., *Acaulospora* sp. and other unidentified forms 15% were taken as mixed VAM inoculum for *D. strictus* seedlings. After six months, data collected on height of the plants showed that mixed VAM inoculum increased the height two times over control.

Verma and Jamaluddin (1995) studied the association and activity of vesicular arbuscular mycorrhizae of teak (*Tectona grandis*) in central India. Vesicular arbuscular mycorrhizal (VAM) fungi of teak were isolated from 20 different sites in Madhya Pradesh, including nursery, plantation and natural forests, and studies were made of the percentage root colonization and of the effect of inoculation on teak seedling growth. Root colonization ranged from 25 to 95% at different sites. Genetically superior (plus trees) showed heavier root colonization than other trees. Sixteen different VAM fungi were isolated and identified belonging to 5 genera of Glomales, *viz*., *Acaulospora, Gigaspora, Glomus, Scutellospora* and *Sclerocystis*. Among these, *Glomus etunicatum* and *Acaulospora scrobiculata* were found to be most widely distributed species, followed by *Glomus intraradices* and *G. mosseae*, while *G. fasciculatum* was least frequent. Inoculation of seedlings with *G. fasciculatum* or mixed VAM fungi showed better height, growth, biomass and percentage root infection in the nursery than uninoculated (control) seedlings. The mixed VAM inoculum was most effective treatment.

Mathur and Vyas (1996c) collected three VA mycorrhizal species (*Glomus fasciculatum, G. mosseae* and *Scutellospora calospora*) from rhizosphere soils of *Zizyphus mauritiana (Zizyphus mauritiana)*, a multipurpose fruit tree species from arid and semiarid regions. The VAM fungi were cultured on *Cenchrus ciliaris*, whose roots were used to inoculate potted seedlings of *Z. mauritiana*. Seedling growth, root colonization and nutrient uptake were recorded 90 days after inoculation. *G. fasciculatum* was significantly better than the other 2 VAM fungi in increasing seedling biomass and uptake of N, P, K, Ca and Mg.

Mehrotra (1996b) grown the tree seedlings of *Acacia catechu* and *Dalbergia sissoo* for 2-2.5 months in root trainers, containing pre-sterilized potting mixture and VAM inoculum, and then transferred to polypots containing unsterilized potting mixture where they grow for the rest of the nursery period. The seedlings were inoculated in root trainers with different amounts of VAM inoculum of *Glomus mosseae* and *G. geosporum* with or without *Rhizobium*. He has given the growth data for *A. catechu*. According to him the seedlings grew best with the lower amount of inoculum (25g), and the addition of *Rhizobium* made very little difference. *D. sissoo* showed a similar response.

Selvaraj *et al.* (1996) also worked on a number of tree seedlings of the multipurpose application e.g. *Acacia leucophloea, A. mangium* , *A. nilotica, Delonix regia* , *Derris indica, Pongamia pinnata* and *Tamarindus indica*. They inoculated the seedlings with 3 arbuscular mycorrhizal fungi *viz*., *Glomus fasciculatum, G. mosseae* and *Gigaspora margarita*. The growth responses of the seedlings were analyzed in terms of height, root collar diameter, root and shoot biomass etc. In addition, accumulation of P in root, stem and leaves was also analyzed. Overall growth response was best with *G. margarita* for *A. mangium, A. nilotica* and *Delonix regia* , *G. fasciculatum* for *A. leucophloea* and *G. mosseae* for *P. pinnata*. He has not observed significant response by *T. indica*.

Subhan *et al.* (1997) carried out the investigations to improve the successful transplanting micropropagated plantlets of *Sesbania sesban* using vesicular arbuscular mycorrhizal fungus *Glomus fasciculatum*. They developed the plantlets from somatic embryos and or adventitious buds induced from various explants on Gamborg's medium supplemented with 6-benzylaminopurine, in presence of Giberellic acid. Subsequent to nodulating the roots with Rhizobia, plantlets were transplanted into sterile garden soil and inoculated with or without *G. fasciculatum.* Only 30% of plantlets transferred to soil without *G. fasciculatum* survived. In contrast, all the plantlets inoculated with *G. fasciculatum* survived. These observations show that mycorrhizal association helped to increase the potential of micropropagated plantlets to successfully withstand transplantation shock.

Mehrotra (1997) observed a fantastic growth response of *Paulownia* to mycorrhization and fertilization. Cuttings were prepared from 1 yr-old seedlings of *Paulownia tomemtosa* at New Forest, Dehra Dun, and planted in pots with sand. On sprouting, the cuttings were transferred to another set of pots containing partially sterilized soil/sand (2:1) and 2% FYM. Half the cuttings were inoculated with 30 kg each of VAM inoculum (sand + VAM infected roots and VAM spores). After 3 weeks half of the VAM inoculated plants were fertilized with superphosphate (500mg), muriate of potash (48 kg) and neem cake (2g). Plants were allowed to grow for a certain period and then growth measurements were recorded. VAM inoculation increased shoot height by 26.4% and biomass production by 54.2% over control, but additional treatment with fertilizer gave growth increases of 87.6 and 151% respectively, compared with controls. On the basis of results, he suggested that the normal practice of direct planting of cuttings be discontinued and appropriate nursery technology be adopted for the production of high quality *Paulownia* planting stock.

Ananthapadmanabha *et al.* (1998) studied comparative effect of inorganic and biofertilizer on growth of teak plants. A trial was conducted in a 0.5 ha block teak (*Tectona grandis*) plantation of 1-yr-old plants of more or less uniform height at 2x2m spacing. Six treatments were applied: application of calcium nitrate (CN), diammonium phosphate (DP) or CN + DP (at rates equivalent to 250 kg/ha), inoculation with *Glomus caledonium* VAM (vesicular arbuscular mycorrhizal) spores or composite teak rhizosphere VAM (250-300 spores/ 100 g soil), and no treatment. Foliar N and P were increased in all treatments, but not K. Best growth of the plants was recorded in VAM treatment. Population of VAM spores increased in the VAM treatment. Available P in soil reported to have also increased in VAM inoculated plants.

Mehrotra *et al.* (1998) studied the effect of VAM, fertilizer and deoiled neem cake application on the growth of *Paulownia tomentosa* raised from root cuttings. Root cuttings of *Paulownia tomentosa* prepared from 1-yr-old polypot seedlings raised at New Forest, Dehra Dun, were planted in root trainers in sterilized sand and on sprouting transferred to more root trainers containing partially sterilized potting mixture of soil/sand (2:1), 2% FYM and 2 g deoiled neem seed cake. Half the plants were inoculated with VAM (FRI-MI) comprising *Glomus mosseae* and *G. geosporum*, and after 3 weeks they were fertilized using a nominal dose of inorganic fertilizers (500 mg superphosphate, 40 mg muriate of potash). After 2.5 months the plants were transferred to polypots using the same medium except that the inoculated and fertilized plants were given a supplement of deoiled neem cake of 10 g per pot. The inoculated and fertilized plants showed increase in shoot height, collar diameter and plant biomass of 83.2, 50.6 and 333%, respectively, compared with control plants. Deoiled neem cake checked root-knot nematode infestation completely (control plants showed 69% infestation by root-knot nematodes). Roots inoculated with VAM fungi exhibited about 20% incidence of VAM formation, whereas the uninoculated roots showed complete absence of VAM fungi. Mycorrhizal roots exhibited lower SRL, lower root/soil respiration and about 10% lower fine root mortality than non-mycorrhizal roots after 15 weeks of exposure to dry surface soil. Results demonstrated the feasibility of examining mycorrhizal effects on the fine roots of adult trees in the field using simple inexpensive methods.

Nema *et al.* (1998) studied the effect of VAM on growth and development of bamboo planted under agroforestry system. The growth performance of 3 bamboo species (*Bambusa bambos, Dendrocalamus strictus and Bambusa nutans*) after inoculation with mixed VAM (vesicular arbuscular mycorrhizas) was studied at 4 agricultural field sites (I-IV) in Madhya Pradesh. Two of the sites (II and III) were very similar in soil texture and nutrient status. The VAM fungi used to inoculate the seedlings included *Glomus mosseae*, *G. intraradices*, *Gigaspora* spp. and *Acaulospora* spp., and inoculation was done at the time of transplanting to the field. Controls were not inoculated. The growth responses to VAM inoculation was small and generally not significant, but was comparatively better at site I than at the other sites. *Bambusa bambos* grew better than the other 2 bamboo species. Percentage mycorrhizal infection was more in the bamboo species planted at sites I, II and III than at site IV. The VAM spore population in rhizosphere did not exhibit any significant variation in inoculated and uninoculated plants.

Sumana and Bagyaraj (1998) in a greenhouse experiment inoculated the *Dalbergia latifolia* seedlings with 8 vesicular arbuscular mycorrhizal (VAM) fungi. Inoculated plants had greater plant height, stem girth, dry weight and P content than non-inoculated plants. They also had greater mycorrhizal root colonization and higher spore number in rhizosphere soil. The best VAM fungi were *Glomus leptotichum* and *G. fasciculatum* .

Naqvi and Mukerji (1998) investigated the influence of VA mycorrhizas and *Rhizobium* during ex vitro hardening on the growth and survival of micropropagated *Leucaena leucocephala*. Ex-vitro inoculation with *Glomus fasciculatum* increased growth and survival of *L. leucocephala*. Dual inoculation with *G. fasciculatum* and *Rhizobium* resulted in increased root and shoot growth, number of leaves, number of nodules and mycorrhizal colonization and 100% survival of micropropogated plants up to 5 months, compared with 91.7 and 62.2% survival in plants treated with *G. fasciculatum* or *Rhizobium* alone, respectively. The non-inoculated plants grew poorly and died after 5 months. They have concluded that this technique could prove useful in micropropagation of *L. leucocephala* during the nursery stage.

Puther *et al.* (1998) also reported the improved establishment of micropropagated *Leucaena leucocephala* plantlets. They obtained the shoots of *L. leucocephala* from hypocotyl callus grown in 10-1 M BAP, on Gamborg's B5 medium. They were rooted on half strength

B5 medium supplemented with 5 x 10-6 M IBA. Subsequent to the nodulation of their roots with *Rhizobium* (strain PRGL 001) in soilrite, plantlets were transferred to sterilized garden soil inoculated with the VAM fungi, *Glomus fasciculatum* or *G. macrocarpum* around their roots. Only 20% of the plantlets survived in soils lacking VAM fungus. In contrast, 100% of the plantlets established very well and showed good growth in VAM inoculated soil. Inoculated roots had both external and internal hyphae with well-formed arbuscules and vesicles confirming the establishment of good mycorrhizal association to help in successful establishment of tissue culture raised plantlets of *L. leucocephala* in field conditions by alleviating the transplantation shock.

Gurumurthy and Sreenivasa (1998) carried out an investigation to screen and select an efficient VAM fungus that improves growth, nutrition and dry matter production of teak in unsterilized soil. Different VAM fungi tried in this investigation were *Glomus fasciculatum, Gigaspora margarita, Acaulospora laevis, Sclerocystis dussii*, Local isolate-1 and Local isolate-2. Per cent root colonization, mycorrhizal spore counts, population of free living N_2 fixers and P-solubilizers were significantly higher in the rhizosphere of teak plants inoculated with *Sclerocystis dussii* as compared with the other VAM fungi tested. A matching trend was observed in plant height, stem diameter, leaf area, root length, shoot dry weight, root dry weight and the concentrations of P, Zn, Cu, Mn and Fe in shoots which also greatest in plants inoculated with *S. dussii*.

Garg *et al.* (1999) conducted experiments to identify an efficient mycorrhiza for promoting the growth of *Eucalyptus tereticornis* in nursery. Two species of VAM, *G. fasciculatum* , *G. mosseae* and three VAM consortia dominated by *G. mosseae, G. mosseae 85, G. mosseae 88* and *G. mosseae* 92 were screened. The screening procedure involved (i) prior to sowing, mycorrhizae application on elite seeds of *E. tereticornis* in root trainer cups at three different geographical locations of India namely Patiala (North India), Choudwar (South East India) and Sewa (Central East India) containing a potting mix (ii) allowing germination and growth at 32 $\pm$5% moisture in a green house and (iii) monitoring the rate of germination, root ramification, shoot biomass and general vigour of the seedlings as a function of mycorrhizal infection, for 100 days (from sowing until harvesting/transplantation). It was observed that in nursery studies vis-a-vis control *G. mosseae* 88 promoted maximum plant height at Patiala while *G. mosseae* 85 at Choudwar. None of the VAM showed positive response at Sewa. Mycorrhizal strains affected differently at different sites on the basis of plant height and % root infection.

Sharma *et al.* (1999c) studied the mycorrhizal dependency of *Eucalyptus tereticornis* on vesicular arbuscular mycorrhizal fungi in a semi-arid alfisol. *Eucalyptus tereticornis* was grown in a greenhouse in low phosphorus (0.67 ppm Olsen's P) soil (typic Haplustalf) inoculated with mixed indigenous vesicular arbuscular mycorrhizal (VAM) fungi. Soil was amended to achieve P levels of 10, 20, 25, 30 and 40 ppm to evaluate the growth response and dependency of *E. tereticornis* to inoculation with VAM fungi. A positive response to mycorrhizal inoculation was evident at first two levels of soil P, *i.e.*, at 0.67 and 10 ppm, but not at the higher levels of soil P. Dry matter yield of inoculated plants beyond 20 ppm soil P was similar or even less compared to their uninoculated counterparts. Inoculated plants produced maximum dry matter (root and shoot) at 10 ppm soil P, whereas uninoculated plants did not produce maximum dry matter until the level reached 20 ppm. The percentage root length colonized by VAM fungi decreased the VAM colonization significantly. Inoculated plants had higher shoot P and N contents compared to uninoculated plants at all levels of soil P. However, at the first two lower levels of soil P, inoculated plants showed significantly higher shoot P and N contents over their respective uninoculated counterparts. The increasing shoot P accumulation beyond 10 ppm did not enhance dry matter yields. Inoculated plants had lower values of phosphorus utilization efficiency (PUE) and nitrogen utilization efficiency (NUE) at all levels of soil P except at the unamended level (0.67 ppm) where the inoculated plants showed higher values of NUE compared to uninoculated control plants. Taking dry matter yield into consideration, *Eucalyptus* plants were found to be highly dependent on mycorrhiza for maximum dry matter production at 10 ppm of soil P. Any further amendment of P to soil would not benefit either VAM symbiosis or plant growth.

Mohan (1999) studied the effect of VAM inoculation on the growth of *Prosopis* spp. in soils of arid zone. In the present study, plant growth responses of *Prosopis cineraria* and *P. juliflora* to inoculation with different VAM fungal isolates *viz., Glomus fasciculatum* and *G. mosseae* was carried out under nursery condition. Inoculation of both the VAM fungal isolates resulted in better plant height, shoot and root biomass, percent colonization of roots and number of VAM fungal spores in soils in unsterilized nursery soil over control. *G. fasciculatum* performed better than the other isolate and the control plants.

Ismail *et al.* (1999) conducted pot culture experiments to assess the influence of different biofertilizers on biomass production of *Azadirachta indica* and *Eucalyptus tereticornis*, grown in ordinary and sludge amended soil and irrigated with sewage water. In general, biofertilizer application significantly enhanced biomass production both in ordinary and in sludge amended soil, but showed significant variability among themselves and with tree species. The pattern of influence also showed variability with soil and with tree species indicating specificity. The results suggested that higher biomass production was often due to additive effect of biofertilizer and sludge. From this experiment it was concluded that mycorrhizal and other biofertilizers in conjunction with sewage water and sludge can be used for forest biomass production, the most ideal environment friendly way to waste disposal.

Paroha *et al.* (1999) studied the growth performance of *Bambusa arundinacea* in relation to VA-mycorrhizal inoculation. One-month-old seedlings of *Bambusa arundinacea* were transplanted to mixed nursery soil/sand (1:1) in polybags and inoculated with mixed vesicular arbuscular mycorrhizal (VAM) fungi (*Acaulospora, Glomus mosseae* and *Glomus intraradices*, 500 spores per bags) or mixed fungal spores + infected roots. Seedlings growth (as dry matter) and mycorrhizal infection were measured 3, 4 and 9 months after inoculation; the results are tabulated. All the inoculated seedlings grew better than uninoculated controls but growth and mycorrhizal infection were best in the spore + infected roots treatment. After 9 months growth was increased by 59.4% in the spore treatment and 71.6% in the spore + infected roots treatment; corresponding figures for root infection were 82.35 % and 87.37% (compared with 50.88% in the control).

Garg *et al.* (1999) studied the effect of mycorrhiza on the growth of *Eucalyptus tereticornis* in the nursery in different geo-climatic conditions. In order to identify suitable mycorrhizas for promoting the growth of *Eucalyptus tereticornis* in the nursery, 6 species of mycorrhizas *Laccaris fraterna* EM-1983, *Glomus fasciculatum, G. mosseae* and 3 VAM (vesicular arbuscular mycorrhizal) mixtures dominated by *G. mosseae (G. mosseae* 85, 88 and 92) were screened. The screening procedure involved the application of mycorrhizas to elite seeds of *E. tereticornis* prior to sowing in root trainer cups at 3 different geographical locations in India (Patiala in north India, Choudwar in southeast India, and Sewa in central east India). The growing medium was pasteurized soil/FYM/sand (2:4:4), and used the local soil at each site (sandy loam from Patiala, and clay loams from Choudwar and Sewa). Germination and growth was carried out at 32+2^{0}C in a polyhouse, with periodic irrigation of seedlings to maintain 30+5% moisture. The rate of germination, root system development, shoot biomass vigour of the seedlings was monitored as a function of mycorrhizal infection over 100 days (from sowing until harvesting/transplanting). *L. fraterna* promoted maximum plant biomass at Patiala and Sewa, while *G. mosseae* 85 was best at Choudwar. The test mycorrhizal strains influenced height growth, chlorophyll content, and percentage mycorrhizal infection differently at the different sites.

Giri and Mukerji (1999) observed significant increase in growth and biomass of the *Sesbania grandiflora* plants with VAM fungus, *Glomus macrocarpum*, in saline soils. The percentage VAM colonization of roots, production of VAM fungal spores in the rhizosphere soil and number of root nodules were significantly higher. *Sesbania grandiflora* was highly dependent on VAM, with maximum at early stages of plant growth. *Glomus macrocarpum* presented *Sesbania grandiflora* against salinity stress by increasing its establishment and survival in saline soils.

Srinivas *et al.* (1999a) studied vesicular arbuscular mycorrhizal dependency of *Albizzia lebbeck* (L.) Benth. Occurrence of mycorrhizal roots on *Albizzia lebbeck [Albizia lebbeck]* was studied in samples collected from 12 sites in Godavari belt forests, India. The percentage of infection varied from 26 to 61%. Species of *Glomus* dominated the rhizosphere of *A. lebeck* in all the soils. Other mycorrhizal species were *Sclerocystis, Gigaspora, Scutellospora* and *Acaulospora* (which was infrequent). Mycorrhizal dependency was determined by growing seedlings in sterilized and unsterilized soil. Height, biomass, chlorophyll and phosphorus content were all greater in mycorrhizal than non-mycorrhizal plants. Mycorrhizal dependency was greatest in September and least in November and June.

Ponnambalam *et al.* (1999) studied the growth and biomass production of *Casaurina equisetifolia* and *Casuarina junghuhniana* in Quartz sand dump from cement manufacturing company mining for limestone at Madukkarai near Coimbatore in Tamil Nadu to determine species suitability for afforestation. The dump consisted of rock debris mixed with some finer particle of sand (85.2%), silt (6.8%) and clay (8.0%): the pH was 8.9; organic matter content 0.3 and moisture content 0.39%. Six-month-old *Casaurina equisetifolia* and *Casuarina junghuhniana* seedlings and rooted cuttings of *Casaurina equisetifolia* were planted in 30 cm 3 [30x30x30] pits spaced at 1 x 1 m. Decomposed coir pith (2 kg) was added for improving the soil texture and water holding capacity. Growth measurements -

height, BDH (diameter at breast height), biomass production, accumulated litter and vesicular arbuscular mycorrhizal colonization - were recorded 2 years after planting. *Casaurina equisetifolia* seedlings performed better in growth, biomass and litter production than the *Casuarina junghuhniana* seedlings and rooted cuttings of *Casuarina equisetifolia.*

Gurumurthy and Sreenivasa (1999) studied the effect of inoculation of efficient VAM fungus at different levels of P on nutrition and growth of Silveroak (*Grivillea rabusta*). One month-old seedlings of silveroak (*Grivillea rabusta*) transplanted to woven sack bags (25 x 30 cm) filled with unsterile red sandy clay soil, farmyard manure and sand (1:1:1) amended with different doses of single superphosphate (equivalent to 0, 50, 75 and 100 kg P_2O_5ha). The soil already contained phosphate equivalent to 10 kg P_2O_5/ha. Inoculation with *Gigaspora margarita* (a VAM fungus) was performed using a pot culture maintained on rhodes grass [*Chloris gayana*] by placing a thin uniform layer 2 cm below the soil. Percentage mycorrhizal root colonization and mycorrhizal spore counts increased significantly with increase in the level of P up to 75 kg P_2O_5/ha. Stem diameter, leaf area, root length, root dry weight and shoot dry weight increased with increase in the level of P. However, there was no significant difference between the 75 and 100 kg P_2O_5/ha, except for stem diameter, leaf area and root length which were significantly higher with the application of 100 kg P_2O_5/ha. A matching trend was observed with respect to shoot P, Ca and Mn concentrations which also increased with increase in P level up to 75 kg $P_2O_{5/}$ha, beyond which they increased slightly or remained practically constant. Shoot Zn and Fe concentration were significantly higher at 100 kg P_2O_5/ha. All these parameters were least in the uninoculated seedlings at all levels of P.

Rajan *et al.* (2000) conducted a greenhouse study under nursery conditions to study the efficacy of 9 arbuscular mycorrhizal (AM) fungi on *Tectona grandis*. Stumps were raised in polythene bags containing soil inoculated with isolates of different AM fungi. Teak seedlings raised in the presence of AM fungi generally showed an increase in plant growth and plant nutritional status (as measured by phosphorus, zinc and copper contents) over those grown with no soil inoculation of AM fungi. The extent of growth and nutritional status enhanced by the AM fungi varied with the species of fungus. Considering the various plant growth parameters and nutritional status of the plants, it was observed that *Glomus leptotichum* is the best AM symbiont of those tested in this experiment. The other fungi tested were *Acaulospora laevis, Gigaspora margarita, Glomus caledonicum, Glomus fasciculatum, Glomus intraradices, Glomus macrocarpum, Glomus mosseae* and *Scutellospora calospora.*

Gurumurthy *et al.* (1999a) screened four standard VAM fungi *viz.*, *Glomus fasciculatum, Gigaspora margarita, Acaulospora laevis, Sclerocystis dussii* and two local isolates from Dharwad, *viz.*, local isolate-1 and local isolate-2, for their symbiotic responses on growth, nutrition and dry matter production of shisham (*Dalbergia sissoo*) in unsterile red sandy clay soil. Shisham plants responded well to inoculation with VAM fungi. However, percent root colonization, spore count, the population of free living N_2-fixers and P-solubilizers were significantly higher in plants inoculated with *Glomus fasciculatum* as compared with other VAM fungi. A similar trend was observed in plant height, stem diameter, leaf area, root length, shoot dry weight, root dry weight and the concentrations of P, Zn, Cu, Mn and

Fe in shoot which were significantly higher in shisham plants inoculated with *G. fasciculatum* as compared to other VAM fungi. All these variables were least in the uninoculated plants.

Gaur and Adholeya (1999) inoculated micropropagated plantlets of *Syngonium podophyllum* and *Dracaena* sp. during an early weaning stage of acclimatization with a mixed indigenous consortium of arbuscular mycorrhizal (AM) fungi. Both species were colonized but a significantly higher colonization was observed (38%) in *Dracaena* than *S. podophyllum* when it was harvested after 20 weeks. *Dracaena* plants showed little difference in the extent to which they were colonized, when examined either at the weaning stage or hardening stage; however, *S. podophyllum* plants at the weaning stage were better colonized, than at 20 weeks. Survival was high in inoculated plants at lower fertility in both hosts. Moreover, *S. podophyllum* showed better stolon production than uninoculated controls at both fertility levels, though the increase was higher at lower fertility. *Dracaena* showed no response in shoot height to any treatment. Chlorophyll accumulation in both hosts was significantly influenced by inoculation, fertility and stage (weaning and hardening). A significant increment in shoot P uptake was also observed in both hosts related to inoculation, stage and fertility. Inoculation with the AM consortium had a significant and favorable effect on acclimatization of micro propagated *S. podophyllum* and *Dracaena* at the weaning stage, saving almost 15 days in the total hardening process.

Kumar *et al.* (2000) investigated the mycorrhizal dependency (MD) of 29-agro forestry tree species representing 14 families. The tests were done in pot experiment using P deficient sandy soil either sterilized with no addition, or sterilized and inoculated with VAM fungi (mixed *Glomus mosseae, G. fasciculatum* and *G. epigaeum* [*G.versiforme*]). Mycorrhizal infection, seedling biomass and P content of roots and shoots were recorded 90 days after sowing. On the basis of the results, the species were divided into 3 groups: those with high MD (which included *Azadirachta indica, Leucaena leucocephala, Gliricidia maculata, Sesbania grandiflora, Cassia siamea [Senna siamea]* and *Acacia melanoxylon)* those with moderate MD (the majority), and those with no MD *(Diospyros melanoxylon, Mangifera indica, Murraya koenigii, Polyalthia longifolia, Psidum guajava, Saraca indica* and *Ziziphus mauritiana)*. No correlation was observed between percentage mycorrhizal infection and MD.

Bhattacharya *et al.* (2000) studied arbuscular mycorrhizal dependence of *Eucalyptus tereticornis* in a low-phosphorus alfisol. It is concluded that *E. tereticornis* may not be an arbuscular mycorrhizal dependent plant at early developmental stages, although arbuscular mycorrhizal fungi infect it freely, improving its phosphorus acquisition efficiency in low-phosphorus soil.

Sharma and Adholeya (2000a) studied the response of *Eucalyptus tereticornis* to inoculation with indigenous AM fungi in a semi arid Alfisol achieved with different concentration of available soil P. *Eucalyptus tereticornis* was grown in a greenhouse in a lower phosphorus (0.67 ppm Olsen-P) soil (Typic Haplustalf) inoculated with mixed indigenous arbuscular mycorrhizal (AM) fungi. Soil was amended to achieve P levels of 10, 20, 25, 30 and 40 ppm to evaluate the growth response and dependence of *E. tereticornis* to inoculation with AM fungi. A positive response to mycorrhizal inoculation was evident at

0.67 and 10 ppm soil P but not at the higher levels of soil P. Dry matter yield of inoculated plants beyond 20 ppm soil P was similar or even less compared to their uninoculated counterparts. Inoculated plants produced maximum dry matter (root and shoot) at 10-ppm soil P, whereas uninoculated plants did not produce until the level reached 20 ppm. The percentage root length colonized by AM fungi decreased from 31% to 3% as the concentration of P increased beyond 10 ppm soil P. Higher levels of soil P depressed the AM colonization significantly. Inoculated plants had higher shoot P and N contents compared to their uninoculated counterparts at all levels of soil P. However, at the first two lower levels of soil P, inoculated plants showed significantly counterparts. The increasing shoot P accumulation beyond 10 ppm did not enhance dry matter yields. Inoculated plants had lower values of phosphorus utilization efficiency and nitrogen utilization efficiency (NUE) at all levels of soil P except at the unamended level (0.67 ppm), where the inoculated plants showed higher values of NUE compared to uninoculated control plants. Taking dry matter yield into consideration, *Eucalyptus* plants were highly dependent on 10 ppm of soil P for maximum dry matter production. Any further amendment of P to soil was not beneficial either for AM symbiosis or plant growth.

Giri *et al.* (2000) studied *Sesbania aegyptiaca* pers. seedling response to VA mycorrhization in two types of soil. Seven-day-old seedlings of *Sesbania sesban* were grown in garden soil (nutrient-sufficient) and stressed soil (nutrient-deficient), which had been sterilized, in the presence or absence of the VAM fungus *Glomus macrocarpum.* A 100 g inoculum containing about 90 viable VAM fungal spores was applied as a thin layer below 1.5 kg of soil in polyethylene bags. Growth parameters of the plant (nodule number, root and shoot length and biomass production) were determined 30 and 60 days after seed germination and nutrient (P, K, Na and Ca^{2+}) and concentrations after 60 days. Biomass, root length and number of nodules increased in mycorrhizal plants in both garden and stressed soil. P, K and Na concentrations also increased in mycorrhizal plants compared with non-mycorrhizal plants. In contrast, Ca^{2+} concentration decreased in mycorrhizal plants. Mycorrhizal dependency of *S. sesban* was higher (84.4%) in nutrient deficient soil than in cultivated soil (79.3%).

Muthumukar and Udaiyan (2000b) studied the role of seed reserves in arbuscular mycorrhizal formation and growth of *Leucaena leucocephala* (Lam.) de Wit. and *Zea mays L.* Seed reserves in *Leucaena leucocephala* and *Zea mays* are important for mycorrhizal formation and seedling growth. Seed reserve removal reduced mycorrhizal formation markedly in *Leucaena* but not in maize, except at 15 and 45 days after seed reserve removal. Partial or total removal of seed reserves decreased plant growth and tissue nutrient concentrations in both hosts. Nodule number in *Leucaena,* which was related positively to plant biomass and mycorrhizal infection levels, was reduced when one or both cotyledons were severed. *Leucaena* seedlings without or with partial seed reserved had higher nutrient use efficiencies throughout seedlings growth. But such an effect was observed only initially in maize. Seed reserve removal increased the specific observation rates of nutrients in both hosts. Phosphorus absorption rate was significantly and positively related to root infection levels in both *Leucaena* and Maize. Though the growth rates of plants without seed reserves were low initially, these plants had higher growth rates during later stages. We conclude that seed reserves are not only important for seedling growth, but also for mycorrhizal formation and nodulation.

Rajeshwari *et al.* (2001) evaluated seedling growth and vigour of casuarina (*Casuarina equisetifolia*) raised in polybags after inoculating nursery soil mixture with 3 cultures of arbuscular mycorrhizas (AM) along with 3 levels of phosphorus fertilizer (0, 300 and 500 mg/polybags). AM inoculated seedlings grew faster and healthier than uninoculated seedlings. Among AM cultures, *Glomus fasciculatum* was found to be efficient. When combined with P at 300 mg/polybag, its efficiency was greater and it produced robust seedlings with longer roots and shoots as well as higher dry matter and P uptake. However, when P was applied at 500 mg/polybag in combination with cultures, AM colonization appreciably decreased and no improvement in growth of seedlings was observed.

Jamaluddin *et al.* (2001b) inoculated the seedlings of *Bambusa nutans* with soil containing *Glomus mosseae, Glomus intraradices* and *Gigaspora* sp. and infected roots (root + spore + soil inoculum). Spore inoculum (200 chlamydospores each of the fungi) and root inoculum. Plants were harvested 7 months after inoculation. Development of vesicular arbuscular mycorrhizas (VAM) and plant growth were observed. A significant increase in growth and biomass was observed in plants inoculated with VAM compared to control. Root colonization increased with increasing biomass and was highest in soil-based inoculum-treated plants.

Trate and Srivasuki (2001) inoculated the seedlings of *Acacia nilotica, Pterocarpus santalinus, Tamarindus indica* and *Peltophorum pterocarpum* with *Glomus fasciculatum* and *Peltophorum pterocarpum* were inoculated with *Glomus fasciculatum* and *Pisolithus tinctorius.* The plants were subjected to unstressed (well-watered) and drought (water stressed) conditions. In *P. pterocarpum* and *T. indica* both mycorrhizal inoculations stimulated plant growth. P and N uptake compared to their non-mycorrhizal (NM) plants under both watering regimes. However, in *A. nilotica* and *P. santalinus*, these parameters were only stimulated by either *P. tinctorius (A. nilotica)* or by *G. fasciculatum (P. santalinus).* This was attributed to competition for carbon between *Acacia* and *G. fasciculatum* and a parasitic association of *P. santalinus* and *P. tinctorius.*

Sharma *et al.* (2001c) studied mycorrhizal dependency and growth response of *Acacia nilotica* and *Albizia lebbeck* to inoculation by indigenous AM fungi as influenced by available soil P levels in a semi-arid Alfisol wasteland. A series of available phosphorus levels ranging from 10 to 40 ppm were achieved in a semiarid soil, a Hyperthermic Typic Hallustaff collected from a wasteland site in Haryana State. The influence of the levels of phosphorus on the symbiotic interaction between two subtropical tree species, *Acacia nilotica* and *Albizia lebbeck* and a mixed inoculum of indigenous arbuscular mycorrhizal (AM) fungi was evaluated in a greenhouse study. The extent to which the plant species depended on AM fungi for dry matter production decreased as the levels of soil P increased, but the degree of this decrease differed in the two species tested. *Acacia nilotica* colonized by AM fungi showed a significant increase in shoot P and dry matter at a soil P level of 10 ppm, whereas in *Albizia lebbeck*, such increased occurred at 20 ppm. Mycorrhizal inoculation response disappeared beyond soil P levels of 25 ppm in *Acacia nilotica* and 30 ppm in *Albizia lebbeck.*

Levels of soil P greater than 25 ppm suppressed AM fungus colonization in both species. Soil P levels of 30 and 40 ppm and 40 ppm caused negative mycorrhizal dependencies (MD) in *Acacia nilotica* and *Albizia lebbeck* respectively. Values of MD for both species were negatively correlated with soil P levels. Based on the MD values, regression equations were developed to predict MD for given level of available P.

Goswani and Jamaluddin (2001) used *Albizia procera* as a test species to study the role of vesicular arbuscular mycorrhiza (VAM) in root trainers (300 cc). Four media were used: (a) pure compost, pH 5.8, (b) compost + sand (80:20), pH 6.4, (c) compost + sand (50:50), pH 6.4, and (d) sand, pH 6.8. The compost contained Glomus, *Scutellospora* and Gignspora in low quantities. VAM inoculum (*Gigaspora* spp., *Scutellospora* spp., *Glomus mosseae, Glomus intraradices* and *Acaulospora scrobiculata*) was added at 25g/root trainers, 10 days after seed germination. After inoculation, the number of VAM fungal spores was increased in the potting media containing compost + sand (80:20, 50:50) and pure compost, but low spore development was observed in pure sand. The percent colonization of roots by VAM was highest in plants grown in compost and sand (80:20) and pure compost- Spore population and number of root nodules were lower in pure compost, compared with the compost mixtures. The height of the plants varied in the different treatments.

Vijayakumar and Abraham (2001) observed the frequency of arbuscular mycorrhizal (AM) fungi in the rhizosphere of the endemic trees *Gluta travancorica, Bentinckia condapanna* and *Myristica malabarica* were isolated. Attempts were made to identify the rapidly infecting fungi on these hosts and *Glomus fasciculatum* was found highly colonized with both *Gluta travancorica* and *M. malabarica,* while *Glomus intraradices* was observed as the rapidly infecting fungi on *B. condapanna*. The effects of phosphorus addition on AM root colonization and spore production on these endemic trees were determined. To find out the effect of pesticides on AM fungal development, a few fungicides and insecticides (Dithane [mancozeb], Bavistin [carbendazim], endosulfan and malathion) at two concentration levels (recommended level and one-half the recommended level) with and without application were tested. Data on the effects of phosphorus and pesticides on AM fungal root colonization and on the AM fungal spore count are tabulated.

Srivastava *et al.* (2001) studied the use of biofertilizer, particularly vesicular arbuscular mycorrhizal (VAM) fungi which have a great importance in forestry as it provides minerals, wide absorption area of root zone, water uptake and tolerance to water stress conditions etc. Rajasthan covers approximately 61% of the geographical area of the total desert area of the country. In these harsh conditions, VAM fungi can play a significant role in survival of plants. In the present study, some economically important tree species of Jaipur region namely *Cordia myxa, Artocarpas integer, Dalbergia sissoo, Pongamia pinnata, Mangifera indica* and *Alestronia* sp. were selected for the study. VAM inoculated seedlings of above tree species performed better in term of shoot height, root height, biomass and percentage of colonization. The shoot height increase was maximum (32.2%) in *A. integer* and minimum (11.2%) in *Mangifera indica.*

Kaushal (2001) conducted a study to test the efficacy of VAMF [vesicular arbuscular mycorrhizal fungi] in *Acacia nilotica* at various levels of phosphorus fertilizer and to determine how far VAM in the plants compensate for phosphorus fertilizer requirement. Results showed that mycorrhizal plants showed an increase in the intensity of mycorrhizal colonization as well as VAMF spore population when treated with different rates of P fertilizer. The intensity of infection and spore population decreased when the phosphorus fertilizer rate was more than 50% of the recommended level. Biomass production in terms of dry weight increased with the application of different rates of P fertilizer. Application of VAMF with 50% recommended level of P fertilizer brings about an equal growth when the non-mycorrhizal plants were supplied with 100% recommended level of P.

Bhattacharya *et al.* (2002) conducted a greenhouse study to determine the changes in the root development pattern of *Dendrocalamus strictus* and *Citrus sinensis* plants upon inoculation with arbuscular mycorrhizal (AM) fungi. Seedlings of both plants were grown in pot cultures in low-nutrient soil, with or without AM (*Glomus mosseae, G. fasciculatum* or *Gigaspora margarita)* inoculation. Results showed that *D. strictus* is a moderately AM responsive plant. The AM fungi did not infect its primary roots. Mycorrhization significantly increased the lateral root development particularly in term of the number of 3rd-order branches per cm of primary root. The total number of lateral roots per cm. of primary root was significantly higher in mycorrhizal than non-mycorrhizal plants, the difference being individually significant for the 3rd-order lateral only. The mycorrhizal plant root diameters were marginally wider than the non-mycorrhizal plant root diameters. The total surface area of the 3rd and 2nd-order laterals increased in mycorrhizal over non-mycorrhizal plants. However, *C. sinensis*, a highly mycorrhiza-dependent plant, produced progressively finer Ist, 2nd, 3rd order lateral branches from the taproot. The laterals were devoid of root hairs and were infected by the AM fungi. Mycorrhizal elements were higher in the 3rd-than 2nd- and Ist -order laterals. Taproots of mycorrhizal plants were longer and heavier than those of non-mycorrhizal plants. The number of lateral roots, both total per plant and per unit length of taproot, was higher in mycorrhizal than non-mycorrhizal plants. The average lengths of lateral roots of mycorrhizal plants of all orders were significantly lower than those of the non-mycorrhizal plants.

Prasad (2002) carried out a study to assess the effect of arbuscular mycorrhizae on biomass yield, uptake and translocation of nitrogen, phosphorus and potassium in *Azadirachta indica* L. Data showed that AMF increased the yield and uptake of NPK nutrients in a period of 120 days of plant growth. The plants inoculated either with mixed inoculum of AMF or with *Glomus fasciculatum* did not differ significantly in their growth. It appears that *G. fasciculatum* could compete with other AMF present in mixed inoculum and establish an effective relationship with *A. indica*. However, *G. fasciculatum* has emerged as the best strain suited to this plant.

Sempavalan *et al.* (2002) conducted a pot culture experiment to investigate the effect of different vesicular arbuscular mycorrhizal fungi inoculation on nodulation, growth (shoot and root dry weight) and N and P uptake of *Casuarina equisetifolia.* The soil (sandy loam soil) used in the study was very low in P status (3.31 kg P_2O_5/ha) and having a pH level of 8.2. VAM fungal cultures were: *Glomus mosseae, Glomus fasciculatum, Glomus etunicatum, Gigaspora calospora* and *Acaulospora laevis.* Plants were harvested after 60, 90 and 120 days after inoculation (DAI). Among the inoculants, *Glomus fasciculatum* recorded the maximum biomass and resulted 2.21 g shoot weight/plant and 0.91 g root dry weight/plant as against 1.06 and 0.52 g in control at 120 DAI. *Glomus fasciculatum* recorded 2.01-14.2 nodules/plant followed by *Glomus mosseae,* which produced 2.0-14.2 nodules/plant as against control (1.6-8.18 nodules/plant) during 60-102 DAI. Maximum colonization was observed with *Glomus fasciculatum* (71.2%) and was significantly higher than *Glomus mosseae* (62.0%) and *Gigaspora calospora* (60.6%) at 120 DAI. Maximum uptake was noticed at 120 DAI with the inoculation of *Glomus fasciculatum* as 27.42 mg N and 9.20 mg P/ plant, followed by 22.80 mg N and 7.98 mg P/ plant.

Panwar and Vyas (2002b) conducted pot experiments to study the effect of 8 arbuscular mycorrhizal (AM) fungi (*Acaulospora mellea, Gigaspora gigantea, Gigaspora margarita, Glomus deserticola, Glomus fasciculatum, Sclerocystis rubiformis, S. calospora* and *Scutellospora nigra*) on the biomass production and nutrient uptake of *Tamarix aphylla.* All the AM fungi increased the biomass production (plant height and dry weight) and nutrient uptake of *T. aphylla.* Of the AM fungi studied, *Scutellospora calospora* was the most effective in increasing plant biomass and nutrient uptake of *T. aphylla.*

Chandra and Ujjaini (2002) studied the interaction between AMF and organic matter and its influence on biomass and root infection on six forest tree species. Mycorrhizal fungi were inoculated with 3 levels of soil organic matter in *Tectona grandis* L., *Dendrocalamus strictus* Nees, *Emblica officinalis* Gaert., *Albizia lebbeck* Benth., *Gmelina arborea* L., and *Anthocephalus chinensis* Miq. Seedlings in a factorial randomized block design experiment. Mean responsiveness of plant biomass was increased up to 36.85% at OM1 (lowest OM level) with the inoculation and decreased with the increasing OM levels in soil *A. lebeck* exhibited highest response to inoculating AMF whilst *E. officinalis* rendered lowest response. Addition of OM in soil was found to enhance the pant biomass but it was found to decrease the responsiveness to AM fungi. Root infection percent of plants was also enhanced with AMF introduction. The improvement in mean root infection was found maximum up to 3.29 fold at OM1 and minimum up to 1.79 fold with OM3 compared root infection by native endophytes. Mycorrhizal infection was increased in non-inoculated plants, while decreased in inoculated plants with increasing level of OM in soil.

Prakash and Kasera (2003) evaluated the effects of different nutritional treatments on the growth and biomass of *Salvadora persica* in a field experiment conducted in Jodhpur, Rajasthan during 1999-2001. The treatments comprised application of 100 g vesicular arbuscular mycorrhiza (VAM)/plant, 8-10 t farmyard manure (FYM)/ha, 40 q Hexameal/ ha, full (60:40:30) or half (30:20:15) of the recommended dose for NPK, applied alone or in combination with Hexameal. Plant height, collar diameter, and fresh and dry weight of leaves, stem branches and bark, and roots were highest with the application of VAM. Treatment with Hexameal resulted in the lowest moisture loss of the plant.

MISCELLANEOUS

Kandasamy *et al.* (1986a) inoculated the nursery beds of *Chrysanthemum cinerariaefolium* with a mixed culture of *Glomus fasciculatum, G. caledonium* and *Gigaspora margarita*. They reported increased seedling height, number of branches/seedling, root length and seedling dry weight, compared with untreated controls. When transplanted in the field, the inoculated plants showed a 7-10 days advancement in flowering.

Kale *et al.* (1987) studied the influence of worm cast (Vee. comp. E UAS'83) on the growth and mycorrhizal colonization of two ornamental plants. The 3 treatments consisted of (1) Equal quantities of fine sand garden soil and FYM (control) (2) sand: garden soil: air dried worm casts at 1.0:1.5:0.5 and (3) as in (2) but equal quantities of each component. One week old seedlings of dwarf salvias and asters (*Callistephus chinensis*) were transplanted into pots containing the 3 mixtures. When worm casts were used instead of FYM, the mycorrhizal root colonization of both species was higher and this had a beneficial effect on plant vegetative characteristics and flowering.

Potty (1988b) studied the response of cassava (*Manihot esculenta*) to VAM inoculation in acid laterite soil. Influence of VAM fungi on growth, dry matter, tuber production and soil microflora and soil enzymes was studied in acid laterite soil. Rooted infected cuttings responded favourably in both sterile and natural soils. Significant difference was noticed in dry matter production and tuber production in natural soil. Sterilization had very little influence over natural soil. Half recommended dose of P significantly enhanced tuber production and dry matter in the presence of VAM fungi. Quantitative difference in soil microflora around root region of VAM infected and control plants was observed but not to a significant level. Rhizosphere microflora favours the VAM association and vice versa in the cassava root region. Soil enzyme activity, *i.e.,* urease, phosphatase and dehydrogenase showed higher activity as the age of the plant increased in presence of VAM fungi. Micronutrients like Cu and Zn accumulated in leaf tops of mycorrhizal cassava plants. But difference was observed between infected plants in natural and partially sterile soil. Not much difference existed between the starch content of mycorrhizal and non-mycorrhizal plants in tubers. However, higher P uptake coincided with high starch content in tubers.

Reddy and Bais (1990) studied the prevalence of native VAM fungi and their relative performance in infectivity of local crops *viz.* chickpea (*cv. Annigeri*), pigeonpea (GSI), sorghum (*cv. Neerjola*) and onion (*cv. Desi*) at Gulbarga. Three VAM fungal species, all belonging to the genus *Glomus viz., G. constrictum, G. fasciculatum* and *G. aggregatum* were recorded in the present survey of native soils. All the four local crops responded to infectivity by all five VAM fungi tested with characteristic vesicles and arbuscules. *G. aggregatum*, though infected sufficiently, the formation of vesicles and arbuscules were scarcely seen. Native species *viz. Glomus fasciculatum, G. aggregatum, G. constrictum* were tested on 4 local crops along with two other VAM fungi, *G. fasciculatum* and *Gigaspora margarita* in pot culture experiments using local soil. Variability was observed in infectivity and % mycorrhiza formation.

Sivaprasad *et al.* (1990) studied the comparative efficiency of different mycorrhizal fungi on cassava (*Manihot esculenta*, Crantz). In a pot experiment with cassava inoculated with *Glomus fasciculatum* gave the highest plant height, shoot and root dry weight and VA mycorrhizal infection, compared with inoculation with *G. mosseae, G. constrictum, G. etunicatum, Acaulospora morroweae* or no inoculation.

Varma and Schuepp (1994) studied positive influence of arbuscular mycorrhizal fungus on *in vitro* raised hortensia plantlets. Micropropagated plantlets of hortensia (*Hydrangea* [*macrophylla*] *cv. Leuchtfeuer*) were inoculated with *Glomus intraradices*. Roots were heavily colonized at the acclimatization stage. There was 100% survival, shoot apices were active and no apparent "transient transfer shock" was noticed. Infected plants showed better growth and larger leaf area than the controls. Eighteen-week-old infected, potted plants exhibited considerable resistance towards induced stress environmental conditions in the phytotron. In contrast, non-infected plants were severely wilted and the number of apical buds was reduced. The general bacterial population was several fold higher in the rhizospheres of colonized roots than in non-infected roots, with *Pseudomonas fluorescens* biovar.1 being predominant.

Srivastava and Mukerji (1995) studied the field response of mycorrhizal and non-mycorrhizal *Medicago sativa* var. *local* in the F1 generation. They collected the seeds from plants of *M. sativa* (lucerne) var. *local* inoculated separately with *Glomus macrocarpum* and *G. fasciculatum* in (pot experiments). These seeds were grown in garden soil and the percentage germination, general health and yield of subsequent plants were recorded (F1 generation). The percentage germination was highest in seeds of *G. macrocarpum* - inoculated parents followed by those inoculated with *G. fasciculatum* . Seeds of uninoculated parent plants showed the lowest germination. Vegetative yield of the progeny was highest in plants inoculated with *G. fasciculatum* followed by those inoculated with *G. macrocarpum* and lowest in uninoculated plants. However, reproductive yield was highest for plants whose parents were inoculated with *G. macrocarpum* followed by *G. fasciculatum* and lowest for seeds of uninoculated parent plants.

Naik *et al.* (1995) conducted a field experiment to measure the effects of different VAM fungi at different P rates on the growth, yield and P content of China aster (*Callistephus chinensis* (L.) Nees.) cv. "Ostrich plume". *Gigaspora margarita*, *Acaulospora laevis* and *Sclerocystis dussii* were used with 0, 50, 75 or 100% of the recommended 'P' rate (120 kg/ha). *G. margarita* inoculation gave the greatest increase in shoot dry weight, 'P' uptake and flower yield at 75% of the recommended 'P' rate suggesting a net saving of 25% of the recommended P with VAM inoculation.

Sulochana *et al.* (1995) studied the effect of inoculation with VAM fungi on growth and tuber yield of cassava in comparison with inorganic fertilizer application under field conditions. They used the mixed inoculum of VAM fungi *Glomus fasciculatum* and *G. etunicatum* and three fertilizer doses: (i) recommended dose of NPK (50:50:50) (ii) full dose of NK and 50% of P_2O_5 and (iii) no fertilizer. The study reveals that VAM inoculation in cassava with or without application of NPK considerably increased the biomass production, tuber yield and P content of plant tissue. VAM inoculation with full dose of NK and 50% P application was found to be an appropriate treatment.

Bojan *et al.* (1995) investigated the role of VAM fungi in the successful establishment of sugarcane seedlings regenerated through tissue culture. Calli were obtained in MS medium amended with 2,4-D (5 mg)-1 and BAP (0.1mg-1) from young juvenile leaves. After successful establishment of calli, they were transferred to shoot induction medium amended with growth adjuvants such as CW (25% v/v) + adenine - sulphate 6 mg-1. The shoots were transferred to half strength MS liquid medium containing IBA, 5 mg-1 for rhizogenesis supported by sterilized filter paper. The fully established sugarcane seedlings were transferred to vermiculite soil and mixture (1:1) inoculated with sterilized spores of *Glomus mosseae*, *G. aggregatum* and *Gigaspora margarita*. Synthesis of mycorrhizae in the roots was assessed on 15th, 30th and 45th day. The survival rate, growth parameters such as height, shoot length, root length, shoot: root ratio, collar diameter, number of green leaves were assessed on 30th and 45th day. All the 3 species of VAM fungi were successfully synthesized. The percentage of infection ranged from 40-60%. Highest infection (60%) was found in the seedlings inoculated with *Glomus aggregatum*. The survival rate was higher (60-80%) in the VAM inoculated seedlings over control. *G. aggregatum* supported the highest survival rate compared to *G. mosseae* and *G. margarita*. *G. aggregatum* supported all the growth parameters except the collar diameter. However, the seedlings inoculated with *G. margarita* were dwarf and the collar diameter increased.

Rajgopal and Ramarethinam (1997) studied the interaction between vesicular-arbuscular mycorrhiza and tea roots. In pot trials, the modified slide technique was used to inoculate tea cuttings with spores of *Glomus fasciculatum*. Examination of tea roots 100 days later showed that the cortical cells were penetrated by *G. fasciculatum* and that the penetration was mainly through the lateral hairs. Slight invagination was noticed at the point of penetration. In a separate trial, P fertilizers (superphosphate and rock phosphate) were mixed with the potted soil at 25, 50, 50 or 150 mg P/g soil and the soil inoculated with *G. fasciculatum* by addition of 1 cm lengths of mycorrhiza-colonized tea roots to give the inoculum dose of 2 x 104 cells / pot. The P content of shoots and roots were measured after 30 days of growth by spectrophotometry. P uptake was significantly increased by mycorrhizal colonization, the effect being greater for superphosphate than for rock phosphate. P uptake also increased with increasing rate of P application, except in shoots of non-inoculated plants given rock phosphate.

Jain *et al.* (1998a) studied the interactive effects between the VAM fungus *Glomus fasciculatum* and phosphorus levels (P0, P20, P40 and P 80) on three forage legumes (*Trifolium alexandrinum, Medicago sativa* and *Vigna unguiculata*) in pots containing P deficient sterilized soil. Growth and P uptake of the plants were improved when inoculated with mycorrhizas. However, the biomass and P uptake further increased in inoculated plants when the soil was supplemented with different P levels, being highest at P80 application. A desirable level of biomass improvement was achieved at P 20 with VAM inoculation, suggesting that P rate can be reduced by the use of VAM.

Krishnamurthy *et al.* (1998) studied the VAM association in sandal- a hemi-root parasite. Observation of the root squash and sections confirmed the establishment of VAM in sandal. This is the first report on the VAM infection in the roots of *Santalum album*. The comparison in the growth parameters of the control and VAM inoculated plants showed that the treated plants registered greater growth than the control plants. The effect of VAM fungi (*G. fasciculatum*) on the initial growth and establishment of *Santalum album* was studied. On the basis of various growth parameters studied, it was concluded that the VAM inoculated plants registered a greater growth than the control plants.

Naik *et al.* (1998) conducted a field trial in India to determine the role of vesicular-arbuscular mycorrhizal (VAM) fungi (*Gigaspora margarita, Acaulospora laevis* and *Sclerocystis dussii*) in the production of *Callistephus chinensis* grown with P at 0, 50, 75 or 100% of the recommended dose (120kg P_2O_5/ha). Among VAM fungi tried, *G. margarita* was the best at improving plant height, number of branches, flower yield, P-uptake and returns (Rs 44 465/ha and a cost benefit ratio of 1:6.3) at 75% of the recommended P dose.

Palipane (1999) studied the influence of vesicular arbuscular mycorrhizal fungi on the growth of coffee plants. He found that vesicular-arbuscular mycorrhizae (*Glomus* spp.) were associated with vigorously growing unfertilized ten years old coffee plants. Artificially inoculated coffee seedlings had a significantly higher growth rate than that of uninoculated seedlings. No competition was exerted on the mycorrhizal fungi by the indigenous microorganisms on artificial inoculation of soil. There was no significant difference between the growth of artificially inoculated coffee plants grown in the field with application of fertilizer. Unhealthy stunted plants were observed in the uninoculated coffee plants grown without application of fertilizer, hence, it was concluded that inoculated coffee plants can be grown successfully in the field without application of fertilizer in the first year after field planting.

Sivaprasad *et al.* (1999) studied the influence of vesicular arbuscular mycorrhizal fungi (VAMF) *Glomus fasciculatum*, *G. etunicatum* and a native isolate *Glomus* sp. on establishment, growth and nutrient uptake of tissue culture plantlets of banana and *Alocasia* sp. VAMF colonization significantly increased the establishment rate from 68.5 to 92 per cent in banana and from 73.6 to 100 per cent in *Alocasia*. Both the plant species showed significant increase in growth, vigour and biomass production during acclimatization and after transplanting to plots, when inoculated with *G. fasciculatum* and *Glomus* sp. significant increase in total and OD phenol content was noticed at the end of acclimatization phase in banana plantlets, while the phenol content was more with *G. fasciculatum* colonization in *Alocasia*. In both the plant species, *G. fasciculatum* colonization remarkably increased P and Zn content.

Sharma *et al.* (1999d) studied the effect of mycorrhiza on mushroom and silkworm cocoons yield. Red and white stem castor varieties were inoculated with native VAM fungi alone as well as in combination with other treatments *viz. Pleurotus sajor-caju*, vermicompost and FYM and cultivated on normal soil in Micromodel, IIT Delhi. Inoculation with VAM fungi, *P. sajor-caju*, and vermicompost in combination markedly influenced the growth and mycorrhizal development. Mycorrhizal inoculation enhanced mushroom and 'cocoon' yield quantitatively. Red castor being more mycorrhizal in nature than white castor, when cultivated in plots treated with VAM, *P. sajor-caju* and vermicompost, yielded maximum growth of mushroom with biological efficiency 112%. On the other hand, paddy straw a traditionally used substrate for mushroom growth produced fruit bodies with biological efficiency 80%.

Shashikala *et al.* (1999a) conducted an experiment under glasshouse conditions to study the effect of *Glomus mosseae* on the growth and nutritional status of micropropagated *Syngonium podophyllum cv. White Butterfly* and *Spathiphyllum floribundum cv. Petite*. The results revealed that both species responded positively to *G. mosseae* inoculation. However, the extent of plant growth benefits conferred by the fungus on each host species varied to a great extent depending on soil P. To elicit maximum plant growth in the presence of *G. mosseae* 100% of the recommended P rate should be applied to *Syngonium*, whereas only 50% of the recommended P rate was necessary for *Spathiphyllum*.

Naik *et al.* (1999) studied the efficacy of different vesicular-arbuscular mycorrhizal (VAM) fungi on growth and yield of China aster (*Callistephus chinensis*) flowers under field conditions (Dharwad, Karnataka) at 4 different P rates. Three VAM fungi, *Glomus fasciculatum, Acaulospora laevis* and *Sclerocystis dussii* and an uninoculated control (UIC) were tested at 4 P rates (0, 50, 75 and 100% of the recommended dose: 120 kg P_2O_5 ha). Among the VAM fungi tried, *G. fasciculatum* resulted in the highest dry matter accumulation rates for leaves, stems and flowers, flower yield, P-uptake and mycorrhizal parameters (percentage root colonization and spore count) at 75% of the recommended P rate.

Gnanadevi and Haripriya (1999) conducted an experiment to screen the efficient VAM fungi for chrysanthemum at Annamalainagar, Tamil Nadu. Three *Chrysanthemum* cultivars (*Bangalore White, Local Yellow and Shyamal*) were planted in pots and inoculated separately with 3 vesicular arbuscular mycorrhizae (VAM) fungi (*Glomus mosseae, G. fasciculatum* and *Acaulospora laevis*). All VAM fungi increased growth and yield of *Chrysanthemum* cultivars compared with the control. *G. fasciculatum*-inoculated plants had the greatest average plant height, highest number of laterals, number of flowers per plant, flower yield (136.62, 130.95 and 137.54 g/plant for *Bangalore White, Local Yellow and Shyamal*, respectively), and shortest duration before first flowering (118.75, 108.00 and 107.83 days for *Bangalore White, Local Yellow and Shyamal,* respectively) occurred.

Prasad *et al.* (2000) studied influence of *Glomus fasciculatum* inoculation on growth and phosphorus uptake in *Gladiolus* sp. *Gladiolus* plants inoculated with *G. fasciculatum* showed enhanced growth compared with the controls. Phosphorus uptake was significantly higher than in the uninoculated controls.

Lakshmpathy *et al.* (2000) screened nine different VAM fungi, *viz., Acaulospora laevis, Gigaspora margarita, Glomus caledonium, G. fasciculatum, G. intraradices, G. leptotichum, G. macrocarpum, G. mosseae and Scutellospora calospora* for their ability to enhance the growth and P uptake in cashew rootstock Ullal-1 under glass house conditions. Seeds were sown in plastic bags containing soil inoculated with infective propagules (12 500 /bag). Effects on growth were recorded 120 days after sowing. Among the fungi studied, *A. laevis* and *Glomus mosseae* significantly increased plant height, stem girth and total biomass of cashew rootstock compared to uninoculated plants. Similarly, the percent root colonization and P uptake were also significantly higher in *A. laevis*- and *G. mosseae*- inoculated plants compared to uninoculated plants. These results suggest that *A. laevis* and *G. mosseae* are better symbionts for inoculating cashew rootstock.

Gosal *et al.* (2001) studied *in vitro* establishment of vesicular-arbuscular mycorrhiza in roots of micropropaged sugarcane. Vesicular-arbuscular mycorrhiza (VAM) an obligate symbiont has been found to develop association *in vitro* with roots of micropropagated sugarcane. The co-cultivation of VAM and roots of micropropagated sugarcane plants exhibited up to 61% infection in sugarcane var. Co.J. 83. The microscopic examination showed the presence of hyphae and few vesicles. The VAM infection caused excessive root growth and increased root dry weight. Besides, the VAM infected plants showed better survival upon their transfer to the soil. These findings could be exploited for multiplication of VAM *in vitro* to overcome the mortality rates of micropropagated sugarcane plants in soil. This can also serve as a potential biofertilizer and bioprotector for sugarcane.

Thammaiah *et al.* (2001) conducted a glasshouse trial to determine the response of coffee (*C. canephora)* cv. Sln. 3R (C x R) to different vesicular arbuscular mycorrhizal (VAM) fungi. The VAM species used were: *Gigaspora margarita, Glomus caledonium, Glomus fasciculatum, Glomus intraradices, Glomus leptotichum, Glomus macrocarpum* and *Glomus mosseae.* Results indicated that the percentage root colonization, mycorrhizal spore count, plant height, leaf area index and plant dry weight were highest in plants inoculated with *Glomus leptotichum* , in plants inoculated with *Glomus leptotichum,* compared to those inoculated with other VAM fungi.

Haripriya and Sriramchandrashekharan (2002) conducted the pot culture experiment to study the effect of vesicular arbuscular mycorrhizal (VAM) fungi inoculation on the growth and yield of *Chrysanthemum.* The treatments consisted of three varieties of chrysanthemum *viz., Bangalore White, Shyamal* and *Local Yellow,* and three species of VAM fungi, namely *Glomus mosseae, G. fasciculatum* and *Acaulospora laevis*. Among the species of VAM, *G. fasciculatum* was found to be the best colonizer, followed by *G. mosseae* and *A. laevis*. *Chrysanthemum cv. Local Yellow* was found to perform better under Annamalainagar(Tamil Nadu) conditions and was best colonized by *G. fasciculatum* recording the highest yield.

Roy *et al.* (2002) conducted an experiment to study the VA Mycorrhizae in relation to growth of different tea varieties. Five plants of the same age group in each cultivars and five samples (rhizosphere soil and roots) of each plant were collected from the germplasm bank and brought to the laboratory. The soil samples were processed to screen the VAM spore population by the wet sieving and decantation method. A good correlation was established between VAM population root colonization and plant growth for the tea cultivars of section A. A root colonization of 100% and 15-30 spores/ 10g dry soils were recorded in TV-23 and TV-26, where a good growth of individual plants was also observed. Teen Ali-17 and B-777, which attained poor growth, had 50-75% root colonization and 10-15 spores/10g dry soil. In section B, 80-100% root colonization and 20-30 spores/10g dry soil were observed in UPASI-2 and UPASI-26. These cultivars also recorded a good growth of individual plants. The six-year old TV-26 had 100% root colonization and good growth, while the three-year-old TV-26 showed only 35% root colonization and very poor growth. *Glomus macrocarpum, Glomus fasciculatum and Glomus ambisporum* were invariably recorded in association with roots of all the cultivars while the distribution of *Glomus aggregatum, Gigaspora margarita, Sclerocystis* and *Scutellospora* was not uniform. *Gigaspora margarita* and *Glomus* species associated with plants showing good growth.

Varshney *et al.* (2002) studied the enhanced growth of micropropagated bulblets of *Lilium* sp. inoculated with arbuscular mycorrhizal fungi at different P fertility levels in an alfisol. The effect of three-arbuscular mycorrhizal (AM) inoculum at four available levels of phosphorus (8.41, 12.53, 13.63 and 14.6 ppm) in non-disinfected soil was studied on the growth, flowering, P uptake and root colonization in micropropagated bulblets of *Lilium* sp. (Asiatic hybrid 'Gran paradiso'). The inoculated bulblets fared significantly better then the uninoculated ones in terms of all the growth variables, namely size, weight, shoot length, number of leaves and leaf area, and in P uptake. However, bulblets inoculated with different AM inocula had optimum growth at different P levels. Bulblets inoculated with indigenous mixed vesicular-arbuscular mycorrhiza species (VAM I) and *Glomus intraradices* isolate 2 (VAM III) showed the best growth and early flowering at available soil P of 13.6 ppm, whereas those inoculated with *Glomus intraradices* Isolate I (VAM II) showed higher growth at 2.5-ppm available soil P. Amongst the three-tested inoculated, VAM I promoted maximum shoot length, bulblet size and weight at 13.6 ppm P. The bulblets under this treatment also flowered earlier, nearly a month before the uninoculated control ones.

Bhattacharya and Bagyaraj (2002) conducted a glasshouse investigation to study the effectiveness of eight arbuscular mycorrhizal (AM) fungal isolates (*Acaulospora laevis, Gigaspora margarita, Glomus caledonium, G. fasciculatum, G. intraradices, G. leptotichum, G. macrocarpum* and *G. mosseae*) on *Coffea arabica cv.* Sln. 5. Coffee seedlings raised in the presence of AM fungi in polythene bags generally showed an increase in plant growth (height, number of leaves, leaf area and biomass) and plant nutritional status (P and Zn) over those grown in the absence of soil inoculation with AM fungi. The extent of growth and nutritional status enhanced by AM fungi varied with the isolates of AM fungi inhabiting the roots of coffee seedlings. Of the eight isolates, *Gigaspora margarita* and *A. laevis* were the best AM symbionts for arabica coffee compared with the others used under this experiment.

Gaur and Adholeya (2002) inoculated five fodder crops, *Zea mays, Medicago sativa, Trifolium alexandrinum, Avena sativa,* and *Sorghum vulgare* with consortia of indigenous arbuscular mycorrhizal (AM) fungi in non-sterile PO_4^{3-} deficient sandy loam soil amended with organic matter under field conditions. Shoot and root dry weigths and total uptake of P and N of all the test plants were significantly increased by AM inoculations. Mycorrhizal inoculation increased yield in terms of shoot dry weight by 257% in *T. alexandrinum* followed by 50% in *A. sativa,* 28% in *Z. mays,* 20% in *M. sativa* and 6% in *S. vulgare.* Variation in dependence on mycorrhiza was observed among the fodder crops. *T. alexandrinum* showed a maximum dependence of 72% in contrast to 5.7% dependency in *S. vulgare.* Plant species showed differences in percentage AM colonization, with a high root infection recorded in *Z. mays* (76%). Spores/IP g^{-1} and 103-spores/IP g^{-1} in *S.vulgare.* This study clearly indicates the potential of using indigenous AM inoculations in fodder crops grown in marginal soils along with in situ large-scale production of inocula.

Hemalatha and Selvaraj (2003) determined the symbiotic response of Indian borage (*Plectranthus amboincus*) to 8 vesicular arbuscular mycorrhizas (VAM; *Acaulospora morrowae, Glomus aggregatum, Gigaspora margarita, Glomus ambisporum, Glomus deserticola, Glomus geosporum, Sclerocystis sinuosa* and *Scutellospora calospora*), inoculated at 10 500 infective propagules per planting point during transplanting of the crop in pot experiments. Alongside this, a field survey was conducted to determine the association of Indian borage with VAM. Microscopic examinations of the roots of Indian borage in the field showed extensive (92%) colonization by VAM. Based on morphological characters, the VAM isolate was identified as belonging to the genus *Glomus*. *Glomus aggregatm* and *Glomus ambisporum* were predominant, whereas *Gigaspora, Acaulospora, Sclerocystis* and *Scutellospora* were rarely isolated. Inoculated plants generally had higher values for plant height, biomass production, and nutrient uptake and content compared to the uninoculated controls, with plants inoculated with *Glomus aggregatum* recording the highest values for the parameters measured. Mycorrhizal inoculation also resulted in an increase in the number of spores in the rhizosphere. No significant increase in the essential oil content of Indian borage due to VAM inoculation was recorded.

Murugan and Selvaraj (2003) inoculated eight vesicular arbuscular mycorrhiza (VAM; *Acaulospora scrobiculata, Gigaspora margarita, Glomus aggregatum, Glomus geosporum, Glomus macrocarpum, Glomus mosseae, Sclerocysts pakistanica* and *Scutellospora calospora*) at 12,500 infective propagules per planting point on kashini (*Cintybus*) plants during transplanting in a filed experiment conducted in Tamil Nadu, India to determine the effects of VAM on the growth and chemical composition of the crop. Microscopic examination of the roots of kashini showed intensive VAM development during the active growth period of the crop. All the 8 VAM species were recorded on the roots of kashini. Intramatrical vesicles, arbuscles, mycelial coils and spores were observed in the cortical tissue of the roots of the crop. Inoculated plants generally had greater values for plant height, shoot and root biomass, number of leaves and length and width of leaves compared to the uninoculated controls, with *Gigaspora margarita* inoculation resulting in the highest values for all the parameters measured. Inoculation of kashini with VAM also resulted in increased root volume, dry matter yield and leaf yield although no positive correlation between plant growth and mycorrhizal colonization was established.

Chandra *et al*. (2004) conducted experiments to select an efficient VA mycorrhizal inoculant for *Gladiolus cv. Aldebaran*. Different species of *Glomus* recovered from the roots of flowering plant *viz., G. fasciculatum, G. aggregatum, G. mosseae, G. multicaule* and *G. macrocarpum* were separately evaluated for their efficacy in improving the overall performance of the plants. In general, all the VAM inoculants except *G. macrocarpum* favoured the growth of shoots and roots. Initiation of budding and flowering, number of buds and flowers, fresh and dry weight and keeping quality of flowers. However, *G. fasciculatum* proved to be the most potential inoculant in improving the performance of *Gladiolus*.

Srivastava *et al.* (2005) conducted an experiment to select an efficient VAM inoculant for *Chrysanthemum cv. Chizuka*. This cultivar was selected due to their high export potential in the International market. In order to select an efficient VAM inoculant for *Chrysanthemum*, different species of *Glomus* recovered from the roots of flowering plants *viz., G. fasciculatum, G. aggregatum, G. mosseae* and *G. macrocarpum* were separately evaluated for their efficacy in improving the performance of *Chrysanthemum* especially the quality and number of flowers, duration of flowers etc.

CONCLUDING REMARKS

Out of 5 cereals and 13 millets known to be cultivated in various parts of the country only few including wheat, barley, and paddy amongst cereals and maize, sorghum, finger millet and pearl millet amongst millets have been investigated with reference to the effect of VA mycorrhiza on growth, productivity and nutrient uptake. Although the work on these crops opens up the possibility of improving their productivity through VAM technology, however, the work should be extended to other crops. Millets, which are not of less economic, value and constitute the food of poor community of the country need special attention.

Production of pulses in the country has gone down in recent past because of a variety of reasons. It is heartening to record that the mycorrhizasts of the country have paid due attention to pulse crops. Commendable work has been done on nearly all the important pulse crops. Their investigations show that VA mycorrhizal fungi may be helpful in boosting the production of pulse crops.

The country has the proud privilege of producing more than fifty types of vegetables. Although attempts have been made to explore the possibility of employing VAM technology in improving the production of vegetables but they are restricted to only few vegetables including potato, brinjal, tomato and lady's finger. There is need to extend this exploration to other vegetables which are not of less importance in view of their huge consumption in the country.

There is an urgent need to boost the production of oilseed crops in view of the shortage of vegetable oil in the country. Oils are the main ingredients required for making a number of edible preparations. In addition, they have a variety of other uses. However, the published reports show that Indian mycorrhizasts have concentrated their attention mainly on two oilseed crops namely groundnut and sunflower. There is need to expedite the use of VAM technology in other important oilseed crops being cultivated in the country.

A variety of spices are cultivated in the country but the Indian mycorrhizasts have paid attention mainly to chilli. There is need to investigate the benefits of VA mycorrhiza in other spices. Medicinal plants, aromatic plants, dyes and narcotic plants have also not been investigated extensively with reference to their growth response to VA mycorrhiza.

Out of more than three dozen fruit crops of the country, only few including citrus, papaya, orange, mulberry, apple, strawberry and banana have been investigated for their response to inoculation of VAM fungi. A perusal of published reports clearly shows that mycorrhizal inoculation is beneficial to them with reference to their growth and productivity. However, response of other fruit crops to inoculation of VAM fungi has to be investigated by the mycorrhizasts of the country.

Attempts are going on in the country to provide green cover to wastelands caused due to industrialization, urbanization and afforestation. A number of plantation programmes are being actively executed and attempts are being made to utilize VAM technology to curtail the rate of mortality of the transplants. Indian mycorrhizasts have done commendable work in this direction and a large number of reports are available related to the response of nursery seedlings of trees to inoculation of VAM fungi. A careful perusal of these reports clearly shows that application of VAM technology in revegetation programmes will ensure a better success to such programmes. However, in view of the fact that extent of benefit of mycorrhizal association depends not only on the host and symbiont but also on many factors related to soil and environment, the mycorrhizasts of the country should give due consideration to constraints related to such factors while testing the response of nursery seedlings of trees to mycorrhizal inoculation.

4

VA MYCORRHIZA : PHYSIOLOGY

Favourable effect of VAM association in plants has been attributed to an increased nutrient uptake by mycorrhizal roots. The transfer of ion from the soil to the roots by mass flow or by diffusion depends on the mobility of that particular ion in the soil. Ions such as NO_3^-, SO_4^{--} and Ca^{++} may move rapidly through the soil to the roots by mass flow, and uptake would be limited by the absorbing capacity of the roots. On the other hand, ions such as $H_2PO_4^-$, NH_4^+, Zn^{++} and Cu^{++} are poorly mobile and move to the roots by diffusion, the rate limiting steps is likely to be the movement to the root surface. Under such conditions, it is expected that uptake would be increased if ions could be moved rapidly through extramatrical hyphae of the VAM fungi rather than diffusing slowly through the soil. It is now well established that VA mycorrhizae enhance the uptake of phosphate by the roots but it is likely that they can also assist in the uptake of other ions. The following account provides an information on the findings of Indian mycorrhizasts on the effects of VA mycorrhizal association on nutrient uptake in the plants and its physiological implications.

It seems likely that VA mycorrhizal fungi may have a role in improving the water relations of plants. However, it is not clear whether this effect is a direct result of fungal invasion and attributable to improved water flow through hyphae or a secondary response due to improved nutrition or physiological alteration of the host. If the benefits to the plants with reference to water are directly attributable to the mycorrhizal association, then the site and mechanism of action needs to be determined. Unfortunately, not much information is available on the physiological implications related to this matter. The work of Indian mycorrhizasts related to this matter is also mentioned in this chapter.

It is commonly accepted that in the mycorrhizal association carbohydrates are transferred from the host to the symbiont. However, information on the carbon nutrition of VAM fungi is still far from complete. This is unfortunate that knowledge about the mechanism of transfer of carbon from host to the fungus, its storage in the symbiont or the cause and effect of the loss of carbon of the host is only fragmentary.

The mycorrhizal plants usually have lower concentration of nitrogen in their tissues probably due to their increased growth. Higher N content in the mycorrhizal plants has also been reported and it has been suggested that VAM may affect the nitrogen nutrition of plants directly. It seems likely that extramatrical hyphae absorbs nitrogen from the soil and translocate it to the plants. Physiological implications of the nitrogen nutrition and metabolism of mycorrhizal associations have not been properly worked out as yet and need thorough investigation. The attempts made by Indian mycorrhizasts related to N and C nutrition / metabolism of host or symbiont are also included in this chapter.

In order to understand the physiology and metabolism of mycorrhizal associations many workers have attempted to gather information about biochemical constituents of the mycorrhizal plants. Such works by Indian mycorrhizasts are also mentioned in this chapter.

Uptake of Nutrient

Krishna and Bagyaraj (1985) studied the phosphatases in the rhizosphere of mycorrhizal and non- mycorrhizal groundnut. In sterilized soil, acid and alkaline phosphatase activities in the rhizospheres of mycorrhizal and non-mycorrhizal plants were higher than in non-rhizosphere soil. No significant differences were noted between mycorrhizal and non-mycorrhizal plants. In unsterile soil, differences in alkaline phosphatase activity among the 3 treatments were not significant. With acid phosphatase, the activity was highest in the rhizosphere of mycorrhizal plants, followed by uninoculated rhizosphere and then non-rhizosphere soil.

Sharma *et al.* (1988) studied the effect of VA mycorrhiza on uptake of phosphorus and zinc in rice (*Oryza sativa*). Inoculation of potted greenhouse-grown rice plants (*cvs. Jaya and Ratna*) with *Glomus caledonium* led to increase in shoot and root dry weight and in P and Zn levels. Shoots and roots of mycorrhizal plants contained 2-3 times more Zn than in non-mycorrhizal control plants and it is suggested that VAM inoculation may prove promising for the management of Khaira disease of rice caused by Zn deficiency, which is widespread in the Tarai belt of Uttar Pradesh.

Jyrwa and Sharma (1988) studied the VAM phosphatase, Dehydrogenase and nitrification in root region of paddy under different agricultural practices. The intensity of infection and spore population of mycorrhizal fungi were high in case of jhum compared to the valley and terrace cultivation. The number of spores and intensity of mycorrhizal infection increased with the age of the plant and were maximum at the grain development stage of the crop. A negative correlation existed between mycorrhiza and moisture content of the soil. The phosphatase and dehydrogenase activities were higher in valley and terrace systems than the jhum cultivation system. Dehydrogenase activity decreased with increased age of the plant in jhum and valley system. The enzyme activities showed a positive correlation with moisture content and organic carbon of the soil. The rate of nitrification was higher in case of jhum system than valley and terrace cultivation systems. Nitrification was more in old plants than in the young ones and it showed a negative correlation with water content and organic carbon of the soil. Significant variation did not exist in the level of organic carbon content of the soil throughout the growing period. It was high in case of terrace followed by valley and jhum cultivation. The jhum cultivation showed a minimum moisture content of the soil ranging from 34-39%, whereas in valley and terrace cultivations, it was 50% or more throughout the growing season of paddy. The pH of the soil was acidic in all the three systems.

Konde *et al.* (1988) studied bulb yield and phosphorus uptake in onion as influenced by *Glomus, Gigaspora* and microphos inoculations under graded phosphorus levels. A field experiment was laid out to study the effect of inoculation of *Glomus fasciculatum, Gigaspora calospora* and carrier based inoculant of microphos at diffent levels of P (25, 50, 75, 100 kg P as single superphosphate per ha) in split plot design consisting of 20 treatments replicated

4 times on the bulb yield and P-uptake by onion (*Allium cepa*) in P-deficient soil (2.3 ppm Olsen P) having pH 8.3 and chlamydospore count 58 spores/50 ml soil suspension. Increasing P levels upto 75 kg/ha conjugated with inoculations significantly enhanced the bulb yield and P-uptake over the corresponding control treatments. However, results at 75 kg/ha plus inoculation were at par with those at 100 kg phosphorus/ha plus inoculations. The *Glomus* inoculation was significantly superior by *Gigaspora* and microphos respectively over the control.

Santhi *et al.* (1988) studied the efficiency of VAM inoculation with different sources of phosphorus on the availability and uptake of nutrients and the yield of green gram. A pot experiment with 3 sources of P, *i.e.*, superphosphate (SP), di-ammonium phosphate (DAP) and rock phosphate (RP) at 50 and 100% recommended dose with and without VAM was conducted on the availability and uptake of nutrients and yield of green gram. The results revealed that irrespective of the sources and levels. of P, VAM inoculation significantly increased the available P, uptake of P and other nutrients and grain yield of the crop over control. Among the 3 sources of P, rock phosphate was more efficiently utilized along with VAM than the other two sources. VAM inoculation along with 50% of RP was as good as full dose of DP. On an average, VAM inoculation increased the grain yield by 54% over uninoculated control irrespective of the sources of P. Inoculation with VAM fungi also increased the total uptake of N, K and Mg by green gram. RP at 50% along with VAM increased total Ca uptake.

Kapoor *et al.* (1988) studied the phosphatases of mycorrhizal and non-mycorrhizal *Trigonella* roots. The experiment was conducted in the plots of Botanical gardens of Delhi University. The soil used was sandy loam (pH 7.8; available-K, 245 Kg/ha; available-P, 880 Kg/ha). Cultures of *Glomus fasciculatum* made on soybean roots were used as inoculum in 3 replicates. Twenty-five replicates of *Trigonella* plants from both mycorrhizal and non-mycorrhizal treatments were harvested at 15, 30, 45 and 60 days after planting. Fresh weights of the plants were taken and roots were washed in ice cold distilled water for the analysis of phosphatases. Acid phosphatase activity was observed greater in mycorrhizal roots than in the non-mycorrhizal roots during the four harvests. On the 15th day, the increase in the activity of alkaline phosphatase in the mycorhizal roots over the control was not significant. But from 30 day onwards, it was significantly higher in mycorrhizal roots. The increased activity of alkaline phosphatase in mycorrhizal roots could be due either to a specific alkaline phosphatase or to induction of higher activity in the roots. The enhanced enzymatic activities were closely related to the increased growth rates and percentage mycorrhizal infection, which suggests that mycorrhizal root infection probably, helps in efficient utilization of P from soil by employing the non-specific acid and alkaline phosphatases. Spetrophotometric scanning of polyacrylamide gels showed 7 distinct bands of the isoenzymes of acid phosphatases. The general trend of the electrophoretic profiles showed differential appearance and disapperance of isoenzymic bands. Non-mycorrhizal 15 days old roots exhibited 4 distinct bands whereas the number of isoenzymes increased to 6 in the corresponding mycorrhizal roots. They have shown the different isoenzymes of the acid phosphatase associated with different stages in the development and function of fungal hyphae, which are of fundamental importance in mycorrhizal P nutrition.

Mohandas (1990a) observed enhanced phosphatase activity in mycorrhizal papaya plants (*Carica papaya*). Inoculation with VA mycorrhizal fungi increased the activity of both acid and alkaline phosphatases on the root surface in enzyme extracts from root and surrounding root region. Acid phosphatase activity was greater than alkaline phosphatase activity. Plants inoculated with *Glomus mosseae* showed better activity than those inoculated with *G. fasciculatum*.

Chaturvedi and Singh (1990) studied the ^{32}P uptake and translocation in chickpea (*Cicer arietinum*) inoculated with VA mycorrhiza. Test plants of the cultivars L-550 and C-235 exhibited higher ^{32}P uptake in pot experiments when inoculated with *Glomus caledonicum (G. caledonium)* and *Rhizobium* together than with either symbiont alone. Uptake was least in uninoculated plants.

Kumari *et al*. (1990) worked on characterization of mycorrhiza-specific alkaline phosphatase from french bean. Alkaline phosphatase was purified from non-mycorrhizal and mycorrhizal treated plants on 45th day. The enzyme preparation from non-mycorrhizal and mycorrhizal french beans was purified using sephadex gel chromatography and DEAE 52 to achieve 71, 103 and 73, 121 fold purification. The purified preparations gave a single band on 5 to 10% gradient PACE. A single protein peak was observed by UV spectrophotometry with an absorption maxima at 275 nm. The Vmax and Km differed slightly for the two preparations but they had similar pH (8.5) and temperature (40°C) optima. The mocecular weight of the alkaline phosphatase from non-mycorrhizal french beans was found to be 139,000 daltons. Whereas subunit molecular weight was 67,000. This enzyme preparation showed a wide substrate specificity as compared to the alkaline phosphatase purified from plants infected with VA mycorrhiza. Both the enzyme preparations showed end product inhibition towards pi and mycorrhizal association showed modification of alkaline phosphatase. Mycorrhizae are known to enhance the phosphate uptake by enhancing phosphatases activity. Phosphatases play an important role in the solubilization of phosphorus from non-available phosphorus sources. Alkaline phosphatase play an important role in catalyzing the hydrolysis of complex phosphate esters. Alkaline phosphatase is known to help in the solubilization of phosphorus sources and thus it can provide additional phosphorus for important metabolic processes which lead to improved production and yield.

Sharma *et al*. (1992) studied the kinetics of zinc uptake by mycorrhizal (*Glomus macrocarpum*) and non-mycorrhizal corn (*Zae mays* L.) roots at pH 6.0 and Zn concentration of 75 μ mol to 1.07 mol/m^3. Five concn-dependent phases of Zn absorption were recognized; phase 0 (1.5-4.0 m mol/mp^3 was linear but the other 4 phases (4.0 mmol to 1.07 mol/m^3), obeyed Michaelis-Menten kinetics. At low concn. <4 mmol/mp^3), sigmoidal kinetics of Zn absorption were observed. The absorption of Zn by mycorrhizal maize was greater at low concn but decreased at higher levels. This appeared to be a result of a higher maximal uptake rate in phase 1 and lower K_m values in the subsequent phases. Kinetic models yielding continuous isotherms could not account for the observed multiphasic pattern.

Rao and Tarafdar (1993) studied the role of VAM fungi in nutrient uptake and growth of clusterbean in an arid soil. In pot experiments in a net house under natural sunlight, a thin layer of various VAM fungi (400 spores / 50 g) was placed just below the loamy sand

(Camborthid) soil surface before sowing *Cyamopsis tetragonoloba*. The soil contained efficient *Rhizobium* strains. The number of N-fixing root nodules / plant ranged from 15.5 for the control to 20.6 for *Glomus fasciculatum*. Nitrogenase activity increased from the control level of 386.2 to 683.4 n mol C_2H_4 / plant/ hour in *G. gilmorei*. Alkaline phosphatase was enhanced with inoculation from 13.0 n Kat/ 100g in the control to 18.8 n kat/100g in *G. epigeum*. Dry matter production and seed yield were significantly improved by inoculation although the different inoculants gave different results. They concluded that VAM inoculation may enable the exploitation of a large volume of soil water and nutrients.

Jha *et al.* (1993) studied the phosphate uptake by a tripartite association involving *Frankia, Glomus mosseae* and *Alnus nepalensis*. The response of non-mycorrhizal plants to P fertilization was more pronounced than that of their mycorrhizal counterparts. The yield of nodulated mycorrhizal plants was greater at low than at high P doses. At the highest P application, however, reduction in nodulation as well as nitrogenase activity was observed. Mycorrhizal infection was reduced by phosphate application, but markedly stimulated nodulation in *N. nepalensis* seedlings. The nitrogen content of the nodules was higher than that of the shoot of nodulated mycorrhizal plants. Nodule dry weight, nitrogenase activity and nodule nitrogen content increased with increase in P application to quantities typical of those in soil, after which it declined.

Tewari *et al.* (1993) studied host genotype dependency and growth enhancing ability of VA-mycorrhizal fungi for *Eleusine coracana* (finger millet). Inoculation of *E. coracana* plants with 6 different VAM fungi (*Glomus caledonicum, Glomus mosseae, Glomus fasciculatum, Glomus epigaeum [G. versiforme], Gigaspora calospora* and *Gigaspora margarita*) increased plant biomass, height, leaf area and absolute growth rate, but the effectiveness of various VAM fungi varied significantly. Maximum root colonization and mycorrhizal efficacy was observed in plants inoculated with *G. caledonicum*. Out of 5 host genotypes (HR-374, PES-110, Pe-4, PES-400 and PES-176) tested for mycorrhizal dependency against *G. caledonicum*, HR-374 gave the highest plant biomass, mycorrhizal efficacy and root colonization and inoculation resulted in increased mineral (phosphate, nitrogen, Zn^{2+} and Cu^{2+}) content and uptake in shoots.

Pai *et al.* (1994) studied the calcium uptake by cowpea as influenced by mycorrhizal colonization and water stress. They raised cowpea plants in pots containing soil inoculated with *Glomus fasciculatum*. The soil, maintained at different moisture levels, received known amounts of radioactive calcium. The radioactivity present in various parts of the plants was assessed after 60 hours. High Ca levels were present in all parts of VAM inoculated plants compared with non-mycorrhizal plants at all levels of moisture stress.

Mehrotra *et al.* (1995) studied the movement of ^{32}P in sunflower plants inoculated with single and dual inocula of VAM fungi. Sunflower plants were inoculated with single and dual inocula of *Glomus intraradix* and *Glomus mosseae*. ^{32}P-labelled phosphate was applied after sufficient infection was obtained in all treatments. Plants inoculated with dual inocula of the two VAM fungi showed significantly higher shoot weight than those inoculated with single inocula and control treatments. Of the ^{32}P transported to the shoot was 59.9 in *G. intraradix* + *G. mosseae*, 25.12 in *G. intraradix*, 33.31 in *G. moseae* and 35.2 in uninoculated control. The presnt study reveals that the increased growth in plants inoculated with dual inocula compared to those inoculated with single inocula of VAM fungi may be due to the increased transfer of P to the shoot from the root.

Mohandas *et al.* (1995) studied the uptake of ^{32}P labelled superphosphate by endomycorrhizal papaya (*Carica papaya*). In greenhouse-grown payaya *cv. Coorg Honey Dew* seedlings, inoculation with a mycorrhizal fungus (*Glomus mosseae* or *G. fasciculatum*) increased plant P uptake and total P content. The percentage of P derived from fertilizer was low in mycorrhizal plants than in controls, while soil-derived P in mycorrhizal plants, percentage utilization of applied P in mycorrhizal plants and soil available P in mycorrhizal soil were higher, indicating that mycorrhizal plants had utilized forms of P not available to non-mycorrhizal plants.

Tarafdar (1995) studied the visual demonstration of *in vivo* acid phosphatase activity of VA mycorrhizal fungi. Filter papers treated with a mixture of 1-naphthyl phosphate as substrate and diazonium salt Fast Red TR as an indicator were used for the nondestructive visual demonstration of the release of acid phosphatase by vesicular-arbuscular mycorrhizal (VAM) fungi. Wheat was grown for 6 weeks in sterilized soils in specially designed pots with 5 compartments with or without hyphal and root barriers. Plants were inoculated with *Glomus mosseae* and then treated filter paper was placed at the outer surface of the hyphal compartment. Acid phosphatase activity was visualized as a red coloured hyphal print on the filter paper.

Kumar and Arockiasamy (1995) studied the distribution of zinc in various parts of *Sorghum bicolor* seedlings in the presence of VAM fungi. Seedlings of Sorghum grown in a sterilized soil sand mixture were soil inoculated with *Glomus fasciculatum* After ensuring the establishment of VAM infection, zinc sulfate at -5 or 100 mM was applied. In inoculated seedlings more Zn was present in the roots and leaves. Root protein content was significantly greater in the presence of *G. fasciculatum*. Synthesis of new proteins that could bind the Zn and retain it in the roots may explain increased Zn tolerance in the presence of VA mycorrhizas.

Kumar *et al.* (1997) studied the alkaline phosphatase levels as indicator for screening VAM fungi in spring wheat. In screening and differentiating efficient symbiotic VAM for use under field conditions, experiments were conducted using six VAM fungi i.e. *Glomus versiforme, G. mosseae, G. monosporum, G. intraradices, G. aggregatum, Acaulospora* and wheat variety '*Moskovskaya*-35'. *Glomus versiforme* showed higher alkaline phosphatase activity and also gave higher mycorrhizal index (35.1) compared to the other VAM isolates.

Sitaramaiah and Khanna (1997) studied the effect of *G. fasciculatum* on the growth and chemical composition of the maize grown in low and high phosphate soils maintained under sterilized and unsterilized conditions, in a greenhouse. The results revealed that mycorrhizal plants produced greater vegetative growth with higher level of chemical constituents than non-mycorrhizal control plants. The endomycorrhizal fungus significantly increased root and shoot weight, root volume, and chemical constituents (nitrogen, phosphorus, potassium, calcium and magnesium) when maize was grown in sterilized, low phosphate soil compared with unsterilized high phosphate soil.

Purkayastha *et al.* (1998) studied the growth and iron nutrition of broccoli (*Brassica oleracea* var. *italica*) grown in a Typic Ustochrept, as influenced by VAM fungi in the presence of pyrite and farmyard manure. Individual applications of either VAM or pyrite with NPK fertilizer significantly enhanced both the Fe^{2+} content of leaves and total uptake of Fe and resulted in increased yield compared with those observed with NPK alone. Though the application of FYM decreased the Fe^{2+} content of leaves relative to plants supplied with NPK alone, this result was not statistically significant. The available Fe content of the soil, after harvest of broccoli, was lower in the presence of VAM than in the control.

Sharma *et al.* (1999a) studied the uptake and translocation of phosphorus (^{32}P) by mycorrhizal (*Glomus macrocarpum*) and non-mycorrhizal *Sesbania aculeata* L. in a short-term solution culture experiment at low (2×10^{-4} k mole P m^{-3}) P concentrations. Uptake in mycorrhizal *Sesbania* was more sensitive to 2,4 dinitrophenol, especially under low P supply. Total P uptake in plants at different time intervals conformed to a Langmuir type equation. The calculated maximum P uptake capacity of the mycorrhizal plants was 1.83 and 1.61-fold higher than the non-mycorrhizal plants under low and high P supply, respectively. During the early duration (0.25 h), P accumulation by root and its translocation to stem was greater and more metabolically-dependent in mycorrhizal than non-mycorrhizal plants. This effect was more evident under low P supply. At a later phase (8h), P accumulation in roots of mycorrhizal plants remained highly metabolic but the translocation to stem and leaves became much less dependent on cell metabolism.

Sharma *et al.* (1999b) investigated the kinetics of Zn uptake by mycorrhizal (*Glomus macrocarpum)* and nonmycorrhizal roots of French bean (*Phaseolus vulgaris* L.) employing ^{65}Zn. At extremely low concentration (below 0.05 mmol m^{-3}), Zn uptake followed a sigmoidal pattern. In the concentration range, 0.1 mmol to 100 mmol m^{-3}, four other distinct uptake phases each one following Michaelis-Menten kinetics were observed. Mycorrhizal roots exhibited generally higher Zn absorption rate than the nonmycorrhizal roots; this was attributed to higher maximal uptake rate (V_{max}) for phases 1, 2 and 3. The greater specificity of phase 4 was result of lower k_m value.

Harikumar and Potty (1999) studied the performance of mycorrhizal and non-mycorrhizal sweet potato plants in the uptake of P in a P fixing acid laterite soil pots. The study revealed that the fixation and removal of P in the laterite soil due to iron and aluminium induced P deficiency could be rectified to certain extent by inoculation of mycorrhiza.

Khaliq *et al.* (1999) investigated the influence of *Glomus mosseae* on growth and uptake of mineral nutrients and essential oil contents of menthol mint. The association of vesicular-arbuscular mycorrhizal fungi *Glomus mosseae* was observed to influence the growth of menthol mint (*Mentha arvensis*) when grown on sterilized as well as unsterilized soil amended with or without farm yard manure (FYM). Benefits from mycorrhization was invariabilily favoured by sterilized soil without FYM amendment and showed marked increase in shoot length (64.3%), shoot fresh weight (172.9%), shoot dry weight (165.9%) and essential oil content (6.8%) as compared to other treatments. Maximum root colonization (91%) and chlamydospore population in rhizosphere soil (676 spores/ 100 g soil) was found in the inoculated plants grown in sterilized soil with FYM amendment. Uptake of P, K and

Zn was found maximum in inoculated plants grown in sterilized soil without FYM amendment. The investigations thus indicated the potential use of efficient VA-mycorrhizal fungi for improving growth and productivity of mints.

Bhadraiah *et al.* (1999) studied the effect of VAM fungi and rock phosphate on phosphatase activities in *Terminalla arjuna*. Seedlings of *Terminalla arjuna*, grown in polythene bags in sterilized soil were treated with *Glomus mosseae, G. fasciculatum* and rock phosphate separately and also in various combinations. The root and shoot tissues of plant were analysed for the acid and alkaline phosphatases. Acid phosphatase activity increased to a maximum in Gf treated roots followed by Gm+P and Gm+Gf treated roots while in shoots the activity was maximum only in Gm+P treated plants. All other combination reduced acid phosphatase activity. Alkaline phosphatase activity was considerably lower than the acid phosphatase activity. Alkaline phosphatase activity was maximum in Gm+P treated shoots followed by P treated plants. Positive correlation was noted between acid phosphatase activity and P concentration.

Senthilkumar *et al.*(1999) studied the role of vesicular arbuscular mycorrhiza in nitrate reduction system and zinc uptake in *Zea mays.* Maize (*Zea mays*) seedlings were inoculated with *Gigaspora margarita* and heavy metal stress was induced with zinc sulphate at 50 and 100 mm concentrations after ensuring the establishment of the fungus. In general, nitrate reductase activity and protein content were higher in all the mycorrhizae colonized seedlings. Zinc content was estimated both in leaves and roots of infected and non-infected seedlings. 30.5% of heavy metal accumulation was found in the roots and about 69.49% in the leaves in the absence of Vesicular Arbuscular Mycorrhizae (VAM), whereas in the VAM colonised seedlings about 40.1% was accumulated in the leaves and 59.85%in the roots point out to the retention of the observed heavy metal in the roots. This may indicate the synthesis of newer proteins that could bind the heavy metal and retain it in the root, a possible mechanism of heavy metal resistance.

Ramesh *et al.* (2000) elucidated the chemical compositions of inter and intracellular hyhae in *Pennisetum pedicellatum* roots colonized by *Gigaspora margarita* by employing energy dispersive X-ray microanalysis. Maximum amount of phosphate was estimated in polyphosphate granules present in the roots. Amount of phosphate was low in fungal hyphae, arbuscular trunks and degenerated arbuscules. Elements S, K and Ca were present in the hyphae, arbuscular trunk and degenerated arbuscules.

Rupnawar and Navale (2000b) studied the effect of VAM inoculation to pomegranate layers with respect to N and P uptake. In a pot culture experiment conducted using sterile, P-deficient soil and 150 g vesicular arbuscular mycorrhizal fungi per pot, observations made 90 and 180 days after planting revealed that mycorrhizal treatments (3 species applied singly and in combination) were superior to non-mycorrhizal treatments of pomegranate layers. *Glomus mosseae* alone and a *Glomus epigaeum [G. versiforme]* + *Glomus mosseae* + *Gigaspora calospora* mixture recorded the highest values for root colonization and N and P uptake.

Pukakayastha and Chhonkar (2001) studied the influence of vesicular-arbuscular mycorrhizal fungi (*Glomus etunicatum* L.) on mobilization of zinc in wetland rice (*Oryza sativa* L). In a greenhouse trial, rice (*Oryza sativa*) cultivar Pusa Basmati-1 was grown in a Zn-deficient Typic Ustochrept soil from IARI farm, New Delhi. The experimental design included two rates of inoculation with vesicular arbuscular mycorrhizal fungi (VAMF), *Glomus etunicatum*(nil and inoculated) and different combination of organic (farmyard manure, FYM) and inorganic (NPK, $ZnSO_4$) fertilizers. The results revealed that a high intensity of root colonization in rice inoculated with *G. etunicatum* could be achieved by raising seedling in P-and Zn-deficient soil in the nursery under aerobic conditions. Moreover, the VAMF that infected rice seedling in the nursery also survived when the seedlings were transplanted into pots under waterlogged conditions. The application of $ZnSO_4$ significantly increased the inflow of Zn to rice roots at the panicle-initiation stage (40 days after transplanting) relative to NPK. The former treatment increased root length, root weight, root volume and total uptake of Zn and thereby increased the grain and dry matter yields. Alternatively, these variables were substantially enhanced by inoculating rice with the VAMF, *G. etunicatum*. The VAMF-colonized rice plants were more active in acquiring Zn from either added or native sources than non-colonized plants, and consequently the available-Zn content in soil was lower after the harvest of rice.

Uptake of Water

Ramakrishnan *et al.* (1988b) studied the influence of VAM fungus *Glomus caledonius* on free proline accumulation in water stressed maize. They planted the maize seeds in sterilized or unsterilized soil and inoculated with *G. caledonius* or a mixture of indigenous inoculum, prepared from P-deficient soil, and *G. caledonium* to give the same inoculum potential. Uninoculated and P-amended pots served as controls. With sterile soil, 5 week-old plants exhibited 48% indigenous infection compared with 32% in the VAM + *G. caledonium* treatment. In unsterile soil, root colonization was 85% and 26% for the two treatments, respectively. At osmotic potential of 0, -2, -5 and -10 bar, leaf potentials were C -3, -5, -8 and -12 bar, respectively for both treatments. They reported a marked increased free proline content in all the cases, with a decrease in leaf water potential. This was greatest in P-amended plots, followed by VAM and then indigenous VAM +*G. caledonium* treatments. According to them all these 3 treatments may confer greater drought resistance on maize than the non-VAM control.

Ramakrishnan *et al.* (1988a) examined the influence of introduced vesicular arbuscular mycorrhizal fungus *Glomus caledonius* (Nicol. and Gerd.) Gerdemann and Trappe and that of the indigenous population (IVAM) in osmotically stressed maize grown in sterile/unsterile soil, with and without P-additions. The leaf water potential at various osmotic stress was nearly of the same order in both mycorrhizal and non-mycorrhizal maize plants. Photosynthetic and photorespiratory activities decreased with the increasing osmotic stress; the chlorophyll content, however, did not yield any significant change. There was no significant increase in photosynthesis under different treatments and soil conditions in unstressed maize but it increased significantly in VAM and P-amended stressed plants. Under non-stressed conditions photorespiration was considerbly greater in VAM and IVAM

+ *G. caledonius* (Gc) treated plants over other treatments. Some increase in chlorophyll a and b contents was observed in VAM, IVAM +Gc and P amended unstressed plants.

Ramakrishnan *et al.* (1990) studied the response of mycorrhizal maize plants to variations in water potentials. 5-week-old mycorrhizal and non-mycorrhizal plants were exposed to osmotic potentials ranging from -3 to -12. Net photosynthesis increased insignificantly in both mycorrhizal and non-mycorrhizal plants at 0 to -5 osmotic potentials. At very low osmotic potential (-10 bar) there was a significant increase in CO_2 assimilation by VAM plants. Although stomata of VAM plants began to close earlier if leaf water potential reache the threshhold level for closure, no change was observed in leaf water potentials in the two groups of plants to indicate a mycorrhizal influence on stomatal regulation.

Panwar (1993) studied the response of VAM and *Azospirillum* inoculation to water status and grain yield in wheat under water stress condition. In a pot experiment, wheat cv. HD-2428 seeds were inoculated with *Glomus fasciculatum* and /or *Azospirillum brasilense* and were water stressed between 70 and 80 days after sowing or not stressed. The relative leaf water content, leaf area, total grain yield, chlorophyll and nitrate reductase activity increased and the leaf ion leakage decreased with the combined inoculation with *G. fasciculatum* and *A. brazilense* compared with single inoculations, resulting in higher biomass production and grain yield under water stressed conditions. Lower grain yields were obtained from the non-inoculated control treatments. Yield from water stressed and non-water stressed plants were similar when seeds were inoculated with both *G. fasciculatum* and *A. brasilense* while in the uninoculated plants, water stress decreased yield.

Nitrogen metabolism

Sunderesan *et al.* (1988) studied the nitrate-reducing capacity of 10 VAM isolates from Tamil Nadu, and of *Glomus fasciculatum* and *G. aggregatum*. Reduction was highest in isolates MKU7, KK2 and *G. fasciculatum*, moderate in MKU4, MKU7 and *G. aggregatum*, least in CK2, CK3 and CK4 while isolates Tm2, Tm6 and Tm 8 showed minimal activity.

Varma (1988) studied the growth and some aspects of nitrogen metabolism of *Gigaspora gigantea*. Hyphal growth from spores of *Gigaspora gigantea* was stimulated by glycine, lysine, glutamine and cystine but repressed by methionine. GS/GOGAT (glutamine synthetase/synthase) pathway which appeared to be operative in contrast to the fact that yeast and fungi do not utilize G3/GOGAT pathway for nitrogen assmilation. The regulatory role of GS and cystine has been discussed.

Tilak *et al.* (1989) studied the occurrence of nitrogen-fixing *Azospirillum* in vesicular-arbuscular mycorrhizal fungi. Nitrogenase activity measured by acetylene reduction was detected when surface-sterilized spores of vesicular-arbuscular (VA) mycorrhizal fungi (*Glomus fasciculatum, G. intraradices, G. scientillans, G. mosseae, Gigaspora gilmorei* and *Endogone dusii*) were inoculated into nitrogen-free liquid medium containing malic acid and incubated under microaerophilic conditions (99% N_2 + 1% O_2) at 30^{0}C. *Azospirillum* spp. were isolated from the nitrogenase-active cultures.

Singh (1990) studied the role of VAM fungi inoculation, in association with *Rhizobium* sp. on symbiontic process, especially on N_2 fixation, in *Cajanus cajan* legume, using ^{15}N isotopes dilution technique under soil potted condition. Nodulation, shoot and grain yield of pigeon pea (*C. cajan*) was significantly increased due to *Rhizobium* sp. and *Glomus fasciculatus* (VAM) inoculation especially at 50 kg P_2O_5 ha^{-1} application. VAM alone enhanced the nodulation of pigeon pea. Using ^{15}N dilution technique, total N yield of the crop, utilization of soil nitrogen and biologically N_2 fixation was more with VAM inoculation, which was at par with *Rhizobium* sp. inoculation. N_2 fixed in grain and straw was higher due to combined effect of *Rhizobium* sp. and VAM, as compared to their individual effect. It was more pronounced in presence of 50 kg P_2O_5 ha^{-1} and 20 kg N ha^{-1} in the form of superphosphate and ammonium sulphate respectively. Increase in total N yield of shoot and grain could be due to increase in translocation of soil nitrogen to plant mediated by VAM, resulting in improvement in nutrition of the crop plant. The higher amount of translocation of soil N and P is positively correlated with the intensity of VAM root colonization and symbiotic parameters.

Tilak and Dwivedi (1990b) studied the nitrate reductase activity of VA-mycorrhizal fungi. Extramatrical chlamydospores of *Glomus fasciculatum, G. mosseae, G. intraradices, G. caledonium, Gigaspora margarita, G. calospora, Endogone discii* and *Acaulospora* sp. were collected from single species pot cultures maintained with *Cenchrus cilliaris* using wet seiving and decanting technique. The results revealed that all the VAM fungal spores exhibited the property of nitrate reducing ability and it varied from 1.5 μ moles to 3.8 μ moles/ tube 24 hours. The maximum nitrate reductase activity was noticed with *Glomus caledonicum* and the lowest activity was recorded with *Acaulospora* sp.

Thapar *et al.* (1990) studied the effect of foliar application of urea on nitrogen metabolism of mycorrhizal moong plants (*Vigna radiata*), under varying phosphorus levels. In pot experiments, the plants inoculated with *Glomus fasciculatum* showed increases in leaf nitrate reductase, glutamine synthetase activities and ammonia concentration at P0 and P_10 levels equal to those at the P_20 level in non-mycorrhizal plants. The increased glutamine synthetase activity and total amino acid concentrations suggested that foliar applied urea was assimilated into amino acids.

Mathur and Vyas (1995d) studied the changes in enzymes of nitrogen metabolism by VA mycorrhizae in *Zizyphus nummularia*. They have shown that mycorrhizal plants had increased activities of glutamine synthetase, nitrate reductase and glutamine dehydrogenase, and increased soluble protein concentrations in leaves and roots. Among the 5 VAM species used *Glomus fasciculatum* had the greatest effect on *Z. nummularia* metabolism, resulting in a leaf soluble protein concentration of 14.12 g/ 12 kg fresh weight. It has been suggested that *Z. nummularia* could be useful for producing highly proteinaceous leafy fodder.

Carbon Metabolism

Singh and Tilak (1990a) studied the translocation and partitioning of assimilated carbon using radioactive CO_2 in mycorrhizal and non-mycorrhizal sorghum plants. According to them, the establishment of VA mycorrhizal infections of *Glomus versiforme* in sorghum roots altered the distribution of C assimilates in the ethanol soluble and acid digested fractions of leaves and roots compared to non-mycorrhizal plants. Greater amounts of C were translocated from shoot to root of mycorrhizal plants without decreasing the leaf area.

Potty and Indira (1990) studied the influence of vesicular-arbuscular mycorrhizae on the photosynthesis and photorespiration of sweet potato (*Ipomoea batatas*). They reported that the growth of mycorrhizal plants was better than that of non-mycorrhizal ones under conditions of low or medium P content in a pot experiment. Mycorrhizal plants showed an increase in both leaf area and dry matter and higher rates of CO_2 fixation.

Prasad and Bilgrami (1995) studied the impact of VA mycorrhizae and phosphates on biomass, yield and sucrose content of *Saccharum officinarum*. Soil inoculation was done with *Glomus fasciculatum*. They recorded more plant biomass, i.e. dry weight, stem height, stem thickness, number of leaves and nodes in VAM inoculated plants. The data also reflected that the metabolic activities associated with juice and sucrose synthesis was higher in VAM inoculated plants than the addition of different sources of phosphates. The joint effect of VAM and fertilizer was always superior to the effect of VAM or fertilizer alone.

Mathur and Vyas (1995c) studied the influence of VA mycorrhizae on net photosynthesis and transpiration of *Zizyphus mauritiana*. Five VAM species *Acaulospora morrowae, Gigaspora marganita, Glomus deserticola, Glomus fasciculatum* and *Scutellospora calospora* were selected for the greenhouse experiments under sterilized soil. According to them, VAM inoculation resulted in a significant increase in net photosynthesis, total chlorophyll and carotenoid concentrations and sugars and starch concentrations. The VA mycorrhizas also increased stomatal resistance, thereby reducing the rate of transpiration.

Singh and Tilak (1995) studied the effect of *Glomus versiforme* on the distribution of photoassimilates in sorghum. A fully expanded leaf of six-week old plant of sorghum with $7.4x10^2$MBg of $^{14}CO_2$, the plants were transferred inside a greenhouse for a chase period of 24 hours. After 24 hours, plants were harvested and grouped into (i) labelled leaf (ii) leaves other than labelled leaf (iii) stem and (iv) roots. They observed that ^{14}C-assimilates were distributed from the labelled leaf to other leaves, stem and roots of plants within 24 hours-chase period. Mycorrhizal plants translocated 6% more amount of ^{14}C-assimilation from shoots to roots as compared to the non-mycorrhizal plant. Their observations on the distribution of ^{14}C-activity in different chemical fractions of plant parts showed that maximum amount of activity was present in the ethanol-soluble fraction of the labelled leaf (62 to 46% of the total activity present in the leaves of mycorrhizal and non-mycorrhizal plants). In roots, there was 51% and 62% of the total ^{14}C activity present in the ethanol-soluble fraction of mycorrhizal and non-mycorrhizal plants.

Thakur and Panwar (1997) studied the response of *Rhizobium*- VA mycorrhizal symbionts on photosynthesis, nitrogen metabolism and sucrose translocation in greengram (*Phaseolus radiatus*). In a field experiment, greengram cv. PS 16 and 105 were seed inoculated with *Rhizobium* and / or soil inoculated with *Glomus fasciculatum* or not inoculated. *Rhizobium* inoculated plants had larger area (9.10%), higher chlorophyll content (11.4%) and photosynthesis activity (21.6%), transpiration (15.89%) and stomatal conductance

(19.35%) than the uninoculated plants. Inoculation increased nodulation, nitrate reductase and nitrogenase activities. These parameters were highest in the plants inoculated with *Rhizobium* + *G. fasciculatum*. During the vegetative stage, the dual inoculated plants translocated a significantly higher percentage of the assimilates to roots and nodules compared with the other treatments. Pusa 105 had higher activities of the parameters studied than PS 16.

Muthukumar and Udaiyan (2000) examined the interaction of plant nutrients, root-soluble carbohydrate availability and arbuscular mycorrhizal (AM) fungi in field-grown crowpea (*Vigna unguiculata*) at a site in Tamil Nadu. Plant nutrients were altered through application of cow dung, sheep manure, and green manures (sunnhemp, *Pongamia*). Organic amendments increased plant growth, AM fungal colonization, soluble carbohydrate concentration in roots, and spore numbers. Percent total colonization, root length with vesicles and spore numbers in soil were negatively correlated with the concentration of soluble carbohydrates within roots, which in turn were related to tissue nutrient levels. However, a positive correlation existed between soluble carbohydrate concentrations within root and root length with arbuscules. But the mycorrhizal parameters were related more to plant nutrient level and their ratios, indicating that tissue nutrients have another level of control in addition to their effect on soluble carbohydrate concentration in roots. Increased AM colonization due to organic amendment significantly reduced nutrient imbalances. The strong relationship between colonization and root-soluble carbohydrate concentration levels validates the basic assumption that mycorrhizal fungi act as a 'strong sink' for photosynthate. This study indicates that the host influences AM colonization by regulating the formation of AM fungi structures and spore formation via availability of root carbohydrate.

Gautam and Prasad (2000) conducted a field experiment to determine the effect of *Glomus macrocarpum* on the fiber strength properties and production of holocellulose, hemicellulose, and cellulose by *Dendrocalamus strictus* under field conditions. VAM-inoculated plants had higher growth than the control (uninoculated). VAM-inoculated plants had increased fibre diameter (48.99%), fibre length (40.00%), and concentrations of holocellulose(21.59%), hemicellulose (28.24%) and cellulose (19.46%) in 4-year-old plants.

Miscellaneous

Krishna and Bagyaraj (1986) studies the phenolics of mycorrhizal and uninfected groundnut var. MGS-7. Soil was inoculated with *Glomus fasciculatum* chlamydospores and infected segments of *Panicum maximum* before sowing. Total phenol in mycorrhizal groundnt roots was generally greater than that found in uninfected roots. On the 10th day after inoculation total phenol concentration was very high and decreased during 10-20 days.

Doss *et al.* (1988) studied the anatomical and histochemical changes in the roots and leaves of finger millet colonized by VA mycorrhiza. They reported that the number of starch grains as well as levels of RNA and insoluble proteins in the root cap cells of finger millet plants infected by *Glomus fasciculatum* were higher than in the root cap cells of non-mycorrhizal plants. *G. fasciculatum* colonized only cortical region of the root and was not observed in the steler portion. Leaves of mycorrhizal plants showed an increase in the thickness, size of midrib veins, major veins, minor veins, mesophyll cells, motor cells and the number of plastids.

Tilak *et al.* (1988) made characterization of nitrogen-fixing *Azospirillum* isolated from within sporocarps of three endomycorrhizal fungi associated with Douglass - fir: *Hebeloma crustuliniforme, Laccaria laccata* and *Rhizopogon vinicalor*. Each strain possessed characteristics typical of *Azospirillum* and all closely resembled *A. brasilense*. They exhibited maximum nitrogen fixing ability (acetylene-reducing activity) at pHs ranging from 6.5 to 8.5 and at temperatures between 30 and 35°C when grown in N-free, semi-solid sodium malate medium. The presence of combined nitrogen (0.1% NH_4Cl) suppressed acetylene reduction by all three strains.

Selvaraj and Subramanian (1990b) studied phenols and lipids in mycorrhizal and non-mycorrhizal roots of *Sesamum indicum*. Mycorrhizal inoculation resulted in a significant enhancement of the percentage infections, dry matter and phenolic content of sesamum plants in sterilized soils. Histochemical studies revealed stimulatin of different types of lipids and phenolic compounds in VAM structures; particularly neutral lipids and catechal tannins in vesicles. It is suggested that reports on increased levels of lipids and phenols in mycorrhizal plants could be due to the contribution of fungal structures.

Mada and Bagyaraj (1993) studied the root exudation from *Leucaena leucocephala* in relation to mycorrhizal colonization. According to them, lower amounts of root exudates (13 mg/g dry root) were produced by leucaena plants inoculated with *Glomus fasciculatum* than by uninoculated plants (21 mg/g dry root). Mycorrhizal plants exuded less K^+, Pi and sugar (mainly glucose) but more protein, nitrogen, phenolics and gibberellins than uninoculated plants. Glyceine, alanine, cysteine, arginine, tryptophan and valine occurred only in the root exudates of the former. Uninoculated plants exuded more of a root elongation inhibitory substance than the inoculated ones, while the concentration of giberrellins was lower.

Bansal and Mukerji (1994) studied positive correlation between VAM-induced changes in root exudation and mycorrhizosphere mycoflora. The influence of vesicular-arbuscular mycorrhizal (VAM) fungi (*Glomus macrocarpum*) on rhizosphere mycoflora was studied together with the possible mechanism involved in this process. Six combinations of VAM fungi and phosphorus fertilizer treatments were applied to *Leucaena leucocephala* roots and quantitative and qualitative observations were made periodically of the rhizosphere mycoflora and constituents of root exudates. The results obtained indicated that the presence of specific mycoflora in the rhizosphere of mycorrhizal roots was mediated through root exudates rather than being an outcome of improved P nutrition.

Harinikumar and Bagyaraj (1995) studied the spread of vesicular arbuscular mycorrhizal fungal hyphae in soil. Active spread of *Glomus fasciculatum* hyphae from the roots of inoculated plants through soil was studied using rectangular battery box filled with soil and partitioned into 3 compartments. Mycorrhizal *Leucaena leucocephala* and non-mycorrhizal *L. leucocephala* were raised in the side compartments. No plant was raised in the middle compartment. Care was taken to prevent root growth from one compartment to the other.

The active spread of the external hyphae from inoculated plant through soil was monitored at periodical intervals by taking soil cores different distances and testing for mycorrhizal colonization. The results showed that mycorrhizal hyphae grew to a distance of 300 mm in 180 days. The uninoculated plants became mycorrhizal after 180 days, confirming the active spread. It was deduced that VAM hyphae spread in soil (without roots) at the rate of 1.66 mm/day.

Mathur and Vyas (1995a) studied the changes in isozyme patterns of peroxidase and polyphenol oxidase by VAM fungi (*G. fasciculatum*) in roots of *Zizyphus* species *viz., Z. mauritiana, Z. nummularia* and *Z. xylopyra* grown under greenhouse conditions. After 60 days of inoculation, when compared with untreated plants, all inoculated plants showed two additional peroxidase isoenzyme bands and one additional polyphenol oxidase band, both showed increased activities also.

Panwar and Thakur (1995) carried out the physiological and biochemical studies in VA-mycorrhizae, *Rhizobium* inoculated mungbean plants under field conditions. Data collected on different aspects related to leaf area, chlorophyll content, photosynthetic rate and nitrogenase activity showed that higher activity was in dual inoculated plants as compared to any other single inoculation. In addition to P-uptake the enhanced uptake of K, Fe, Mn, Zn and Cu were found in the VAM inoculated plants. The host microbial interaction was found better in dual inoculated plants.

Kanakadurga and Rama Rao (1995) studied the effect of VA mycorrhizae on root and shoot phenol concentrations of *Terminallia arjuna*. They used *Glomus mosseae* and *G. fasciculatum* as VAM inoculants, alone as well as in combination. Rock phosphate was also supplied alone and in combination with VAM fungi. Data showed that phenols decreased in the root system of the efficient P-transporter-Gm+P. There was increase in the shoot phenol concentrations in Gm+P and Gf which were efficient P-transporters and also had high infection rates. The shoot/root levels of phenols were higher in all the treatments when compared to control.

Raman and Elumalai (1995) studied the effect of mycorrhizal and frankial inoculation in *Casuarina equisetifolia* and production of indole-3 acetic acid. They inoculated the seedlings with *Glomus fasciculatum*, ectomycorrhizal fungus - *Pisolithus tinctorius* and *Frnkia* separately and in combination. They recorded a higher amount of IAA in the nodules of *G. fasciculatum,* and *Frankia* inoculated seedlings than that of the plants inoculated with *Frankia* alone. The VAM and *Frankia* combinatioin gave a tremendous increase in the dry weight over the frankial inoculation.

Mathur and Vyas (1996a) studied the influence of VAM fungi namely *Acaulospora morrowae, Gigaspora margarita, Glomus fasciculatum, G. mosseae, Scutellospora calospora* etc. on the physiological changes in *Prosopis cineraria*. Their investigation revealed that in general, the nutrient contents were higher in VAM plants as compared to the control. However, different VAM species varied in their efficacy to increase nutrient level in *P. cineraria*. Total protein contents in VAM plants ranged from 3.12 to 6.92 mg/ 100 mg dry weight as compared to 2.16 mg/100 mg of control. Out of five VAM species used, *S. calospora* increased the protein content most efficiently. The increased total chlorophyll contents and carotenoids were also noticed in VAM inoculated plants. Likewise, increase in total sugar contents and starch content was recorded more in VAM inoculated plants. *S. calospora* increased all the parameters two to three folds than other VAM fungi, hence, they recommended it for *P. cineraria* for producing highly nutritive leafy fodder.

Mathur and Vyas (1996b) studied the physiological changes in *Zizyphus mauritiana* by different VAM fungi. They investigated the relative efficiency of seven VA mycorrhizal species *viz., Glomus fasciculatum, G. mosseae, G. constrictum, Scutellospora calospora, Gigaspora margarita, Acaulospora morrowae* and *S. rubifermis*, collected from rhizosphere soils of *Z. mauritiana* from arid and semiarid regions of the country, as inoculum for potted seedlings. Measurements were made of seedling biomass production, nutrient uptake and contents of sugar, starch, chlorophyll, protein and carotenoids after 150 days. Root colonization and soil spore contents of the mycorrhizae were also recorded. They reported *G. fasciculatum*, the best inoculant, followed by *G. mosseae*. These VAM fungi gave the highest biomass production and nutrient uptake and colonized the roots most effectively. They suggested that they may be used in the production of highly nutritious leaf folder by *Z. mauritiana*.

Ratti and Janardhanan (1996a) studied the effect on growth, phosphorus and cytokinin contents of palmarosa (*Cymbpogon martinii*) by *Glomus* inoculation *viz., G. aggregatum, G. fasciculatum, G. intraradix* and *G. mosseae*. The inoculation significantly increased shoot dry weight, P content of root and shoot tissue and cytokinin production in palmarosa (*Cympopogon martinii*) roots as compared to uninoculated control plants. Per cent root colonization and number of chlamydospores were positively correlated with the dry weight of shoot, root and shoot P content and cytokinin content. Variation in infectivity and effectiveness among the *Glomus* species was observed in this study. Of the four, *G. aggregatum* and *G. fasciculatum* had superior symbiotic relationship to palmarosa than others.

Mathur and Vyas (1997) studied the enzymatic changes in cluster bean by different VAM fungi. *Cyamopsis tetragonoloba* was soil inoculated with *Glomus fasciculatum, G. mosseae, Gigaspora margarita* or *Scutellospora calospora*. Enzymes were assayed in homogenized roots. Mycorrhizal inoculation increased activity of peroxidase and polyphenol oxidase, particularly in plants inoculated with *Glomus mosseae*.

Raman and Elumalai (1998) studied the Gibberellic acid in mycorrhizal and actinorhizal inoculated *Casuarina equisetifolia. C. equisetifolia* seedlings were raised and inoculated with *Frankia, Glomus fasciculatum* and *Pisolithus tinctorius*, individually and in combination. The roots, nodules and cladodes were collected and analysed for the presence of giberrellic acid. Significantly increased GA was detected in the triple inoculated plants of *C. equisetifolia*.

Joseph *et al.* (1999) studied the effect of different isolates of VAM on growth nutrient uptake and biochemical constituents of *Pueraria phaseoloides* Benth. *P. phaseoloides* seeds inoculated with *Bradyrhizobium* species were sown in polythene bags containing 11 mycorhizal inoculants isolated from rubber growing soils. After 50 days of inoculation, plants were uprooted and studied for various growth parameters, nitrogenase activity and biochemical constituents. Among the 11 mycorrhizal inoculants, *G. fasciculatum* recorded maximum percent of root colonization and this was at par with *A. laevis, G. fasciculatum* and *A. laevis*. Though all the isolates augemented the shoot lenght and biomass, *G. fasciculatum* was significantly superior to other species followed by *A. leavis*. *G. fasciculatum* also significantly recorded an increase in nodule number, nodule weight and nitrogenase activity. Higher levels of phenols, amino nitrogen, sugars and starch were recorded in *G. fasciculatum* treated plants followed by *A. leavis*.

Gautam and Prasad (1999) studied the impact of *Glomus macrocarpum* VAM fungus on proximate chemical composition of *Dendrocalamus strictus* under field conditions. The VAM inoculated plants were superior over uninoculated control. VAM inoculated and uninoculated *D. strictus* plants were transplanted in experimental field in 1993 and data were recorded after 32 months of plant growth under field conditions. The proximate chemical composition of *D. strictus* increased significantly in VAM inoculated plants than control plants. The per cent increase was recorded for cold water solubiles 35.60%-47.16%, hot water solubiles 13.22%-42.86%, alcohol solubiles 20.00%-30.65%, alcohol benzene solubiles 10.77-15.49% and NaOH solubiles 12.94%-14% over control plants.

Tandon and Prakash (1999) studied the role of VAM fungi in distribution and uptake of heavy metal in *Sesamum indicum* seedlings. Indigenous mycorrhizal inoculum was used to colonize *Sesamum* seedlings. After ensuring the establishment of infection, heavy metal treatment was provided at different concentrations. The uptake and distribution of the heavy metal in roots and leaves was estimated in the presence and absence of mycorrhiza. The mycorrhizal roots showed a significant increase in the heavy metal accumulation in comparison to the non-mycorrhizal roots. There was a significant increase in root protein content in mycorrhizal plants as compared to the non-mycorrhizal plants at all the concentrations of heavy metal treatment. Heavy metal content in the root increased significantly in *Sesamum* due to the association with mycorrhizal fungi thus suggesting the role of the root protein in the retention of heavy metal in roots of infected plants.

Thippeswamy and Sreenivasa (1999) conducted a pot experiment at Dharwad during kharif 1996 to determine the physiological, nutritional and yield responses of tomato to the inoculation of *G. fasciculatum* under water stress conditions (withholding water for 5 days) imposed at 25, 50 and 75 days after transplantation (DAT). It was observed that percentage root colonization and spore counts were higher in the rhizosphere soil of inoculated tomato plants. Inoculation of *G. fasciculatum* resulted in higher relative leaf water content, chlorophyll content and a decrease in proline content. The dry weights of shoots and roots, plant height, number of fruits and fruit weight were higher in inoculated plants as compared with uninoculated plants under water stress conditions. Water stress imposed at 50 DAT was found to be critical and detrimental to growth and yield compared with stress imposed at an early (25 DAT) or later (75 DAT) stage of crop growth.

Ramesh *et al.* (2000) elucidated the chemical compositions of inter and intracellular hyhae in *Pennisetum pedicellatum* roots colonized by *Gigaspora margarita* by employing energy dispersive X-ray microanalysis. Maximum amount of phosphate was estimated in polyphosphate granules present in the roots. Amount of phosphate was low in fungal hyphae, arbuscular trunks and degenerated arbuscules. Elements S, K and Ca were present in the hyphae, arbuscular trunk and degenerated arbuscules.

Mathur and Vyas (2000) studied the effects of *Gigaspora margarita, Glomus constricutm, G. fasciculatum. G. mossea, Sclerocystis rubiformis* and *Scutellospora calospra* on nutrient uptake and physiology in *Ziziphus mauritiana* seedling under water stress were studied in a pot under experiment. Increase in amino acids, protein, chlorophyll and sugar contents resulted from mycorrhizas. Mycorrhizas had a marked effect on *Z. mauritiana* under water stress conditions.

Muniyamma *et al.* (2000) tested VAM fungal spore extract and VAM root extract as substitutes to the rooting hormone NAA in cultures of sugarcane variety Co.419 in Murashige and Skoog (MS) liquid medium. The results showed that rooting response was enhanced with either VAM fungal spore extract of VAM root extract, compared to NAA.

Panwar and Vyas (2000) investigated the effects of inoculating 8 arbuscular mycorrhizal (AM) species *(Acaulospora mellea, Gigaspora margarita, Gigaspora gigantean, Glomus deserticola, Glomus fasciculatum, Sclerocystis nibiformis, Scutellospora nigra* and *Scutellospora calospora*) on the peroxidase and polyphenol oxidase activities in the roots of *Tamarix aphylla* under greenhouse conditions. The activities of these enzymes were higher in AM-inoculated plants than in the uninoculated plants. The highest increase in enzyme activities was observed in plants inoculated with *Scutellospora calospora.*

Devi and Reddy (2002) reported that mycorrhiza and *Rhizobium* inoculation resulted in a significant increase in the quantities of phenolic compounds in roots and shoots of groundnut (cv. TMV2) plants compared to uninoculated plants. The production of some new compounds, and the increase in the activity of peroxidase and polyphenol oxidase [catechol oxidase], following inoculation was observed. The accumulation of phenolics and the activation of oxidative enzymes were more pronounced in the dual-inoculated plants.

Bhadraiah *et al.* (2002) studied the effect of AM fungi and rock phosphate on phosphatase activities in *Tectona grandis* L. Seedlings of *T. grandis* grown in polythene bags in sterilized soil were treated with *Glomus mosseae, G. fasciculatum* and rock phosphate (P) separately and also in various combinations. The root and shoot tissues of treated plants were analysed for the acid and alkaline phosphotases. Acid phosphatase activity greatly increased in Gm + Gf + P- treated roots followed by Gf and Gf- treated roots. While in shoots the activity was maximum in P-treated plants followed by Gf + Gm + P- and Gm+ Gf- treated plants. All other combinations have reduced acid phosphatase activity. Alkaline phosphatase activity was considerably lower than those of acid phosphatase activity in both roots and shoots of all VAM-treated *T. grandis* plants. Alkaline phosphatase activity was maximum in GF-treated roots and P-treated shoots, followed by P and Gm+P Alkaline phosphatase activity was increased to the maximum in P followed by Gf and Gm+Gf+P- treted shoots of teak. A positive correlation was noted between acid phosphatase activity and P concentration in the root of treated teak plants.

Panwar and Vyas (2002a) evaluated the efficiency of eight arbuscular mycorrhizal species, *Acaulospora mellea, Gigaspora margarita, Gigaspora gigagiganlea, Glomus deserticola, Glomus fasciculatum, Sclerocystis rubiformis, Scutellospora calospora* and *Scutellospora nigra,* collected from rhizosphere soils of *Moringa concanensis,* for nutrient uptake and enhancement of acid phosphatase, nitrate reductase, peroxidase and polyphenol oxidase [catechol oxidase] activities in this endangered tree of the Indian Thar desert. Culturing was done under greenhouse conditions and analyses were performed 180 days after inoculation. All fungi showed beneficial effects, with *G. margarita* being the most efficient in promoting all biochemical parameters.

CONCLUDING REMARKS

A perusal of the published works on the effect of VAM fungi on the nutrient uptake in plants clearly shows that a huge bulk of data has been collected by Indian mycorrhizasts on the uptake of P, and its effect on the growth and yield of some economically important plants. Some work is related to the effect of different sources of phosphorus or graded levels of phosphates on the performance of the plants. Work on related physiological implications is negligible. However, few reports on the involvement of phosphates in the process are available. Likewise, few workers have attempted to explore the translocation of absorbed P using ^{32}P. Only fragmentary information is available on the effect of VAM association on absorption or translocation of other nutrients and it is restricted to only Zn^{++}, Ca^{++} and Fe^{++}.

One of the main aims of researches on VA mycorrhiza is to manipulate the symbiosis to improve plant growth under different conditions and different habitats. However, this aim may be fulfilled more authentically if physiological implications related to improved uptake of water as well as carbon and nitrogen nutrition / metabolism of host and symbiont are properly understood. Although Indian mycorrhizasts have made some contribution with reference to this, however, more is expected from them, since VA mycorrhiza has great prospect as biotool for the welfare of the plants in the country. A perusal of the literature shows that contribution of mycorrhizasts of the country includes only 4 reports on water relations, 6 on carbon metabolism and 6 on nitrogen metabolism and the information provided by these reports is very fragmentary and inconclusive. The reports on the biochemical constituents of the mycorrhizal plants by Indian workers are also very few.

5

VA MYCORRHIZA VS OTHER MICROBES

Of the various microorganisms present in and around roots, VAM fungi occupy unique ecological position as they are partly inside and partly outside the host. Their internal phase encounters competition and antagonism from other beneficial symbiotic microbes while their external phase faces competition and antagonism from beneficial non-symbiotic microbes present around the roots. Since they have an ensured source of nutrients from the host, they have comparatively better conditions for their growth and may exert greater effect on plants with other symbiotic and non-symbiotic microbes in and around their roots. It has been pointed out that investigations are needed on the interactions between VAM fungi and these beneficial microbes, and their effect on growth and productivity in plants. A bulk of information has been collected related to this aspect; however, authentic generalization has not been possible as yet. Following description gives an account of the work done by Indian mycorrhizasts on the interactions between VAM fungi and other beneficial symbiotic/non-symbiotic microbes in and around the roots.

VAM + Rhizobium

Godse *et al.* (1978) studied the response of cowpea to *Rhizobium* and VA mycorrhizal dual inoculation. VA mycorrhiza increased nodulation and root and shoot development, not only in plants inoculated with *Rhizobium* but also in the naturally nodulated plants. The increased shoot and root development were presumably due to increased phosphorus uptake.

Bagyaraj *et al.* (1979a) suggested on the basis of results of dual inoculations on plants in a phosphorus deficient sandy loam that *Glomus fasciculatum* could greatly increase nodulation and nitrogen fixation in field grown soybeans with *Rhizobium japonicum*.

Sivaprasad *et al.* (1983) reported that although inoculation of *L. leucocephala* with *Rhizobium* alone and to a greater extent *Glomus fasciculatum* alone significantly increased nodulation, growth and P-uptake over uninoculated controls. Best results were obtained with dual inoculation.

Manjunath and Bagyaraj (1984) studied the response of pigeon pea and cowpea to dual inoculation with *Glomus fasciculatum* and / or *Rhizobium* sp. in a P-deficient non-sterile soil. Application of P (22 kg/ha) increased nodulation and the number of endomycorrhizal spores in the root zone of both legumes. Plants inoculated with both organisms and supplemented with P recorded the highest shoot dry weights and N and P contents, indicating the need for the addition of small amounts of P to derive maximum benefit from dual inoculation with *Rhizobium* and VA mycorrhiza.

Parvathi *et al.* (1985b) reported that in two cultivars of groundnut, combined inoculation with a *Rhizobium* sp. and *Glomus mosseae* resulted in a synergistic effect on the formation of mycorrhiza and nodulation when compared with separate inoculations. The combined inoculum resulted in a fairly good response of the plants in terms of dry weight and uptake of N and P.

Punj *et al.* (1987) studied the dual inoculation with *Rhizobium* and *Glomus fasciculatum* and its effect on growth, nodulation and nitrogen fixation in *Leucaena leucocephala*. Two efficient and competitive *Rhizobium* strains gave significant increase in all the parameters when inoculated in combination with *G. fasciculatum*.

Adholeya *et al.* (1988a) investigated the effect of VAM-*Rhizobium* interaction on productivity, N-P uptake, dry weight and percent infection in mungbean (*Vigna radiata*). *Glomus caledonium* and *Rhizobium* strain KM-1 were used. Treatments included, control (C), Mycorrhizae (M), *Rhizobium* (R) and Mycorrhizae & *Rhizobium* (M & R). Yield in M & R treatment was higher than in other treatments. Similarly N-P uptake, infection percentage and dry matter production were also higher in M & R treatment, which suggested a positive effect of VAM-*Rhizobium* interaction on *V. radiata*.

Adholeya *et al.* (1988b) studied the VAM *Rhizobium* interaction and its effect on colonization and nitrogen fixation in mungbean (*Vigna radiata*) under field trials. Field inoculation of VAM fungus, *Glomus caledonium* and *Rhizobium* strain KM.1 was applied. Treatments included, control (C), Mycorrhizae (M), *Rhizobium* (R) and Mycorrhizae & *Rhizobium* (M & R). Infection rating, nodule number, nodule dry weight, nitrogenase activity was maximum when the crop had dual inoculation (M & R). Positive results indicated important contribution of *Glomus caledonium* and *Rhizobium* sp. when used in combination.

Rangarajan *et al.* (1988) investigated the synergistic effect of rhizobia and VAM on biomass production of certain tree legumes. Inoculation of 2 rhizobia and 2 VAM fungi (*Glomus fasciculatum* and *Gigaspora calospora*) individually and in combination on tree legumes such as *Acacia mellifera, A. nilotica* and *Leucaena leucocephala* resulted in an increased biomass production ranging from 3 to 25% over uninoculated control. Of the 3 tree legumes, maximum response to rhizobial inoculation (13%) was observed in *Acacia mellifera* and to mycorrhizal inoculation (16%) in *L. leucocephala*. Although dual inoculation gave more biomass production than individual inoculation in all the 3 tree legumes, maximum response (25%) was observed in *L. leucocephala*. Synergistic effects of dual inoculation were observed also in respect of nodule dry weight and mycorrhizal colonization.

Rai (1988) conducted a field experiment in calcareous soil (pH 8.4, 34.2% free $CaCo_3$ and 4.2 ppm DTPA - extractable Fe) to assess the role of *Glomus albidus* in Fe nutrition and N fixation of *Cicer arietinum* cv. Pant G 114 and BR - 77 with and without *Rhizobium* strains IC-76 and RG-3. Inoculation with *G. albidus* as well as *Rhizobium* increased number of nodules / plant, nodule dry weight/plant, nodule leghaemoglobin content, nitrogenase activity and total nodule Fe. Seed yield, P, Ca and Fe content in the seeds also showed an increase. However, dual inoculation proved to be superior to single inoculations.

Chaturvedi and Sharma (1988) investigated the interaction of *Glomus caledonium* and *Rhizobium* in relation to uptake and translocation of labelled phosphatic fertilizers in chickpea (*Cicer arietinum*). Maximum concentration of P was recorded in chickpea plants provided with a dual inoculum of VAM and *Rhizobium* followed by plants receiving VAM or *Rhizobium* inoculum alone irrespective of phosphatic fertilizers of different solubility's used. Uninoculated check plants showed minimum level of P. Similar results as indicated above were observed when the treated plants were exposed to X-ray films to record the distribution of P. Irrespective of P source, very little radioactivity was recorded in uninoculated plants. Plants co-inoculated with *Rhizobium* and VAM exhibited highest radioactivity followed by plants inoculated with VAM or *Rhizobium* alone. Radioactive P was distributed throughout the plants indicating good translocation within the plant system.

Champawat (1990b) studied the effect of dual inoculation of pea plants (*Pisum sativum*) with *Gigaspora calospora* and *Rhizobium,* which resulted in greater increases in shoot and root DW, P and N uptake, nodulation and mycorrhizal infection than single inoculations.

Champawat (1990c) demonstrated that dual inoculation with VAM fungi, *Gigaspora calospora* and *Rhizobium* bacteria in chickpea, considerably enhanced the plant growth, uptake of phosphorus and nitrogen and root nodulation.

Mallesha and Bagyaraj (1990) reported the results of greenhouse trial using *Leucaena leucocephala* var. K 67. All the seeds were treated with *Rhizobium* before sowing. Shoot and root growth increased more by inoculation (as seeds) with *Gigaspora margarita* than with *Glomus mosseae*, while combined inoculation further increased root growth. All the treatments increased the P content of shoots (with *Glomus mosseae* alone giving the highest value), while combined inoculation produced highest root colonization.

Ramaraj and Sharmugam (1990) reported that in both sterilized and unsterilized soil conditions, dual inoculation with *G. etunicatum* and *Rhizobium* increased the rhizobial nodules and mycorrhizal colonization in cowpea and consequently the biomass to a greater extent than when either was inoculated alone.

Santhanakrishnan and Oblisami (1990) studied the effect of interaction between *Rhizobium* and VA mycorrhizal fungi inoculation on the growth of groundnut applied with different levels of gypsum. According to them there was an increase in colonization of VAM fungi up to 150-kg/ha gypsum with a decrease in VAM colonization with further increase up to 450-kg/ha gypsum. The magnitude of increase of growth, VAM colonization and nutrient content in groundnut plants was maximum with dual inoculation of both organisms together with 150-kg/ha gypsum.

Shivaram and Rai (1990) studied the influence of *Rhizobium* and VAM on *Cenchrus ciliaris* grown in greenhouse. ELISA studied their associative effect in terms of forage yield, accumulation of N and P in the tissue and also competition of rhizobia. Combined inoculation resulted in a significant increase in dry matter production, nitrogen uptake and phosphorus uptake. *Rhizobium* used as the inoculant accounted for 48-60% of nodulation.

Sivaprasad and Rai (1991) studied the synergistic association between *Glomus fasciculatum* and *Rhizobium* species and its effect on pigeon pea. In field and greenhouse trials in pigeon pea on a P-deficient tropical soil, soil inoculation with *G. fasciculatum* and *Rhizobium* sp. increased nodulation, nitrogenase activity, growth and N and P contents compared with inoculation with *Rhizobium* alone. The association with *G. fasciculatum* increased nodulation due to native rhizobia. The response of dual inoculation was less in the field than in the greenhouse. In the field, dual inoculation gave 15.2% more yield than inoculation with *Rhizobium* alone.

Chaturvedi and Kumar (1991) studied the nodulation and nitrogenase activity in gram as influenced by *Rhizobium* and VA mycorrhiza interaction. Nodulation and nitrogenase activity in chickpea cultivars Pant G-114 and BG-209 raised in P-deficient unsterilized soil and inoculated with *Rhizobium* and *Glomus caledonium* were higher compared with plants given either inoculum alone. Uninoculated cultivars formed poor nodules and had lower levels of nitrogenase activity.

Rawat *et al.* (1991) studied the effect of dual inoculation of *Rhizobium* and mycorrhiza on nodulation and yield in chickpea. In a pot experiment, seeds of chickpeas cv. JG 315 were inoculated with *Rhizobium* strain H-45 and /or strain F-75 and mycorrhiza strain M1 and /or M2, Number of nodules/ pot were highest with *Rhizobium* strain H-45 and did not differ significantly between mycorrhiza treatments. Nodule dry weight and seed yield did not differ significantly between different *Rhizobium* mycorrhiza treatments.

Dubey (1993) studied the effect of mycorrhiza; *Bradyrhizobium* inoculation and phosphorus level on nodulation yield and yield attributes in soybean. In a field experiment on clay loam soybeans cv.JS 75-46 were seed inoculated with *Glomus fasciculatum* (control - noninoculated) and given 0,13.2 or 26.4 kg P/ha as single superphosphate or seed inoculated with *B.japonicum* or both inoculants. Seed yields from these treatments were 1.55, 1.63, 1.87, 1.94, 1.72 and 1.84 t/ ha respectively. Seed inoculated plants had higher nodule dry weight and pods/plant and higher seed N and P contents than the uninoculated controls.

Dhingra *et al.* (1994) studied the response of lentil to VA mycorrhiza (*Glomus fasciculatum*) and *Rhizobium leguminosarum* inoculation in relation to phosphorus levels. Lentil cv. LL56 was raised in loamy sand soil with seeds untreated or inoculated with *R. leguminosarum* and / or *G. fasciculatum*. 0.60 kg P_2O_5/ha was also applied. Seed yields were not significantly different between inoculation treatments or P treatments but a significant interaction between inoculation and P rate was recorded.

Kehri and Chandra (1994b) studied the interaction between VA-mycorrhizae and *Rhizobium* and their effects on chickpea in unsterilized soil. Effect of dual inoculation of VA-mycorrhizal fungi and *Rhizobium* on plant biomass, nodulation, per cent mycorrhizal colonization and yield was examined in chickpea raised in unsterilized soil. All the three VA-mycorrhizal fungi singly or in combination with *Rhizobium* improved growth, nodulation and yield in chickpea, however, two of them exhibited comparatively better efficiency when used alone, while the third one when used in combination with *Rhizobium*.

Verma *et al.* (1994) studied the effect of inoculation of VAM fungi and *Rhizobium* on growth and biomass production in *Acacia nilotica* in nursery. Seedlings of *Acacia nilotica* were inoculated with *Rhizobium* strain isolated from nodules of the same species. VAM fungi and a single dose of nitrogen and phosphorus in unsterilized soil under nursery conditions. A single species of VAM fungus (*Glomus mosseae*) and a mixture of locally isolated VAM fungi (consisting of *Glomus* species and a *Gigaspora* sp.) were used to inoculate the seedlings. Height, growth, collar diameter, nodulation, percentage root infection by VAM fungi and fresh and dry weights of shoots and roots were determined 4 months after inoculation. Inoculation of *Rhizobium* and VAM fungi and application of nitrogen and phosphorus singly or in different combinations enhanced the growth and biomass production prepared with values for uninoculated (control) seedlings. Maximum seedlings volume (height x root collar diameter) was found in seedlings inoculated with *G. mosseae* and *Rhizobium* in combination, and in *Rhizobium* + Phosphorus combination, followed by the mixed inoculum of VAM fungi. Maximum total dry biomass was recorded in seedlings treated with a combination of mixture of VAM fungi, followed by *Rhizobium* + mixed VAM and *Rhizobium* + phosphorus. It is concluded that inoculation of VAM fungi and *Rhizobium* singly or in combination, is equally effective or more beneficial than use of chemical fertilizers in raising seedlings of *Acacia nilotica*.

Kaushik and Kaushik (1995) studied the interaction between VAM fungi and *Rhizobium* and their influence on *Albizia lebbek*. Three VAM fungi, *Glomus mosseae, Gigaspora gilmorei* and *Glomus constrictum* in combination with *Rhizobium* were tested for their influence on dry matter production, nodulation, N_2 fixation and nutrient uptake by *A. lebbek* plants in pots under greenhouse conditions. On the basis of their results they reported significantly more biomass of the plants in combination of *Rhizobium* and VAM inoculants as compared to either that of *Rhizobium* or VAM alone. Dual infection of *Rhizobium* and VAM resulted in a higher number and weight of nodules.

Kumutha and Santhanakrishnan (1995) studied the effect of insoluble phosphate and dual inoculation on soybean. They conducted a pot experiment using *Glomus fasciculatum* as VAM inoculum, recommended doses of N and K, Superphosphate or Mussoorie rock phosphate to meet the P requirement at the rate of 60 kg P_2O_5/ha and *Rhizobium japonicum*. VAM and *Rhizobium* were inoculated individually as well as in combination along with Mussoorie rock phosphate and superphosphate. Dual inoculation was found to be superior to individual inoculation of either VAM or *Rhizobium* in enhancing the nodulation, dry matter and nutrient uptake of soybean. Superphosphate was the most compatible source of P and this combination registered maximum values. Crop response to Mussoorie rock phosphate application was improved due to application of bioinoculants.

Mandal and Kaushik (1995) studied the interaction between VAM fungi and *Rhizobium* and their effects on the growth parameters of *Acacia nilotica*. They carried out pot experiments using *A. nilotica* in nutrient deficient river sand (pH 7.4) inoculated at sowing with 7 VAM treatments (1) *Glomus mosseae* (2) *Glomus constrictum* (3) *Gigaspora gilmorei* (4) 1+2 (5) 2+3 (6) 1+3 and (7) not inoculated. *Rhizobium* was inoculated by drenching the seedling with suspension, non-*Rhizobium* treatments were also included. Seedlings growth was measured at 45 and 90 days after sowing. All inoculation treatments increased seedling growth with the best performances from seedlings inoculated with VAM and Rhizobia and the most effective treatment (6) + *Rhizobium*, and the least effective (2) without *Rhizobium*.

Neeraj and Verma (1995) studied the interaction between VA mycorrhizae and *Rhizobium* in *Cyamopsis*. *Gigaspora margarita* and *Glomus fasciculatum* were taken for tripartite studies. The combinations of these two VAM fungi with *Rhizobium* and/or orthophosphate were taken and mineral nutrients, namely calcium, magnesium, manganese and iron were estimated. Chloroplast pigments estimation was also done. They reported that VAM fungi increased the biomass of guar plants and elevation in mineral nutrient concentration. Higher chloroplast pigmentation and enhanced photosynthetic activity was also reported.

Saxena *et al.* (1995) carried out an investigation to study the effect of *Bradyrhizobium* and different VAM fungi *(Glomus mosseae, G. fasciculatum, G. versiforme, G. macrocarpum, Gigaspora gilmorei, G. margarita, G. calospora, Endogone dussii* etc.) on green gram. They reported that *Bradyrhizobium* sp. strain S-24 interacted differentially with VAM fungi and caused significant variations in nodulation and growth parameters in the crop. Co-inoculations of all the VAM fungi except *E. dussii* produced significantly higher nodules fresh weight and root/shoot dry weights when compared to single inoculation of strain S-24. The nodule occupancy of strain S-24 increased significantly in presence of VAM fungi, *G. mosseae, G. fasciculatum and Gigaspora calospora*.

Selvaraj *et al.* (1995) studied the response of tree seedlings to variable combination of VAM and N_2 fixer under N and P depleted acid soil under nursery conditions. They included in their study *Acacia haloserícea, Leucaena leucocephala* as N_2 fixing tree legumes and non-N_2 fixing *Cassia siamea*, three different VAM fungi *viz. Glomus fasciculatum, G. mosseae* and *Gigaspora calospora*, with and without N_2 fixing *Rhizobium* for *Acacia* and *Leucaena* and *Azospirillum* for *Cassia*. The results revealed that two leguminous N_2 fixing tree species responded well to the combined inoculation of *Rhizobium* sp. and *G. fasciculatum* by recording more plant biomass, root nodulation and VAM infection than the combinations of other VAM, *Glomus mosseae* and *Gigaspora calospora* - in the acid soils (pH 5.6%). In *Cassia siamea* also the seed inoculation of *Azospirillum* with co-inoculation of endomycorrhizae registered better plant growth and establishment over untreated controls.

Shinde and Nair (1995) studied the effect of *Glomus mosseae* and *Rhizobium* on growth and nodulation of *Phaseolus radiatus* by using MNP method. Maximum growth with more nodules was observed in 2^1 dilution. This dilution showed more number of spores and cent percent root infection as compared to other dilutions and control.

Thakur and Panwar (1995) studied the effects of *Rhizobium* - VAM interactions on growth and yield in mungbean (*Vigna radiata*) under field conditions. They reported that, inoculation, either singly or combined, increased plant height, leaf area, photosynthetic rate and total dry matter production compared with no inoculation. A combination of the two inoculants gave higher dry matter production and seed yield compared with either single inoculant. Pusa - 105 had higher leaf area, photosynthetic rate, total dry matter production, pod number, seed number/ pod, seed size and seed yield than PS-16.

Tilak *et al.* (1995a) evaluated different VAM fungi for soybean cv. Pusa 22. A soil and sand mixture, containing extramatrical chlamydospores and infected root segments of *Cenchrus ciliaris* with either *Glomus fasciculatum, G. mosseae, Gigaspora calospora, G. margarita, G. macrocarpum* or *G. versiforme* served as inoculum which contained 1000-

1250 chlamydospores/100 g. Soil inoculation with various VAM fungi in general resulted in significantly higher root infection than that of uninoculated control although the differences among VAM fungi were not significant. However, *Rhizobium* application alone did not show any significant increase in root infection by VAM over uninoculated control. Soil inoculation with *G. fasciculatum* and *G. mosseae* in presence of *G. japonicum* produced significantly higher bean yield than their corresponding controls.

Vyas and Srivastava (1995a) investigated the effect of dual inoculation with *Rhizobium japonicum* and *Glomus fasciculatum* on yield and growth of cluster bean. They observed significant changes in one-month old plants in terms of plant growth, dry weight, P-uptake, root colonization etc. when inoculated with dual inoculants.

Shinde and Nair (1996) studied the effect of *Rhizobium* alone, VAM alone (*Glomus fasciculatum*) and *Rhizobium* and VAM on growth and nodulation of *Phaseolus radiatus* Rox. Dual symbionts resulted in better growth and nodulation compared to *Rhizobium* or VAM alone. VAM alone was comparatively better than *Rhizobium* alone.

Thapar and Uniyal (1996) studied the growth response of *Acacia nilotica* to inoculation with a VAM fungus *Glomus macrocarpum* and/or *Rhizobium* in sodic and (normal) New Forest (Dehra Dun, Uttar Pradesh) soils (singly and at ratios of 1:1, 2:1 and 3:1 sodic/normal) under greenhouse conditions. Inoculations with *Glomus macrocarpum* resulted in height increase, from 17.8% in pure sodic soil to 50.5% in mixed sodic soil (3:1) over 7 months compared with control, non-inoculated plants. A significant increase of 166.7% in biomass (dry) was observed with VAM. *Rhizobium* inoculation did not improve seedling growth either alone or with VAM. Significant improvements in growth were also observed after inoculation in normal soils with VAM alone.

Greep *et al.* (1997) studied the response of *Macroptilium atropurpureum* Urb. to vesicular-arbuscular mycorrhizal fungi and *Rhizobium* sp. in sterilized soil. The effect of inoculation of six species of vesicular-arbuscular mycorrhizal fungi, *Gigaspora margarita, Glomus mosseae, G. monosporum, G. versiforme, G. fasciculatum, G. deserticola* and *Rhizobium* sp. on growth and NPK content in *Macroptilium atropurpureum* was investigated. The mycorrhizal species were isolated from the rhizosphere forest soils and multiplied in pot culture with sorghum as the host plant. *Rhizobium* sp. was isolated from fresh nodules of the plants and maintained on yeast extract mannitol broth. The influence of VAM fungi and *Rhizobium* sp. either singly or in combination increased growth, nodulation and nutrient contents of *M. atropurpureum*, with the best results given by inoculation with *Rhizobium* together with multiple VAM fungi.

Dubey (1998) studied the response of soyabean (*Glycine max*) to biofertilizers with and without nitrogen, phosphorous and potassium on swell-shrink soil. In a field of Sehore, Madhya Pradesh in the 1991-92 rainy seasons, different combinations of inoculation with *Bradyrhizobium*, vesicular-arbuscular mycorrhizas (VAM) and phosphate-solubilizing bacteria and application of NPK fertilizers were compared for soyabeans cv. PK472 grown on a vertisol. Data on nodulation, growth, yield components and crop yield are tabulated. Yield was 0.95t/ha in controls and 1.05-1.21 t in the other 14 treatments, being highest from combined use of the 3 inoculants without NPK.

Pandey *et al.* (1998) studied the response of cowpea (*Vigna unguiculata* Linn. Walp.) to VAM fungi *Glomus fasciculatum* and *Rhizobium* inoculation under different sources and levels of phosphors. In a pot experiment, cowpeas *cv. Bundel Lobia-2* were not inoculated or soil inoculated with *Glomus fasciculatum* (Gf) and / or seed inoculated with *Rhizobium* (Rh), and given 0, 40 or 80 kg P/ha as Mussoorie rock phosphate (MRP), diammonium phosphate or single super phosphate. There were highly significant increases in plant growth, nodulation, % root colonization, P content in the shoot and available P in the soil with inoculation with Gf or Rh, or with P fertilizer, compared with the control. MRP at 40 kg/ha with Gf + Rh gave the best results.

Kumar and Potty (1998) conducted a pot experiment to study the dual effect of vesicular arbuscular mycorrhiza (VAM) fungi *Glomus microcarpum* and *Bradyrhizobium* in the biomass production of yam bean. Plants treated with VAM fungi and *Bradyrhizobium* produced more than 3 times the weight of the control. In addition, the production of nodules, soil spores and percentage of VAM colonization also increased. *Rhizobium* treated plants have shown VAM colonization and VAM inoculated plants produced less number of nodules, combined inoculation of *G. microcarpum* and *Bradyrhizobium* showed significant increase in biomass production, nodule number and weight, soil spore count and fungal colonization percentage compared with individual inoculations.

Khan and Uniyal (1999c) studied the growth response of some forest trees to VAM and rhizobium inoculations and fertilizer application. The growth response of *Dendrocalamus strictus, Bambusa nutans, Acacia nilotica* and *Populus deltoides* to VAM inoculation alone and in combination with fertilizer application was investigated. In *A. nilotica,* dual inoculation with VAM and *Rhizobium* and in Bamboo and *Populus* VAM in combination with fertilizers was given. Inoculation with the above endophytes in combination with the fertilizers proved more effective in promoting growth of plants than when fertilizers and endophytes were applied alone.

Prasad (1999a) studied the effect of indigenous VAM fungus (*Glomus fasciculatum*) and *Rhizobium* sp., on the growth of *Acacia nilotica* plant *viz.*, root and shoot length, dry weight, basal area, number of leaves/ plant, nodulation, NPK concentration, and VAM infection in root and infestation in soil. Native VAM *G. fasciculatum* isolated from the rhizosphere of *A. nilotica* was multiplied and maintained on *Paspalum notatum, Rhizobium* sp. isolated from nodules of *A. nilotica* was cultured and maintained in yeast extract mannitol broth. The results indicated that VAM and *Rhizobium* sp. inoculated *A. nilotica* plants showed much better growth than untreated plants. The differences were significant for control plants and also between mycorrhizae and mycorrhizae plus *Rhizobium* sp. treated plants. The maximum growth, of NPK uptake and nodulation was observed in double inoculated plants.

Kumar and Chandra (1999) studied the response of mungbean (*Vigna radiata*) to *Rhizobium* and VA mycorrhiza inoculation with varying phosphorus levels. Combined inoculation of Rhizobium + VAM (*Glomus caledonium*) resulted in significantly more nodule number at 30 and 50 DAS and grain yield of field grown mungbean (*Vigna radiata*) with numerical increase in plant biomass over single inoculation with either *Rhizobium* or VAM in a soil of medium P status. Inoculation of VAM also significantly increased infection in plant roots over without inoculation treatment. The applied P doses (0, 25, 50 and 75 kg P_2O_5 ha^{-1}) significantly increased the plant dry matter at 50 DAS and grain yield over no P application.

Gupta and Rahangdale (1999) evaluated the effects of dual inoculation with *Rhizobium* species and arbuscular fungi on the growth of two forest tree species *Albizia lebbeck* and *Dalbergia sissoo* in pot experiment in a net house. Seedlings were raised in sterilized soil inoculated separately with two *Rhizobium* -Al and R-AcN) and two arbuscular species (*Glomus* spp.). The effects of various inoculations on shoot height, root length, number of leaves, dry weight of shoot, root and leaf, number and biomass of nodules were recorded. The biochemical effects of these symbionts in relation to protein and sugar levels were also determined in plant tissue. In general, *Rhizobium* and /or arbuscular inoculations enhanced growth with respect to height, biomass and nodule status. Higher protein and sugar levels were also recorded in the inoculated seedlings. The advantages of preinoculation with these symbiotic organisms at the nursery stage of *D. sissoo* and *A. lebbek* in plantation programmes are discussed.

Das *et al.* (1999) studied the effects of *Rhizobium* and vesicular arbuscular mycorrhiza (VAM) on green gram (*cv. Nayagarh local*) in acid lateritic soil (Alfisol) of Central Agricultural Research Farm, Bhubaneswar, India. Three P-sources and organic matter were included in the experiment with and without lime. The dry matter yield and N, P and K uptake of green gram harvested at 65 days of growth were significantly higher due to dual inoculation, and were more pronounced when amended with lime. The investigation suggested that dual inoculation of *Rhizobium* and VAM in the presence of organic matter along with application of Mussorie rock phosphate improved the growth and yield of green gram.

Balachandar and Nagarajan (1999) studied the dual inoculation of green gram (*Vigna radiata*) with vesicular arbuscular mycorrhizae (*Glomus mosseae*) (as a soil inoculant) and *Rhizobium* (as seed inoculation) in comparison with single inoculation and fertilizer nutrition. Among the various treatments tested, the dual inoculation with *Glomus mosseae* and *Rhizobium* with 50 per cent of recommended nitrogen and phosphorus fertilization recorded the maximum plant height, root length, plant biomass, number of nodules per plant, nodule weight and grain yield of green gram variety Vamban 1. Mycorrhizal infection of green gram positively influenced rhizobial colonization for better nodulation and nitrogen fixation. The results revealed that dual inoculation of VAM and *Rhizobium* with 50 per cent of recommended nitrogen and phosphorous fertilizer gave higher yield in Vamban 1 green gram compared to single inoculation or 100 per cent of the recommended fertilizer application.

Khan and Uniyal (1999a) reported the effects of inoculating *Acacia nilotica* seedlings with VAM and *Rhizobium* and *Populus deltoids* cuttings with VAM alone and in combination with NPK fertilizer. The trials were done in polypots under field condition Uttar Pradesh. The growing medium was soil/sand (2:1)+2%FYM .Two VAM preparations were used in each case – a mixture of *Glomus mosseae* and *G. geosporum*(M1),and one of *Glomus mosseae* and *G. aggregatum*(M2). *Acacia nilotica* responded positively to VAM and *Rhizobium* inoculations both separately and alone with an increase of up to 40.94% in biomass (in the M1+ *Rhizobium* treatment). In *P. deltoids* a maximum increase of 37.64% was achieved with VAM (M2) inoculation in combination with NPK. NPK alone increased growth very little and the 2 VAM treatments alone were not as effective as those with added NPK.

Raju and Reddy (1999) studied the effect of biofertilizers and inorganic nutrition on blackgram (*Phaseolus mungo*) grown in rice fallows. Black gram cv. LBG 20 seeds *Rhizobium* were broadcast 7 days before rice harvest and treated with the following 0.5 kg/ha, 5 kg vesicular arbuscular mycorrhiza (VAM)/ha *Rhizobium*+ VAM 50 kg P_2O_5/ha, VAM +50kg P_2O_5/ha 2% urea spray at pre-flowering. 2% urea spray at pod filling or 4% diammonium phosphate (DAP) spray at pre-flowering, in a field experiment conducted in Maruteru. Andhra Pradesh during the summer season of 1997 and 1998. The first flush of floral initiation appeared 26 days after sowing. However, spraying of 2% urea at pre-flowering caused another flush of floral initiation after 41 days. Spraying urea at pre-flowering, or pod-filling stages and DAP spray at pod filling, increased the seed yield by 40.2, 34.2 and 31.1%, respectively. Application of 50 kg P_2O_5, alone or in combination with VAM, increased root length by 47.8% and 50.7% respectively, and seed yield by 25.2 and 30.8%, respectively. The highest number of fruiting branches per plant (8.14) number of pods per plant (44.1) haulm yield (4325kg/ha), phytomass (185 g/plant), seed yield (1204 kg/ha) and benefit: cost ratio (71:3:1) were recorded in plots sprayed with 2% urea at pre-flowering stage.

Prasad (2000) studied the growth response in *Acacia nilotica* (L.) Del. inoculated with *Rhizobium* and *Glomus fasciculatum*. The native vesicular arbuscular mycorrhiza (VAM) *Glomus fasciculatum* was isolated from the rhizosphere of *Acacia nilotica* and multiplied and maintained on *Paspalum notatum, Rhizobium* sp. isolated from nodules of *A. nilotica* was cultured and maintained in yeast extract mannitol broth. The symbionts were used individually and in combination to inoculate seeds *A. linotica*, and seedlings growth measured after 45 and 90 days. *Rhizobium* and VAM inoculated *A. nilotica* plants showed much better growth than uninoculated plants; with significant differences both between control and inoculated plants and between individually and jointly inoculated plants. Both maximum growth and uptake of NPK was observed in combined inoculated plants.

Kumar *et al.* (2001) when inoculated cowpea with *Glomus fasciculatum* (a vesicular arbuscular mycorrhiza), *Rhizobium* (CP-3) and *Azotobacter,* exhibited highest seed germination, nodule number and dry weight, plant height, mycorrhizal colonization and nutrient content higher with combinations of the 3 above mentioned organisms compared to inoculation with either one organism or two organisms. Interactions between host and symbiont also varied with cultivars. Vesicular arbuscular mycorrhiza or *Azotobacter* or *Rhizobium*, or their combinations can have important effects on nodulation, nitrogen fixation and P uptake in legumes.

Sarvanan and Natrajan (2000) grown *Acacia nilotica* seedlings in the nursery in steam sterilized and unsterilized soils inoculated with vermiculite inoculum of *Pisolithus tinctorius*, soil inoculum of *Glomus mosseae* and liquid inoculum of *Bradyrhizobium* sp., and various growth parameters determined at 7 months old. Seedlings grown in both sterilized and unsterilized soil with a combination of *P. tinctorius. G. mosseae,* and *Bradyrhizbium* sp.

inocula generally showed greater total dry weight of shoots and roots, number of nodules, weight of nodular tissue, and nodule nitrogenase activity (measured by acetylene reduction) than the seedlings grown in soils with other combinations of organisms and uninoculated seedlings. Mycorrhizal development was not much different between the triple inoculation treatments and treatment involving *Bradyrhizobium* sp. and one mycorrhizal fungus (dual treatments). Dual inoculation treatments generally gave better results than inoculation with *Bradyrhizobium* sp. alone, and all inoculation treatments gave better results than the control-uninoculated treatment. The response to inoculation was generally greater in the sterilized than unsterilized soil.

Hazarika *et al.* (2000) studied the effect of vesicular arbuscular mycorrhizal (VAM) fungi and *Rhizobium* on growth and yield of green gram [*Vigna radiata* (L.) Wilczek]. *V. radiata* cv. K-851. Seeds were inoculated with *Glomus mosseae, G. fasciculatum* or *Rhizobium,* or *Rhizobium* combined with either of the two *Glomus* spp. before sowing. All treatments significantly altered the growth, dry matter and nodulation of *V. radiata.* Shoot and root length, shoot and root dry weight and the number/plant and dry weight of nodules were highest in treatments with *Rhizobium* combined with *G. mosseae* or *G. fasciculatum.* The latter treatment resulted in an 80, 64, 110 and 110% increase in shoot length; root length, shoot dry weight and root dry weight, respectively. Sporulation, root colonization, number of pods/plant, seed weight, yield and the N, P and K concentration and uptake in plants were higher in treatments with VAM fungi and *Rhizobium* compared to uninoculated plants and those treated with *Rhizobium* or VAM fungi alone. *Rhizobium* + *G. fasciculatum* showed a 74, 27.8, 24.3, 25.3 and 26.6% increase in spore count, root colonization, pods/plant, seed weight/plant and yield. This treatment also resulted in the highest nutrient concentration and uptake in plants.

Kumar *et al.* (2001) when inoculated cowpea with *Glomus fasciculatum* (a vesicular arbuscular mycorrhiza), *Rhizobium* (CP-3) and *Azotobacter,* exhibited highest seed germination, nodule number and dry weight, plant height, mycorrhizal colonization and nutrient content higher with combinations of the 3 above mentioned organisms compared to inoculation with either one organism or two organisms. Interactions between host and symbiont also varied with cultivars. Vesicular arbuscular mycorrhiza or *Azotobacter* or *Rhizobium,* or their combinations can have important effects on nodulation, nitrogen fixation and P uptake in legumes.

Devi and Reddy (2001) studied the influence of *Rhizobium* species and the vesicular arbuscular mycorrhizal (VAM) fungus, *Glomus mosseae* on the growth, nodulation, phosphorus content and acid and alkaline phosphatase activities of groundnuts. Groundnuts were inoculated with VAM (250 spores per pot) only, *Rhizobium* (10^6 spores per pot) only, VAM + *Rhizobium,* and uninoculated (control) under sterilized soil conditions. Seeds of groundnut were sown in 6 or 9-inch pots containing different inoculum. Observations were made 15.25 and 45 days after inoculation. Groundnuts inoculated with VAM + *Rhizobium* showed the highest increase in growth, number of nodules, mycorrhizal colonization, phosphorus content and phosphatase activity compared with sole applications of VAM and *Rhizobium.*

Tarafdar and Rao (2001) conducted a field experiment under rainfed conditions at a research institute in Jodhpur, Rajasthan to investigate the effects of biological management practices in loamy sand on clusterbean (*Cyamopsis tetragonoloba)* production. The 10 treatments were: control; N (20 kg N/ha as urea); P (40 kg P_2O_5 /ha); farmyard manure (FYM, C 14.3% and N 1.12%) with a C: N ratio of 13:1 (4 tonnes/ha), *Rhizobium,*AMF (*Glomus mosseae*); FYM + *Rhizobium*; FYM + AMF, AMF + *Rhizobium;* and FYM + *Rhizobium* + AMF. Results showed that inoculation of clusterbean with *Rhizobium* and AMF coupled with the application of FYM significantly improved nodule numbers, nitrogenase activity, root infection, and the number of AMF spores. In the rhizosphere of treated cluster bean biomass C increased over control except with *Rhizobium.* Maximum enhancement was found in the treatment FYM +AMF + *Rhizobium.* Dual inoculation with FYM also resulted to maximum acid and alkaline phosphatases and dehydrogenases (oxidoreductases] activities. N concentration was found highest in dual Rhizobium inoculation coupled with FYM application while improvement of P concentration was observed in the inoculation of AMF. Increase in K, Ca and Mg concentrations was found in the AMF + *Rhizobium* + FYM treatment. Sodium and manganese remained unaffected. Dry matter production and seed yield of cluster bean were increased with inoculation in both years with maximum enhancement with dual inoculation coupled with FYM application. Results indicated that soil fertility was improved by inoculation or with the application of FYM.

Singh *et al.* (2001) studied the effect of *Rhizobium,* vesicular arbuscular mycorrhiza and phosphorus on the growth and yield of lentil (*Lens culinaris*) and field pea (*Pisum sativum*). Lentil cv. LL 147 and field pea cv. PG 3 were treated with 20 and 40 Kg P_2O_5/ha, respectively (T1). *Rhizobium leguminosarum* (T2), *Glomus fasciculatum* (T3), T2+T3, (T4), T2+20 Kg P_2O_5/ha (T5), T3+20 kg P_2O_5/ha (T6), T4 + 20 kg P_2O_5/ha (T6), T4 + 20 kg P_2O_5/ ha (T7) and control (untreated, T8) in a field experiment conducted in Punjab, during the rabi season of 1992-93. The number of pods per plant was higher in the T1, T5 and T7 treatments than in T8. Highest grain yield of lentil and pea was recorded from T7 (1546 and 1643 Kg/ha, respectively), followed by T6 (1521 and 1381 kg/ha, respectively) and T5 (1467 and 1524 kg/ha, respectively), which were on a par with T1 (1283 and 1586 kg/ha, respectively).

Chauhan and Pokhriyal (2002) made an attempt to study the effect of biofertilizers i.e., *Rhizobium* and mycorrhiza on nodulation and nitrogen fixation in *Albizia lebbeck* seedlings. Inoculated plants alone or combined performed better than uninoculated control plants. Plants performed best when inoculated with both *Rhizobium* and mycorrhiza (R+M) as compared to plants, which received either *Rhizobium or* mycorrhiza, and even plants, which were not at all inoculated. Plants with *Rhizobium* inoculation had an edge over mycorrhizal inoculated plants.

Rajasekhar *et al.* (2002) inoculated together *Rhizobium* isolated from *Pterocarpus santalinus* and VA mycorrhizal fungi *viz, Glomus mosseae, G. fasciculatum* and *G. macrocarpum* on *P. santalinus.* The individual inoculation of the rhizobia and VA mycorrhizae enhanced plant growth and biomass production over control, where as the dual inoculation showed a much significant increase in plant biomass, number of nodules, phosphorus and nitrogen centents in the plants.

Raverkar and Tilak (2002) conducted a pot culture experiment consisting of two cultivars (Kalitur and Lee), four different strains of *Bradyrhizobium japonicum* (S 103, SM 113, SB 117 and SB 119) and an uninoculated control using unsterilized sandy-loam soil under greenhouse conditions. In both cultivars vesicular arbuscular mycorrhizal (VAM) colonization by indigenous propagules improved significantly ($P<0.05$) due to inoculation with various bradyrhizobial strains. Significant enhancement in synthesis of number of nodules, nodule dry biomass and nitrogenase activity was recorded due to bradyrhizobial inoculation. Amount of N_2-fixed in Kalitur and Lee due to inoculation with *B. japonicum* strains varied between 150.16 and 329.14 and 79.60 and 200.13 mg plant^{-1}, respectively. Improved VAM colonization due to introduced bradyrhizobia enhanced P uptake substantially in both the cultivars, thus complementing the process of N_2-fixation and dry biomass production.

Pandey *et al*. (2003) conducted a study at Ranichauri, Tehri Garhwal, to evaluate the effects of vesicular arbuscular mycorrhizas (VAM), rhizobium (*Rhizobium leguminosarum*) and phosphorus sources, i.e. single super phosphate (SSP) and Mussoorie rock phosphate (MRP), on the nodulation, growth and yield of pea cv. VL-7. The experiment comprised 10 treatments with different phosphorus sources and biofertilizers either alone or in combinations in acidic soil (pH 5.0 and P fixation capacity 54.8%) under rainfed cool humid temperate conditions. The results revealed that phosphate sources (SSP and MRP) and biofertilizers individually or in combinations significantly increased the nodulation, growth and yield of the crop. Among the treatments, the combinations of dual inoculation, VAM+rhizobium with a mixture of phosphorus sources, i.e. SSP: MRP (50:50), recorded the maximum nodulation, growth and yield.

Reddy and Ahlawat (2001) conducted a field experiment in New Delhi during 1989-90 and 1990-91 to analyze the effect of phosphorus, zinc and biological fertilizers (*Rhizobium leguminosarum* and /or vesicular arbuscular mycorrhizal (VAM) fungi *Glomus fasciculatum*, applied singly or in combination) on dry matter production, grain yield and nutrient uptake in lentil cultivars JL 1 and Lens 4076. JL 1 produced a higher grain yield than Lens 4076 in 1990-91, but both cultivars were similar in dry matter production and nutrient uptake. Treatment with 17.2 Kg P/ha (as single super phosphate) +5 Kg Zn/ha (as $ZnSO_4$) produced a higher dry matter and grain yield than the control (no fertilizer), N, P and Zn uptake were improved by the application of P and Zn. The combined inoculation with Rhizobium and VAM fungi resulted in increases in dry matter, grain and straw yield and nutrient (N, P and Zn) uptake when compared with their individual inoculation.

VAM + Azotobacter/ Azospirillum

Mohandas (1987) reported that inoculation of tomato with *Glomus fasciculatum* and *Azotobacter vinelandii* singly and in combination in the field resulted in significantly increased leaf area, shoot dry weight, nitrogen content, phosphorus content and yield in relation to uninoculated controls. While fungal treatment alone substantially increased growth, nitrogen content, phosphorus content and yield, its combination with *A. vinelandii* produced additional effects on leaf area, shoot dry weight, phosphorus content and yield.

Kandasamy *et al.* (1988) screened 60 sweet potato cultivars which showed root infection by VAM fungi varying from 13.89 to 46.3% depending on genotypes. In pot trials, inoculation of sweet potato cuttings with VAM fungi (*Glomus fasciculatum* or *G. mosseae*), *Azospirillum brasilense* and especially with VAM fungi + *A. brasilense* significantly increased growth, plant N and P contents, tuber weight and starch content.

Nagarajan *et al.* (1989) studied the effect of combined inoculation of *Azospirillum brasilense* and *Glomus fasciculatum* on mulberry. Mulberry cuttings were dipped in a slurry of *A. brasilense* (10^8/g soil) for 15 minutes and then dried in shade. A 100 g *G. fasciculatum* inoculum (260 spores/100 g soil) was placed 2.5 cm below the cuttings in the pots. Treatment with *A. brasilense* and *G. fasciculatum* in combination resulted in increased shoot biomass and leaf weight compared with the control whereas the individual treatments had little effect on these parameters. Effects on plant height and root biomass were not statistically significant.

Indi *et al.* (1990b) studied the yield and nutrient uptake by brinjal influenced by *Azospirillum brasilense* and / or *Glomus fasciculatum* inoculations under graded phosphorus level. The dual inoculations with these organisms combined with applications of 75% of the recommended P fertilizer dosage gave the highest yield in field trial on a P deficient soil.

Negi and Tilak (1990) studied the interaction between *Glomus versiforme* and *Azospirillum brasilense* in barley. Plants inoculated with both the organisms in pot experiments fixed more N than when single inoculations were used. Maximum fixation of 335.3 mg N/plant after 120 days was obtained in the dually inoculated treatment.

Negi *et al.* (1990) studied the influence of soluble phosphorus fertilizer on the interaction between a vesicular-arbuscular mycorrhizal fungus and *Azospirillum brasilense* in barley (*Hordeum vulgare* L.). In pot trials, barley was inoculated with *Glomus versiforme*, alone or with *Azospirillum brasilense*, and given no fertilizer, 50 kg N/ha as ^{15}N-labelled ammonium sulphate, 30 kg P_2O_5/ha or N + P. Inoculation with both endophytes increased growth and grain production compared with the uninoculated control. Inoculation with *G. versiforme* alone increased grain yield only when P was applied. Dual inoculation also caused the greatest increases in N and P uptake by roots, shoots and grains. The amount of N_2 fixed increased with increases in plant growth and was highest after dual inoculation, reaching 58.12 and 79.01 mg/plant in shoots and grain, respectively at 120 d after sowing without applied P and 46.94 and 143.8 g/plant with applied P. Acetylene reduction rates were very low in roots of inoculated plants and were not corrected with plant N content.

Singh *et al.* (1990) studied the interaction effect of *Glomus fasciculatum* and *Azospirillum brasilense* on yields of various genotypes of wheat (*Triticum aestivum*) *viz.*, WZ 771, HD-2329, Kalyansona, WZ 147, HD 2181, Arjun HD 2285, Sonalika, in pots. The colonization among various genotypes was highest with HD-2329 and lowest with Sonalika. Seed inoculation with *A. brasilense* did not significantly increase yield of HD-2285 and WZ 771 wheat, whereas *G. fasciculatum* inoculation significantly increased grain and straw yield of all the varieties. Increase in yield was higher when *G. fasciculatum* and *A. brasilense* were coinoculated. Nitrogen and phosphorus contents in plants were also higher after soil inoculation with *G. fasciculatum*.

Chokraborty and Sharma (1993) studied leaf nutrient composition of sweet orange as affected by combined use of bio- and chemical fertilizers. Application of the bacterium *Azospirillum brasilense* (3 g/plant, as a root dip at transplanting) to sweet orange cv. Mosambi plants substituted for at least 1/4 of the N fertilizer requirement. Application of the vesicular arbuscular mycorrhiza *Glomus fasciculatum* (50 g of culture in a sand medium placed 15 cm below the soil near each plant) completely substituted for the P fertilizer requirement and also increased leaf Zn, Cu, Fe and B content, compared with untreated controls.

Kumari and Balasubramanian (1993) studied the effect of combined inoculation of VAM fungi (*Glomus fasciculatum, Gigaspora margarita* and *Acaulospora laevis*) with *Azospirillum brasilense* on growth and nutrient uptake by coffee seedlings in a nursery experiment. Inoculation with *G. margarita* + *A. brasilense* significantly increased shoot length, root length and total dry weight of the plants compared with control and resulted in maximum VAM colonization of roots (44%). Both N and P uptake were significantly increased by inoculation with *G. margarita* + *A. brasilense* and the uptake of micronutrients Fe, Cu, Zn and Mn were also increased. It has been suggested that the use of these biofertilizers may significantly enhance the growth and vigour of coffee seedlings grown in the nursery.

Subbiah (1994) conducted experiments with chilli (*Capsicum* sp.) (Co. 1) and bellary onion (NP.53) at the University Orchard, Tamil Nadu Agricultural University, Coimbatore, during Sep.-Oct. 1987 and June-July 1989, respectively. The soil, a vertisol, was low in available N (75-107 kg/ha) and P (2.8-3.5 kg/ha), and high in available K (280-300 kg/ha). Seeds were treated with *Azospirillum brasilense* (50-100/100 g) and sown in nursery beds which had been inoculated with VAM fungi (*Glomus fasciculatum* (*G. fasciculatum*) at 1 kg/m^2 and *A. brasilense* at 2 kg/ha. Seedlings (45 days old) from beds inoculated with the biofertilizers, or from untreated beds, were transplanted onto ridges and received fertilizer treatments comprising 50, 75 or 100% of the recommended N dose + 100% of the recommended dose of P (as superphosphate). Data on the yields of dry chilli pods and onion bulbs, and on N, P and K uptake are tabulated. No significant effects on yields were observed. Application of biofertilizers had some significant effects on nutrient uptake.

Singh *et al.* (1995) studied the effects of vesicular arbuscular mycorrhizal fungi, phosphate solubilizing microbes (PSM) and *Bradyrhizobium japonicum* on nodulation and DM accumulation at 45d in soybeans cv. Bragg. Inoculation with mycorrhizal fungi caused about 56% root infection. Addition of rock phosphate (RP) and inoculation with *B. japonicum* or PSM stimulated root infection by native as well as exotic mycorrhizal fungi. Addition of RP and inoculation with *B. japonicum*, mycorrhizal fungi and PSM individually increased the number and DW of nodules/plant and shoot DM accumulation. The number and DW of nodules/plant were significantly higher than the control due to RP x *B. japonicum*, RP x mycorrhizas, *B. japonicum* x PSM, mycorrhizas x PSM, *B. japonicum* x mycorrhizas and RP x PSM interactions. Among the 4 factor interactions, the number and DW of nodules/ plant were highest with RP x *B. japonicum* x mycorrhizas x *P. straite* (? *Pseudomonas striata*).

Tilak (1995) studied the interaction between VA mycorrhiza and *Azospirillum brasilense* in pearl millet. A field experiment was conducted using a mixed culture of the most efficient strains of *Azospirillum brasilense* B1 and B2 and *Glomus versiforme* as a sand and soil (1:1) mixture containing extramatrical spores and root segments of *Cenchrus* sp. The amount of N_2 fixed by the plant was higher in inoculated plants and was maximum in the dually inoculated treatment. This treatment also brought a significant increase in grain yield over all other treatments. Plants inoculated with both VAM and *Azospirillum* fixed more N_2 and produced higher grain yield than singly inoculated plants. This observation was attributed to the collective effects of both mycorrhizae and *Azospirillum*. The effect was, however, more pronounced in the presence of P, showing that P was required for N fixation. The P made available by the VAM fungus might have contributed to the N_2 fixation in the presence of *Azospirillum*.

Gowda *et al*. (1995) investigated the role of synergistic effect of VA mycorrhiza and *Azotobacter* in mulberry. They used eight month old cuttings of S-54, K_2 and S-34 mulberry varieties, six *Azotobacter* strains and different VAM inocula *viz., Glomus fasciculatum, Glomus mosseae* and their mixture. The observations were made for survivability percentage, fresh and dry biomass of leaf, NPK content of leaf etc. They reported that all the parameters of three mulberry varieties studied were improved due to dual inoculation of VAM fungi and *Azotobacter* when compared with single or no inoculation. They suggested dual inoculation technique of mulberry with VAM fungi and *Azotobacter* to be, an ideal technology for improved mulberry production.

Sharma and Mukerji (1995) studied the interaction between VAM fungus (*Glomus macrocarpum*) and free-living nitrogen fixers (*Azospirillum brasilense*) and its effect on growth of jute (*Corchorus allitorius*). The plant height, dry weight, and yield increased in order of *G. macrocarpum* and *A. brasilense* in combination > *A. brasilense* alone > *G. macrocarpum* alone > without *G. macrocarpum* and *A. brasilense*.

Prabakaran *et al*. (1995) studied the effect of endomycorrhizae and *Azospirillum* on growth and yield of maize in acid soil. They conducted a field experiment with bio-inoculation of VAM fungi namely *Glomus fasciculatum, G. mosseae, Gigaspora calospora* and *Acaulospora laevis* and *Azospirillum brasilense*. They recorded significant enhancement in growth and yield attributes over their N and P controls.

Ratti and Janardhanan (1996b) studied the response of dual inoculation with VAM *Glomus aggregatum* and *Azospirillum* on the yield and oil content of Palmarosa (*Cymbopogon martinii* var. *motia*). Dual inoculation increased the growth, yield and oil content of palmarosa significantly over uninoculated control as well as inoculations with *G. aggregatum* or *A. brasilense* alone. *A. brasilense* also stimulated the VAM colonization and increased VAM spore population in the rhizosphere soil of palmarosa. Nitrogen content of leaf tissue was higher in the plants inoculated with *A. brasilense*, while leaf P content was greater in *G. aggregatum* inoculated plants than in dual inoculation.

Murumkar and Patil (1996) investigated the effects of VAM fungi *Glomus fasciculatum, Gigaspora margarita, Acaulospora* sp. and their combination, and the diazotrophs *Azotobacter chroococcum, Azospirillum lipoferum* and their mixture on *Capsicum annum* var. *grossum*. In general, shoot and root length, dry weight, fruit yield, nutrient uptake and mycorrhizal establishment were significantly increased by mycorrhizal inoculation together with different diazotrophs. The most effective VAM treatment was a mixture of all the 3 species, followed by individual treatments with *G. fasciculatum, G. margarita* and *Acaulospora* respectively. The combination of VAM fungi and mixed diazotrophs treatment gave the highest yield and VAM root colonization values.

Mandhare *et al.* (1998) carried out a green house pot experiment with onion (cv. N.-53) showed that application of *G. fasciculatum* + *A. chroococcum* + 50% of the recommended P rate resulted in the greatest root length (30.50 cm), plant height (57.90 cm), bulb girth (19.20), bulb fresh weight (321.66 g/bulb), root colonization 78.00%) and P uptake (38.60 mg/plant). The recommended P rate consisted at 50 kg/ha applied as single superphoshate.

Sansamma *et al. (*1998) studied the influence of biofertilizers on productivity of guinea grass intercropped in a coconut garden. In a field experiment from September 1993 to August 1995 in Vellayani, Kerala, guinea grass [*Panicum maximum*] intercropped in a coconut garden was inoculated with *Azospirillum, Azotobacter* or vesicular arbuscular mycorrhiza (VAM) alone or with different rates of N, P and K. Yield was increased by all biofertilizer treatments, compared with uninoculated controls. Yields were generally greatest with VAM, followed by *Azotobacter*, and were generally increased by addition of mineral fertilizers. Biofertilizers could reduce the N fertilizer requirement by 25%. VAM inoculation combined with N and K fertilizers gave better results than the recommended NPK treatment.

Sugavanam *et al.* (1999b) conducted an experiment for the selection of an efficient vesicular arbuscular mycorrhizal fungi and *Azospirillum* sp. for inoculating *Tectona grandis.*The influence of 6 vesicular-arbuscular mycorrhizal (VAM) fungi (*Gigaspora margarita, Glomus deserticola, Glomus fasciculatum, Glomus monosporum, Glomus mosseae, Glomus versiforme*) on the growth of 30 day-old seedlings of *Tectona grandis* was tested both individually and in combinations with *Azospirillum* sp. Seedling measurements were made 150 days after sowing [120 days after inoculation]. Inoculation with individual VAM fungi or *Azospirillum* sp. increased growth, biomass production, root colonization and tissue N, P, and K concentration, and reduced root shoot ratio in all treatments, compared with uninoculated control\values, but some treatments were not significantly different from the control. Root colonization was successful in all cases. The most significant responses were observed with *Gigaspora margarita*or *Glomus versiforme* inoculation in conjunction with *Azospirillum* sp.

Rajput (1999) studied the effect of bio-fertilizers application on yield of bajra (pearl millet). A field experiment was conducted during 1988 at the JNKVV Zonal Agricultural Research Station, Morena, to study the effects of biofertilizers on yield of bajra [*Pennisetum glaucum*]. Application of 40 kg N/ha gave significantly higher grain yield (2492 kg/ha) than other treatments. Application of biofertilizers (*Azospirillum* + VAM) gave 13% higher yield than in controls, and similar yield to application of 20 kg N/ha.

Pushpa Kumari and Geethakumari (1999) determined the effect of the inoculation of *Azotobacte*r and vesicular arbuscular mycorrhiza (VAM) alone or in combination with different N and P levels (100, 75 and 50%) on sweet potato. The N: P: K recommended rate was 75:50:75 kg/ha. Results revealed that treatments involving microbial inoculation were superior to those without microbial inoculation at all levels of nutrients. A higher efficiency was observed at lower levels of nutrients with the inoculation of VAM alone. Economic evaluation revealed that 50% N and P applications along with VAM inoculation gave the maximum net return and cost benefit ratio.

Mengistu and Singh (1999) studied the effects of biofertilizers on growth, yield and nutrient uptake of onion (*Allium cepa* L). *Azospirillum brasilense, Gigaspora calospora* (vesicular arbuscular mycorrhiza, VAM), N and P fertilizers (50 and 25 kg/ha, respectively), alone or in combination, were applied on onions (cv. Pusa Red) grown in New Delhi. The highest bulb horizontal (5.38 cm) and vertical (4.06 cm) diameters were obtrained using *A. brasilense* + VAM + 50 kg N. The treatment VAM + *A. brasilense* + 25 kg P produced the highest number of bulbs. Shoot dry weight (8.28g) and leaf number (8.1) were highest using *A. brasilense* + 50 kg N + 25 kg P. *A. brasilense* + VAM + 50 kg N + 25 kg P resulted in the highest dry bulb weight (5.50 g); total soluble solids content (13.11%); and bulb P ad N uptake. Plants treated with VAM matured earlier (116 days).

Paroha *et. al.*(2000) studied the synergistic role of VAM and Azotobacter inoculation on growth and biomass production in forestry species. One month old seedlings of *Bambusa arundinacea, Gmelina arborea, Tectona grandis* and *Dalbergia sissoo* in the nursery of the State Forest Research Institute at Jabalpur, Madhya Pradesh, were inoculated with 20 g VAM (200 chlamydospores of *Glomus intraradices, Glomus mosseae* and *Acaulospora scrobiculata*, along with 10 g infected roots of the trap species *Panicum maximum*) and / or with 10 g of culture of *Azotobacter chroococcum* (10 cells/g culture). The inoculated seedlings exhibited improved growth and biomass, but there were variations among treatments and species. In general root development was more positively affected than shoot development in all species except *T. grandis*. In all 4 species root length development was maximum either with VAM or with VAM + *Azotobacter*, *T. grandis* and *Gmelina arborea* doubled their biomass yield just by mycorrhization. The response to *Azotobacter* in *T. grandis* was comparatively low. The effect of the biofertilizers was least in bamboo. The increase in biomass yield with VAM decreased with the age of the plant, and it is suggested that supplementary doses of chemical fertilizers with VAM and *Azotobacter* and may be helpful in enriching the microbial population and thereby plant growth.

Sreeramalu *et al.* (2000) studied the effect of individual and combined inoculation of *Azotobacter chroococcum* and *Glomus fasciculatum* on growth and yield of sunflowers *cv.* KBSH-1 at different levels of NPK fertilizers under field conditions in Karnataka. Dual inoculation in conjunction with 75% of the recommended NPK rate of 35:50:35 kg/ha gave the best growth and yield. However, half the recommended NPK rate with inoculation was economically optimum due to saving 50% of NPK fertilizer application.

Sugavanam *et al.* (2000) studied the effect of different mycorrhizal fungi and *Azospirillum* on *Eucalyptus tereticornis* seedlings. Transplanted 30 - days- old seedlings of *Eucalyptus tereticornis* in polybags were inoculated with 6 arbuscular mycorrhizal (AM) fungi (*Gigaspora margarita, Glomus deserticola, G. fasciculatum, G. monosporum, G. mosseae and G. verslformae*) individually as well as in combinations with *Azospirillum.*

The result indicated that inoculation of AM fungi or *Azospirillum* increased the growth, biomass, root colonization and tissue N.P. and K concentrations. Significant responses were observed in plants inoculated with *Glomus versiforme* or *Gigaspora margarita* in conjunction with *Azospirillum.*

Rajadurai and Beaulah (2000a) studied the effect of *Azospirillum* and VAM on nutrient uptake and available nutrients in post harvest soil of African marigold (*Tagetes erecta* L.). In pot experiments, African marigolds (*Tagetes erecta*) were inoculated with *Azospirillum* and /or vesicular arbuscular mycorrhizas (VAM) before sowing and given 30.0, 37.5 or 45.8 mg N. 30 or 45 mg P and 37.5 mg K/kg soil. N, P and K uptakes were highest in marigolds inoculated with both *Azospirillum* and VAM at the highest N and P application rates. Soil nutrient status was highest in soils given N and P at 45.) mg/kg soil. The lowest soil nutrient status was lowest in treatments without VAM or *Azospirillum* inoculation.

Rajadurai and Beaulah (2000b) studied the effect of inorganic fertilizers *Azospirillum* and VAM on yield characters of African marigold (*Tagetes erecta*). A pot experiment was conducted to investigate the effects of single or combined treatments of *Azospirillum*, VAM (*Glomus fasciculatum*) and NPK fertilizers on African marigold (*Tagetes erecta*) The fertilizer treatments contained 37.5 mg k/ka-45, 37.5 and 30 mg N/kg and 45 and 30 mg P/kg. Increasing levels of NPK fertilizers resulted in earlier flowering. Most and heaviest flowers, as well as longest stalks, were produced with the highest fertilizer rate combined with Azospirillum and VAM. Combined treatments were better than single treatments.

Sharma *et al.* (2001b) studied the survival rate of *Azotobacter chroococcum* (*Azo* + rhizobacteria) at periodic intervals in the rhizosphere of three F1 hybrids of bread wheat (*Triticum aestivum*) involving four varieties as parents under low input field conditions with and without the inoculation of arbuscular mycorrhizal fungi (AM fungi, *Glomus fasciculatum*) and *Azo*+ AM fungi. Effects of wheat genotypes and AM inoculation on the proliferation and survival of *Azotobacter* were prominent. Maximum *Azo* counts were observed 80 and 120 days after sowing in *Azo*+AM treatment in the cross WH 147 x PBW 175. *Azo* survival rate was associated with increase in the grain and biological yield. Inoculation by AM fungi enhanced proliferation and survival of *Azo*.

Singh *et al.* (2000a) conducted a field experiment during the rainy (kharif) season in Himachal Pradesh to assess the response of guinea grass (*Panicum maximum*) to inoculation with biological fertilizers (*Azotobacter, Azospirillum* and vesicular arbuscular mycorrhiza (VAM) and N fertilizer. N was applied at 0, 25, 50 and 75 kg/ha. Seed inoculation with VAM produced the highest green forage, dry matter and crude protein yields. Green forage, dry matter and crude protein yields increased significantly up to 75 kg N/ha.

Rajadurai *et al.* (2000) conducted a pot experiment to study the effect of *Azospirillum* and VAM on growth characters of African marigold (*Tagetes erecta* L.) with various rates of NPK fertilizers. Application of NPK at 45:45:37.5 mg/kg along with combined inoculation of *Azospirillum* and VAM exhibited increased growth with respect to plant height (144.50 cm), number of leaves (156.20) and laterals per plant (28.30). The plants treated with 45 mg N and 45 mg P/kg with combined inoculation of biofertilizers recorded a leaf area of 106.70, 55.70 and 96.70 cm^2 at 30.60 and 90 days after planting, respectively.

Sharma and Bhutani (2000) studied the leaf nutrient status of apple seedlings as influenced by VAM, *Azotobacter* and inorganic fertilizers. Apple seedlings were grown from sterilized seeds in sterile mixture inoculated with 4 species of vesicular arbuscular mycorrhizas *viz., Glomus mosseae, Glomus fasciculatum, Glomus macrocarpum* and *Glomus versiforme* and *Azotobacter.* A number of inorganic fertilizers were also applied. The leaf samples were harvested after 6 months and ground samples were digested in a diacid mixture. The highest leaf N and P contents were recorded from the dual inoculation with Azotobacter and G. fasciculatum. In general Zn, Cu and Fe contents of the leaves increased with inoculation of the biofertilizer and further increased with biofertilizer + inorganic fertilizer treatments.

Behl *et al.* (2003) conducted a filed experiment in a randomized block design with three replications over 2 years to evaluate the effect of wheat cultivars and dual inoculation of *Azotobacter chroococcum* (Azc) and arbuscular mycorrhiza fungi (AMF, *Glomus fasciculatum*) on root characters and AMF infection in three crosses of wheat. The experimental material comprised four wheat parents, WH-147, WH-157, WH-542 and PBW-175, and three F_1 crosses, WH-147 × WH-157, WH-147 × WH-542 and WH-147 × PBW-175. Comparison of treatment averages, *i.e.* control (mineral nutrients 60 kg N+30 kg P_2O_5 + 12.5 kg $ZnSO_4$ ha^{-1}, as in other two treatments), AMF and AMF + Azc, revealed that inoculation of Azc led to an increase in AMF infection in roots. Maximum root biomass was obtained in F_1 hybrids WH-147 × WH-157 in the AMF treatment and in WH-147 × PBW-175 receiving AMF + Azc. Total root length and AMF infection of roots was maximum in WH-147 × PBW-175 for all the treatments during both years. A positive association between AMF infection in roots and *Azotobacter* survival in the rhizosphere was apparent. Similarly, maximum *A. choroococcum* counts were observed 80 and 120 days after sowing in the AMF + Azc treatment in cross WH147 × PBW175.

Kathiresan and Venkatesha (2002) conducted a field experiment during the kharif season of 1998-99 in Bangalore, Karnataka, to study the effect of *Azospirillum* and vesicular arbuscular mycorrhizas (VAM) combined with different rates of N and P on the growth, yield, and quality of gladiolus cv. White Prosperity. Tabulated data on the effect of *Azospirillum* and VAM in combination with N and P on gladiolos plant height, leaf area, number of leaves, days for spike emergence days for 50% flowering, spikes/ha, corms/ha, cormels/ha florets/spike, rachis length, vase life, and cost benefit ratio are presented.

Shashidhara and Gopinath (2002) conducted studies on calendula (*Calendula officinalis*) *cv. Red Orange* to determine the effects of bioinoculants (*Azotobacter* at 200 g/ha and VAM at 15.6 g/plant) with different levels (100, 75, and 50%) of nitrogen and phosphorus on growth and flowering in Bangalore, Karnataka. during 1998-99. The one month-old seedling raised in seed pans was transplanted at 30×20 cm spacing. Significantly tall statured plants (19.60 cm), high number of branches (8.87/plant) with high number of flowers (19.67) and weight (72.20 g/plant), improved flower quality, high number of leaves (49.30/plant) and advancement of flowering (by 2 days), extension of longevity of flowers on the plant (by 2.5 days) and duration of flowering (by 4 days) were recorded with the application of 135:90:60 kg N, P_2O_5 and K_2O/ha.

VAM + Frankia

Vasanthakrishna *et al.* (1994) studied the response of *Casuarina equisetifolia* to inoculation with the vesicular-arbuscular mycorrhiza, *Glomus fasciculatum* and / or the actinorrhiza *Frankia*, in a phosphorus-deficient unsterile soil. Mycorrhizal root colonization and sporulation was greatest in plants inoculated with *G. fasciculatum* alone compared with other treatments. Single inoculation with either *G. fasciculatum* or *Frankia* significantly improved plant growth compared with uninoculated control plants. The effects of *G. fasciculatum* were greater than those of *Frankia*. Dual inoculation with both the symbionts enhanced the growth of *C. equisetifolia* more than single inoculation with either organism.

Joshaline and Balasubramanian (1995) investigated the response of *Casuarina equsetifolia* under tripartite interaction between *Frankia, Glomus mosseae* and *Casuarina* at two different phosphorus levels. The experiment was conducted in pots and plants were harvested after 90 days and data on shoot length, root length, dry weight, NPK content, nodule nitrogenase activity, nodule number and dry weight were recorded. The growth of *C. equisetifolia* improved considerably due bioinoculants, namely *Frankia* and *G. mosseae* even with half the recommended dose of P.

Tiwari (1995) investigated the effect of tripartite association between VAM fungi, *Frankia* and *Alnus nepalensis*, on seedling growth, nitrogenase activity, and biomass production. The root lengths, nodule dry matter production and acetylene reduction rates were significantly greater when the seedlings were inoculated in combination with VAM fungus, *G. mosseae* and *Frankia* than when inoculated separately.

Udaiyan and Sugavanam (1995) studied the interaction of VAM fungi and *Frankia* sp. on growth and nutrient uptake in *Casuarina equisetifolia*. They reported that single inoculations with *Glomus fasciculatum* resulted in higher growth, biomass, nodulation and spore number, whereas *Gigaspora margarita* increased tissue nutrients concentration. Root colonization increased in *G. monosporum* inoculations while mixed inoculations of selected, compatible VAM fungi in conjugation with *Frankia* sp. aid seedling growth despite the presence of native symbionts.

Balasubramanian *et al.* (1995) studied the interactive effect of vesicular-arbuscular mycorrhizae and *Frankia* on *Casuarina* seedling growth. The compatible strain of *Frankia* for VAM fungus, *Glomus fasciculatum*, was evaluated in a pot culture experiment conducted at Tamil Nadu Agricultural University in 1992. Surface sterilized *C. equisetifolia* seeds were raised in a steam-sterilized soil; aseptically the seedlings were inoculated separately with four different *Frankia* strains, namely ORS 021001, ORS 020607, ORS 022601, and CCI_3. The *Frankia* strains were obtained in the form of calcium alginate beads from the Department of Botany, University of Glasgow, UK. It is interesting to note that CCI_3 + *G. fasciculatum* combination yielded higher shoot length (146.31 cm), root length (43.78 cm) and ultimately resulted in higher biomass production over the control and *Frankia* inoculation. Further, higher nodulation, higher nitrogen build-up both in nodule (18%) and plant (7%) were also obvious in the above combination treatment. The result also explained the better performance of all *Frankia* strains over control. However, the dual inoculation with *G.*

fasciculatum had a notable performance, especially with CCI_3 and ORS 020607. On the other hand there was not much difference between dual inoculation and inoculation of *Frankia* alone for ORS 021001 and ORS 022601 strains.

Ravichandran and Balasubramanium (1999) determined the efficacy of dual inoculation of *Frankia* and VAM fungus (*Glomus fasciculatum* on the growth of *C. equisetifolia* at the Sugarcane Research Station, Cuddalore, Tamil Nadu. The growth of *Casuarina equisetifolia* seedlings can be enhanced significantly through inoculation with suitable strain of *Frankia* and also with a host specific VAM (vesicular arbuscular mycorrhizal) fungus. The host specific *Frankia* strain (CCI_3) and VAM fungus (*Glomus fasciculatum*) were identified through earlier separate pot culture experiments. In this experiment, the efficacy of dual inoculation was compared with that of sole inoculation of either microsymbiont and with uninoculated controls. Dual inoculated plants grew better than sole inoculated plants, and significantly better than uninoculated controls 105 days after inoculation; shoot and root length by was 49 and 88% more, respectively, in dual than uninoculated plants. The root volume, mycorrhizal infection percentage and nodulation characteristics (nodule numbers, nodule dry weight and nodule nitrogenase activity) of dual inoculated seedlings were also significantly higher than in the uninoculated controls.

Sugavanam *et al.* (1999a) studied the influence of tripartite interactions of VAM fungi, *Frankia* and *Casuarina equisetifolia* on growth and nutrient accumulation in host under nursery conditions. Pot culture experiments were carried out in a sandy loam, non-pasterized soil to determine the effects of six different vesicular arbuscular mycorrhizal (VAM) fungi such as *Gigaspora margarita, Glomus deserticola, G. fasciculatum, G. mosseae, G. monosporum* and *G. versiforme* individually as well as in combinations with *Frankia* sp. On the seedling growth of *Casuarina equisetifolia.* Inoculatiion of VAM fungi or *Frankia* sp. increased the growth and biomass. Among the different VAM fungi, *Glomus fasciculatum* followed by *G. versiforme* significantly enhanced the growth, stem girth and biomass. Inoculation with *G. manosparmum* increased mycorrhizal infection and spore number in the soil. Co-inoculation with *Gigaspora margarita* and *Frankia* sp. Significantly increased root colonization and spore number. Nodule number and nodular biomass were increased with the inoculation of VAM fungi and were further increased by inoculations with *Frankia* sp. with a maximum in *Glomus mosseae* and *Frankia* sp. inoculated seedlings. Inoculation with *Gigaspora margarita* individually or in combination with *Frankia* sp. significantly increased tissue P and K concentration. Significant increase of N concentration was observed in *Glomus monosporum* inoculation.

Sempavalan *et al.* (2001) investigated the relationships between the host plant and the microorganisms involved in the multiple symbiosis and determine the degree of colonization and the relative effect of each combination on plant growth, nodulation and nitrogen fixation of the host plant *Casuarina equisetifolia.* Tabulated data showing the combined inoculation of *Frankia* strains (Ce2, ORSO20607 and CC13) and *Glomus fasciculatum* on plant dry weight, nodule dry weight and mycorrhizal colonization in *C. equisetifolia* are given. Inoculation of either *Frankia* or *G. fasciculatum* increased the biomass by 70% over control. For nodulation of *Frankia* strains. Inoculation of Ce2 produced an average of 0.467 nodules/

plant during the study period as against uninoculated control (0.068 nodules/plant). Ce2 and *g. fasciculatum* recorded the maximum colonization of 65.6% at 120 days after inoculation (DAI), against the inoculation of *G. fasciculatum* alone (58.5%). Nitrogen uptake of dually colonized plants ranging from 23.2-48.72 mg N/plant as maximum compared to 14.6 mg N/plant in uninoculated control at 120 DAI. Phosphorus uptake was also higher in dually colonized plants, ranging from 3.56-10.03 mg P/plant as compared to 21.6 mg P/ plant in the control.

VAM + Phosphate Solubilizing Microbes (PSM)

Gopalkrishna *et al.* (1990) studied the interaction between *Glomus fasciculatum* and two phosphate solubilizing fungi, *Penicillium funiculosum* and *Aspergillus niger* in finger millet. The inoculations produced a synergistic action and resulted in improved growth and nutrient uptake.

Gaur and Rana (1990) studied the role of VA mycorrhizae, phosphate solubilizing bacteria and their interactions on growth and uptake of nutrients in wheat. Inoculations with PSB's such as *Pseudomonas striata* and *Agrobacterium radiobacter* and VAM fungi such as *Glomus fasciculatum* and *Gigaspora margarita* improved dry matter yield. Higher yield was obtained with an application of microbial inoculation and fertilizer together.

Sreenivasa and Krishnaraj (1992) reported increased fruit production in chillies following inoculations with *Glomus fasciculatum* or *G. macrocarpum*. The percentage root colonization by the mycorrhizal fungi, sporulation and the number of infective propagules increased with increasing amounts of rock phosphate present in the treatment. Native endophytes of VAM fungi were not stimulated by rock phosphate alone, but mycorrhizal development increased when *Pseudomonas striata* was present. Dual inoculation with *P. striata* and *G. fasciculatum* or *G. macrocarpum* increased VAM development, with *G. fasciculatum* showing better response. Similarly dual inoculation with *G. fasciculatum* and *P. striata* resulted in the greatest increase in shoot dry mass, fruit yield and shoot P content of chilli. Micronutrient content also increased. It was concluded that P solubilizing bacteria increased the amount of soluble P and the mycorrhizas increased P uptake of the plant and that a synergistic effect occurred when both were present.

Jones and Sreenivasa (1993) studied the effect of inoculation of a VAM fungus *Glomus fasciculatum* and / or a phosphate solubilizing bacterium *Pseudomonas striata* on the population of free living nitrogen fixing *Azotobacter* and phosphate solubilizing bacteria in the rhizosphere of sunflower (*Helianthus annuus*) at 30, 60 and 83 days after sowing. Phosphate was given as rock phosphate (0.521 - 2084 mg/pot) or super phosphate (0.2344 mg pot). Each pot contained 10 kg of a P-deficient black clayey soil. The populations of *Azotobacter* and phosphate solubilizing bacteria increased with increasing P concentration in host plants inoculated with both *G. fasciculatum* and *P. straita* up to 60 days after sowing and then decreased. Superphosphate was superior to rock phosphate in increasing the *Azotobacter* population. Population increases were greater following dual inoculation than individual inoculation.

Chandraghatgi and Sreenivasa (1995) studied the possible synergistic interactions between *Glomus macrocarpum* and *Bacillus polymyxa* in chilli. Their study clearly brought out that the combined inoculation of *G. macrocarpum* and *B. polymyxa* in chilli was better and yielded highest fruit weight. According to them, it was possible to replace 25% of P in form of Superphosphate with Rock phosphate following combined inoculations as compared to individual inoculations.

Tilak *et al.* (1995b) investigated the synergistic effects of phosphate solubilizing bacterium - *Pseudomonas striata* and *Gigaspora margarita* in bean. Uniform seed treatment with *Bradyrhizobium japonicum* was also given. They reported a significant increase in root infection, root biomass and yield with *G. mosseae* or *G. fasciculatum* than the soil inoculation with VAM or *P. striata* alone.

Potty and Harikumar (1995) studied the interaction between VAM fungi and phosphobacterium in sweet potato rhizosphere. They reported that effectiveness of combined inoculation of VAMF and phosphobacterium within a lower range of phosphorus was beneficial and mainly mycorrhiza-mediated, whereas further addition of P release by phosphobacterium, and the uptake mechanism of host were not only mycorrhizae mediated, but also had other mechanisms which may be operative. The contribution of phosphobacterium may be the reduction of plant growth substance rather than phosphate solubilizers in their study.

Karthikeyen *et al.* (1995) studied the effect of dual inoculation of phosphobacterium (*Pseudomonas striata*) and VAM fungi (*Glomus fasciculatum*) on the growth of neem under potted condition. They reported that combined inoculation enhanced the dry matter production, VAM colonization, and plant nutrient uptake significantly over uninoculated control and individual inoculation.

Das *et al.* (1995b) studied the effect of coinoculation with VAM fungus *Gigaspora* sp. and phosphate solubilizing *Penicillium* sp. PSB1 and PSB2 on tomato and greengram grown in sterilized soils. They reported varied growth responses of crop to *Gigaspora* sp. and *Pencillium* sp. depending upon nutrient levels of growth medium, amount of growth medium and stage of crop growth. *Gigaspora* sp. and *Pencillium* sp. had no effect on tomatoes grown in 4 kg of soil-sand mixture with low level of nutrients. *Gigaspora* sp. increased grain yield of greengram at half the recommended dose of P applied to Assam soils whereas PSB isolate increased grain yield at recommended doses. They suggested the testing of *Gigaspora* sp. and PSB isolate 1 for field scale application for their use as biofertilizer in future.

Bayan and Talikdar (1997) studied the effect of VAM fungus (*Gigaspora* sp.) and a phosphate solubilizing bacterium (*Bacillus* sp.) on greengram grown in a sterilized soil: sand mixture containing two levels of Mussoorie rock phosphate (78 and 156 mg 12 kg soil: sand mixture). *Gigaspora* sp. significantly increased grain yield of greengram at both the levels of applied rock phosphate, while *Bacillus* sp. increased it at the higher level of rock phosphate addition. Coinoculation resulted in the elimination of the beneficial effect of *Gigaspora* sp.

Tandon and Prakash (1998) studied the influence of soil inoculation with VA mycorrhiza and phosphorus solubilising microorganism on growth and phosphorus uptake in *Sesamum indicum*. The effect of microphos, a culture containing *Pseudomonas striata, Bacillus polymyxa* and *Aspergillus awamori*, individually and in combination, on the solubility and availability of insoluble phosphates and the growth response in sesame, were investigated in pot experiments during the kharif season. Rock phosphate (0 to 75 kg P_2O_5/ha) was used as a phosphate source in a phosphorus deficient soil. The inoculation with VAM and/or PSM resulted in a significant increase in the dry matter accumulation and phosphorus uptake by the plants. Synergistic effects of VAM and PSM on the dry mater accumulation and phosphorus uptake by the plants were observed with applications of 30 kg/ha rock phosphate. The number of capsules produced per plant increased by 41% over the respective control in the mycorrhizal plants given 45 kg/ha rock phosphate, thus resulting in an increase in seed yield per plant.

Singh and Kapoor (1998) studied the effects of inoculation of phosphate-solubilizing microorganisms and a vesicular arbuscular mycorrhizal fungus on mungbean grown under natural soil conditions. *Bacillus circulans* and *Cladosporium herbarum* as PSM and *Glomus fasciculatum* as VAM, with or without Mussoorie rock phosphate was used in a P-deficient natural non-disinfected sandy soil. The mycorhization levels increased following inoculation with PSM or *G. fasciculatum*. In general, the increase in yield was higher in the presence of rock phosphate and inoculation with a combination of PSM and VAM fungus. Highest N and P uptake by *V. radiata* was recorded after treatment with a combination of *B. circulans, C. herbarum* and *G. fasciculatum* in the presence of rock phosphate. Generally the PSM population increased after fungus inoculation.

Singh and Kapoor (1999) studied the effect of inoculating wheat (*Triticum aestivum*) with the phosphate solubilizing microorganisms (PSM) *Bacillus circulans* and *Cladosporium herbarum* and the vesicular-arbuscular mycorrhizal (VAM) fungus *Glomus* sp. with or without Mussoorie rock phosphate (MRP) amendment in a nutrient-deficient natural sandy soil. In the sandy soil of low fertility root colonization by VAM fungi was low. Inoculation with *Glomus* sp. improved root colonization. At maturity, grain and straw yields as well as N and P uptake improved significantly following inoculation with PSM or the VAM fungus. These increases were higher on combined inoculation of PSM and the VAM fungus with MRP amendment. In general, a larger population of PSM was maintained in the rhizosphere of wheat in treatments with VAM fungal inoculation and MRP amendment. The results suggested that combined inoculation with PSM and a VAM fungus along with MRP amendment can improve crop yields in nutrient-deficient soils.

Gurumurthy *et al*. (1999b) investigated the effect of inoculation of an efficient VAM fungus *Sclerocystis dussii* and a P-solubilizer, *Pseudomonas striata* on growth and nutrition of teak (*Tectona grandis* L.) seedlings at optimum level of P tried in two forms as single superphosphate (SSP) or rock phosphate (RP). Per cent mycorrhizal root colonization, mycorrhizal spore count, population of free living N_2-fixers and P-solubilizers, activities of acid phosphatase and alkaline phosphatase were significantly highest in the dual inoculated seedlings as compared to seedlings inoculated with either *S. dussii* or *P. striata* when given

P as SSP or RP. Similarly, plant height, stem diameter, leaf area, root length, root and shoot dry weights were also significantly highest in the dual inoculated seedlings as compared to individual inoculations. A matching trend was observed with respect to shoot P, Zn, Cu, Mn, and Fe concentrations in the dual inoculated seedlings given P as SSP or RP. All these parameters were least in the uninoculated seedlings given P as SSP or RP.

Chandra *et al.* (2000) explored the possibility of employing VAM and PSF technologies in curtailing the cost of production of *Chrysanthemum* and *Gladiolus*. Through their preliminary screening of a number of VAM fungi and PSM, they identified *Glomus fasciculatum* and *Aspergillus niger* to be the most potential bioinoculants for both the crops. Response of both the crops to these inoculants with different doses of P fertilizer (full, half and quarter of the recommended dose) in terms of growth of plants as well as quality and number of flowers under pot/microplot conditions alongwith other inputs used in their cultivation was evaluated. The performance of both the crops with quarter dose of P fertilizer and bioinoculants was found to be as good as with full dose of P fertilizer alone. The findings indicated that the application of VAM and technologies in cultivation of *Chrysanthemum* and *Gladiolus* may curtail the expenses incurred on P fertilizer to one fourth and make their production less expensive.

Mukherjee and Rai (1999) studied the sensitivity of P uptake to change in root growth and soil volume as influenced by VAM, phosphate solubilizing bacteria (PSB) and P levels in wheat and chickpea. In pot and field trials in New Delhi, India, wheat cv. Kundan and chickpeas [*Cicer arietinum*] cv. Pusa 372 were given 0 or 60 kg P/ha and were not inoculated or soil inoculated with the VAM fungi *Glomus fasciculatum* or *G. macrocarpum* or seed inoculated with the PSB *Pseudomonas striata*. P fertilizer in both pot increased Root dry weight and P uptake of both crops and field trails. P uptake was significantly higher under field conditions. Under field conditions, root dry weight and P uptake were generally highest with *P. striata*, followed by *G. fasciculatum* followed by *P. striata* in treatments without P fertilizer, with no significant difference between biofertilizers in treatments with P.

Meshram *et al.* (1999) compared different P sources, mycorrhizal species and P-solubilizing bacteria for chickpeas cv. Phule G-12. Yield was 2.38 t/ha without added P, and among P sources ranged from 2.68 t with superphosphate to 3.52 t with rock phosphate in a field study at Rahuri, maharashtra in rabi 1991/92. Inoculation with a mixture of 3 VAM species (*Glomus mosseae, Gigaspora calospora* and *Acaulospora* sp.) gave a yield of 3.78 t/ha, compared with 3.54, 3.22 and 3.17 t from the individual species as listed.

PSM

Chandra *et al.* (2000) explored the possibility of employing VAM and PSF technologies in curtailing the cost of production of *Chrysanthemum* and *Gladiolus*. Through their preliminary screening of a number of VAM fungi and PSM, they identified *Glomus fasciculatum* and *Aspergillus niger* to be the most potential bioinoculants for both the crops. Response of both the crops to these inoculants with different doses of P fertilizer (full, half and quarter of the recommended dose) in terms of growth of plants as well as quality and number of flowers under pot/microplot conditions alongwith other inputs used in their cultivation was evaluated. The performance of both the crops with quarter dose of P fertilizer and bioinoculants was found to be as good as with full dose of P fertilizer alone. The findings indicated that the application of VAM and technologies in cultivation of *Chrysanthemum* and *Gladiolus* may curtail the expenses incurred on P fertilizer to one fourth and make their production less expensive.

Kalavathi *et al.* (2000) examined the effect of inoculation with VA-mycorrhizal (VAM) fungi (*Glomus fasciculatum*) and/or phosphorus solubilizing bacteria (*Pseudomonas straiata* or *Bacillus polymyxa* [*Paenibacillus polymyxa*] on the growth and nutrient uptake of neem (*Azadirachta indica*) seedlings under nursery conditions. Data on dry matter production, VAM colonization and N, P and K uptake are tabulated for 30, 60 and 90 days after sowing [and inoculation]. The results clearly indicated that combined inoculation of VAM and phosphobacteria markedly increased the neem seedling growth when compared with the individual inoculants alone or with the uninoculated control. The increase in the growth was attributed to the accompanying increase in nutrient uptake (N, P, K) by the neem seedlings. The results indicated that the 3 types of organisms have a synergistic effect.

Mukherjee and Rai (2000) conducted a field experiment in New Delhi to study the effect of vesicular arbuscular mycorrhizas (VAM) and phosphate-solubilizing bacteria (PSB) on the wheat (*Triticum aestivum* cv. Kundan) and chickpea (*C. arietinum cv. Pusa 372*) growth, yield and phosphorus uptake. Main plot treatments comprised: 0 and 60 kg P/ha and subplot treatments comprised: inoculation with *Glomus fasciculatum, G. macrocarpum, Pseudomonas striata* or no inoculation (control). Application of phosphorus exhibited perceptible influence on yield of the crops. Biofertilizers and phosphorus nutrition interaction showed significant influence on growth and phosphorus uptake by wheat and chickpea compared with either of the components applied separately. Among the biofertilizers *P. striata* registered the highest grain yield in both crops at both levels of phosphorus, closely followed by *G. fasciculatum*. Among the VAM fungi, *G. fasciculatum* had positive effects on yield and phosphorus uptake, unlike *G. macrocarpum*.

Korikanthimath *et al.* (2000) studied the status of vesicular-arbuscular mycorrhizae and phosphate solubilizers in robusta coffee and cardamom mixed cropping system. An alternate row of robusta coffee, which was planted in 1947, removed in May 1985 and replaced with cardamom in June 1995 in a field experiment conducted in Kodagu, Karnataka. The population of vesicular arbuscular mycorrhizas (VAM) and phosphate solubilizer bacteria and fungi were determined from soil samples collected from the rhizosphere of both mixed and monocropped coffee and cardamom. Cardamom and coffee recorded higher number of VAM spores in the mixed (32 and 34 spxes/10 e soil, respectively) compared to the monocropping (21 ad 32 spores/10 g soil, respectively) system. The population of stabilizing bacteria and fungi was higher in coffee and cardamom, respectively in both cropping systems. An increase in the population solubilizing fungi was observed in the mixed cropping.

Subbiah and Narayanaswami (2001) conducted an experiment to study the interaction effect of phosphate and biofertilizers on P use efficiency, yield and quality of fruits in bhendi (*Abelmoschus escuentus* L. Moench) in Coimbatore with treatments, consisting of two sources of P (mussooriephos and superphosphate), three levels of phosphorus (20, 35 and 50 kg P_2O_5/ha), two methods of inoculation of biofertilizers: without biofertilizers and with biofertilizers (vesicular arbuscular mycorrhizas (VAM) fungi at 100 g/m^2 as soil application and phosphobacterin at 2 kg/ha as seed treatment and soil application), and two types of soils, red (noncalcareous) and black soil (calcareous). Results revealed that application of 35 kg P_2O_5/ha as mussooriephos in association with dual inoculation of VAM and phosphobacterin in black soil had beneficial effect not only on enhancing the fruit yield and P-use efficiency but also on improving the quality of fruit in bhendi, besides saving the cost of 15 kg P_2O_5.

Singh and Tilak (2001) studied the phosphorus nutrition through combined inoculation with phosphorus solubilising rhizobacteria and arbuscular mycorrhizae. The synergistic effect of combined phosphorus fertilizers sources of natural P (varisite, strengite, fluorapatite, hydroxyapatite, and tricalcium phosphate) along with bioinoculants (*viz*., phosphate solubilizing rhizobacteria and arbuscular mycorrhizas) is discussed.

Vijaya and Srivasuki. (2001) inoculated seedlings of *Azadirachta indica* with the vesicular arbuscular mycorrhizal fungus *Glomus fasciculatum* and/or the phosphatase-producing fungus *Aspergillus fumigatus* in a pot experiment. Treatments used were: (i) sterilized control; (2) *G. fasciculatum* inoculation (#) *A. fumigatus* inoculation; (4) *G. fasciculatum* + *A. fumigatus* inoculation; and (5) unsterilized control. Dual inoculation increased shoot and root biomass, root length, concentrations of P, and to a lesser extent of P and Mg, and nutrient uptake of *A. indica* seedlings, compared with single inoculation. Cu and Zn concentrations in shoots were enhanced *by G. fasciculatum*. At harvest, depletion of P in the rhizosphere soil increased in the order of sterilized soil < *A. fumigatus* < *G. fasciculatum* < *A. fumigatus* + *G. fasciculatum*, which corresponds with the enhanced ion concentration in the plants. Inoculation with *A. fumigatus* increased phosphatase activity in the rhizosphere.

VAM + PSM + Nitrogen Fixers

Poi *et al*. (1989) studied the response of chickpea *Cicer arietinum* L. to combined inoculation with *Rhizobium*, phosphobacteria and mycorrhizal organisms. *Rhizobium, Bacillus polymyxa* and *Glomus fasciculatum* were inoculated to *Cicer arietinum* in pot experiments. Inoculation with *Rhizobium* alone markedly increased N uptake by the plants but a significantly greater response was obtained by inoculation with all 3 test organisms. Triple inoculations also resulted in significantly greater dry matter production and phosphate uptake as compared with single or double inoculation with any of the test organisms. The results suggest that *G. fasciculatum* and phosphobacteria can greatly assist symbiotic N fixation as well as phosphate uptake by *C. arietinum* particularly when the crop is grown in soils containing insoluble phosphate. Use of combined inoculation was suggested for establishing leguminous crops with minimum supplies of N and P fertilizers.

Singh and Singh (1993) studied the interaction of rockphosphate, *Bradyrhizobium*, vesicular-arbuscular mycorrhizae and phosphate-solubilizing microbes on soybean grown in a sub-Himalayan mollisol. The interactive effects of rockphosphate (RP), *B. japonicum*, vesicular arbuscular mycorrhizae (VAM) and phosphate-solubilizing microbes (PSM) on soyabeans (cv. Bragg) was investigated in a mollisol in greenhouse experiments. Two levels of Mussoorie RP with or without VAM fungi and *B. japonicum*, and 4 treatments of PSM (uninoculated or inoculated with *Pseudomonas striata, Bacillus polymyxa* or *Aspergillus awamori*) were used. Inoculation with endophytes alone resulted in 70% root colonization. Addition of RP or inoculation with PSM, except *B. polymyxa*, stimulated root infection of native as well as introduced VAM endophytes. Application of RP or inoculation with *B. japonicum*, mycorrhizal fungi or PSM increased nodulation, N uptake, available soil P and the PSM population in the soil after crop harvest. The grain and straw yields did not increase following RP addition or mycorrhizal inoculation but increased after inoculation with *Bradyrhizobium* or PSM. In general, the application of RP, *Bradyrhizobium*, VAM and PSM in combinations of any 2 or 3, resulted in increases in nodulation, plant growth, grain yield and uptake of N and P. Among the 4 factor interactions, RP, *Bradyrhizobium*, and *P. striata* in the absence of VAM resulted in maximal nodulation, grain and straw yields and N uptake by soyabean. The highest P uptake by soyabean grain was recorded with *Bradyrhizobium* and *A. awamori* in the absence of RP and VAM. Generally, available soil P and PSM population after crop harvest were not increased by the treatment combinations giving the maximal uptake of nutrients. However, they increased in response to PSM, which produced no significant increase in total uptake of nutrients.

Singh (1994) studied the response of dual inoculation with *Bradyrhizobium* and VA mycorrhiza or phosphate solubilizers on soybean in Mollisol using inoculants of *Pseudomonas striata, Bacillus polymyxa*, indigenous VA mycorrhiza, *Glomus fasciculatum, Bradyrhizobium japonicum* and their few main dual combinations. As a result of increased activities of microorganisms in the rhizosphere region either directly or indirectly, inoculation with *Bradyrhizobium*, VAM and phosphate solubilizers individually increased the number and dry weight of root nodules and organic matter accumulation in plant. Increase in the grain yield due to individual inoculation with *Bradyrhizobium*, VAM and phosphate solubilizers ranged from 8.66 to 13.82%, 0.0 to 46.55% and 0.0 to 41.38% respectively. The efficacy of multistrains inoculum of phosphate solubilizers was almost similar to the single strain inoculum. It was concluded that, even in high phosphate soils, inoculation with single-strain inoculum of *Bradyrhizobium* or VAM or phosphate solubilizers was an essential technique for the successful cultivations of soybean.

Singh (1995) studied the effect of VAM fungi, phosphate solubilizing microbes and *Bradyrhizobium japonicum* on nodulation and dry matter accumulation at 45 days in soyabeans *cv. Bragg* under greenhouse conditions. Inoculation with VAM fungi caused about 56% root infection. Addition of rock phosphate and inoculation with *B. japonicum* or PSM stimulated root infection by native as well as exotic mycorrhizal fungi. Addition of rock phosphate and inoculation with *B. japonicum*, mycorrhizal fungi and PSM individually increased the number and dry weight of nodules per plant and shoot dry matter accumulation. The number and dry weight of nodules per plant were significantly higher than the control due to rock phosphate.

Mishra *et al.* (1995b) studied the response of maize to chemicals and biofertilizers. In a field trial at Indore, Madhya Pradesh, Maize cv. Chandan - 3 was given no fertilizers or 40 kg N + 18 kg P + 20 kg K/ha with or without seed inoculation with *Azotobacter* and application of VA mycorrhiza or phosphorus solubilizing bacteria (PSB). *Azotobacter*, VAM and PSB increased grain yield by 37, 6 and 31% respectively, in plots which received no fertilizers compared with the control. Grain yield was highest with application of N,P, K + *Azotobacter* + VAM or *Azotobacter* + VAM + PSB each giving a 25% increase in yield compared with application of NPK alone. Net returns were highest with a combination of NPK + *Azotobacter*.

Prathibha *et al.* (1995) studied the effects of seed inoculation with *Azospirillum* and / or *Pseudomonas striata* on cotton cv. DDH-2 in pot experiment. Soil was inoculated with *Glomus fasciculatum* and 0 or 5 tonnes biogas spent slurry(BSS), 0, 50 or 100% of the recommended rate (40 kg/ha) of N as urea and 20 kg P/ha as superphosphate or Mussoorie rock phosphate were also applied. In general, application of BSS stimulated the establishment of inoculated organisms in the rhizosphere. The population of *Azospirillum* and P solubilizing bacteria were highest when combined inoculation with *Azospirillum* + *P. striata* was given and 50% N as urea and 100% P as rock phosphate was applied. Mycorrhizal spore numbers and percentage root colonization were highest in plants inoculated with *G. fasciculatum* + *Azospirillum* with no application of N or P fertilizers. Plant dry matter, N and P uptake and seed yield were highest when *Azospirillum* and *P. striata* were applied and 50% N as urea and 100% P as rock phosphate was applied.

Kehri *et al.* (1997) studied a package of microbial inoculants for linseed. A package of triple inoculants consisting of a highly efficient vesicular arbuscular mycorrhizal fungus (*Glomus aggregatum*), a potential phosphate solubilizing fungus *(Aspergilus niger*) and an appropriate nitrogen fixer (*Azotobacter chroococcum*, SCSA 6) identified on the basis of preliminary screening was evaluated for its efficacy. All the three microbial inoculants improved the shoot biomass, nitrogen and phosphorus uptake as well as yield when used separately. Similar improvement was recorded when they were used in combination of two as dual inoculants. However, maximum improvement in all the parameters were recorded when all the three were used together as triple inoculation. Possibly, the superiority of triple inoculation was due to uninhibited supply and improved uptake of both the important nutrients *viz.* nitrogen phosphorus to the crop, a situation which was not available to it when it was offered single or double inoculations.

Sekar *et al.* (1997) carried out experiments with 6-month-old seedlings of 3 forest tree species (*Syzygium montanum, Syzygium arnotianum* and *Elaeocarpus oblongus*) to elucidate the influence of biofertilizer inoculation on the seedling growth. Seedlings were inoculated with *Azospirillum* (50 g/plant), phosphobacteria (50 g/plant) and VA-mycorrhizal fungi (20 g/plant) singly and in combination, and seedling growth parameters were recorded 0, 3 and 6 months after inoculation. Combined inoculation with *Azospirillum,* phosphobacteria and VA-mycorrhizal fungi significantly enhanced the total dry weight and shoot root ratio compared with control. Amongst single inoculations, the VA-mycorrhizal fungal inoculation was the best.

Swarupa *et al.* (1997) investigated the effects of biofertilizers (*Azospirillum* - N fixing) and/or phosphobacteria or VA mycorrhiza (P-mobilizing) applied 1 day prior to sowing, on germination of coffee. The best germination (> 93% after 60 days) was observed in seeds treated with phosphobacteria, either alone or in combination with VAM and/or *Azospirillum*.

Babu *et al.* (1998) studied the effect of the interaction between *Glomus fasciculatum, Azospirillum*, Phosphobacterium applied alone or in different combination on maize in a field trial. Triple inoculation showed a synergistic effect.

Dubey (1998) studied the response of soyabean (*Glycine max*) to biofertilizers with and without nitrogen, phosphorous and potassium on swell-shrink soil. In a field of Sehore, Madhya Pradesh in the 1991-92 rainy seasons, different combinations of inoculation with *Bradyrhizobium*, vesicular-arbuscular mycorrhizas (VAM) and phosphate-solubilizing bacteria and application of NPK fertilizers were compared for soyabeans cv. PK472 grown on a vertisol. Data on nodulation, growth, yield components and crop yield are tabulated. Yield was 0.95t/ha in controls and 1.05-1.21 t in the other 14 treatments, being highest from combined use of the 3 inoculants without NPK.

Singh *et al.* (1999) conducted a greenhouse experiment to find out uptake and concentration of N in different plants parts as influenced by rockphosphate (R), *Bradyrhizobium japonicum* (I), VA mycorrhizal fungi (M) and phosphate solubilizing microbes (S), in soybean. Almost all the main, two three and four factor interaction significantly increased N concentration in shoot, nodule and grain over the control. The maximum N conc. in shoot (1.27%), nodule (5.35%) and grain (6.27%) resulted due to RxI MxS_3, RxI x M_0xS_3 or RxIXMxS_0 and RxIxM_0xS_1 interaction, respectively. The highest total N uptake (210.4% more than control) was recorded with RxIxMxS_1 followed by RxIxS_1 interaction (193.13%).

Udaiyan and Muthukumar (1999) investigated the effect of applying indigenous vesicular arbuscular mycorrhizal (VAM) fungi, *Rhizobium* and phosphate solubilizing bacteria (PSB) to a tree nursery. *Acacia auriculiformis* and *A. holosericea* were grown for 120 days in unsterile soil inoculated with indigenous VAM fungi, *Rhizobium* and PSB and their combinations. Inoculation of microbes either individually or in combinations improved the growth of the acacia species significantly. The growth stimulation was accompanied by higher mycorrhizal colonization and tissue nutrient levels. The results demonstrated that application of indigenous AM fungi with PSB and *Rhizobium* could enhance the growth of acacias effectively in tree nurseries.

Kumar *et al.* (1999a) investigated the effect of artificial inoculations of VAM fungi and other bioinoculants like *Rhizobium, Azotobacter* and PSB singly and in combination on the growth of nursery seedlings of two agro-forestry tree species *viz*., *Sesbania grandiflora* and *Pongamia pinnata*. VAM colonization and resting spore population showed an increasing tendency till the age of 180 days. Dual inoculations were found to be superior for both VAM colonization and biomass production. *Rhizobium* inoculation in combination with mycorrhizae helped in better uptake of phosphorus than mycorrhizae alone. Among different VAM fungi *Glomus fasciculatum* formed better colonization and induced more growth. However, a mixture of different VAM fungi was found to be better inoculum than the individual VAM fungi.

Kennedy and Chellapillai (1999) conducted an experiment to determine the effect of inoculation with vesicular arbuscular mycorrhizal fungi (mixed *Glomus fasciculatum* and

G. mosseae), *Azospirillum*, and phosphobacteria, either individually or in combination, or seedling growth and nutrition uptake of 4 shola tree species, *Elaeocarpus munroii*, *Gordonia obtusa, Ligustrum roxburghii* and *Sideroxylon tomentosum*. Shola is a distinctive type of vegetation found in Kerala and Karnataka (India), in which grassland and evergreen trees grow in juxtaposition. The shola tree seedlings were inoculated at about 6 months old in polybags in shola soil, and measurements were made 180 days later. All the treatments increased seedling growth over control non-inoculated values, with the seedlings, which received the combined inoculation of VAM fungi; *Azospirillum* and phosphobacteria had showing the highest increased height, total dry weight, VAM colonization, and total N and P uptake in all 4 species tested. Among the individual inoculation, the VAM treatment was the best. The magnitude of response to the biofertilizers was greater in *Gordonia obtusa* and *L. roxburghii* than in the other species.

Nirmala *et al.* (1999) subjected cucumber (*Cucumis sativus* Linn) *cv. Local.* to 12 treatment consisting of farmyard manure (FYM) at 30 t/ha, with or without N (35 kg/ha), neem cake (0.25 kg/pit), groundnut oil meal (0.25 kg/pit), *Azospirillum* (2 kg/ha), phosphobacteria (2 kg/ha) and vesicular arbuscular mycorrhizas (VAM; 2 kg/ha), alone or in various combinations, in an experiment conducted in Tamil nadu. The first female flower in earlier node, closest sex ratio, largest fruit size and the highest number of tender fruits were recorded from FYM + *Azospirillum* + phosphobacteria + VAM. The controls produced the first female flower in the later node, widest sex ratio, smallest fruit size, and the lowest number of tender fruits. FYM+ *Azospirillum* + phosphobacteria + VAM produced the highest fruit yield, 2053.13 g/vine, while the controls had the lowest fruit yield, 810.25 g/vine.

Chezhiyan *et al*. (1999) investigated the effect of inorganic (NPK fertilizers), organic (farmyard manure), and biofertilizers (*Azospirillum*, phosphobacteria, and VAM) on growth and yield of banana *cv. Virupakshi* in Thadiyanudisai. Application of biofertilizers and organic manure along with 75% NPK was effective in increasing bunch weight.

Padma *et al.* (1999) conducted field trials to study the effect of the vesicular arbuscular mycorrhizal fungus *Glomus mosseae*, N_2 fixing *Azotobacter chroococcum* and a commercially available phosphate solubilizing microbial inoculant (Biophos) on mulberry (*Morus alba*) sapling in the nursery. Combined inoculation with all three inoculants enhanced the biomass and P yields. The scope for utilizing these organisms to improve the growth and vigour in commercial mulberry, nursery was discussed.

Nirmala and Vadivel (1999) studied the effect of combined application of organic manure and biofertilizers on growth and productivity of cucumber. Cucumber was supplied with combined application of organic manure and biofertilizers in a field trial conducted in Periyakulam. The treatments were: control (T1); FYM at 30 t/ha +35 at kg N/ha, applied 30 days after sowing (T2); FYM at 30 t/ha (T3); T3 + neem cake at 0.25 kg/plant (T4); T4 + groundnut cake at 0.25 kg/plant (T5); T3 + *Azospirillum* at 2 kg/ha (T6); T3+ phosphobacteria (T7); T3 + VAM at 2 kg/ha (T8); T6 + phosphobacteria (T9); T6 + VAM (T10); T7 + VAM (T11); and T9 + VAM (T12). T12 gave the longest vine (330cm.) and tender fruit (22cm.), highest number of leaves per plant (170), leaf area (29 396 cm^2), dry matter production (335 g/plant), tender fruit yield per vine (2053 g), and the shortest number of days to 1st female flower appearance (28). The highest girth of the tender fruit (18 cm) was recorded with T6. In all the treatments the number of fruits per vine were on a par with each other compared to the control.

Kandiannan *et al.* (2000) evaluated the effect of inoculation of biofertilizers (*Azospirillum,* phosphobacteria (*Bacillus megaterium* var. *Phosphaticum*) and VAM (*Glomus* sp.) along or in combination) on the growth and nutrient content of black pepper in an experiment conducted during 1996-97 in Calicut, Kerala. Plant height, leaf area, biomass, dry matter and nutrient contents were higher in inoculated plants than the uninoculated control. Growth was significantly higher when all the biofertilizers were inoculated compared to individual inoculations and the control. Among individual inoculations, maximum growth was observed in VAM followed by phosphobacteria and *Azospirillum.* In the dual combinations, phosphobacteria + VAM was the most effective followed by *Azospirillum* + VAM and *Azospirillum* + phosphobacteria.

Ratti *et al.* (2001) studied the interactive effects of phosphate solubilizing bacteria, N_2 fixing bacteria and arbuscular mycorrhizal fungi (AMF) in a low phosphate alkaline soil amended with tricalcium insoluble source of inorganic phosphate on the growth of an aromatic grass palmarosa (*Cymbopogon martinii*). The microbial inocula consisted of the AM fungus *Glomus aggregatum*, phosphate solubilizing rhizobacteria *Bacillus polymyxa [Paenibacillus polymyxa*] and N_2 fixing bacteria *Azospirillum brasilense*. These rhizobacteria behaved as "mycorrhiza helper" and enhanced root colonization by *G. aggregatum* in presence of tricalcium phosphate at the rate of 200 mg kg^{-1} soil (P1 level). Dual inoculation of *G. aggregatum* and *B. polymyxa* yielded 21.5 g plant dry weight (biomass), while it was 21.7 g in *B. polymyxa* and *A. brasilense* inoculated plants as compared to 14.9 g of control at the same level. Phosphate content was maximum (0.167%) in the combined treatment of *G. aggregatum, B. polymyxa* and *A. brasilense* at P1 level, however acid phosphatase activity was recorded to be 4.75 μ mol mg-^{1}min-^{1} in *G. aggregatum, B. polymyxa* and *A. brasilense* treatment at P0 level. This study indicates that all microbes inoculated together help in the uptake of tricalcium phosphate, which is otherwise not used by the plants, and their addition at 200 mg kg^{-1} of soil gave higher productivity to palmarosa plants.

Tomar *et al.* (2001) conducted a field experiment to study the effect of *Rhizobium,* vesicular arbuscular mycorrhiza (VAM, *Glomus caledonium*) and phosphate solubilizing bacteria (PSB, *Pseudomonas striata* strain P-27) inoculation, with and without P on black gram (*Vigna mungo*) seed yield. Phosphorus application in soil with medium P content (5.4 mg kg1) increased nodulation, gram yields, N and P in plant and grain over no phosphorus control. Forty kilograms of P_2O_5 each ha recorded an increase of 20.6% in nodule dry weight, significant increases of 0.35 g kg^{-1} in concentration and 1.28 g kg^{-1} in P concentration of plant over 20 kg P_2O_5 ha^{-1}. Similar significant increases of 0.59 g kg^{-1} in grain yield and 0.54 and 0.23 g kg^{-1} in N and P concentrations of the grain, respectively, over 20 kg P_2O_5 ha^{-1} were also obtained with higher dose. Inoculation of *Rhizobium* +VAM +PSB at all stages of plant growth recorded maximum increase in all the parameters studied. Dual inoculation of *Rhizobium* with either VAM or PSB was generally significant in the effect and better than that of VAM + PSB however, P accumulation in plant and grain was more with VAM +

PSB. Among single inoculations, *Rhizobium* gave the highest and 21.0% more nodule number, 34.7% more nodule dry mass, 0.73 g kg^{-1} more N in grain and 4.2% higher grain yield over PSB. PSB, however, registered significant increases in P concentration in plant and grain over VAM and *Rhizobium*.

Rajalingam and Kumar (2001) conducted an experiment to study the effect of digested coir pith compost (DCC) and biofertilizers on 3 cultivars of tea, *i.e.* Assam Jat, China Jat and ATK clone. Eighteen treatments were imposed involving 100, 75, 62.5 and 50% of estate practice, along with DCC at 3 and 6 tones/ha, and biofertilizers, namely *Azospirillum lipoferum*, phosphobacteria (*Bacillus megatherium*) and vesicular arbuscular mycorrhizal fungi (*Glomus fasciculatum*), each at 40 kg/ha. Results revealed that the different cultivars recorded higher green leaf yield in different treatments containing DCC and biofertilizers either alone or in combination with full or reduced level of estate practice. This study indicates that DCC and biofertilizers are suitable as amendments in improving the physical, chemical and biological properties of tea soil and also the green leaf yield of tea.

Muthukumar *et al*. (2001) studied the response of neem (*Azadirachta indica* A. Juss.) to indigenous arbuscular mycorrhizal fungi, phosphate solubilizing and asymbiotic nitrogen-fixing bacteria under tropical nursery conditions in Coimbatore, Tamil Nadu. Seedlings were harvested at 60 and 120 days after transplanting. Microbial inoculation resulted in increased mycorrhizal colonization, greater plant height, leaf area and number, root collar diameter, biomass, phosphorus, nitrogen and potassium content, and seedling quality. Inoculated seedlings also had low root: shoot ratios and low nutrient utilization efficiencies. Populations of PSB declined with seedling growth; contrarily populations of *A. brasilense* increased. *A. brasilense* and PSB population were related to each other and influenced root colonization by AM fungi. Microbial inoculation effects were greatest when seedlings were inoculated with a combination of microbes rather than individually. This clearly indicates that these microorganisms act synergistically when inoculated simultaneously; with maximum response being when both AM fungi were coinoculated with *A. brasilense* and PSB. The results emphasize the importance of microbial inoculations for the production of robust, rapidly growing seedlings in nurseries and illustrate the advantage of inoculating soils of low microbial population with indigenous microbes.

Rajendran *et al.* (2000) studied the influence of biofertilizers on the biomass production of *Casuarina equisetifolia* in farm forestry. *Casuarina equisetifolia* seedlings inoculated with different biofertilizers such as *Azospirillum*, phosphobacterium, vesicular arbuscular mycorrhizal (VAM) fungi and *Frankia* and their combinations were planted at a farmland, in Tamil Nadu, India. Growth parameters were collected at bimonthly intervals up to 30 months. Biomass was estimated 24 months after planting. Maximum height, girth at breast height (gbh) and total biomass were obtained in the combined application of *Azospirillum*, Phosphobacterium, VAM and *Frankia*. The combinations of VAM + *Frankia* among double inoculation, and the combination of *Azospirillum*, VAM and *Frankia* in triple inoculation also proved to be the best treatment in promoting the significant total height, gbh and total biomass production.

Kamalakannan and Manivannan (2002) conducted a field experiment to study the influence of *Azospirillum,* phosphobacterium and vesicular arbuscular mycorrhizas (VAM) on growth parameters of radish *cv. Pusa Chetki*. The result revealed that maximum shoot length, number of leaves and leaf area were recorded in treatment which received NPK + *Gigaspora margarita.* The maximum shoot: root ratios were recorded in the treatment, which received the full compliment of NPK along with *Azospirillum+Gigaspora margarita.*

Sajan *et al.* (2002) conducted a field experiment during 1999 kharif season in Bangalore to study the effect of biofertilizers combined with different N, P and K fertilizer rates on the growth and yield of chilli *cv. Byadagi Dabba.* NPK fertilizer (75 or 100%) was applied alone (100%, the control) or in combination with *Azotobacter, Azospirillum,* phosphate solubilizing bacteria (PSB), and vesicular arbuscular mycorrhizas (VAM). The plants inoculated with *Azotobacter, Azospirillum,* PSB, and VAM in combination with 75% NPK + 100% K recorded the maximum plant height (100.03 cm), leaf area (28.79 dm^2/plant), and dry matter (187.36 g/plant) compared to the control. The plants also produced 36% more fruits per plant (111.38) and 45% more dry fruit yield (2.27 t/ha) compared to the control (81.68 and 1.56 t/ha, respectively). Inoculation of biofertilizers alone recorded the lowest dry fruit yield (1.06 t/ha) compared to the control (1.56 t/ha).

Vijaykumari and Janardhanan (2003) conducted a pot culture experiment in red loamy soil at the Avinashilingum University, Coimbatore to assess the effect of biofertilizers on seed germination and seedling growth of silk cotton (*Ceiba pentandra*). Soil inoculation using *Azospirrillum,* phosphobacteria and vesicular arbuscular mycorrhiza (VAM) singly and in different combinations was tried. Germination percentage was recorded 30 days after sowing and biometric and biochemical changes such as chlorophylls, total soluble carbohydrates, reducing sugars, total free amino acids, buffer-soluble proteins and phenolics were recorded on 60, 90 and 120 days after sowing. The result indicated that combined inoculation of biofertilizers viz., *Azospirillum* + phosphobacteria + VAM improved the investigated biometric as well as biochemical attributes.

Miscellaneous

Krishna *et al.* (1982) studied the interaction between a VAM fungus and *Streptomyces cinnamomeous* and their effects on finger millet. The growth and P nutrition of the crop in a sterile P-deficient soil was improved by inoculation with either *Glomus fasciculatum* or *S. cinnamomeous*. The microorganisms interacted antagonistically when added simultaneously or with a two week interval between them. *S. cinnamomeous* reduced spore production and development of infection by *Glomus* while *Glomus* reduced the multiplication of *S. cinnamomeous*. Because of this antagonism dual inoculations stimulated plant growth less than single ones.

Singh *et al.* (1991) studied the interaction between VAM fungi and *Saccharomyces cerevisiae* in legumes. In greenhouse trials, nodule number, shoot and root dry weight, root infection (native VAM) and formation of vesicles, arbuscules and spores were enhanced in *Leucaena leucocephala* cv. K-63, soyabean cv. PK-416, pigeonpea cv. Pusa 78, *Phaseolus mungo* (*Vigna mungo*) cv T-9, mungbean cv. Pusa Baisakhi and cowpeas cv. C-152 when inoculated with live yeast cells. The increase in parameters varied with legume and type of yeast culture. Live yeast cells were more effective than dead cells in increasing the nodulation, VAM root colonization and formation of vesicles and arbuscules.

Krishnaraj and Sreenivasa (1992) studied increased root colonization by bacteria due to inoculation of vesicular-arbuscular mycorrhizal fungus in chilli (*Capsicum annuum*). More bacteria were present in the endorhizosphere of chilli plants inoculated with *Glomus fasciculatum* or *G. macrocarpum* than in the rhizosphere of uninoculated plants, with the number of bacteria increasing with percentage root infection by mycorrhizal fungi. Increased colonization by bacteria capable of growing on hydroxyapatite tricalcium phosphate and nutrient agar was observed on plant roots with VAM fungi, with a corresponding increase in the percentage P in shoots and in shoot biomass. It has been suggested that these increases might have resulted from the increased rhizosphere bacteria which liberate organic acids and solubilize otherwise insoluble P in the rhizosphere, which might have been taken up and transferred to the plant by the VAM fungi.

Kumar *et al.* (1995b) studied the effect of mushroom-mycorrhizal fungal interactions on yield and quality of seeds in pearl millet. *Volvariella valvaceae* infested paddy straw was mixed with unsterilized non-cultivated soil in the ratio of 1:9 and *Glomus fasciculatum* inoculum was layered three inch below the soil. Seeds of pearl millet var. BJ 104 were sown in pots. Control series without mushroom paddy straw and VAMF inoculum were also maintained. The authors recorded more available nutrients at the time of harvest than at the time of seed sowing. There was a tremendous increase in available phosphorus in non-cultivated soils. It was 325% in paddy straw amended soil, 525% in the *Volvariella* inoculated soil, and 850% when the VAMF *G. fasciculatum* was inoculated in addition to the mushroom fungus.

Kumar *et al.* (1995a) studied the influence of plant growth promoting rhizobacteria on mycorrhizal associations in wheat. They planted the seeds of wheat cv. HD-2204, bacterized separately with each of ten PGPR (out of 10 PGPR, 3 isolates were of *Bacillus subtilis*, two of *B. polymyxa* two of *Pseudomonas putida* two of *P. fluorescence* and *P. aeruginosa*) in 30 cm pots filled with field soil containing a mixture of mycorrhizal fungi (*Glomus aggregatum, G. fasciculatum, G. intraradix, G. invermayanum, G. magnicaule, G. mosseae, Gigaspora* spp., *Acaulospora* spp., *Sclerocystis* sp.). Pots planted with uninoculated seeds served as control. They collected data on mycorrhiza formation, root and shoot weight, P-uptake, earheads/ plants, grain wt./ earhead etc. In general, *Pseudomonas fluorescence* increased grain yield and *Bacillus subtilis* and *P. pultida* increased dry matter production, through enhanced root colonization by VAM fungi.

Machado and Bagyaraj (1995) studied the effect of the interaction between mycorrhiza and helper bacteria on growth of cowpea. They used *Glomus fasciculatum* as VAM fungus and three helping bacteria. All the three helping bacteria were effective in enhancing per cent mycorrhizal colonization, total plant biomass and total phosphorus content of the plant. However, their performance was variable. The results illustrated that helping bacteria improve the ability of mycorrhizal fungi to colonize plant roots and in turn plant growth.

Pal and Singh (1995) studied the effect of different bacteria *viz.,Pseudomonas, Enterobacter, Beijerinckia, Azospirillum* and *Azotobacter,* isolated from sporocarp of mushroom on VA mycorrhizal root colonization by *Glomus mosseae* in pearl millet. On the basis of their observations they concluded that these bacteria may be used as biofertilizers to produce VAM inocula blended with the bacteria/bacterium to boost up the plant growth.

Rangarajan and Santhanakrishnan (1995) conducted an experiment to investigate the effect of plant growth promoting rhizobacteria and biofertilizers on the fresh leaf yield and nutrient content in *Morus alba.* They conducted the experiment using efficient strain of *Azospirillum brasilense*, VAM, *Glomus fasciculatum* and growth-promoting rhizobacterium, *Pseudomonas fluorescens* alongwith the recommended level of NPK for mulberry. The trial had nine treatments with three replications. Three prunings were done for each year and the leaf fresh weight of mulberry was recorded. They reported that the combined inoculation of all the three inoculants-PGPR *(Pseudomonas fluorescens) Glomus fasciculatum* and *Azospirillum* to be superior to twin, or single inoculants, or uninoculated control with two levels of N and P. The combined inoculation of biofertilizers also enhanced the quality of mulberry leaves. This resulted in the increase of all economically important characters of silkworm larvae and silk production.

Pahwa (1996) conducted an experiment using red sandy loam soil (pH 7.2) in which *Leucaena* seeds were sown after single and combined inoculations with *Rhizobium* LL-28-2, *Bacillus* polymyxa H_5 and *Glomus fasciculatum.* The pots were given NPK fertilizer and micronutrients, and observations were made on nodulation, root length, forage yield (green, dry and crude protein) and plant N and P contents after 80 days. Inoculation with *Rhizobium* + *Glomus* gave the best yields and nodulation, followed by inoculation with all the 3 microbes.

Sreenivasa and Majjigudda (1996) studied residual influence of VA mycorrhizal inoculation on rhizosphere microflora and growth of subsequent crop. Wheat cv. DWR-162 was grown for 90 days in pots where soil was inoculated with *Glomus fasciculatum, Gigaspora margarita, Acaulospora laevis* or *Sclerocystis dussii* or not inoculated. Thirty days after the harvest of the first crop, a second crop was grown in the same pots. Recommended NPK rate was applied to the first crop. Percent root colonization, spore count, populations of rhizosphere free living nitrogen fixing and P-solubilizing microflora were lower during the growth of the second crop. Plant height, dry matter and shoot P concentration were significantly increased by soil inoculation; *G. fasciculatum* was the best inoculant.

Rani *et al.* (1998) examined the impact of soil supplemented with VAM, *Rhizobium* sp. and *T. harzianum*, single and in combination with one another on the biomass of *Prosopis cineraria.* The biomass yield of different treated and untreated plants was determined by observing various parameters. Dual inoculation with VAM and *Rhizobium* sp. resulted in maximum plant growth and nodulation.

Kapoor and Mukerji (1998) studied the effect of VAM (*Glomus macrocarpum*) inoculation of the soil on rhizosphere mycoflora of *Anetham graveolens* at different growth stages (15, 45 and 75 days after germination). Although the total number of fungal colonies decreased, a relative increase was noticed for saprophytes and biocontrol agents. The root-borne pathogens, however, were suppressed on mycorrhization.

Kichadi and Sreenivasa (1998) conducted a pot experiment to study the interaction between the vesicular arbuscular mycorrhizal (VAM) fungus, *Glomus fasciculatum,* and the fungal antagonist, *Trichoderma harzianum*, on the rhizosphere of tomato *cv.* L-15 plants grown in soil supplemented with spent biogas slurry. The percentage mycorrhizal root colonization and mycorrhizal spore counts were highest in the rhizosphere soil of plants inoculated with both fungi. Disease severity index was least in the dual inoculated plants. The presence of spent slurry also enhanced VAM colonization and increased spore counts. The interaction effects of the fungi not only increased plant growth and yield but also improved P nutrition.

Bhatt *et al.* (1999) studied the influence of pseudomonas-vesicular arbuscular mycorrhizal fungal interaction on growth of maize (*Zea mays* L.). An indigenous consortium of VAM fungi and a growth promoting fluorescent pseudomonad strain PRSq were evaluated in an interactive manner for growth behaviour of maize in unsterile soil. Single inoculation of VAM and PRSq improved fresh and dry weight of root and shoot at 15, 25 and 35 days of harvest, combined inoculation exhibited an additive influence. Phosphorus content was significantly greater in VAM plants, followed by VAM + PRSq or single inoculation of PRSq. Similar behaviour was observed for nitrogen content, however, plants inoculated with PRSq alone showed the least N level. Irrespective of the harvest time and treatment, significantly lower C content was recorded in plants inoculated with PRSq. Maximum alkaline phosphatase was recorded in plants treated with PRSq. However, enzyme level declined with successive harvest period in control, PRSq alone and VAM + PRSq treated plants. Acid phosphatase was maximum in control plants after 15, 25 and 35 days, enzyme activity increased in all but for control and reached a maximum for VAM treated plants. Root infection was significantly higher in plants inoculated with VAM singly followed by the combined inoculation with fluorescent *Pseudomonas*.

Rani *et al.* (1999) conducted experiments to determine the influence of vesicular arbuscular mycorrhizal (VAM) fungi, *Rhizobium* sp. and *Trichoderma harzianum* individually as well as in combination on *Acacia nilotica* seedlings. *G. mosseae* was isolated from the rhizosphere of *A. nilotica, Rhizobium* sp. was isolated from nodules of *A. nilotica* and *T. harzianum* was isolated from the rhizoplane of *A. nilotica*. Co-inoculation of seedlings with *G. mosseae* and *Rhizobium* sp. resulted in maximum plant growth and best nodulation.

Lilly and Santhanakrishnan (2000) studied the effect of interaction between *Glomus fasciculatum* and *Trichoderma viride* on the growth of sunflower. Combined inoculation of *Glomus fasciculatum* and *Trichoderma viride* increased the growth, percentage root colonization by *Glomus fasciculatum* and uptake of phosphorous by sunflower when compared to individual inoculation or uninoculated controls at different stages of sampling. The results show a synergistic effect of *Glomus fasciculatum* and *Trichoderma viride* on the growth of sunflower and indicate the potential benefits of using combined inoculation.

Sastry *et al.* (2000) studied the effect of an AM fungal consortium and *Pseudomonas* on the growth and nutrient uptake of *Eucalyptus hybrid. Eucalyptus* is an important tree species used for afforestation of large tracts of marginal and wastelands. *Eucalyptus*-arbuscular mycorrhizal fungal (AMF) interactions in seedling establishment and growth promotion have been inadequately dealt with. Efforts were made to assess the role of AMF-pseudomonad (PRS9, plant growth promontory fluorescent *Pseudomonas*) interactions in growth promotion and nursery establishment of *E. hybrid.* Seedlings were subjected to six different treatments: (i) uninoculated control, (ii) 400 AM spores. (iii) 800 AMF spores+ PRS9, 10,20 and 30 ppm P. Root and shoot length, and root and shoot fresh and dry weights were highest at 400 AMF spores followed by 400 AMF spores and 400 AMF spores + PRS9. In general, plant growth was high at 20 ppm P. Root P content increased significantly with 400 AMF spores followed by 800 at 20 ppm P. Independent of soil P levels, the quality index of mycorrhizal treatments without PRS9 was significantly higher than the treatments including PRS9 or PRS9 alone. Mycorrhizal inoculation efficiency was superior at 10 ppm Irrespective of the treatment. AM alone 9400 spores) significantly improved the inoculation efficiency. PRS9 in association with AM fungi inhibited growth promotion and nutrient uptake.

Pandey and Gupta (2000) studied the scope of biofertilizers in agroforestry. Biofertilizers (vesicular arbuscular mycorrhiza, *Rhizobium, Azotobacter, Azospirillum*, blue green algae, *Azolla,* phosphate solubilizing microorganism and *Frankia*) have an important role to play in improving the supply and availability of nutrients to crops in India. With the quantity of nutrients removed from the soil being almost double what is added through fertilizers, there is growing nutrient deficit causing a major threat to soil health and crops productivity. The recent 60% increase in the prices of inorganic fertilizers has resulted in a decline in their use, underlining the need for integrated nutrient management involving biofertilizers. Biofertilizers hold a promise to improve the crop yield and nutrient dynamics in agroforestry mainly in the trees as conduits through which the nutrient cycle flows, making them sites for the accumulation of nutrient in an agroecosystem.

Mohan (2000) studied endomycorrhizal interaction with rhizosphere and rhizoplane mycoflora of forest tree species in Indian arid zone. Microorganisms are present in great numbers on and near the feeder roots of trees and they play vital roles in numerous physiological processes. Associations of microorganisms participating in saprophytic, pathogenic and symbiotic root activities mediate these dynamic processes. The major symbiotic associations on tree species are mycorrhizal fungi. Among the mycorrhizal fungi, vesicular arbuscular mycorrhizal (VAM) fungi are ubiquitous and distributed in most ecosystems. These fungi act as biocontrol agents against root and soil borne pathogens and improve plant growth and vigour. In this investigation, an experiment was conducted to study the interaction of VAM fungi with rhizosphere and rhizoplane mycoflora of 4 important arid zone species, viz. *Acacia nilotica, A. Senegal, A. tortilis* and *Prosopis cineraria.* Seedlings were raised under sterile conditions, and then transplanted either to rhizosphere soil enriched with mixed VAM inocula (*Glomus fasciculatum* and *G. aggregatum*) or raised under ordinary nursery conditions without inoculation. After 60 days, the mycoflora of the rhizosphere and rhizoplane was very much less in VAM inoculated seedlings than in the uninoculated (control) plants, in all the tree species studied. Most of the pathogenic fungi, such as species of *Cylindrocladium, Fusarium* sp. *F. oxysporum, Rhizoctonia bataticola [Macrophomina phaseolina]* and *R. sonalni* were suppressed or completely eliminated from VAM inoculated plant roots and the rhizosphere soil samples. The significance of these findings is discussed.

Rani *et al.* (2001) investigated the performance of *Prosopis cineraria* on soil amended with vesicular arbuscular mycorrhizas (*Glomus fasciculatum, Rhizobium* sp., *Rhizobium* sp. + *G. fasciculatum, Trichoderma harzianum* and *T. harzianum* + *G. fasciculatum)* in a pot experiment. At 45 and 90 days after treatment, the greatest shoot length, shoot dry weight, root length, root dry weight, nodule number, nodule weight and infection rate were obtained with *G. fasciculatum* + *Rhizobium* sp.

Earanna *et al.* (2001) studied the response of *Coleus aromaticus* to *Glomus fasciculatum* and other beneficial soil microflora. Stem cuttings of *Coleus aromaticus [Plectranthus amboinicus]* were treated with vesicular arbuscular mycorrhiza (VAM; *G. fasciculatum)* and beneficial microorganisms, viz. *Bacillus coagulans, Pseudomonas fluorescens, Azotobacter chroococcum,* and *Trichoderma harziarium* either singly or in combination of organisms. The numbers of leaves and branches as well as the toal biomass, nitrogen, and phosphorus contents were higher in plants treated with *G. fasciculatum* alone. Synergistic effect was noticed especially in plants treated with *G. fasciculatum* + *P. fluorescens.* Mycorrhizal colonization in roots was also maximum when VAM was applied alone.

Chinnamuthu and Venkatakrishnan (2001) conducted a study in Coimbatore, Tamil Nadu, during Kharif (June-September) 1994, 1995 and 1996 to determine the effect of integrating inorganic fertilizer with vermicompost (phosphate solubilizing bacteria) and vesicular arbuscular mycorrhizas (VAM) on the productivity of sunflowers (cultivars Co 2 and Co 3). Treatments included farmyard manure at 5 t/ha, vermicompost at 2 t/ha (Factor M), inorganic fertilizers, NPK at 100% (40:20:20 kg N: P: K/ha). 50% and)%. Application of 100% NPK gave the highest seed yield. NPK +FYM, VAM produced better seed yield than vermicompost.

Mamatha *et al.* (2002) investigated the effect of soil inoculation with arbuscular mycorrhizal (AM) fungi and a mycorrhiza helper bacterium (MHB) on mulberry and papaya plants (cv. M5) inoculated with *Glomus fasciculatum* and 1.5-year-old pawpaw plants (var. Solo) were inoculated with a mixed culture of *G. mosseae,* and *G. caledonium* with or without *Bacillus coagulans* at two levels of P fertilizer. Growth, P uptake, yields and AM colonization levels were monitored. Leaf yield in mulberry and AM colonization levels were minimal in uninoculated plants given 50% recommended P. However, crop yield of both mulberry and papaya inoculated with AM fungi alone or together with the bacterium and given 50% recommended P were statistically on a par with that of uninoculated plants given 100% recommended P. As inoculated of *B. coagulans* increased mycorrhiza levels in AM fungal-inoculated plants, this may be included in the class of MHB. Thus, mulberry and papaya already established in the filed may respond to AM inoculation and MHB may increase symbiosis development by efficient AM fungi.

Srinath *et al.* (2003) conducted a greenhouse investigation to study the influence of the arbuscular mycorrhizal (AM) fungus *Glomus mosseae* and the plant growth-promoting rhizomicroorganisms (PGPRs) *Bacillus coagulans* and *Trichoderma harzianum* on the growth and nutrition of micropropagated *Ficus benjamina* plantlets. The AM fungus was inoculated either singly or in combination with the PGPRs. Plants showed maximum plant height, biomass, P content, mycorrhizal root colonization, spore numbers and populations of *T. harzianum* and *B. coagulans* in root zone soil when all the three organisms were inoculated together. Thus, when *G. mosseae* is co-inoculated with PGPRs, the growth and nutrition of *Ficus benjamina* is enhanced. *T. harzianum* and *B. coagulans* are thus designated as mycorrhizal helper organisms.

Sailo and Bagyaraj (2003) conducted a glasshouse investigation to study the influence of the arbuscular mycorrhizal (AM) fungus, *Glomus mosseae,* and plant growth promoting rhizomicroorganisms (PGPRs), *Bacillus coagulans* and *Azotobacter chroococcum,* alone and in combination, on the growth and nutrition of *Simarouba glauca* [*Quassia simarouba*]. Individual inoculation of these organisms significantly enhanced plant biomass, but biomass was significantly greater with inoculation of *G. mosseae* than with inoculation with PGPRs. Dual inoculation of *G. mosseae* plus *B. coagulans* further significantly enhanced plant biomass compared with all other treatments. This was also true for other parameters studied; plant height, number of leaves and plant P-content. The percent mycorrhizal colonization in the root and spore numbers in the root zone soil was also highest in seedlings inoculated with both *Glomus mosseae* and *Bacillus coagulans.*

Sood (2003) determined the chemotactic responses of the plant-growth-promoting rhizobacteria *Azotobacter choroococcum* and *Pseudomonas fluorescens* to roots of vesicular-arbuscular mycorrhizal (*Glomus fasciculatum*) tomato plants. A significantly (P=0.05) greater number of bacterial cells of wild strains were attracted towards vesicular-arbuscular mycorrhizal tomato roots compared to non-vesicular-arbuscular mycorrhizal tomato roots. Substances exuded by roots served as chemoattractants for these bacteria. *P. fluorescens* was strongly attracted towards citric and malic acids, which were predominant constituents in root exudates of tomato plants. *A. chroococcum* showed a stronger response, towards sugars than amino acid, but the response was weakest towards organic acids. The effects of temperature, pH, and soil water matric potential on bacterial chemotaxis towards roots were also investigated. In general, significantly (P=0.05) greater chemotactic responses of bacteria were observed at higher water matric potentials (0, -1, and –5 kPa), slightly acidic to neutral pH (6, 6.5 and 7), and at 20-30°C (depending on the bacterium) than in other environmental conditions. It is suggested that chemotaxis of *P. fluorescens* and *A. chroococcum* towards roots and their exudates is one of the several steps in the interaction process between bacteria and vesicular-arbuscular mycorrhizal roots.

CONCLUDING REMARKS

A number of economically important leguminous crops including soyabean, pigeonpea, groundnut, bengal gram, pea, lentil, mungbean, green gram and cowpea have been investigated by Indian workers with reference to their response to dual inoculation with *Rhizobium* and VAM fungus. Nearly in all the cases synergistic effects of dual inoculation were observed in respect of nodulation, mycorrhization, as well as growth and yield in crops. Some workers with respect to some leguminous forest trees reported similar observations. Some workers have investigated the response of *Casuarina* and *Alnus* with reference to interaction between a symbiotic nitrogen fixer - *Frankia* and VAM fungi. Dual inoculation with both the symbionts enhanced the growth more than single inoculation with either organism.

Inoculation of non-leguminous crops with non-symbiotic nitrogen fixers like *Azotobacter* and *Azospirillum* has been reported to improve the performance of these crops. The practice is in regular use in many crops of the country. A number of mycorrhizasts of the country have attempted to investigate the effect of interaction between VAM fungi and *Azotobacter*/ *Azospirillum* on the growth and yield of a number of important crops including cereals and millets (barley, wheat, pearlmillet, maize etc.), vegetable crops (tomato, brinjal), fruit crops (mulberry, sweet orange), spices (chilli), fibre crops (jute) and plantation crops (coffee). Nearly all the crops showed better performance when coinoculated with both VAM fungi and nitrogen fixer than when inoculated with either of them. It clearly indicated that the two inoculants exerted a synergistic effect on the growth and yield of the crops and dual inoculation technique may be used for improved productivity in non-leguminous crops.

Many soil microorganisms solubilize unavailable forms of P and these microorganisms called phosphate-solubilizing microbes (PSM) have been shown to improve the growth and productivity of different crops not only under laboratory conditions but also in fields. Some Indian mycorrhizasts have studied the interaction between VAM fungi and PSM on the performance of different crops. They have reported a synergistic effect on the growth and yield of crops due to dual inoculation. Majority of them concluded that synergistic effect was observed because PSM rendered more P soluble, while mycorrhiza enhanced P uptake.

A number of workers have investigated the effect of triple inoculation with VAM fungi, PSM and nitrogen fixers. Nearly all of them have reported that triple inoculation resulted in significantly better performance of the crops they investigated in terms of growth and productivity as compared with single or double inoculation with the test organisms. It has been suggested that mycorrhizal fungi and PSM assisted the nitrogen fixation as well as P uptake by the plants resulting in improved performance of the crops.

Majority of interaction studies between VAM fungi, nitrogen fixers and PSM have been done under pot conditions with a limited number of crops. However, synergistic effect shown by double or triple inoculants on the performance of the crops studied encourage intensifying field studies which would make double / triple inoculation a practical proposition not only in agricultural crops but also other plant communities, which are being used for other purposes like reclamation and afforestation. If the synergistic interactions could be repeated under field conditions, it would cut down the cost of fertilizers in practical agriculture.

6

FACTORS AFFECTING VA MYCORRHIZA

Factors which influence the colonization and life cycle of VAM fungi have important bearing on the extent to which they may benefit their host. As we move closer to using mycorrhizal fungi to increase growth and productivity of different plant communities, a thorough understanding of the factors which influence their survival, germination, root colonization an d sporulation will be critical. Seasons, physico-chemical characters of the soil, agronomic practices, plant protection chemicals, pollutants and genotypic characters of the host may affect VAM fungi and their potentiality to colonize and provide benefit to the host. In view of this, the effect of such factors on the VA mycorrhizal symbiosis and its benefit to diverse plant communities has been extensively investigated worldover.

A variety of plant protection chemicals are employed to fight the ailments of plants due to various biological entities in India. A knowledge of the interaction of these chemicals with VA mycorrhizal association appears to be essential in order to use these chemicals as well as VAM technology for benefit of plants in an integrated manner. Indian mycorrhizasts have done commendable work in this direction.

India has a number of agroclimatic regions. At the same time it has a variety of soils differing in their physicochemical characters. Season and soil specific crops are raised in different parts of the country. Thus, there appears to be a need to investigate the effect of seasons and soils on the growth and productivity of the crops with reference to the benefit which they may acquire from the VA mycorrhizal association. A number of agronomic practices are employed to ensure better performance of the crops. In order to ensure benefit from both VA mycorrhizal associations as well as agronomic practices, a clear understanding of the effect of agronomic practices on the mycorrhizal associations is a prerequisite to integrate agronomic practices and VAM technology for welfare of plants. Mycorrhizasts of the country have published a number of reports on this subject.

Pollutants affect not only the VAM fungi but also the magnitude of mycorrhizal association. In view of this, a number of Indian workers have investigated the effect of different pollutants on the VAM fungi and their potentiality to benefit the plants. Mycorrhizal association and its extent of benefits is expected to vary not with the hosts but also with their genotypes. Indian mycorrhizasts have provided some information also on this aspect. The following account consolidates the work done by the mycorrhizasts of the country with reference to the effect of various factors on the mycorrhizal fungi and their potentiality to colonize and benefit diverse communities of plants.

Plant Protection Chemicals

Reddy and Rajasekhar (1988) studied the antibiotic induced VAM stimulation in peanut and microbial interaction. Seed and foliar treatment with 9 antibiotics at 100 mg/l stimulated VAM development and plant growth. With Oxytetracycline and Doxycycline at 500 mg/l enhanced VAM development in unsterile soil only but not in sterile soil. OT treatment showed an enhanced rhizosphere microflora with a predominance of gram-negative bacteria over gram-positive bacteria, an increase in *Aspergillus* and *Rhizopus* sp. and elimination of *Trichoderma* spp. The bacterial population increased up to 10 days while fungal and actinomycete population increased up to 15 days. VAM stimulation by OT was assigned to a change in microbial ecology.

Padmavathi *et al.* (1990a) investigated the effect of pesticides on endomycorrhizal symbiosis in foxtail millet and sorghum. The soil used as the potting medium (3.5 kg pot $^{-1}$) was a black soil with an indigenous population of 87 spores of VAM fungi per gram soil. Pesticides chosen for the study included three fungicides, *viz.*,agrosan, dithane Z-78 and sulphur, and four insecticides, BHC, carbofuran, endosulfan and quinalphos. Agrosan, at all the concentrations tested, did not affect the growth of foxtail millet. But 100 ppm alone was inhibitory to the mycorrhizal development. Plant growth was not affected by the sulphur treatment. However, the colonization by the VAM fungi significantly increased with all the concentrations. The enhancement in mycorrhizal development was more pronounced with 5 and 25 ppm levels. The effect of dithane Z-78 was similar to that of sulphur. BHC at 25 ppm level caused maximum mycorrhizal development. Significant inhibition in the per cent mycorrhizal colonization was observed with 100 ppm BHC. Higher concentrations of carbofuran (10 and 100 ppm) were toxic to sorghum. Carbofuran application at 25 to 100 ppm levels, significantly inhibited the root colonization by VAM fungi. Similar trend was observed even with two successive applications of carbofuran. Though 5 ppm of endosulfan was not phytotoxic, there was significant reduction in mycorrhizal development at this level. All the other concentrations caused reduction in plant growth and mycorrhizal colonization. This was true for single application or two repeated applications of the insecticide. Quinalphos application also resulted in a similar response as carbofuran. The results showed that application of pesticides at recommended levels had no adverse effect on mycorrhizal colonization in sorghum and foxtail millet.

Tiwari and Shukla (1991) studied the influence of pesticides on *Glomus fasciculatum* and growth, development and nutrient uptake of soybean. In greenhouse experiments carbendazim had no inhibitory effect on the root colonization and chlamydospore population of the VAM fungus, or on the N,P,K content, plant height, branching, number of *Rhizobium* nodules, number of pods, plant dry weight or seed weight. Chlorothalonil did not affect VAM colonization but inhibited all the other components. All the other fungicides, insecticides and herbicides tested were inhibitory to all the aspects studied. In particular, oxadiazon was highly inhibitory to all the yield components through its effect on *G. fasciculatum*.

Jalali and Chhabra (1991) investigated the effect of Bavistin (carbendazim), captafol, thiram, BHC on *Glomus fasciculatum* infection of *Pennisetum americanum*. Thiram was reported as the most inhibitory to plant growth and mycorrhizal development. Captafol and Bavistin reduced total dry matter production by shoots and roots. All the pesticides adversely affected nutrient transport in the host plants.

Nair *et al.* (1991) studied the effect of certain plant protection chemicals on the incidence of VA mycorrhiza in black gram (*Vigna mungo*). In pot experiments with fungicides and insecticides, applied earlier to the soil or the seeds, most of the treatments reduced the incidence of VA mycorrhizas in black gram roots. Vitavax was more harmful to mycorrhizal development than Bavistin.

Kehri and Chandra (1993a) studied the effects of bavistin spray on soil microorganisms and VAM formation in greengram in relation to its yield. Bavistin spray (200, 300 and 500 ppm) was tfor its effect on VAM formation in roots and microbial population in and around roots of greengram. The fungicide suppressed the population of *Macrophomina phaseolina* (dry root rot pathogen of the crop) and improved the yield. However, it caused an unfavourable efect on mycorrhization, nodulation and population of non-symbiotic nitrogen fixers. According to them, to make the treatment more beneficial to the crop, there is a need to exploit the possibility of employing some method which may nullify its adverse effect on beneficial microorganisms.

Vijaylakshmi and Rao (1993) studied the effect of six insecticides and one fungicide on the development of VAM fungi in peanut (*Arachis hypogaea*). Monocrotophos at 5 to 50 ppm stimulated VA mycorrhizal (*Glomus* spp.) colonization while dimethoate and cypermethrin reduced it. Dichlorvos stimulated colonization at low concentration but was inhibitory at 50 ppm. A slight increase in VAM colonization was observed following application of 5 or 10 ppm fenvalerate, but endosulfan significantly inhibited *Glomus* growth even at 5 ppm. Ziram reduced colonization by 10-20% and plant growth was reduced by application of 20 or 50 ppm Ziram.

Sugavanum *et al.* (1994) studied the effect of fungicides on VA mycorrhizal infection and nodulation in groundnut (*Arachis hypogaea*). They investigated the effects of 6 fungicides *viz*., carbendazim (as Bavistin), mancozeb (as Dithane, M-45), 2-methoxyethyl-mercury chloride (as Emisan), fytolan, thiram and Ziram, applied (at field application rates) both as seed dressisng and soil drench, on the formation and function of VAM and Rhizobial nodular systems in groundnut. Fytolan treatment produced a range of positive responses in VAM infection and spore production, rhizobial nodulation and growth as well as yield. The other fungicides restricted mycorrhizal infection to different levels. The rhizosphere spore count was increased in the early growth stages of the host by fungicides, but declined in the later stages. Their effects on sporulation varied depending on VAM fungal species. Root nodulation, growth and yield of the host plant were not significantly affected by the application of fungicides, except the interference of 2-methoxyethyl mercury chloride in the early growth stages. The inhibitory effcts of 2-methoxyethylmercury chloride on the VAM and *Rhizobium* colonization resulted in a significant decrease in the growth and yield of the host plant.

Anusuya (1995a) studied the introduction of carbendazim and vesicular-arbuscular mycorrhizal fungi on growth and yield of chilli (*Capsicum annum* L.). The percentage of root colonization, spore number, and number of infective propagules increased with increase in carbendazim concentration (500 and 1000 mg l^{-1}) and the age of the host. Carbendazim significantly increased shoot dry matter in both 60 and 90 day old plants. Shoot P-concentration increased considerably over control plants when applied at 500 mg l^{-1} concentration of carbendazim in 40 and 90 days old plants. Carbendazim application showed a beneficial effect on the uptake of Zn, Cu, Fe, and B in the early stages of plant growth (40 and 60 days) at 500 and 1000 mg l^{-1} concentration over control, but the effect differed markedly as the age advanced. At 90 days, the level of Cu, Mn, and Fe were significantly reduced.

Anusuya (1995b) studied the interaction of carbendazim and VAM fungi on growth and yield of tomato. They reported no significant difference in the per cent mycorrhizal colonization and mycorrhizal spores in soils due to carbendazim treatment. However, carbendazim significantly increased the shoot dry matter in 60 and 90 days old plants.

Anusuya and Shobha (1995) studied the effect of glyphosate on VAM association in greengram. They applied the pesticide glyphosate (N-Phosphonomethyl glycine) to soil as solution to obtain a concentration of 5, 10, 100 mg and VAM inoculum of *Glomus fasciculatum*. They revealed that pesticide application helped the host plant in enhancing phosphorus concentration of root and shoot in the early stage of plant growth (30, 45 and 60 days), yield and nutrient uptake especially phosphours.

Udaiyan *et al.* (1995) studied the biostatic effect of fumigation and pesticide drenches on an endomycorrhizal-*Rhizobium*-legume tripartite association under field conditioins. They assessed the effects of formaldehyde fumigation and pesticide drenches with carbendazim (as Bavistin), Ziricle (as Cuman) [Ziram], copper oxychloride (as Copperthom), sulphur (as Sulfex), carbofuran (as Furadon) and chloradane (as Termix) [chlordane] at recommended rates on VAM association and *Rhizobium* sp. nodules, regularly for a period of 90 days in pigeonpea, *Dolichos biflorus,* ested *Vigna mungo* and cowpea under field conditions. The fumigants and the pesticides initially reduced VAM fungal colonization and the number of spores in all the crops. Following the initial decrease there was slow recovery, but by 90 days after emergence, root colonization was either parallel to or still lower than the control, and the number of spores was still well below control levels for all species except pigeonpea, which had more VAM spores than the control in all the treatments except fumigation and carbofuran. Although the number of nodules did not differ from control levels at 30 days after emergence, differences were evident during the later stages of plant growth for all species except cowpea. The effect of pesticides on VAM fungi and root nodulation varied with the associated host plant species. Plant tissue P and VAM colonization were significantly correlated in all host plants. The pesticide treatments had no marked effect on plant growth, but accumulation of nutrients in pesticide-treated plants was lower than that in untreated plants. Growth and nutrient status of the legumes varied with VAM fungi colonization.

Ammani and Rao (1995) studied the effect of RD9 a potential neem-based formulation, namely, RD 9 (REPELIN) developed by ITC and quinolphos on vesicular arbuscular mycorrhizal colonization in cotton. Aqueous solution of quinolphos (0.05%) was sprayed one month after sowing on one set of the pots. On the other set, RD 9 (0.5%) was sprayed. Potted plants receiving no pesticide or RD 9 served as controls. Five pots with a minimum of three healthy plants were maintained for each treatment.Crop growth was luxurious with RD 9 application. Its application was on par with quinolphos in terms of pest population, damage, and yield. From the data it wass evident that application of quinolphos resulted in reduction of per cent colonization and in number of vesicles/ spores when compared to the control. In contrast, application of RD 9 resulted in slight increase in per cent colonization, arbuscules, and average number of vesicles/spores.

Robert *et al.* (1995) studied the effect of fungicides on vesicular-arbuscular mycorrhizal fungi under natural conditions. Data indicated that the application of Ridomil-72WP, either in combination with the Bordeaux mixture or Akomin-40, resulted in maximum mycorrhizal colonization and spore production. Evidently, the time of application was also critical in spore production. Ridomil-72 WP applied during the second week of September showed maximum spore count than when applied at the onset of monsoon (T4). The study envisages the potential merit of Ridomil-72 WP application in stimulating the colonization and spore production of *Sclerocystis coremioides*. This suggests the better utilization of Ridomil-72 WP in the VAM based bio-rational disease control practices against foot rot incidence in black pepper.

Venkateswarlu and Daft (1995) made an attempt to determine the toxic effects of insectide on plant growth and mycorrhizal colonization by a promising VAM fungus, *Glomus clarum* Nicolson and Schenck, in groundnut. The plants, both non-mycorrhizal and mycorrhizal, that received carbofuran at concentrations up to 5 kg ha^{-1} were significantly taller than the untreated controls. Increasing concentrations of carbofuran resulted in a decrease in leaf number. Carbofuran applications had no inhibitory effect on initial flowering of non-mycorrhizal plants. However, there was a significant increase in the carbofuran. Interestingly, no variation in relation to dry weights of shoot and root, total root length, and specific root length was observed with the insecticide treatment either in the presence or absence of mycorrhizal inoculum. The concentrations of 0.5 and 2 kg ha^{-1} levels of carbofuran were innocuous to mycorrhizal development of *G. clarum* in groundnut. The application of carbofuran only at 5 kg ha^{-1} significantly inhibited the percentage of root colonization by the VAM fungus.

Anusuya and Dhaneswari (1995) studied the effect of fungicides bavistin (carbendazium) and metalaxyl on nodulation and vesicular-arbuscular mycorrhizal development in groundnut. The number of nodules on the 30th and 45th day were more in plants sprayed with 500 and 1000 mg^{-1} carbendazim and metalaxyl, than in control plants. Data showed that on 30th day dry weight of shoot and nodules were reduced considerably in plants sprayed with 250 and 500 mg^{-1}, carbendazim over control. However, in plants sprayed with carbendazim and metalaxyl at 1000 mg l^{-1}, there was increase in dry weight of nodules. The number of vesicles and arbuscules were reduced significantly at 500 and 1000 mg l^{-1} in carbendazim and metalaxyl-sprayed plants. These observations clearly indicate that VAM infection is considerably inhibited by application of different fungicides.

Vyas and Vyas (1995) investigated the influence of pesticides on mycorrhizal infection and chlamydospores formation in soybean (*Glycine max*). Among the pesticides, insecticides were least toxic to MI and chlamydospores production. The data indicated that a definite trend was established based on their toxicity toward AMF. The descending order of their toxicity toward mycorrhiza was : thimet > carbaryl > methyl-parathione > chloropyriphos > quinolphos. Herbicides were found to be more toxic than the other pesticides. The MI and chlamydospores number was significantly reduced in various treatments. This gives a caution to us that their use should be judicious to harness the benefits of mycorrhiza in crop productivity. This study concludes that as the pesticides and mycorrhizal interactions studied through this technique showed varied effects, the compatibility between these should be pre-determined. Among the pesticides, herbicides were more toxic towards mycorrhizal infection.

Udaiyan *et al.* (1996) studied the effects of formaldehyde fumigation and Fytolan drench on VAM fungi and nodulation in some leguminous forest tree seedlings in Tamil Nadu. Seedlings of 12 legume tree spp. *viz., Acacia cassia, A. catechu, A. nilotica, Albizia lebbeck, Dichrostachys cinerea, Leucaena latisiliqua* and *Pterocarpus marsupium* were raised in formaldehyde fumigated or Fytolan drenched beds in nursery in Coimbatore. Seedlings in the formaldehyde fumigated beds had stunted growth and were chloretic, had poor VAM and root colonization (18-25%) and spore density (3.1-10.6/g soil) and lower nodule numbers (3-8/plant) and nodular biomass (100-870 mg/plant). The total biomass (15.5-72.9/plant) and field survival rate (31.2-40.4%) of the seedlings were very low. In contrast, seedlings from Fytolan drenched beds showed normal growth, enhanced biomass (18-83.2g/ plant), had a higher field survival rate (71-86%), intense VAM root colonization (53.4-100%) and higher spore density (36-82.8/g soil) and higher nodule numbers (7.4-17.6/ plant) and nodular biomass (195-950 mg/ plant) than control seedlings. Roots of these plants exhibited extensively developed arbuscular and vesicular structures of the 7 VAM species recorded from the rhizosphere soils of control and Fytolan drenched beds. *Acaulospora bireticulata, G. fasciculatum* and *G. geosporum* were the dominant species.

Rachel *et al.* (1996) studied the effect of six agrochemicals *viz.*, Captan, Dithane M-45, Agrimycin, Urea, Gibberrellic acid and 2,4-D on mycorrhizal infection of sunflower with *Glomus mosseae* and *Acaulospora morrowea* separately and in combinations. Two fungicides showed inhibitory effect on all the mycorrhizal fungi. Foliar spray of urea inhibited the mycorrhizal infection. GA and 2,4-D also inhibited the mycorrhizal infection. Phosphorus content of the plant decreased when treated with two fungicides. Agrimycin had shown positive effect on shoot phosphorus. GA has increased the phosphate absorption of the plant whereas 2,4-D inhibited.

Shetty and Magu (1997) studied the influence of metalaxyl on *Glomus fasciculatum* associated with wheat. Soil incorporation of metalaxyl significantly enhanced the root colonization. The stimulatory response of VA mycorrhizal fungi to low concentrations of metalaxyl resulted in increased plant biomass production, nutrient uptake and grain yield of wheat. However, higher concentrations of metalaxyl, particularly 2.5 ppm of metalaxyl affected mycorrhizal infection and seed yield of wheat.

Jamaluddin *et al.* (1998) studied the effect of fungicides on colonization and development of VAM in *Dalbergia sissoo* under nusery conditions at the Tropical Forest Research Institute, Jabalpur, Madhya Pradesh. Fungicide application and mycorrhizal inoculation were done at the seed stage, and observations were made on 6-month-old seedlings. All the treatments received both VAM (with rhizobia) inoculations and fungicide, and controls were run with VAM (+ rhizobia) alone and with neither VAM nor fungicide treatment. All the 3 fungicides tested (Dithane M-45 [mancozeb], Bavistin [carbendazim] and captafol) enhanced infection by VAM and VAM spore formation in the rhizosphere at 0.1% concentration. At the higher concentration of 0.2% the fungicides suppressed VAM colonization. Seedling growth was also increased with 0.1% fungicial treatment. Mixed and single VAM inoculation using *Glomus fasciculatum* and *G. mosseae* without fungicide enhanced VAM infection and seedling growth, although seedling growth was less than with the 0.1% fungicide treatments.

Mishra (1999) studied the influence of *G. fasciculatum* in combination with 9 pesticides (herbicides and fungicides) on the growth of 4 tomato cultivars indicated that Pusa Ruby supported the highest multiplication of VAM [vesicular arbuscular mycorrhizas], followed by Pusa Early Dwarf, Gaurav and K21. All pesticide treatments reduced plant growth compared with the G. *fascuculatum* treatments. The lowest reduction in growth, percentage VAM colonization and production of chlamydospores was observed in the carbendazim + *G. fasciculatum* treatment. The herbicides under study exhibited a significantly higher degree of toxicity than the fungicides studied. The highest reduction in VAM colonization and spore production was observed in Pusa Early Dwarf treated with oxyfluorfen.

Udaiyan *et. al.* (1999) assessed the influence of formaldehyde fumigation and pesticide drench with Bavistin [carbendazim], Cuman [ziram], Copperthorm [copper oxychloride], Sulfex [sulfur], Furadan [carbofuran] and Termex [chlordane] at recommended rates on vesicular-arbuscular mycorrhizal (VAM) colonizatioin, growth, biomass and nutrient contents in *Eleusine coracana, Panicum miliaceum* and *Paspalum scrobiculatum* under field conditions at every 30 days up to 90 days. Experiments were conducted at Bharathiar University, Coimbarore, Tamil Nadu. The VAM root colonization of *E. coracana* was significantly decreased by all the pesticide treatments in comparison with the control, whereas, root colonization was significantly enhanced in *P. miliaceum* and *P. scrobiculatum* by Cuman and Copperthom treatments, respectively. The effect of pesticides on VAM fungi (*Glomus* spp.) sporulation varied according to plant species and pesticide. The N and P content of *Eleusine coracana* was significantly reduced by all pesticide treatments, whereas, the N content of *P. miliaceum* and the N, P and K content of *P. scrobiculatum* were significantly enhanced in treatments with Sulfex.

Udaiyan *et al.* (2001) investigated the influence of formaldehyde fumigation and pesticide drenching with Bavistin [carbendazim], Cuman [Ziram],Copperthom. Sulfex [sulfur], Furadan [carbofuran], and Termix on the dynamics of root colonization by arbuscular mycorrhizal (AM) fungi, spore number and rhizobial nodule number in an experiment conducted in Coimbatore, Tamil Nadu. Plant growth and nutrient accumulation were assessed regularly. Pesticide reduced plant growth of the legume seedlings at all age levels. Nitrogen and potassium contents were higher in seedlings grown in formaldehyde fumigated beds than in the control, but the accumulation of phosphorus was lower in all pesticide treatment.

Kumar *et al.* (2002b) studied the effects of soil application of four fungicides (*viz.*, captan, thiram, Bavistin [carbendazim] and Apron [metalaxyl]) on vesicular arbuscular mycorrhizal colonization and development in chickpea in pot experiments. None of the fungicides affected seed germination, plant height and number of pods in chickpea except, thiram (50 ppm and 100 ppm), which reduced plant height and number of pods. Mycorrhiza-inoculated plants exhibited significant increase in plant height, seed weight, and number of pods, biomass production and nutrient content compared to control. The fungicides, along with the mycorrhiza, caused reduction in plant height, seed weight, biomass production and nutrient content. Soil application of thiram caused maximum reduction in sporocarp number and mycorrhizal colonization in chickpea whereas Apron had lest the effect on mycorrhizal colonization.

Seasonal variation

Harinikumar and Bagyaraj (1988c) studied the effect of season on VA mycorrhizae of leucaena and mango in semi-arid tropic. Their results showed that maximum colonization and sporulation occurred during winter (November to January) months. Summer months (April to June) were unfavourable for the proliferation of VAM fungi. There was a positive correlation between relative humidity and mycorrhizal activity, while the correlation between temperature (ambient and soil) and mycorrhizal activity was negative.

Mohankumar and Mahadevan (1988a) studied seasonal changes in spore density and root colonization of VAM in a tropical forest. The seasonal variation in VAM spore population and root colonization in Kalakad Reserve Forest (K.R.F.), a tropical forest in Tamil Nadu over a period of 2 years from September 1982 to August 1984 was studied. K.R.F. has 7 different zones of vegetation; evergreen, grassland at low and high altitudes, semievergreen, mixed deciduous, teak and scrub jungle. In all the forest types, the number of spores and per cent infection did not coincide with each other. Generally, the rate of infection was more during summer than in rainy season. Surprisingly, the number of spores was higher during rainy season than in summer. The number of spores decreased markedly with increase in the depth of soil.

Jha *et al.* (1988) studied the status of VAM in degraded forest ecosystem. A seasonal study on the status of VAM in the 2 degraded forest stands in tropical and sub-tropical humid forests was carried out. Plants in more degraded stands were highly mycotrophic compared to the less disturbed site. VAM fungi were dominant in tropical conditions than in the sub-tropical ones. The population of the endogonaceous spores exhibited two seasonal peaks, *i.e.*, late winter and early winter in both the sites. *Glomus* sp. was dominant mycorrhizal fungi. Soil moisture correlated with infection intensity and spore population.

Pandey and Singh (1990a) studied the ecophysiological responses of litchi cultivars to vesicular arbuscular mycorrhizal colonization. They investigated the effect of seasonal variations on mycorrhizal association and root respiration in different varieties of litchi.

Relationship between the intensity of endophyte colonization and pH of roots and rhizosphre soil was also studied. The dry weight, nitrogen, phosphorus, potash and ash content of external mycelium of the endophyte including sporocarps on the root surfaces were high compared to those of litchi roots. But the moisture and organic carbon content of the external mycelium was less than the roots. The mycorrhizal colonization and root respiration was highest in October and minimum in June, suggesting a close relationship between mycorrhizal infection and new root growth. There were also clear differences in mycorrhizal colonization in different varieties.

Rachel *et al.* (1990) studied seasonal variation in the colonization and spore population of VA-mycorrhizal fungi in semi-arid soil of Andhra Pradesh. The percentage infection by VAM fungi varied among 30 species belonging to 19 angiospermic families. According to them root colonization and number of resting spores in the rhizosphere were greater in summer than in winter and least in rainly season. Arbuscules predominated in summer.

Sulochana and Manoharachary (1990) studied the impact of season on the distribution of VAM fungi associated with sesame. A total of 11 VAM fungi belonging to *Acaulospora, Glomus*, and *Gigaspora* were associated with kharif and rabi seasons with 6 sesame cultivars. The *cv. Gowri* showed maximum VAM association and colonization was greater in the kharif season.

Singh *et al.* (1992) studied the effect of soil depth and seasonal changes on spore population and mycorrhizal colonization of kinnow (Mandarins) and rough lemon seedlings. They reported maximum and minimum infection in June and November respectively. The maximum and minimum spore population of endophytes in soil were recorded in June and October respectively. VAM colonization in soil increased root, shoot length and the number of leaves per rough lemon seedling.

Mago and Mukerji (1994) studied the VA mycorrhizae in Lamiaceae with refernce to seasonal variation especially in *Mentha viridis, Ocimum sanctum, O. basilicum, Salvia officinalis* and *Coleus varigatus*. In all the plant species studied, VAM colonization was lowest during winter and highest during late summer and autumn.

Ibrahim *et al.* (1995) studied the seasonal response of groundnut genotypes to arbuscular mycorrhza-*Bradyrhizobium* inoculation. Nine cultivars of groundnut (*Acachis hypogaea*), earlier shown to respond to arbuscular mycorrhiza in a soil poor in available N or P, were studied in the same soil over three subsequent seasons for response to VAM (*Glomus fasciculatum*) and *Bradyrhizobium* (R) singly or in combination (VAM+R). The responses were measured for six traits, namely percentage of root length colonized by AM (VAMC), root P, Biomass (BM), carbon dioxide exchange rate (CER), leaf nitrogen and pod yield. Differences in the traits under 4 treatments, namely control, VAM alone, R alone and VAM+R, were statistically evaluated across cultivars and seasons. High improvements over the control were consistently recorded under VAM+ R in experimental plots. Particularly yield responses were generally genotype and season-dependent. The most responsive to combined inoculation were cultivars 1441 A1, NF P 140, 1423 VB and RB15. A basis for explaining improved yield sustenance by dual inoculation with VAM and *Bradyrhizobium* was provided by the significant correlation between VAMC and root P in addition to the desirable correlation between CER, BM and root P. The results provided ample reason for integrating VAM schedules into breeding and agronomic approaches in programmes on stabilizing groundnut yields.

Iqbal and Mahmood (1998) made a study to judge the seasonal fluctuations in VAM association and VAM fungal spore population on some commonly cultivated crops at four different localities *viz*., Shahjamal, Kasimpur and Badarbagh in Aligarh District. The intensity of VAM association, VAM formation, and VAM fungal spore population showed great seasonal fluctuations, during the different seasons of the year. In all localities, six species of the VAM fungi *viz*., *G. mosseae, G. constrictum, G. fasciculatum, G. aggregatum, Gigaspora gigantea*, and *Acaulospora scrobiculata* were identified. Predominant VAM species found in all the four localities were *G. mosseae, G. constrictum*, and *Glomus fasciculatum*. However, the species of *G. aggregatum, Gigaspora gigantea*, and *Acaulospora scrobiculata* rarely occured and their distribution differed from one locality to the other. The population of VAM fungal spore was seemingly very high during early summer and late winter in all the four localities. Spore population was suppressed during late summer and early winter in all the samples observed. Intensity of VAM colonization also showed a great variation in all the localities during different seasons.

Muthukumar and Udaiyan (1999b) studied the spore-in-spore symdrome in vesicular-arbuscular mycorrhizal fungi and its seasonality in a tropical grassland. Sporocarps and spores of vesicular-arbuscular mycorrhizal (VAM) fungi isolated from a tropical grassland (dominated by *Cymbopogon caesius*) in the Maruthamalai Hills, Coimbatore, Tamil Nadu, were occupied by spores of other VAM fungal species. Species of *Gigaspora, Glomus, Sclerocystis* and *Scutellospora* were occupied by spores of *Acaulospora, Glomus* and *Scutellospora* species. Typically spores of a single VAM fungal species occurred within a single occupied spore and their number varied from 1-35. Seasonal trends in the density of live, dead and occupied spores were analysed during a 2-year period to determine within-year variations in the density of occupied spores and to assess whether the trends in density of occupied spores were same in both years. Seasonal trends in density of occupied and live spores emerged, but dead spore densities exhibited no seasonal patterns. The presence of VAM spores inside dead spores of other VAM fungal species suggested that spores of VAM fungi act as microhabitat when dead, in addition to their normal role as propagules.

Lakshman *et al.* (1999) studied the seasonal fluctuations of vesicular arbuscular mycorrhizal fungi on some commonly cultivated crops of Dharwad District in Karnataka. The results indicated that the seasonal changes in the climatological factors cause marked fluectuations in VAM associated with cultivated crops.

Chandra and Jamaluddin (1999a) studied seasonal variations in rhizosphere vesicular arbuscular mycorrhizal (VAM) spores and in root VAM infection in a teak (*Tectona grandis*) seed production area and a nearby plantation of the bamboo *Dendrocalamus strictus* (established in 1969) in Maharashtra. The soil and root samples were collected in rainy, spring, winter and summer seasons [year not given] at up to 30 cm depth and mixed to form one samples. The mycorrhizal fungi present were *Glomus* spp. (*G. aggregatum, G. mosseae,*

G. invermaium, G. microcarpum), Acaulospora scrobiculata, Gigaspora, Scutellospora and SCL (*Sclerocystis*]; *Glomus* spp. were the most common. In teak, the maximum numbers of rhizosphere VAM spores were found in the summer and least in spring and winter. For the bamboo plantation, and maximum numbers of rhizosphere VAM spores were found in the summer and winter, and the least in the rainy season. Spore populations were higher in the bamboo than in the teak rhizosphere. VAM infection showed very little seasonal variation but was slightly greater in the teak stand then the bamboo stand.

Kumar (2002) studied the seasonal variations in the biodiversity of arbuscular mycorrhizal fungi in forest ecosystems. The spore population of arbuscular mycorrhizal fungi in soils collected from scrub, deciduous, evergreen and Scholas areas at Alagar hills, Maduraiwas studied. In all soil samples, the spores of *Glomus* spp. were most dominant, followed by *Sclerocystis* and *Scutellospora* spp. The highest number of spores was recorded during the winter season (December and January), whereas the lowest was observed during the rainy season (August and September). The sandy clay loam soil contained the highest number of spores compared with clay loam and sandy loam soils.

Physico-chemical soil characteristics

Gunasekaran *et al.* (1987) studied the effect of various pH (5.5, 6.5, 7.5 and 8.5) and temperature (20, 25 and 30⁰C) on the spore germination of *G. fasciculatum*. Spore germination was studied on cornmeal agar, modified Hepper medium, charcoal agar (0.01%) and soil extract agar. Maximum germination (81.3%) occurred at pH 6.5 at 25⁰C.

Mohankumar and Mahadevan (1987) studied the effect of litter leachates on VA mycorrhizae. According to them, litter leachates greatly suppressed the number of mycorrhizal spores in soil and this inhibition increased with incubation time. Extracts of leaves from an evergreen forest inhibited the mycorrhizal activity.

Mohankumar and Mahadevan (1988b) studied the VA mycorrhizal distribution in the tropical Kalakad Reserve Forest, with respect to change in organic matter content in the soil. No definitive relationship existed between organic matter content and the number of spores, nor could any correlation be established between organic matter and infection. No single factor influenced the distribution and survival of mycorrhizas in the tropical forest ecosystems.

Vyas and Srivastava (1988b) studied VAM of moth bean with special reference to ecological implications. Moth bean plants along with rhizosphere soils were collected from 6 different localities. The VAM spore population and percentage root colonization were affected by change in soil pH, soil moisture, P, organic carbon, etc. A variation in the VAM fungal flora was recorded. An increase in soil pH directly influenced the spore population and percentage of root colonization. An increase in soil moisture decreased the spore population and percentage of root colonisation. A decrease in spore population and percentage root colonization was observed with an increase in organic carbon content of the soil. The available soil P directly affected the spore population and percentage of root colonization. Maximum soil P was associated with maximum spore population and percentage of root colonization. The native VAM fungi preferred high soil pH (8.7), low soil moisture (5.0%), low organic carbon (0.10%) and 4.3 mg/100 g available phosphorus for maximum spore population and root colonization.

Sulochana *et al.* (1988) worked on the VAM fungi associated with *Sesamum indicum* in relation to some ecological factors. Rhizosphere soil samples of 6 varieties of *Sesamum indicum* were screened for VAM fungi. Altogether 5 species of *Glomus*, 2 species of *Gigaspora* and 1 species of *Acaulospora* were observed. Moisture, pH, organic matter and available phosphorus of soils affected the VAM association.

Kehri and Chandra (1988) studied the mycorrhizal infection and its relation to the rhizosphere microflora in urd under water stress conditions. A quantitative evaluation of the intensity of mycorrhizal association and the microbial population in the rhizosphere in 2 varieties (T-9 & Bhadeli) of urd (*Phaseolus mungo*) grown under water stress/non-stress conditions was undertaken. The percentage VAM infection in roots, VAM spore population as well as relative proportion of microbes, i.e., fungi, bacteria and actinomycetes in the rhizosphere fluctuated with the age of plants. Water stress affected not only the microbial spectrum of the rhizosphere but also the intensity of root infection and the population of VAM spores in the root region.

Veeraswamy *et al.* (1988) studied the development of VAM on *Sorghum* under field conditions. In a systematic survey, the extent of VAM development was measured in sorghum, up to 80 days after sowing, in agricultural fields of Anantput district, Andhra Pradesh, comprising soils (alluvial, red laterite and black) with varying physico-chemical characteristics. VAM developed differentially on sorghum raised in different soils. In all the cases, 50% root system became mycorrhizal by 25 to 30 days after sowing. A plateau in VAM formation, with more than 90% soil and laterite soil while in black soil about 87% of root length was colonized by VAM fungi only after 75 days. In the plants from alluvial and laterite soil, internal vesicles/spores were noticed after 15 days and few days later, there was initiation of the formation of external vesicles/spores. However, sporulation by VAM fungi was delayed in black soil. External vesicles/spores started appearing only after 55 days while arbuscules were predominant throughout in the interior of the roots.

Iyer *et al.* (1989) studied the effect of fertilizer levels on vesicular-arbuscular mycorrhiza colonization in cardamom. Roots of cardamom plants growing in experimental plots with three fertilizer levels were examined before and after the south-west mansoon. Percent colonization in roots varied from 40-100 percent in pre-mansoon samples and 63-94 per cent in post mansoon samples. Results showed that increasing doses of fertilizer tended to decrease the intensity and incidence of the VAM population in soil and roots.

Singh and Tyagi (1989) studied the occurrence of VAM fungi in the roots of *Aeschynomene indica* under the influence of various ecoloical factors. No VAM fungi infection was detected in the roots of *A. indica* plants, grown under waterlogged conditions, but plants grown under normal moisture conditions were highly infected with VAM fungi. Dry weight of shoots and percent VAM fungal root colonization were progressively reduced due to incorporation of sodium chloride, potassium chloride, and glucose in the soil. However, concentrations of 2.0, 1.0 and 4.0 of sodium chloride, potassium nitrate and glucose respectively, inhibited the growth of *A. indica* plants. Potassium nitrate was more toxic to plant growth. Maximum root colonization by indigenous VAM fungi and dry weight of plants were recorded at 50 kg P_2O_5/ha. 100 kg P_2O_5 / ha reduced VAM fungal root colonization. On exposure of *A. indica* plants to increased light duration, dry weight of shoots and roots and VAM colonization were progressively increased. However, 12 hour light exposure was optimum for the growth of *A. indica* and maximum VAM fungi root colonization.

Kehri and Chandra (1989c) studied the mycorrhizal association in crops under sewage farming. A comparative assessment of VA-mycorrhizal status in 18 crops irrigated with civic sewage and water indicated that sewage irrigation was favourable for the mycorrhiza formation in a number of crops including *Vigna sinensis, Spinacia oleracea, Coriandrum sativum, Solanum tuberosum, Trigonella foenum-graecum*. However, in certain other crops including *Saccharum officinarum*, it exhibited a suppressing effect. Under both the types of irrigation, the intramatrical growth of the symbiont within the roots and its extramatrical sporulation failed to show any relation with each other.

Kehri and Chandra (1990) studied the VAM association in urd as affected by water stress condition in soil and foliar spray. The intensity of VAM infection in roots and sporulation in urd var. T-9 and Bhadeli, raised under normal level of water supply, fluctuated with the age of the plants. Water stress condition in soil affected the VAM association in both the varieties. However, the effect was nullified to a great extent by the application of foliar spray of 250 ppm of $NaNO_3$ to T-9 and of KCl to Bhadeli.

Gokuldas *et al.* (1990) studied the mycorrhizal reproduction as influenced by moisture stress. In pot experiments with cowpea plants grown with *Glomus fasciculatum*, maximum mycorrhizal root colonization occurred under conditions of severe stress with watering once in 5 days to field capacity (FC). When pots were maintained at only 50% FC, the number of spores and infective propagules were less. Watering at FC once in 3 or 5 days favoured sporulation more than when they were maintained at 50 or 75% FC.

Kehri and Chandra (1991) studied the spectrum of the mycorrhizosphere of green gram as affected by non-indigenous VAM fungi under normal and strained water schedules. Out of ten symbiotic fungi used in the study, only four *viz., Glomus fasciculatum, G. macrocarpum*, isolates NS and 4S could cause an improvement in root colonization in *var.* T-44 under normal supply of water. On the other hand, eight including these four could make the mycorrhizosphere more conducive for sporulation. But under strained water supply, only three symbionts could improve root colonization and seven, the spore production. Root colonization under similar water schedule in *var. Pusa Baisakhi* showed an improvement due to the introduction of only two symbionts *viz., G. fasciculatum* and isolate 7MY. However, all the symbionts could cause an increase in spore production in the mycorrhizosphere. However, under strained supply, eight symbionts could improve root colonization but all the ten, the spore production.

Shukla and Vanjara (1992) reported the levels of available NPK in soil in following crops of sesame, sunflower and mustard inversely proportional to the VAM spore density, indicating an inhibitory effect of fertilizers on the mycorrhizas.

Jindal *et al.* (1993) studied the effect of VA mycorrhizae and NaCl salinity on metabolism of moong plants. In pot experiments, the effects of (a) 0 (b) 12.5 or (c) 25 mM NaCl were studied in the presence or absence of *Glomus fasciculatum* mycorrhizal infection on metabolism of mung beans cv. Ml.131 at 40 (pre-flowering, PF) and 62 (flowering, F) days after sowing. Leaf nitrate reductase activity was greater in mycorrhizal than non-mycorrhizal plants with (c) but nitrogenase activity was similar in both types of plants. Proline accumulation was greater in the mycorrhizal plants at both growth stages with (b) but only at F with (c). The concentration of total sugars was greater in the mycorrhizal plants with (c) at both growth stages, but total N was greater only at F.

Kehri and Chandra (1993b) studied the response of mungbean to different VA-mycorrhizal fungi under normal and strained supply of water. Efficacy of eleven VAM fungi in improving the performance of two varieties of mungbean was evaluated under pot conditions with normal and strained supply of water. Marked variations were noted between different VAM fungi with respect to their effect on mycorrhization, growth, nodulation and yield. An isolate of *G. fasciculatum* collected locally showed high versatility and was recommended for field trials under irrigated and rainfed conditions.

Kehri and Chandra (1994a) studied the effect of foliar sprays on the VA-mycorrhiza formation in chickpea. The effect of selected sprays *viz.* 2, 4-D 50 ppm; KNO_3, 50 ppm, $(NH_4)_2HPO_4$, 50 ppm; Urea, 50 ppm and 200 ppm on the mycorrhization in the roots of chickpea var. T-3 was investigated under normal as well as water stress condition. The intensity of mycorrhization varied not only with the age of the plants and watering schedule but also with the foliar spray used. Under normal condition, only urea, 50 ppm caused an improvement in the mycorrhization, others showed a suppressive effect. Under water stress condition, on the other hand, all the chemicals but for 2, 4-D showed an improvement in the mycorrhization. The results of the study have been discussed in the light of known importance of foliar sprays as well as VA mycorrhizae in the welfare of crop plants.

Eranna and Parama (1994) studied the'effect of liming, mycorrhizal inoculation and rock phosphate on phosphorus uptake and growth of soybean in an acid soil. In a pot experiment with soyabean using an acid (pH 5.1) soil, O or 2.64 g lime/ kg soil and 0, 10, 20 or 40 g rock phosphate/kg soil were applied and the soil was inoculated with *Glomus mosseae*. Liming had no significant effects on soyabean dry weight but it decreased P uptake. Mycorrhizal inoculation increased both dry weight and P uptake. Shoot dry weight increased with rate of P application whereas root dry weight was highest with the lowest rate of P. Shoot P uptake increased up to 20 g P/kg soil whereas root P uptake increased up to 40 g P/ kg soil.

Prasad (1995) studied the effect of Physicochemical characteristics of soil on VAM colonization (*Glomus fasciculatum*) in *Saccharum officinarum*. Experimental soil exhibited a sandy loam texture with 62.4 to 66.1% sand, 8.4 to 10.5% clay and 21.1 to 32.2% silt. Temperature variations between 20.5-35.5°C were recorded during diffirent phases of plant growth over a year with maximum infection at 30.5°C. Infection percentage and population density of VAM fungi, *G. fasciculatum* was highest at a moderate level of soil moisture (19.2%). In general, alkaline pH (8.6) was suitable for VAM infection and spore population.

Among other chemical parameters, EC presented direct correlation. Available P and potash contents of soil exhibited an almost inverse relationship between VAM infection and spore population. When P and K contents declined in soil, % infection and spore concentration increased. There was no correlation between available Nitrogen and VAM infection.

Joshi and Singh (1995) studied the interrelationship among VA mycorrhizae population, soil properties and root colonization capacity of soil. The population was determined in different soil types of India using the most probable number technique and correlated with soil properties and VAM root colonization capacity of the soils. The soils were sandy loam, silty clay loam, gravelly loam, loam and sandy loam containing 21, 9, 17, 15 and 23 propagules/g soil respectively. Two of the five soils were rated as high in VAM population. The population of VAM propagules in soil showed a positive correlation with soil properties such as N, organic C, available K, sand content, pH and percentage VAM root infection capacity of soil, while there were negative correlations with CEC, available P, silt and clay content.

Raghavendra and Lakshman (1995) studied the effect of different soil pH values on VA mycorrhizal association in tree species. *Artocarpus heterophyllus, Eugenia bracteata, Pterocarpus marsupium* and *Terminalia bellirica* were included in the study. *Glomus mosseae* was used as inoculum. Data revealed that all the plants grown in natural soils with different pH value exhibited varied responses. The soils having pH 4.5 to 5.5 were found unfavourable for the growth of host as well as endophyte. However, biomass formation, weight of the plants, percentage of VAM association etc. were maximum at soil pH between 6.5- 7.5.

Khan and Kulshreshtha (1995) studied the impact of fly ash on VAM fungus *Glomus caledonium* and root nodule bacterium *Bradyrhizobium* in black gram. Fly ash amended soil influenced root colonization and spore number of the VAM fungus and root nodulation by the root nodule bacterium on black gram. Root nodulation and spore numbers of VAM fungus were enhanced at lower levels (20 and 40%) and suppressed at high levels of fly ash. Root colonization by the VAM fungus was adversely affected at all the levels. Enhancements were greater in dual inoculated plants and suppression in single inoculated plants. Fly ash applications in soil at low levels appear to be helpful for root symbionts and consequently may enhance plant growth and their productivity. According to them, this significant impact can be explored further for agriculture.

Raman and Nagarajan (1996) studied the effect of forest fire on VAM fungi in tropical forest of southern India. Twelve species (including several tree species) were selected for the study. According to them all macroelements apart from nitrogen, as well as microelements, increased following fire. A total of 22 VAM fungal species were isolated; 15 from the burned and 16 from the unburned area with 9 species common to both. VAM infection was considerably lower in the burned area (18-52%) than in the unburned (37-72%) area.

Bagyaraj and Machado (1996) conducted greenhouse experiments to investigate phosphorus concentrations in the soil solution suitable for VAM activity in *Leucaena leucocephala*. They collected soil from an uncultivated field of the University Campus, Bangalore, and liming was done at the rate of 1 kg of $Ca(OH)_2$/kg of soil. The soil was subjected to 3 cycles of wetting and drying to raise the pH to 6.2. Six differnt levels of P were established by applying varying amounts of KH_2PO_4 to the limed soil. (0, 25, 50, 100, 250 and 400 mg/kg soil). Other essential nutrients were supplied at optimum levels for VAM activity. This resulted in soil solution P levels of 0.007, 0.022, 0.049, 0.088, 0.232 mg p/litre respectively. *Glomus aggregatum* inoculum was added at the rate of 25 g/kg of soil. Germinated seeds of *Leucaena leucocephala* were planted in pots @ 2/pot and maintained in a glasshouse for 60 days. Plant growth, root and shoot P content, plant biomass and mycorrhizal colonization was highest (86.1%) at 0.022 mg P/litre. The benefits derived from VAM inoculation were inversely proportional to the soil solution P level.Taking biomass production into account, the level of P soil solution that was optimum for VAM symbiosis was estimated as 0.022 mg P/litre.

Muthukumar *et al.* (1997) studied the influence of native endomycorrhiza, soil flooding and nurse plant on mycorrhizal status and growth of purple nutsedge (*Cyperus rotundus* L.). Above ground (AG) and below ground (BG) dry weight of mycorrhizal plants were significantly lower than the non-mycorrhizal plants. However, except when grown with onion, ratios of below ground:above ground dry weight were the same. Percentage root colonization of *C. rotundus* and onion was significantly higher when the 2 were grown together, although plants growing alone had less infection. Flooding of soil inhibited mycorrhizal formation in *C. rotundus* and mycorrhizal infected plants produced smaller tubers than uninfected plants. Above ground tissue P concentration was lower in mycorrhizal *C. rotundus* but below ground P concentration did not vary significantly between infected and uninfected plants. P utilization efficiency was similar for mycorrhizal and non-mycorrhizal *C. rotundus*. Plants in flooded soil had the maximum AG and BG dry weight, tuber dry weight and P content, but had lower AG:BG dry weight and P utilization efficiency.

Udaiyan *et al.* (1997) studied the possible role of arbuscular mycorrhizal (AM) fungi on drought tolerance in *Vigna unguiculata* subsp. *unguiculata* (L.) Walp and *Leucaena latisiliqua* L. Pot culture experiments were carried out in sterilized soil to determine the effects of root inoculation with *Acaulospora scrobiculata, Glomus aggregatum* and *G. etunicatum,* both individually as well as in combinations, on growth and drought resistance of *Vigna unguiculata* and *Leucaena latisiliqua* (*L. leucocephala*). All plants were inoculated with *Rhizobium* sp. During the first 45 days, all pots were watered to field capacity and the plants were then subjected to drought cycles by withholding water for 7 days. The plants were further subjected to 7 drought cycles. Inoculation with mycorrhizas in general produced higher growth, biomass and root nodules, than the uninoculated controls. Proline content was higher in all the plants inoculated with *G. etunicatum*, whereas inoculation with *G. aggregatum* in *Vigna unguiculata* and *G. etunicatum* in *L. leucocephala* produced the highest plant N, P and K contents. In general, plants inoculated with more than one mycorrhiza species produced higher plant dry matter than those inoculated with one mycorrhiza species only. For *V. unguiculata,* inoculations with *A. scrobiculata* + *G. aggregatum* produced the highest plant dry weight for *L. leucocephala*, the best combination was *A. scrobiculata* + *G. etunicatum*.

Reddy *et al.* (1997) studied the influence of water stress on VAM colonization and growth of sunflower (cv. *Moderan*) in field experiments conducted at the Kakatia University Campus, Andhra Pradesh. Two species of VAM fungi, *Glomus mosseae* and *Acaulospora morrowae* were used. Mycorrhizal infection and number of resting spores were greater in water stress conditions than non-stress conditions. According to them mycorrhizal inoculations ameliorated the water stress conditions to some extent.

Singh *et al.* (1998) studied the occurrence of ecto- and endo-mycorrhizae with the roots of *Casuarina equisetifolia* in various soil types. Out of 14 soil samples collected from various location in India, vesicular arbuscular mycorrhiza formation in seedlings of *Casuarina equisetifolia* grown for 120 days was observed in 9 samples. External mycelial (ecto-mycorrhizal) associations were also recorded in 10 samples. Hartig net and intercelular mycelial colonization with the root under the situation were not observed. In a few soil samples plant roots were colonizaed with both endo-mycorrhizal and ecoto-mycorrhizal fungi, whereas in 5 samples only endo-mycorrhizas and in 3 only ectomycorrhizas were observed. Only in one soil sample, no mycorrhizal association developed, indicating the probable absence of both type of mycorrhizal fungi in the soil. Maximum VAM root colonization was observed in soils from Jabalpur (Madhya Pradesh), and higher ectomycorrhizal colonization was recorded in the Baroda (Gujrat) soil samples.

Gaur and Adholeya (2000b) studied the effect of particle size of soil-less substrates upon AM fungus *Glomus intraradices* inoculum production in a locally available sand graded particle size, planted with *Zea mays* and fertilized with a nutrient solution. Plants in sand with particle sizes of 0.50-0.78 mm had higher root fresh weights, spore production and percent mycorrhizal colonization than with other particle sizes. Production of spores and infectious propagules was enhanced by a nutrient solution without P. Plants were also inoculated with *G. intraradices* in pots containing clay-brick granules. Charcoal,Coalmarl, sand or perlite of the optimal particle size (0.50-0.78mm). Percent root length colonized by *G. intraradices* and production of infectious propagules were 40-50% higher for plants grown in clay-brick granules and sand than in the other media.

Kaushal (2000) studied the influence of edaphic factors on VAMF spore population and root colonization in *Acacia niloica* in Rajasthan. The rhizosphere soil and root samples of *A. nilotica* were periodically collected from six localities of arid and semi-arid region of Rajasthan. Vesicular arbuscular mycorrhizal fungal spore populations as well as root colonization were influenced by abiotic factors like soil moisture, soil pH, soil organic carbon content, soil phosphorus and soil nitrogen. The spore population was negatively correlated with soil moisture, soil carbon and soil phosphorus but positively correlated with soil pH. However, there was no correlation in spore population and percentage of root colonization with regard to changes in soil semen. Seasonal variations in the spore population were also observed. The number of spores was highest in the rhizosphere soil during the months of July and August and lowest during summer (May and June).

Agronomic Practices

Harinikumar and Bagyaraj (1988a) studied the effect of crop rotation on native VA mycorrhizal propagules in soil. Finger millet (*Eleusine coracana*) was grown as the first season crop in 15 plots. In a second season a mycorrhizal host (cowpea) and a non-mycorrhizal host (mustard) were grown in 5 plots each, and the remaining 5 plots were left fallow. In the third season cowpea was grown in all the plots. Leaving the land fallow reduced the number of mycorrhizal propagules by 40% while growing a non-mycorrhizal host reduced it by 13%. Cowpea grown in the third season coincided with a slow build up of mycorrhizal propagules in soil. There was a slow build up of mycorrhizal propagules late in the season irrespective of the treatment in the preceding season.

Nirmala *et al.* (1988) investigated the viability of VAM spores in the fallow fields of Maduravoyal, Tamil Nadu. Soil samples were collected from 6 different rice fields which remained fallow from 1980 to 1984 and from a field under regular rice cultivation. Spores of VA mycorrhizal fungi were collected from the soil and viability of VAM spores was assessed in onion roots. Though the soil samples of 1980 and 1981 had a good number of VAM spores they did not infect onion roots. The per cent mycorrhizal infection was less in the soil collected from the field under regular cultivation. Onions grown in the soil collected from 1982 and 1983 fields which contained low amount of phosphorus had 35-40% infection. Spores collected from uncultivated fields lost their viability to infect roots.

Harinikumar and Bagyaraj (1988b) studied the effect of intercropping and organic soil amendments on native VAM. The effect of mono and intercropping with soybean and maize together with amendments like paddy straw, maize straw and pongamia leaf on native VAM was studied. Intercropping stimulated proliferation of the VAM fungi compared to monocropping with maize or soybean. Of the 3 amendments studied, addition of pongamia leaf increased native mycorrhizal activity to the maximum followed by maize straw. Addition of paddy straw to soil was least stimulatory.

Harinikumar and Bagyaraj (1989) studied the influence of cropping sequence with and without fertilizer and farmyard manure application on VA mycorrhizae over three consecutive seasons. In the first season maize was grown on all the plots. In the second season cowpea (*Vigna ungiculata*), groundnut and finger millet were raised on the same plots and in the third season, sunflower was grown on all the plots. The groundnut grown in the second season stimulated mycorrhizal root colonization, sporulation and infective propagules in the soil and these effects were carried over to the next season. The plots cropped to finger millet in the second season had the lowest number of mycorrhizal spores. The application of farmyard manure stimulated VA mycorrhizae while fertilizers at the recommended level decreased the mycorrhizal propagules.

Harinikumar *et al.* (1990a) studied the effects of monocropping and intercropping of soyabean and maize together with rice straw, maize straw and pongamia leaf soil amendments on native VAM fungi. Intercropping stimulated proliferation of VAM fungi compared to monocropping with maize or soyabean. In the soil cultivated with soyabean there were more mycorrhizal propagules as compared to maize. Addition of organic amendments to soil increased the proliferation of VAM fungi. Of the three organic amendments studied, addition of pongamia leaf showed the highest increase of native mycorrhizal presence, followed by maize straw. Increased numbers of VA mycorrhizal propagules in the soil was correlated with improved plant P uptake.

Gupta *et al.* (1990a) studied the possible role of weeds in maintaining mycorrhizal infectivity of rice soils. It was observed that rice plants were colonized with VAM fungi at all sampling sites. The weeds belonging to Malvaceae, Acanthaceae, Euphorbiaceae and Graminae were colonized with VAM fungi at two sampling sites and the members of Compositae and Polygonaceae were having VAM infection at all the sites sampled. Plants belonging to families Papilionaceae, Scrophulariaceae, Labiatae and Cyperaceae were also colonized with VAM fungi. Mycorrhizal spores extracted from these soils belonged to the genus, *viz. Glomus fasciculatum, G. magnicaulis, G. macrocarpum, G. mosseae G. reticulatum*. This sudy shows that weeds such as *Ageratum conyzoides, Sphaeranthus indicus, Sida acuta, Asteracantha longifolia, Hyptis suaveolens, Leontis nepetifolia* and *cyanodon dactylon* may be responsible for maintaining the mycorrhizal infectivity status of these soils.

Shukla and Vanjare (1990) undertaken the extraction of vesicular arbuscular mycorrhizae spore and studied the effect of fertilizers on their population in oilseeds cropped soils. The effect of nitrogen on VAM population with regards to oilseed crops (Sesamum, Sunflower and Mustard) revealed a unique adverse effect of increasing level of nitrogen. From the initial low level itself, the population decreased and continued gradually till zero at high levels indicating high sensitivity of VAM to increasing level of nitrogen in oilseed cropped soils. A declining trend of VAM population with increase in phosphorus levels has been observed in Sesamum, Sunflower, and Mustard crops. High phosphorus levels in soil are detrimental for the symbiotic establishment of VAM with these oilseeds roots. In oilseed crops, there is an adverse effect of potash as its level is increased. In oilseeds, it seems that there is inverse relationship between VAM spore population and potash level. In all soil samples analyzed, only one genus i.e. *Glomus* spp. was encountered.

Singh *et al.* (1990) investigated the effect of glyphosate N-Phosphonomethyl glycine, a post emergence, non-selective herbicide on VA mycorrhizal association in groundnut. The results indicated that mycorrhizal inoculation increased significantly the root and shoot dry weights. Glyphosate applied at 100 ppm concentration reduced significantly the root dry weight compared to treatment receiving 10 ppm concentration. Interaction effect of mycorrhiza with glyphosate showed that mycorrhiza along with glyphoste at 5 and 10 ppm concentration significantly increased the root and shoot dry weight. In fact the highest root dry weight (2.6 g) was recorded in mycorrhizal tretment receiving 10 ppm glyphosate. Per cent mycorrhizal colonization increased with increase in glyphoste concentration, the maximum being 87% in 100 ppm glyphosate treatment. Glyphosate application did not significantly alter either the P concentration or the P content of the shoot. However, it was noted that both the parameters increased with increase in glyphosate level applied. Per cent mycorrhizal colonization increased with increase in glyphoste concentration. A significant increase in per cent and total P content in mycorrhizal plants was recovered. Application of glyphosate alone (10 ppm) increased per cent P concentration of the root. They attributed it to the enhanced P availability in the soil due to glyphosate application. The results clearly indicated that the VAM fungal association in groundnut was not affected by glyphosate even at unrealistically high concentration.

Mehrotra and Baijal (1992b) studied the influence of certain agronomic practices on mycorrhizal associations in field crops. An assessment of the VA mycorrhizal associations in crops grown under certain agronomic practices revealed that (i) a lower mycorrhizal colonization occurs in crops grown in sewage than in tubewell water fields, (ii) crops belonging to the families Cruciferae and Chenopodiaceae, which are generally sparsely colonized, had higher mycorrhizal colonization when plants belonging to the family Leguminosae have been grown in the same field, (iii) additional phosphorus added during fertilizer applications in tube-well irrigated fields, supported better mycorrhizal colonization than sewage irrigated fields with lower P, (iv) water-logged conditions, under which paddy is cultivated was deterimental to mycorrhizal growth and development and (v) taxa of mycorrhizal species belonging to the genus *Glomus* were predominant over others.

Sreenivasa and Srihari (1994) studied the influence of monocropping and intercropping of different crop plants on native endophytes of VAM fungi in vertisols. Intercropping stimulated proliferation of VAM fungi more than monocropping. More mycorrhizal propagules were present in soil cultivated with legumes than in those with cereals.

Verma (1994) studied the effect of shifting cultivation on the mycorrhizal loss in the forest soils of north-eastern hills of India. Shifting cultivation is practiced in North-Eastern India where it is known as jhuming. Due to population growth the jhuming cycle has been reduced to 3-5 years which is not sufficient to restore soil fertility through natural succession. The effects of shifting cultivation on root production, VAM infection and VAM spore production in the soil, and the effects of the depletion of mycorrhizal propagules on the vegetation were investigated. Two contrasting sites, an upland open site subjected to frequent burning and a closed fallow site with *Pinus kesiya* as the dominant tree species, were compared. The results showed that the VAM spore population has been reduced in the frequently disturbed soils and increased in the undisturbed soils of the valleys. There was a decrease in root biomass and number of VAM spores with decreasing depth at both sites. The author has discussed the implication of the results for soil fertility in the area.

Wani and Lee (1995) studied the effect of crop and soil management practices on mycorrhizal association with crops in the semi-arid tropics. They studied the effect of crop, plant growth stage and cropping system on VAM, effect of tillage on VAM, effect of N fertilizer and effect of organic amendments on VAM. They reported that short fallowing between harvesting the crop and sowing of the next year crop reduced the VAM spores in soil. The deep tillage increased the mycorrhizal colonization in the roots. Higher doses of N fertilizer reduced the VAM colonization. When two levels of FYM and bioearth were applied to sorghum and pearl millet three legumes-soybean, groundnut and pigeonpea, the mean VAM colonization across the crops was not influenced by organic amendments. Application of 1% bioearth significantly reduced the VAM colonization in pigeonpea and soybean whereas in sorghum and pearl millet it was increased over no amendment treatment. FYM application had not adverse effect on root colonizatioin of the crops.

Maiti *et al.* (1996) studied the perpetuation of native VAM fungi under monocropped, rainfed upland agroecosystem. According to them, native VAM fungi could be encouraged in soils used for rice by allowing growth of preferred weeds during the off-season and maintaining established mycelial networks by longterm minimal tillage practices.

Srinivasan *et al.* (1996) studied the influence of effluent irrigation on VA mycorrhizae. They assessed the effect of effluent irrigation on VAM status of three crop species *viz.*, *Capsicum annuum, Pennisetum typhoides, Sorghum bicolor,* 4 tree species *viz., Azadirachta indica, Carica papaya, Musa paradisiaca, Psidium guajava* and 7 weed species *viz., Ageratum conyzoides, Calotropis procera, Croton sparsiflorus, Euphorbia hirta, Parthenium hysterophorus, Sida acuta, Tridax procumbens*. The incidence of VAM infection was greater for weeds than for the crop and tree species. Effluent irrigation markedly altered the nutrient status of the soil and significantly increased the root infection in *Sorghum bicolor* and *Psidium guajava* while significantly reduced spore numbers in all plant species except *Azadirachta indica*.

Harinikumar and Bagyaraj (1996) studied the persistence of an introduced VA mycorrhizal fungus in the field under the influence of a mycorrhizal host, a non-mycorrhizal host and when the field was left fallow. There was a significant greater build-up of the mycorrhizal fungus in plots cropped with mycorrhizal hosts (finger millet) {*Eleusine coracana* or field beans (*Lablab purpureus*) compared to plots cropped with a non-mycorrhizal host (mustard) {*Brassica juncea*} or left fallow. *Glomus intraradices*, the introduced fungus was further monitored by the electrophoretic mobility of the isozymes of malate dehydrogenase extracted from the resting spores of the fungus. The zymogram from *G. intraradices* spores, showed 3 distinct isozyme bands 1, 3 and 7. A zymogram of spores isolated from plots inoculated with *G. intradices* for three seasons exhibited similar bands. Spores isolated from plots inoculated with *G. intraradices* for one or two seasons showed only one weak band, corresponding to isozyme band 7. Their results indicated that the introduced fungus persisted in the field for only one season.

Lakshman (1996) studied the role of weeds in maintaining VA-mycorrhizal colonization in rice soils. Forty five weed species belonging to 19 angiospermic families were collected from paddy fields in the Hubli-Dharwad region of Karnataka and screened for the presence of VAM. All were found to be mycorrhizal but the extent of colonization varied: it was 4-17% in the Chenopodiaceae, Cruciferae, and Cyperaceae, 19-31% in Polygonaceae, Aizoaceae, Scrophulariaceae and Oxalidaceae and 33-53% in the Primulaceae, Commelinaceae and Umbelliferae (Apiaceae). The greatest colonization (54-82) was recorded in the members of the Zygophyllaceae, Leguminaceae (Fabaceae), Malvaceae, Euphorbiaceae, Liliaceae, Labiatae, Compositae (Asteraceae) and Graminae (Poaceae).

Munyanziza *et al.* (1997) discussed the role of two main types of mycorrhiza, ectomycorrhiza which mainly occur in tropical pines, Caesalpinaceae, Dipterocarpaceae, and Myrtaceae, and arbuscular mycorrhiza (AM) which occur in most of the agricultural and horticultural crops and several tree species in tropical ecosystem. The changes in mycorrhizal biodiversity due to the establishment of tropical cropping system on areas previously occupied by species rich forest ecosystems and the detrimental effect of modern, high input agricultural practices are compared with low input sustainable agricultural methods which enhance the population of mycorrhizal fungi. It is concluded that more studies are needed to understand the role of mycorrhizal fungi in soil aggregation and non-destructive methods have to be developed for studying mycorrhizal biodiversity in natural ecosystem.

Shinde and Nair (1998) investigated the effect of continuous cropping on VAM association of maize. Maize was grown both as Kharif and Rabi crops in Maharashtra. Assessment of VAM fungi was done from Baglan area of Nasik district. During Kharif season, when the plants were in the seedling stage, the number of VAM spores and the percentage root infection was low, which gradually increased as the plants matured. This slightly decreased during one month break between Kharif and Rabi crops, probably due to absence of root exudates. However, there was an increase when the rabi crop was in field and reached a maximum at the end of growth of the plants. This steady increase in VAM spores as well as infection was probably because there was a continuous supply of same type of nutrients in the form of root exudates of maize. A correlation was established between VAM and the edaphic factors.

Panja and Chaudhuri (1999) tested four different organic manures for their effects on mycorrhizal relations of *Cajanus* in low nutrient alluvial soil. Root colonization intensity was not changed by the manures except by oilcake which at very high concentration reduced root colonization intensity significantly. Promotion of root development by these manures resulted in significant increase in root biomass yield and also spore production except by oilcake which even at moderate concentration suppressed spore production significantly. Rich husk promoting root development, stimulated spore production very strongly. FYM and green leaf manure caused moderate stimulation of mycorrhizal root biomass and spore production in root association.

Maiti *et al.* (1999) conducted experiments to explore the possibilities of enhancing VAM association for improvement of upland rice (*Oryza sativa* L.) under rainfed agro-ecosystem. A long term experiment was conducted to explore the possibilities of utilising native VAMF flora by enhancing their population in soil, thereby augmenting colonization in rice through different rice based cropping systems (RBCS) and management practices under rainfed upland agroecosystem of south Bihar plateau for improvement of productivity. Among three cropping systems evaluated, rice + pigeon pea (*Cajanus cajan* L.) intercropping was found to be most promising.

Srilakshimi *et al.* (1999) studied the effect of organic amendments *viz.*, pressmud, biogas slurry, farm yard manure and urban compost on the mycorrhizal colonization, growth and yield of MCU - 7 cotton. All the organic amendments were found to cause positive influence. The mycorrhizal fungi inoculated plants recorded higher values of root colonization, growth and yield parameters than uninoculated plants. The positive effect of organic amendments were in the order of farm yard manure, biogas slurry, urban compost and pressmud. In their presence, 56.04, 52.06, 45.05 and 40.27 per cent of root colonization, respectively was recorded. The *Glomus fasciculatum* inoculated plants applied with farm yard manure, biogas slurry, urban compost and pressmud recorded 84.57, 72.82, 68.55 and 62.66 per cent of root colonization, respectively. The increase in growth and yield parameters observed with farm yard manure was the highest followed by the treatments applied with biogas slurry, urban compost and pressmud.

Maiti and Variar (2000) studied the effect of different rice (*Oryza sativa* L.) based cropping systems on population dynamics of native vesicular-arbuscular mycorrhizal fungi under rainfed upland ecosystem. Different rice-based cropping systems were evaluated for the augmentation of native vesicular-arbuscular mycorrhizal fungi, in Hazaribag, India, during wet seasons 1993-97. The 4 systems evaluated were: rice; rice followed by toria during 1993-94, rice followed by horse gram (*Dolichos biflorus*) during 1995, maize followed by niger (*Guizetia abyssinica*) during 1996 and rice alone in 1997; rice intercropped with red gram (*Cajanus cajan*); rotation, rice alone in 1994, fallow in 1995, black gram (*Phaseolus mungo*) in 1996 and only rice in 1997. The population of mycorrhizal fungi peaked during September in all treatments. The 2nd treatment was most favourable for mycorrhizal fungi.

Maiti *et al.* (2002) conducted a filed experiment in 1998 and 1999 to determine the effect of rice-legume intercropping on the native vesicular arbuscular mycorrhizas (VAM) in the soil. Treatments were: rice alone; rice: pigeon pea at 4:1; and rice: green gram at 4:1. The cultivars used were rice cv. Vandana, pigeon pea cv. Bisra Arhar, groundnut cv. BG 3-6, black gram cv. Pant U 19, and green gram cv. Sunayana. Root samples were collected at 20,40 and 60 days after emergence (DAE). All intercropping systems result in higher crop VAM colonization and VAM population compared to sole rice at 60 DAE. The highest colonization percentage was observed in rice: groundnut intercropping.

Rana *et al.* (2002) conducted a field experiment in Hazaribag, Jharkhand during 1998 and 1999 to determine the effect of rice-based intercropping systems on vesicular arbuscular mycorrhizal (VAM) colonization, P uptake, and crop yield. The intercropping systems were rice cv. Vandana (the control), rice-pigeon pea (Birsa Arhar), rice-groundnut (BG 3-6), rice-black gram (Pant U 19), and rice-green gram (Sunayana). Root colonization by VAM fungi in rice at 60 days after seedling emergence (DAE) ranged from 30.9% in the control to 54.7% in rice-groundnut system. P concentration and uptake in rice straw at 60 DAE were influenced by the intercrops. The competition for P by the intercrops was highest with groundnut followed by black gram, pigeon pea, and by green gram. The intercropping systems, in general, increased the yield-attributing characteristics, i.e. leaf area index, tiller number, and panicle number, of rice. In 1998, the tiller number and panicle number were significantly higher in intercropping systems than the control, but in 1999, only leaf area index was higher than the control. Rice grain yield was higher in 1999 than in 1998, except in rice-groundnut system. The total productivity of rice+pigeon pea was significantly and consistently higher than the other cropping systems in 2 years, but was at par with rice + groundnut system in 1998.

Yaragoppa *et al.* (2003a) conducted a field experiment during kharif of 2000-2001 at the Agricultural College Farm, Raichus, and Karnataka on red sandy loam soils to investigate the influence of plant densities and phosphorus management on dry matter and its accumulation pattern in *Sesbania aculeate*. There were 21 treatment combinations comprising of three plant densities viz., low (13 333 plants/ha), medium (20 000 plants/ha) and high (40 000 plants/ha) in main plots. These plant densities were achieved by providing variable intrarow spacing of 75, 50 and 25 cm with a constant interrow spacing of 100 cm. In subplots, four P levels (0, 60, 80 and 100 kg/ha) without vesicular arbuscular mycorrhizas (VAM) and three P levels (60, 80 and 100 kg/ha) with VAM inoculation were used. *S. aculeate* sown under low plant density produced significantly higher dry matter per plant as compared to dry matter produced per plant at medium and high plant densities at all growth stage (30, 60, 90 DAS and at harvest). At harvest, low plant density recorded an increase of 5 and 11% over medium and high plant densities. At harvest, low plant density recorded significantly higher dry matter accumulation in leaves, stems and pods when compared to medium and high plant densities. Application of 60 kg P_2O_5/ha+VAM produced significantly higher dry matter at all growth stages and dry matter accumulation in leaves, stems and pods at harvest when compared to other treatments, except for the treatment which received 8- kg P_1O_5/ ha+VAM and 100 kg P_2O_5/ha+VAM.

Harikumar and Thomas (2003) studied the occurrence of arbuscular mycorrhizal (AM) species under different management in coconut plantations. Mycorrhizal colonization in roots and spore density in the root region soils of coconut in a plantation in Kerala were found to be affected under treatments of tillage, herbicide, fertilizers and a combination of tillage + organic and inorganic manures when compared to that in plots which were not receiving these treatments. Identification of the AM fungi associated with coconut under different treatments revealed the presence of nine species of AM fungi belonging to the genera viz., *Glomus, Acaulospora* and *Gigaspora.* The distribution of various AM species under different management treatments was also determined. *Glomus macrocarpum* was found to be the most abundant species under different treatments.

Singh and Adholeya (2003a) determined the effects of crop rotation *viz.,* wheat-rice, wheat-pigeon pea and wheat-sugarcane on the development and diversity of vesicular arbuscular mycorrhiza (VAM) in a filed experiment conducted in Uttar Pradesh. Wheat-pigeon pea rotation recorded the highest diversity, spore count, root colonization, infectivity and extrametrical hyphal biomass of VAM as well as the highest enzyme active staining. VAM isolates collected from this rotation also recorded a significantly higher dry weight and nutrient uptake compared to those isolated from the other crop rotation systems. Wheat-rice rotation recorded poor VAM development and less efficient VAM isolates.

Yaragoppa *et al.* (2003b) conducted a field experiment during kharif season of 2000 at the Agricultural College Farm, Raichur, and Karnataka on red sandy loam soils to investigate the effect of plant densities and phosphorus (P) management on growth and seed yield of *Sesbania aculeate.* There were 21 treatments combinations, comprising of three plant densities, viz., low (13, 333 plants/ha), medium (20, 000 plants/ha), and high (40, 000 plants/ ha) in main plots. These plant densities were achieved by providing variable intrarow spacing of 75, 50 and 25 cm with a constant interrow spacing of 100 cm. In subplots four P levels (0, 60, 80 and 100 kg P_2O_5/ha) without vesicular arbuscular mycorrhizas (VAM) and three P levels (60, 80 and 100 kg P_2O_5/ha) with VAM inoculation were used. Observations were recorded on growth and yield components are plant height (cm), number of leaflets per plant, number of branches per plant, dry matter production and its distribution (g), number of pods per plant, length of pod (cm), number of seeds per pod, 100 seed weight (g), pod yield per hectare (q) seed yield per hectare and stalk yield per hectare (q). High plant density recorded significantly higher seed; stalk and husk yield as compared to medium and low plant densities. The increase in seed, stalk and husk yield with high plant density was mainly compensated by high plant population per unit are even though the growth and yield attributes like number of leaflets, number to branches, number of pods per plant, number of pods per plant and 100 seed weight were significantly low when compared to growth and yield attribute recorded under low and medium plant densities. However, and inverse relationship was observed with respect to plant height. Application of P_2O_5 at 60 kg/ha+VAM recorded significantly higher seed, stalk and husk yield due to higher growth and yield attributing characters which in turn caused significantly higher B: C ratio as compared to the treatments which received only P and control. However, it was on par with the application of P_2O_5 at 80 kg/ha+VAM and 100 kg/ha+VAM.

Pollutants

Jagpal *et al.* (1988b) studied the distribution of VAM fungi in relation to pollution. Survey of endogonaceae in Delhi revealed that the number of VAM fungi decreased near Indraprastha Thermal Power Station and their number increased with increase in distance, from the Station. The forms encountered were sclerotial and lost their efficiency of forming association with the plant growing in the vicinity.

Jagpal *et al.* (1988a) studied the effect of automobile pollution on VAM spore population of soil. The response of VAM spore population to the automobile exhaust pollution was examined. The soil samples were taken from 6 diffeent traffic intersections of Delhi for 4 months. The spore population in the samples was determined. The traffic density and lead concentration were negatively correlated with spore number.

Sambandan *et al.* (1992) studied the distribution of VAM fungi in heavy metal polluted soils of Tamil Nadu. These soils contained very high concentration of Zn followed by Cu, Pb, Ni and Cd. Of 18 plant species examined, 16 were mycorrhizal. VAM clonization and diversity of spore density in the rhizosphere soils were investigated. Of 15 VAM fungi observed, 10 species were of *Glomus*, 2 were of *Gigaspora*, 2 were of *Acaulospora* and 1 was of *Sclerocystis*.

Babu *et al.* (1995) studied the vesicular-arbuscular mycorrhizal colonization and biomass production in economically important crop plants grown directly over the sludge. The present study critically analyses the effect of application of distillery sludge on VAM infection, biomass production, and crop productivity. Physico-chemical characteristics of the sludge indicated that its water-holding capacity was as much as 122% compared to the garden soil which was about 19%. As regards chemical parameters, pH was very high in the sludge (9.82) and the EC value was 8.63 mS/cm indicating extreme salinity. The sludge contained various organics like sugars, phenols, sucrose, starch, nitrogen and proteins as well as elements such as N P K and Ca. The percentage of germination and seedling vigour index were not at all affected even when the sludge was directly used *in toto* for raising plants. The percentage of VAM infection was more in plants (50-day-old) grown directly over the sludge (21-80%) as compared to the control. The increased pH and EC values of the sludge did not affect the VAM spore occurrence, germination and root colonization. The elements such as N,P, K, and Ca were analysed in the roots of 50-day-old plants which were higher in the plants grown directly in sludge. The phosphorus uptake was more followed by nitrogen, calcium and potassium. Total biomass production was increased significantly in sludge-grown plants (30-73%) and the harvest index was also higher (26-45%) over the control. From the overall perusal of the data, it was surmised that application of sludge was not harmful to plants, and the increased level of pH and EC did not affect the VAM spore populations, spore germination, root colonization, and mycorrhizal effectiveness. The study supports the occurrence of mycorrhiza, most particularly VAM, in any kind of stress environment.

Joseph *et al.* (1995) studied the effect of mycorrhization on two crops raised on land disturbed by petro-effluents. Two crops, maize (cereal) and green gram (pulse), raised in effluent and fresh water irrigated fields were selected for the study. Both effluent and fresh water irrigated maize plants showed efficient mycorrhizal infection (88%). Shoot length of effluent irrigated plants was more than that of fresh water irrigated plants. The number of cobs and the length of cobs were almost similar in both the conditions, but cob weight and number of grains per cob were more in the effluent irrigated plants. Effluent irrigated greem gram plants had higher infection than the fresh water irrigated plants, in both the conditions. The nodules were more in number in the vegetative phase and less during the reproductive phase. The apparent morphological change discernible in the root nodules of the effluent irrigated plants was their comparatively bigger size (about two-fold). At the time of harvest, effluent irrigated plants exhibited comparatively more shoot length, dry weight, higher number of mature pods, more pod length, and higher number of grains per pod than that of fresh water irrigated plants. The fresh weight of the shoot was however a little less in the effluent irrigated plants than in fresh water irrigated plants *Glomus* spp. constituted the dominant species among the VAM fungi associated with the effluent and fresh water irrigated plants of both maize and green gram.

Raman *et al.* (1995) studied the distribution of vesicular-arbuscular mycorrhizal fungi in tannery effluent polluted soils of Tamil Nadu. From the three sites of effluent polluted soils, 15 species of VAM fungal spores were isolated and identified. Out of the 22 plant species screened for VAM fungal colonization, 19 plant species exhibited positive results while the percentage of infection varied between 11 and 64%. The soils of the study sites were found to be clayey loam and alkaline (pH 7.6-8.1). The percentage of organic matter varied between 22 to 29%. Total nitrogen, available phosphorus, (P_2O_5 kg/acre), and available potassium (K_2O kg/acre) were considered as low. Out of 22 plant species examined, 19 were mycorrhizal. Again, among the plant species examined, *Prosopis juliflora, Azadirachta indica*, and *Heliotropium zeylanicum* were the potential tree species growing well in the polluted environment and associated with the maximum number of mycorrhizal fungi. Many VAM fungal species were found to be associated with more than one plant species, among them *Glomus geosporum* and *Gigaspora gigantea*, were found to be associated with the maximum number of plant species, which were incidentally called common VA endophytes in polluted soils examined. In the present study, it was found that polluted soils contain high concentrations of heavy metals and most of the plants were able to survive in the effluent polluted environment with VAM fungal associations.

Reddy *et al.* (1995) studied the association of vesicular-arbuscular mycorrhizal fungi in soils polluted with industrial and sewage effluents. The soil, which was distant from the effluent site and supported many weeds, grasses, and other plant communities, was considered as control. The soil site, which received industrial effluents - rich in caustic soda, acetic acid, soap, sodium silicate, ammonium sulphate, print colour, and other effluents - was considered as the second sampling site. The third sampling soil site received only sewage effluents, and supported only one type of grass. Soil polluted with the sewage effluent supported less VAM population than non-polluted soils. Altogether, 44 VAM fungal species were collected and identified. The non-polluted soils contained 15 VAM species, whereas the soil polluted with industrial effluents harboured 26; only 9 species were found associated with the soil receiving sewage effluents. *Gigaspora* and *Scutellospora,* which were found in non-polluted soils were absent in the soil polluted with industrial and sewage effluents. *Glomus* was completely absent in the non-polluted soil. The soils with below neutral pH favoured more VAM fungal population while those with the pH range from 7.02 to 8.1 supported more VAM fungal species. Soil moisture was inversely related to VAM fungi. Chlorides, bicarbonates, and nitrogen did not effect VAM fungal dynamics. The soils with moderate to less phosphorus content supported more VAM fungal population. Soils with less potassium, 1-3% organic carbon, 53 to 330 ppm of iron, and 4.5 to 17.5 ppm of manganese favoured more VAM fungi. Nickel and lead were found toxic at higher levels.

Jahan and Khan (1996) examined the impact of sulphur dioxide on root nodulation by *Bradyrhizobium japonicum* and root colonization by *Glomus fasciculatum* on mungbean and consequential effects on growth, yield, leaf pigment, seed protein and some elemental contents of the plants in artificial treatment conditions. SO_2 in general, suppressed plant growth and yield characters, leaf pigment, nitrogen and phosphorus contents and seed protein. Root nodulation by root nodule bacterium and root colonization by the VAM fungus and its spore production were also inhibited. Nodulated mycorrhizal and dual inoculated plants showed enhanced growth yield, leaf pigment, seed protein, nitrogen and phosphorus. Suppressions caused by SO_2 were lower in the inoculated plants as compared to uninoculated exposed plants. The root symbionts were thus suppressed by SO_2 but they provided partial protection to the plants against adverse effects of the air pollutant.

Kumar *et al.* (1999) investigated the influence of three pollutants *viz.*, tannery, paperboard and sewage water on VAM fungi with reference to eighteen plant species. With few exceptions, the percentage of infection was suppressed under the influence of effluents. Spore population in the rhizosphere soils varied both with the plant and also with type of pollutant. All the three effects *i.e.* enhancement, suppressing effect and no effect were observed in different plants. VAM fungal species responded differently towards the effluents. *Glomus* and *Acaulospora* species were not significantly affected by the effluents. Paperboard effluents have shown less toxic effect than the other two types of effluents.

Kulshreshtha and Khan (1999a) investigated the impact of sulfur dioxide on *Glomus caledonium* and *Rhizobium* sp. on *Vigna mungo* roots in Aligarh, Uttar Pradesh. It was demonstrated that root symbionts minimize the negative effects of sulfur dioxide.

Kulshreshtha and Khan (1999b) investigated the effect of fly ash from a thermal power plant on *Glomus caledonium* and *Rhizobium* sp. on the roots of *Vigna mungo* in Aligarh, Uttar Predash. It was demonstrated that mycorrhizas and root nodulating bacterium protected the plants from some of the harmful effects caused by fly ash.

Kulshreshtha and Khan (1999c) studied the effect of ozone on root colonization by VAM fungi and root nodulation by *Rhizobium* in black gram. The impact of ozone in two different concentrations (0.02-0. 05-0.02 ppm) and (0.05-0.1-0.05 ppm) for 7 hrs on alternate days on the plant growth (on fresh and dry weights, lengths of shoot and root), yield, chlorophyll, seed protein, phosphorus and nitrogen, root colonization, spores of VAM fungus and root nodulation were studied. Suppression in growth and other above parameters was less in symbiotic plants in comparison to non-symbiotic (uninoculated) plants. Dual inoculated black gram plants had greater plant growth and other above parameters than uninoculated and single inoculated plants.

Jamaluddin *et al.* (2001a) studied the status of vesicular arbuscular mycorrhizal (VAM) fungi in sludge dumps of the Ballarpur paper mill (Maharashtra) where different tree species were planted to beautify the premises and also reclaim the sludge garden. The dominant species at the site were *Glomus* spp., such as *G. deserticola, G. leptotichum, G. intraradices, G. mosseae, G. aggregatum* and *G. invermaium*. Based on the percentage of root colonization by VAM fungi, the planted species did not show any corresponding increase in height. However, the overall correlation with the height and diameter of plant species was positive. The maximum percentage of root colonization was recorded from *Acacia nilotica* (72.39%) followed by *Dalbergia sissoo* (70.36%). The minimum percentage of root colonization (7.36%) was noticed in *Casauarina equisitifolia*. There was abundant VAM spore highest number of 466 spores/ 100 g soil, was observed in the rhizosphere of *Acacia nilotica*, while the lowest, 179 spores/100 g soil, in *C. equisetifolia*. Other species carried over 150 spores in 100 g of rhizosphere soil, but could not express their efficiency for root colonization accordingly.

Ramakrishnan *et al.* (2001) examined the effect of sugar mill effluent polluted soil on the quantitative and qualitative assessment of vesicular arbuscular mycorrhizal (VAM) fungi in Tamil Nadu. Twenty-three species of VAM fungi belonging to five genera *viz.*, *Glomus, Gigaspora, Acaulospora, Sclerocystis* and *Scutellospora* were identified from the soil taken around ten plant species (*Abutilon indicum, Prosopsis, julliflora, solanum nigrum, boerhavia diffusa, Croton sparsiflorus, Ocimum basilicum, Commelina bengalensis, cynodon dactylon, Dacryloctenium aegyptium* and *Cyprus rotundus)* in Eydanur, Sundaravadi, Varakalpattu and Tiruvendipuram, Tamil Nadu.. Among the ten plant species studied, two plant species (*Croton sparsiflorus* and *Cyperus rotundus*) were non-mycorrhizal while the other 8 species were colonized and also the percentage root colonization was less in polluted soil. The number of VAM spores was also less in polluted soil when compared to unpolluted soil. The increased levels of micronutrients and heavy metals were noticed in polluted soil, which caused reduction in the VAM propagules.

Koul *et al.* (2001) Studied the influence of lead in soil on mycorrhizal development and plant growth of *Cyamopsis tetragonoloba* (Linn.). *C. tetragonoloba* (*cv. Pusa Naveen*) plants were treated with 6 levels of lead nitrate (15, 30, 45, 60, 75 and 100 ppm) and an unamended control. Plants were harvested after 70 days without damaging the roots. Growth parameters and total fresh and dry weight were recorded. Lead tissue analysis was performed and mycorrhizal response was estimated. Growth of *C. tetragonoloba* increased with increasing concentration of Pb in soil. Plant biomass declined significantly at concentrations above 60 ppm of Pb. At 100 ppm Pb, root length decreased by 76%, shoot length increased by 58%, and fruit fresh and dry weight decreased by 33 and 52%, respectively, compared to the control. Maximum average Pb accumulation (2.11 μg/g) was recorded for roots treated with 100 ppm Pb. Increasing Pb concentration decreased the number of lateral roots. Root browning, blackening and suberization were also observed in treated plants. At higher concentrations of Pb (60, 75 and 100 ppm), there was a decrease in colonization by mycorrhizal fungi. Vesicular arbuscular mycorrhiza (VAM) hyphae had an irregular size and terminated abruptly in the outer cortex of the roots. The number of VAM spores in the rhizosphere also decreased with increasing Ph concentration. At 100 ppm Pb, colonization was represented only by internal hyphae and few if any vesicles were present.

Gautam and Irshad (2002) collected samples separately for each crop in fields of leguminous crops at polluted sites (Javan and Harduaganj) and pollutions-free sites (Kwarsi and Kharaish war) in Aligarh, Uttar Pradesh. Seven root samples of six commonly cultivated leguminous crops (pea, chickpea, pigeon pea, cowpea, mung bean and clover) were collected to measure root colonization of arbuscular mycorrhizal (AM) fungi. The AM fungal population was affected by pollutants released from the Kasimpur thermal power station. Pollutant-free sites always yielded higher populations of AM spores compared to polluted sites. Among the crops, clover supported the highest production of spores and cowpea recorded the lowest yield in all the four sites investigated. At the polluted sites, mung bean showed the second highest number of AM spores followed by chickpea, pea and pigeon pea. At pollution free sites, the highest and lowest spore numbers were recorded in clover and cowpea, respectively. The second highest number of spores was recorded in mung bean, followed by chickpea and pea. The percentage colonization of AM fungi was affected by pollutants. The percent colonization in clover was highest, 60-64%, 70-74%, at the polluted and non-polluted sites, respectively. The colonization percentage in mung beans was the second highest all the four sites, with an average of 44%. In pea, pigeon pea, chickpea, the average percentage of root colonization was 54.4% and 56.5% respectively.

Kulshreshtha and Khan (2002) studied the effect of ozone (0.02-0.05-0.02 and 0.05-0.1-0.05 ppm applied for 7 h daily at one-day-intervals) on the growth (fresh and dry weights, and lengths of shoot and root),yield, chlorophyll, seed protein, phosphorus, and nitrogen contents, root colonization, spores of vesicular arbuscular mycorrhiza (VAM) fungi and root nodulation in black gram. The suppression of growth and of other aforementioned parameters was less in symbiotic plants than in non-symbiotic (uninoculated) plants. Higher values for growth and other parameters were recorded for dual-inoculated plants than for single-inoculated plants.

Genotypic Characters

Subhashini *et al.* (1988) studied the genotype dependent variation in mycorrhizal colonization and response to phosphorus in *Sorghum bicolor*. The genotype differences among 47 diverse varieties and exotic germplasm for VAM colonization, P uptake in leaf, shoot, root and dry matter production were studied over 2 season in alfisol soil which was low in P_2O_5. VAM infection ranged from 25 to 95% in different cultivars. Restorer lines of the hybrids such as PD-2-1-11, PVR-10, M-148, Rs-29 and Rs-71 showed maximum VAM colonization and high P content in shoot and root. Certain genotypes of sorghum were adapted better than others to P deficient conditions and showed higher efficiency in their uptake. The data on correlation coefficients revealed significant and positive associations of VAM infection with P content of shoot and root as well as with dry matter production.

Ammani *et al.* (1989) recorded increased mycorrhizal colonization in gamma ray-induced green gram micromutants. They treated two green gram cultivars with gamma radiation at 130 kR and 40 kR. At M_6 generation, 4 micromutants showing the greatest yield and the 2 parent cultivars were planted in red laterite soil known to have a high population of VAM fungi. Samples of roots were taken at 15 days intervals up to harvest at 60 days. Average mycorrhizal colonization at 30 days was highest in micromutant LGG 403 (90%) and lowest in the parent cultivars ML-26-10-3 and LGG 127. The amount of external mycelium and the number of vesicles/spores (mainly *Glomus*) were also highest in LGG 403. Phosphorus content, shoot dry weight and yield were all higher in the micromutants than in the parents.

Sreenivasa and Rajashekhara (1989) studied the influence of stage of host development and genotypes on mycorrhizal colonization of field grown wheat. According to them estimates of VAM infection and sporulation showed a positive relationship with the growth of the host until flowering (up to the 60th day), with slight or no increase thereafter. Colonization and sporulation varied between genotypes and was maximum with DWR 39 at flowering stage.

Kumari and Nair (1989) investigated the natural incidence of VA mycorhizas influenced by host varieties. 10 cowpea varieties growing in pot trials at Trivanduram were taken into account. Average mycorrhizal index differed between varieties, with the maximum infection in the variety C 152 followed by Ptb 2 and New Era. Addition of rock phosphate to the soils enhanced mycorrhizal formation, and had similar effect in all the varieties.

Gupta and Krishnamurthy (1994) tested the receptivity of various wheat cultivars NI-8629, CPAN-2005, Raj-1777 and CC-464 for VAM association - *Glomus caledonium*. Symbiotic efficiency varied according to cultivar, CV. NI-8629 benefitted more from the association compared with the other 3 cultivars.

Uniyal and Thapar (1997) studied the status of VAM colonization in *Acacia nilotica* seedlings of different seed origin. The response of *Acacia nilotica* of 13 Indian provinces to inoculation with a mixture of indigenous VAM fungi (*Glomus fasciculatum, G. fulvus, G. mosseae, Scutellospora* sp.) and pure inoculum of *Glomus macrocarpum* was studied under greenhouse conditions. The frequency of VAM infection and percentage of root colonization was variable, ranging from 46 to 96% and 5.14 to 47.57% respectively.

Miscellaneous

Venkataramana *et al.* (1985a) found *Gigaspora nigra* to be the most common VAM fungi in soil samples from Jorhat in North Eastern India. The coefficient of variance for spore number in an undisturbed plot was comparatively low. Culture filtrate of mixed soil microflora and presence of live maize root in sterilized soil favoured the germination of these spores.

Chandra and Zaidi (1987) studied the mycorrhizal association in certain crops as affected by growth stages. The report included the informations regarding the mycorrhizal association in two leguminous (Blackgram and Pigeon pea) and two non-leguminous crops (Pearl millet and Sorghum) commonly grown in trans-Ganges area under Allahabad district. All the samples of roots and rhizosphere soils showed association of only *Glomus* spp. Irrespective of crops, the percent root infection varied with the stages of growth. In spite of minor climatological changes in the area during the cropping period, the spore population of *Glomus* spp. in the non-rhizosphere region remained comparatively low and uniform. In all the four crops, the population in the rhizosphere region fluctuated with the stages of growth. However, irrespective of the crops or their stages of growth, it was relatively higher than that recorded in the non-rhizosphere region.

Deka *et al.* (1990) studied the effect of fuel burning on VA mycorrhizal fungi and their influence on the growth of early plant colonizing species. According to them the root infection in early colonizers by endomycorrhizal fungi was greatly reduced after burning in a forest area. After emergence of seedlings on the burnt soil, infection by VAM increased spontaneously with age of the plants. The population of endogonaceous spores in the soil was also drastically reduced after the fire.

Nair *et al.* (1990) studied the effect of soil solarization on the nodulation and infection by VAM fungi in cowpea. According to them treatment increased nodulation and infection by 20%.

Joshi *et al.* (1991) studied the effect of deforestation on microbes, VAM fungi and their enzymatic activity in eastern Himalaya. Soils of 4 plant communities (forest, undisturbed and disturbed grassland and a highly disturbed (barren) plant community) at Jawai, Meghalaya, were surveyed for microbial population, soil respiration and chemistry, dehydrogenase and urease activities and VAM fungi. Undisturbed grassland soil had the most bacteria and actinomycetes, barren soil had the fewest microbes and symbionts, while the greatest number of microbes was observed in forest soils. Three genera of VAM fungi were isolated from all soils except the barren soil, which had only one. Organic matter, moisture content, CO_2 evolution, and enzyme activities were positively correlated with the microbial population.

A study was conducted by CMCC (1996) to determine the effect of storage condition on infectivity level of VAM fungi. The activity level of inoculum multiplied on roots of pot grown sorghum plants was not significantly changed after storage at 4^0C for 1 yr.

Shailaja *et al.* (1998) studied the influence of phyllody disease on VAM colonization in two cultivars of sesame. The influence of sesame phyllody, an important phytoplasmal disease of sesame, on VAM colonization was studied in the cultivars Rajeshwari and Gowari, uprooted from the same field locations. Both healthy and infected plants of the 2 sesame varieties exhibited mycorrhizal colonization. The percentage VAM colonization was more in Rajeshwari than Gowari and in both cultivars the colonization was greater in infected plants than healthy plants. Vesicles and mycelium were associated with all the types of plants but arbuscules were present only in infected plants of both the cultivars. *Glomus fasciculatum* and *Sclerocystis rubiformis* were observed in all the 4 types of plants. *Glomus mosseae* was associated with healthy and diseased plants of Rajeshwari only and *Acaulospora morrowea* was recorded in the rhizosphere soil of healthy plants of the Gowari cultivar.

Pandey (1998) studied that plant parasitic nematodes reduce VAM infection in dry tropical grasses. Field experiments were conducted in grassland located in the Vindhyan Plateau, Madhya Pradesh to study the effects of plant parasitic nematodes on VAM infection of dry-tropical grasses. The grassland was dominatd by *Dichanthium annulatum, Bothriochloa pertusa* and *Chrysopogon fulvus*. Twenty 8x8 m plots were laid out in the grassland in a randomized block design. Five of the plots were treated with the nematicide DBCP (1,2 dibromo-3 chloropropane), five with the fungicide methyl-1 (butyl-carbamoyl)-2-benzimidazolecarbamate, five with the nematicide and the fungicide, and five plots were untreated. Two ectoparasitic nematodes, *Tylenchorhynchus nudus* and *Helicotylenchus dihystera*, and one endo-parasitic nematode *Meloidogyne* sp. were present in the experimental plots. Plant parasitic nematodes averaged 3000/m^2 in non-treated plots. The populations of nematodes (both parasitic and non-plant parasitic) were reduced by 13.3 and 12.5 %, respectively in the fungicide treated plots. Nematodes were not found in nematicide and fungicide + nematicide treated plants. On average, 10% greater ($P<0.05$) roots were infested by VAM in the nematicide treated plots. The mycorrhizal increase was expressed through significant increase in VAM hyphae, vesicles and arbuscules by 9.1, 10.8 and 10.0%, respectively. None of the roots were infested with VAM in fungicide treated and fungicide + nematicide treated plots. It is concluded that root herbivory by plant feeding nematodes reduces the VAM infection in grasses. It is suggested that the negative relationship between plant parasitic nematodes and VAM may be due to root herbivory limiting the infection sites to VAM.

Mishra *et al.* (2001) studied the status of mycorrhizal fungus in turmeric fields subjected to soil solarization. The mycorrhizal fungi status in solarized and non-solarized fields of turmeric crops was assessed in a study carried out in Gujrat. Soil solarization was carried out using clear low-density polyethylene plastic of 25-mm thickness for 15 days during summer. The control plot showed 62 spores/g of soil and 115-spores/cm root pieces, while solarized soil contained 21 spores/g of soil and 76-spores/ cm root piece. These indicate the detrimental effect of solarizatiion on the density of vesicular arbuscular mycorrhizal fungi spores in the turmeric rhizosphere, and reduction in root colonization, and consequently the yield of turmeric.

Reddy and Garampalli (2002) studied the effect of fly ash amendment at 3 concentrations (10, 20 and 30%) on the infectivity and efficacy of *G. aggregatum* by conducting a pot experiment with sterile low fertile soil using pigeon pea (*Cajanus cajan)* cv. Maruti and chickpea (*Cicer aritenum* [*Cicer arietinum*]) cv. Annigeri, the two major pulse crop cultivars of this region. It is evident from the present investigation that the percent VAM colonization in both the crops significantly decreased with the increase of fly ash content in the soil. The formation of VAM fungal structures (vesicles and arbuscules) inside the host root was also found completely suppressed at higher concentrations of fly ash. The effectiveness of *G. aggregatum* under the influence of fly ash was found significantly affected compared to control, when judged by the growth response of pigeon pea. However, in chickpea, VAM association could slightly increase the growth over its control. Fly ash amendment alone also has shown positive influence on the growth of both the crops over their controls (without VAM association). This influence of fly ash amendment together with the usefulness of VAM fungi, as bioremediation agents can be exploited suitably in the reclamation of wastelands and soils over burdened with fly ash.

Pandey and Trafdar (2002) studied the effect of phosphorus, salinity and moisture on VAM fungal association in neem (*Azadirachta indica* Linn.). Different stains of vesicular arbuscular mycorrhizal (VAM) fungi were screened for their tolerance to varying levels of P, salinity and soil moisture while infecting neem (*Azadirachta indica*). The maximum percent root infection (79%) occurred at 30 mg kg^{-1} P with *Glomus fasciculatum,* which also resulted in the highest biomass production. With increased salinity level up to 3 dS m^{-1}, there was a slight decreased in percent root infection by VAM fungi. However, when the salinity levels were further increased up to 6 dS M^{-1}, a drastic decrease (nearly one third) in percent root colonization occurred. *Glomus mossease* was found to be the most saline-resistant species as compared to *Glomus fasciculatum, Gigaspora maragarita* and mixed inocula. It was observed that VAM infection increased with increasing moisture stress levels. The maximum infection was observed when soil was maintained at 30-60% available water, irrespective of fungal species. *Glomus fasciculatum* sporulated the most and stimulated the highest biomass production, and appeared to be the most efficient VAM fungus for neem and therefore is recommended for propagation of neem seedlings in the nursery.

Kaushal (2002) conducted a study in Rajasthan, to determine the factors influencing the development of vesicular arbuscular mycorrhizas (VAM) and VAM sporulation in the rhizosphere of *Prosopis cineraria.* Periodical surveys were undertaken in ecologically different areas in Rajasthan (Pal, Chopasni, Mandore, University Campus, Mathania, and Chaukha) to collect rhizosphere soil and root samples of plants growing in irrigated, non-irrigated fields and wastelands. The rhizosphere soils were analyzed for pH, percentage moisture, organic carbon, nitrogen and total phosphorus using standardized techniques. The maximum number of spores (280 per 100 g of soil) and root colonization (78%) were observed where soil pH was maximum (9.1). Root colonization and VAM sporulation was much influenced by soil moisture. Excess soil moisture resulted in a decrease in spore population and root colonization. The rhizosphere soil with a low content of carbon (0.09%) had maximum spores, while the soil rich in organic carbon had poor sporulation as in the case of soil collected from Pal. The concentration of nitrogen in the soil ranged form 0.02 to 0.07% in the different rhizosphere soils. The content of nitrogen had little or no direct effect on VAM formation of sporulation.Spore population and percentage of root colonization decreased when the level of phosphorus in the soil increased. Periodical collection of rhizosphere soil and root samples revealed that spore population was minimum in hot summers (April, May and June) and gradually increased after the first rain (July).

Kaushik and Dabas (2003) studied the application of vesicular arbuscular mycorrhiza (VAM), P fertilizer and water stress conditions to record their effect on nodulation and NPK in 120 days old seedlings of *Acacia nilotica* and *Dalbergia sissoo* under greenhouse conditions. The treatments used in the experiment consisted of variations in mycorrhizal inoculation, application of phosphorus fertilizer and repeated water stress. Application of *Glomus mosseae,* P fertilizer and water significantly increased nodule formation in both the tree species. Water stress conditions significantly reduced nodule number and size in *A. nilotica* but did not have significant difference in *D. sissoo.* Interaction between phosphorus × water stress, mycorrhizas × phosphorus, and mycorrhizas × water stress was significant in *A. nilotica.* Average nodule size was comparatively larger in *D. sissoo* than in *A. nilotica.* Individual application either of VAM or P fertilizer statistically increased nodulation in both the tree species, whereas water stress conditions significantly reduced nodulation in *A. nilotica.* VAM inoculation significantly improved phosphorus concentration in shoots of both the tree species, whereas water stress conditions reduced it.

CONCLUDING REMARKS

Utility of VAM technology in improving the performance of a large number of crops cultivated in different parts of the country has been explored by Indian mycorrhizasts. But only few crops including wheat, pearlmillet, sorghum, greengram, pigeonpea, peanut, soyabean, sunflower, tomato and cotton have been studied with reference to the interaction between mycorrhizal association and plant protection chemicals such as antibiotics, fungicides, insecticides, biopesticides, fumigants etc. A perusal of published reports indicate that because of their toxicity the plant protection chemicals exert an unfavorable effect on the VAM fungi and their potentiality to benefit the host. However in many cases, the benefit from both mycorrhizal association as well as plant protection chemicals to the plant may be ensured by manipulating the recommended doses of plant protection chemicals.

The informations collected by mycorrhizasts of the country show that seasons and physico-chemical characters of the soil such as pH, nutrient status, moisture etc. exert a marked effect on the VAM fungi and the benefit, which they provide to the plants. Although the informations relate to only few crops and few soil factors, however it clearly indicates that any recommendation related to the application of VAM technology in crop based on pot trials has no authencity unless it is repeated in fields under diverse agroclimatic conditions and in soils with different physicochemical characteristics.

A number of agronomic practices such as crop rotation, fallowing, monocropping, intercropping, organic amendments etc. are employed for the better performance of the crops in the country. A clear understanding of their influence on the activity and potentiality of native and introduced VAM fungi is imperative for an integrated management of such practices and VAM technology. A number of crops have been investigated by the Indian mycorrhizasts with reference to this and the informations collected show that the agronomic practices affect the diversity and composition of VAM fungi in soil in a variety of ways.

Thus field performance of the VAM inoculum selected on the basis of pot trials should be tested along with recommended agronomic practices before deciding the protocol of the field application of VAM technology for a crop.

Only little work has been done by the mycorrhizasts of the country with reference to the effect of pollutants on the mycorrhizal association. On one hand, the pollutants affect the growth and productivity of the plant and on the other; they have an important bearing on the composition and diversity of VAM fungi in the soil. However, it is clear from the informations collected that mycorrhizal association partially compensates the unfavourable effect exerted by the pollutants on the host. However, more investigations with diverse crops and different pollutants are needed to confirm this assumption.

The VAM fungi do not have marked host specificity, however, the hosts do differ in their extent to which they may derive benefit from a particular symbiont. Same is expected to be true with different genotypes of the same host. Without a clear understanding of the effect of genotypes on the mycorrhizal association and its benefit, ability to manipulate mycorrhizal symbionts to the benefit of crops will be seriously hampered. Limited number of crops, *viz.* sorghum, greengram, wheat and cowpea has been investigated by Indian mycorrhizasts with reference to the influence of genotypes on the mycorrhizal association and its benefit to the host. As expected, these results clearly indicate that genotypes of a host differ in their response to mycorrhization with reference to their nutrient uptake, mycorrhizal colonization, growth as well as productivity.

7

VA MYCORRHIZA AS BIOCONTROL AGENT

Extensive studies have been carried out to explore the possibility of using mycorrhizal model system in inducing suppressions of potential soil-borne pathogens. Its relevance becomes all the more important since most of the soil-borne diseases, by and large, are not easily controlled by conventional methods.

Diseases due to Fungal Pathogens

Most of the studies conducted on the interaction of VAM fungi and fungal pathogens suggest that mycorrhizae decrease the severity of disease, while a few studies suggest that there is either no effect or there is an increase in the severity of disease. It has been pointed out that in view of the fact that VAM fungi deter the severity of disease they may be used as biocontrol agents. Various mechanisms by which VAM could suppress the fungal pathogens have been suggested on the basis of experimental evidences. Morphological, anatomical, biochemical or physiological changes in the host due to mycorrhization may suppress the pathogen and its activity and impart resistance against it. It has also been suggested that better nutrition of the host due to mycorrhization curtails the damage caused by the pathogen resulting in a curtailment of disease expression. Although generalizations regarding interactions between VAM fungi and fungal pathogens is difficult, however, the possibility of biologically controlling the pathogens through VAM fungi looks promising. In view of this many Indian mycorrhizasts have attempted to explore the possibility of and controlling the fungal pathogens through mycorrhization. The following account gives a description of their reports.

Jalali and Thareja (1981) observed suppression of chickpea (*Cicer arietinum*) wilt caused by *Fusarium oxysporum* f. sp. *ciceri* following soil inoculation with vesicular-arbuscular mycorrhizal fungus-*Glomus* sp. The endophyte colonized plant roots extensively which resulted in reduction in wilt incidence from 80% in non-mycorrhizal to 26.6% in mycorrhizal plants. Inoculation with *Glomus* sp. also increased nodule number, total dry matter production and 'P' content in those plants not infected by the pathogen. This suggested that mycorrhizal colonization may provide effective protection against infection with *Fusarium* and perhaps other root pathogens.

Krishna and Bagyaraj (1983) observed that the mycorrhizal fungus *Glomus fasciculatum* reduces the severity of peanut (*Arachis hypogaea*) disease and the number of sclerotia produced by *Sclerotium rolfsii*. Root and shoot dry weight of the host and their 'P' content were highest in plants inoculated with mycorrhiza only and lowest in plants inoculated with pathogen only. Simultaneous addition of mycorrhizal inoculation and the pathogen reduced disease severity.

Hegde and Rai (1984) studied the influence of *Glomus fasciculatum* on damping-off of tomato. In greenhouse pot tests, inoculation with the mycorrhizal fungus *G. fasciculatum* either simultaneously with or 2 weeks before inoculation with *Pythium aphanidermatum* reduced damping-off and increased plant height, and shoot and root weights compared with plants inoculated with the damping off pathogen alone or 2 weeks before mycorrhizal inoculation. Mycorrhizal inoculation alone increased root but not shoot weight over that of uninoculated controls.

Parvathi *et al.* (1985a) studied the influence of root-infecting fungi on the development of *Glomus mosseae* in groundnut. In pot, test plants infected with *G. mosseae* alone exhibited greater mycorrhiza formation than those inoculated with *G. mosseae* + *Rhizoctonia solani* or *G. mosseae* + *Fusarium solani*. However, when the other fungi were introduced separately into the soil after inoculation with *G. mosseae* mycorrhizal development was not affected.

Bisht *et al.* (1985) suggested that vesicular-arbuscular mycorrhiza (*Gigaspora calospora*) exerted an inhibitory effect on the development of pigeonpea (*Cajanus cajan*) blight caused by *Phytophthora drechsleri* f. sp. *cajani*. In studies at ICRISAT (1986), presence of vesicular-arbuscular mycorrhiza reduced *Phytophthora* blight of pigeonpea in glasshouse tests. In pot experiments promising results were obtained with the VAM fungi, *Glomus fasciculatum* and *G. monosporum*, in significantly increasing 'P' uptake.

Chakravarty and Mishra (1986) observed a reduction in *Cassia tora* wilt caused by *Fusarium oxysporum* when roots inoculated seedlings were transplanted to pot containing soil inoculated with chlamydospores of *Glomus fasciculatum* or *G. tenue*. The mycorrhizal fungi stimulated seedling growth and reduced the *F. oxysporum* population in the rhizosphere.

Jalali (1986) showed that mycorrhizal inoculation of mungbean resulted in a measurable reduction in infection by the root pathogen *Rhizoctonia solani*. This effect was more pronounced when seed pelleting with sporocarps of the endophyte was done. This suppressive effect was directly correlated with the ability of the mycorrhizal symbiont to develop in the absence of available soil phosphate. The studies indicated that soil conditions favourable to mycorrhizal colonization were in many respects not conducive to the growth and establishment of the test pathogen.

Chaturvedi *et al.* (1987) studied the occurrence of VA mycorrhiza in *Ascochyta* blight resistant and susceptible cultivars of chickpea. They examined the roots and soil from the root zone of 2 varieties (ICC 8160 and ICC 8161) with resistance to *Ascochyta rabiei* and 2 (ICC 2664 and Annegeri) with susceptibility for VA mycorrhiza at the time of maximum disease development. The soil from the root zone of the resistant cultivars harboured more VA mycorrhizal spores than did that of the susceptible cultivars. However, the difference in spore number between resistant ICC 8161 and susceptible ICC- 2664 was not significant. The roots of the resistant varieties always harboured more VA mycorrhizal spores than did those of the susceptible varieties.

Bali and Mukerji (1988a) investigated the effect of VAM fungi on *Fusarium* wilt of cotton and jute. The investigations were performed in 4 sets of experiments taking mycorrhizae *G. macrocarpum* and pathogens in different combinations. Mycorrhizal plants showed improved plant dry weight and per cent mycorrhizal infection than the non-mycorrhizal plants. Mycorrhizae were beneficial in reducing disease severity in host plant.

Ramraj *et al.* (1988) studied the biocontrol of *Macrophomina* root rot of cowpea and *Fusarium* wilt of tomato by using VAM fungi. Inoculation of *Glomus erunicatum*, besides increasing the biomass and yield of cowpea (*Vigna unguiculata*), induced tolerance to *Macrophomina* root rot infection. The disease incidence was reduced to 16% in the inoculated plants as against 33% in untreated control. In tomato, *G. mosseae* enhanced the growth and also reduced *Fusarium* wilt. The mycorrhizal plants recorded only 11% disease as against 46% in the non-mycorrhizal plants.

Reddy *et al.* (1988) studied the influence of mycorrhiza on chickpea *Fusarium* wilt. Simultaneous inoculation of *Cicer arietinum* with vesicular arbuscular mycorrhizal fungi (*Glomus mosseae, G. constrictum* and *G. monosporum*) and *F. oxysporum* f. sp. *ciceri* had no effect on wilt incidence in the susceptible JG 62 and resistant WR 315 genotypes. However, the wilt tolerant genotype, K 850 had 25% wilt in soil with *Fusarium* + VAM, less than with *Fusarium* alone (35%).

Kehri and Chandra (1989a) studied the interaction between *Glomus fasciculatum* (Thaxter sensu. Gerd.) and Trappe and *Macrophomina phaseolina* causing dry root-rot of chickpea cv. T-3, under normal as well as strained water schedule. Addition of *G. fasciculatum* in the soil caused an increase in mycorrhizal colonization of the roots under both the water schedules. The increase in mycorrhizal colonization was always associated with a decrease in the population of pathogen in the roots and an improvement in the growth and yield in chickpea.

Jalali *et al.* (1990) studied the interaction between vesicular-arbuscular mycorrhizal endophyte and *Macrophomina phaseolina* in mungbean. It was reported that mycorrhizal inoculation significantly restricted the spread of the pathogen in host root tissues. Disease incidence was reduced from 77.9 per cent in pathogen inoculated to 13.3 per cent in mycorrhiza + pathogen inoculated plants. Dual inoculation with mycorrhizal endophyte and the pathogen also resulted in increased total dry matter production, nitrogen, phosphorus and potassium contents of such plants as compared to pathogen-inoculated plants.

Sundaresan *et al.* (1993b) reported that pre-establishment of the vesicular-arbuscular mycorrhizal fungus in cowpea (*Vigna unguiculata*) roots reduced colonization by *Fusarium oxysporum* and severity of the disease as measured by a reduction in vascular discolouration. They observed induction and accumulation of phytoalexins in cowpea roots infected with mycorrhizal fungus. Production of phytoalexins was always higher in mycorrhizal plants than in non-mycorrhizal ones and there appeared to be direct correlation between phytoalexin concentration and the degree of mycorrhizal association. They suggested that the production of phytoalexin compounds in mycorrhizal plants could be one of the mechanisms imparting tolerance to disease.

Jayaraman and Kumar (1995) on the other hand, observed VAM fungi-pathogen interaction in two cultivars of chickpea. In cv. BG 256, VAM fungus *Glomus fasciculatum* enhanced the activity of pathogenic fungus *Sclerotium rolfsii* that causes collar rot disease.

However, in cv. JG 62 the VAM fungus not only nullified adverse effects of the pathogen but also improved plant growth compared with control.

Chandra *et al.* (1995a) evaluated the efficacy of a number of VAM fungi *viz., Glomus fasciculatum, G. mosseae, G. aggregatum, G. claroideum, G. macrocarpum, G. mosseae, G. aggregatum, G. claroideum, G. macrocarpum* and *G. multicaule* recovered from agricultural fields raising leguminous crops in checking the dry root rot of greengram, blackgram and cowpea due to *Macrophomina phaseolina*. The interaction between VAM fungi and pathogen was examined by inoculating the crops with selected VAM fungi alone as well as in combination with selected isolate of pathogen in sterilized as well as unsterilized soil. The rate of mortality, mycorrhizal status, changes in the population of pathogen and yield were recorded. When the inoculum of VAM fungi was added alongwith that of pathogen, a reduction in the rate of mortality and population of pathogen was recorded in all the three crops. This change was coupled with an increase in the mycorrhization in the roots as well as yield. This finding clearly indicated that VAM fungi were instrumental in reducing the damage caused to the crops due to the pathogen. The study under pot conditions was repeated under microplot conditions and more or less similar findings were recorded.

Singh and Singh (1995) studied the effect of *Glomus aggregatum* inoculation and P-amendments on *Fusarium oxysporum* f.sp. *ciceri* causing chickpea wilt. 80 and 20% plants were found healthy and showing yellowing respectively at both 0 and 40 mg P/kg soil after 45 days of sowing in the presence of mycorrhizae and wilt pathogen. All plants, inoculated with *F. oxysporum* f. sp. *ciceri* alone, were found to have wilted after 45 days of sowing at all three levels of P. On the other hand, none of the plants showed wilting symptom in the soil inoculated with VAM fungus. The chickpea plants showed significant resistance against the wilt pathogen in presence of *G. aggregatum*. The 20, 20 and 60% plants showed yellowing symptom (partially infected) at 0, 40, 80 mg P/kg soil after 45 days sowing. The observations revealed that the presence of mycorrhizal fungi significantly delayed infection, as well as reduced its percentage. The root colonization with *G. aggregatum* protected chickpea plants up to 80%, against 100% wilting due to *F. oxysporum* f. sp. *ciceri* in absence of mycorrhizal fungus at variable levels of P. The wilting process in mycorrhizal chickpea plant was found to be significantly slow as compared to non-mycorrhizal plants.

Sivaprasad *et al.* (1995b) studied the vesicular-arbuscular mycorrhizal colonization relation to foot-rot disease intensity in black pepper. A study was undertaken to analyze the relationship, if any, between VAM colonization and foot-rot disease intensity in black pepper. Three major black pepper-growing tracts of Kerala, namely, Idukki, Wynad, and Cannanore districts, were selected for the study. Pepper plants from existing plantations with varying foot-rot disease intensities were selected randomly from different locations. The percentage of vines dead per standard was considered as the index of disease intensity. The observations made during the pre-mansoon period gave a definite negative relationship between disease intensity and VAM colonization percentage and spore count in the rhizosphere soil. The roots were discoloured and the percentage of VAM colonization decreased with increase in disease intensity. For example, sample 14 with 90% dead vines recorded five per cent VAM colonization in the root and a spore count of 15 in rhizosphere soil, as against 51%

colonization and 33 spores recorded in a healthy plant (sample 16). The healthy plants, irrespective of locations, consistently showed higher VAM colonization and spore count compared to the diseased plants. Analysis of the 38 pepper root and rhizosphere samples collected during the period before and after the monsoon indicated a general trend of decrease in foot-rot disease intensity of black pepper with increase in VA mycorrhizal colonization .

Sivaprasad *et al.* (1999) studied the foot rot of black pepper and rhizome rot of ginger as influenced by vesicular arbuscular mycorrhizal fungi (VAMF). Effect of inoculation with different VAMF on foot rot disease of black pepper and rhizome rot of ginger was tested under field conditions. Black pepper cuttings inoculated with *Glomus monosporum* at nursery level and transplanted to the field significantly reduced the foot rot incidence from 100 per cent infection and 83 per cent mortality observed in control to 33 per cent. Similarly, in ginger the rhizome rot infection of 78 per cent and the mortality of 47 per cent was brought down to 32 and 17 per cent by inoculation with *G. mosseae*. Ability of VAMF to stimulate plant growth and impart resistance against pathogen was found as independent traits of VAMF.

Reddy and Hindumathi (1999) conducted experiments in irrigated and non-irrigated soils during kharif and rabi with different varieties of sorghum by using mycorrhizal fungi since sorghum roots appear to be good hosts of mycorrhizae. In all cases charcoal rot was declined to a great extent in mycorrhizal inoculated plots. No disease was observed in the varieties Rs-29 and E-36-1 inoculated with mycorrhizae and pathogen. The disease incidence was reduced upto 80% even in the variety, CSV-8R which is considered as highly susceptible to charcoal rot. The reductioin in disease severity by competition of sites, though the main reason is by strengthening of the host plant so that the damage caused by the pathogen was offset by improved plant growth.

Chandra *et al.* (2000c) tried to manage the soil-borne diseases of two important horticultural crops *viz.*, guava and papaya. An experiment was designed to determine the efficiency of the VAM isolates recovered from various orchards of Allahabad *viz.*, *Glomus aggregatum, G. claroideum, G. fasciculatum, G. mosseae* in improving the growth and biomass of both the crops. VAM inoculants were provided to the crops under pot conditions in sterilized/unsterilized soil. Out of five VAM inoculants, *G. aggregatum* was found to show highest potential in colonizing the roots of guava and papaya and improve their performance. It was also shown to suppress the indigenous population of pathogens in the roots. On this basis, it was selected and its potential to act as biopesticide in controlling the diseases of guava and papaya was evaluated. Dual inoculation experiment was designed to examine the effect of selected VAM and pathogen inoculants on the performance of guava and papaya under pot conditions in sterilized and unsterilized soil. The crops were provided with combination of VAM (*G. aggregatum*) and pathogen {*F. oxysporum* (SA) for guava, *F. semitectum* (NG) for papaya} inoculants. The VAM inoculants *G. aggregatum* was found to be effective in suppressing the root infection in both the crops not only in sterilized but also in unsterilized soil. It was found to be equally effective in counteracting the negative effects of pathogen on the performance of both guava and papaya in terms of growth of roots and shoots.

CONCLUDING REMARKS

Although a number of Indian crops suffer from severe fungal diseases, only few have been investigated with reference to their response to mycorrhizal fungi as biocontrol agents. A perusal of published reports shows that majority of mycorrhizasts have concentrated their attention to some of the important leguminous crops including chickpea, mungbean, greengram, blackgram, pigeonpea, groundnut and cowpea. Wilts and root-rots have been reported to severly affect the productivity in these crops. It is heartening to note that they have been successfully controlled through the use of mycorrhizal fungi. Few other crops e.g. sorghum, guava, papaya, ginger and black pepper have also been investigated and have been shown to give a positive response to mycorrhizal fungi as biocontrol agent for some of the diseases caused to them. It may be mentioned that the published data related to potentiality of mycorrhizal fungi as biocontrol agent is of limited value because the observations are based in nearly all the cases on pot experiments.

DISEASE DUE TO NEMATODES

On the basis of manner in which plant parasitic nematodes feed on roots, they are grouped under three categories: sedentary endoparasites (*Meloidogyne, Heterodera, Tylenchus, Rotylenchus*) migratory endoparasites (*Pratylenchus, Radopholus*) and ectoparasites (*Belonolaimus, Macroposthonia, Paratrichordus, Xiphinema* etc.). Plant parasitic nematodes and mycorrhizal fungi are commonly found inhabiting the same soil and colonizing roots of their host plants. These two groups of microorganisms exert a characteristic but opposite effect on plant growth/ health. The potential role of mycorrhizal fungi as biological agent for the control of nematode diseases has recently received considerable attention. The damage due to the nematode diseases is generally reduced in mycorrhizal plants. In several studies mycorrhizal fungi have shown an antagonistic influence on the population of plant parasitic nematodes. Establishment of VAM in plants generally confers resistance to nematode parasitism or adversely affects nematode reproduction. However, sometimes root colonization by VAM fungi may have no effect on nematode reproduction or may promote it.

The potentiality of VAM fungi to suppress the severity of nematode diseases has been explained on the basis of morphological, anatomical, biochemical and physiological changes in the hosts due to mycorrhization.Better nutrition of the mycorrhizal plants has also been suggested to be the cause of disease suppression. In view of the potentiality of the mycorrhizae in suppressing the disease severity in plants due to nematodes, many Indian workers have tried to explore the possibility of using VAM fungi as biocontrol agents of nematode diseases of the crops cultivated in the country. Following is an account of their published reports related to this.

Suresh and Bagyaraj (1984) reported on the basis of their studies conducted at the University of Agricultural Sciences, Bangalore, that application of *G. fasciculatum* first, followed by *Meloidogyne incognita* reduced nematode infection to tomato better than simultaneous application of VAM nematode or nematode first, followed by VAM.

Suresh *et al.* (1985) reported that root extracts from mycorrhizal plants colonized by *G. fasciculatum* brought about 50% mortality of the larvae of the nematode, *Meloidogyne incognita,* in tomato roots in four days time though mycorrhizal roots did not prevent the penetration by nematode larvae. The number of giant cells formed in mycorrhizal plants was significantly low.

Jain and Hasan (1986) studied the association of vesicular-arbuscular mycorrhizal fungi and plant parasitic nematodes with forage Sorghum (*Sorghum bicolor* L.). Soil/root samples collected from the rhizosphere of forage sorghum in Jhansi, revealed 2 VAM fungi *Glomus fasciculatum* and *G. mosseae* and 3 predominant nematodes. More than 50% VAM infection lowered the number of nematodes, however, the presence of nematodes did not adversely affect VAM sporulation. It was concluded that VAM fungi could be used to manage plant parasitic nematode populations.

Jain and Sethi (1987) showed that the occurrence of *Heterodera cajani*, a nematode and the VAM fungi *Glomus fasciculatum* and *G. epigaeum* in *Vigna unguiculata* were largely independent of each other and the organisms did not modify the effect of each other to any considerable extent. However, an increase in nematode inoculum invariably resulted in reduced root infection and spore production by mycorrhizal fungi. The presence of *G. fasciculatum* showed a profound adverse effect on cyst production and multiplication of nematode while *G. epigaeum* tended to exhibit a reverse trend.

Hasan and Jain (1987) conducted a survey for parasitic nematodes and VAM fungi associated with berseem *(Trifolium alexandrium)* in Bundelkhand region. They collected the samples of berseem roots and rhizosphere soil from 25 locations in 5 districts of the region. Out of 7 genera of plant parasitic nematodes identified, both stunt (*Tylenchorhynchus vulgaris*) and spiral nematode (*Helicotylenchus dihystera*) were the most common and occurred in every sample. However, the population of the former species was observed to be generally high and was also frequently associated with stunted or abnormal growth of the plants. Spores of endomycorrhizal fungi occurred in every sample and out of 3 species identified, *G. mosseae* was the most prevalent. High populations of stunt nematodes were found to be associated with abundant mycorrhizal counts which indicated that both the organisms did not affect the development of each other. A pathogenicity test revealed that 1000 nematodes of *T. vulgaris* caused significant reduction in plant growth of berseem. A positive interaction between VAM and stunt nematode was observed in which the former offset the nematode damage and increased the plant growth and phosphorus content of berseem in greenhouse experiment.

Jain and Sethi (1988) noted the influence of time of inoculation on the interaction between VAM, *Meloidogyne incognita* and *Heterodera cajani* on cowpea. Inoculation of cowpea seedlings with *Glomus fasciculatum* or *G. epigaeum,* one or two weeks prior to inoculation with *M. incognita* or *H. cajani,* alleviated the deleterious effect of both nematodes to a considerable extent. Early establishment of *G. fasciculatum* reduced gall formation by *M. incognita* and cyst formation of *H. cajani*. Prior establishment of *G. epigaeum* resulted in increased number of cyst of *H. cajani*. The presence of nematodes reduced mycorrhizal infection and spore production. Over 60% colonization of the root system by VAM fungi considerably hampered root invasion by *H. cajani*. However, the endophyte did not further influence the development of penetrated jubeniles or the fecundity of the nematodes.

Jain and Sethi (1989) studied the effect of early establishment of *G. fasciculatum* or *G. epigaeus* on the penetration and development of *H. cajani* in cowpea (*Vigna unguiculata*) under pot conditions. Introduction of mycorrhizal fungi 15 days before the nematode, adversely affected root penetration to greater extent than simultaneous inoculations. Over 60% colonization of the root system by VAM fungi considerably hampered root invasion. However, the endophytes did not further influence the development of penetrated jubeniles or the fecundity of the nematode.

Umesh *et al.* (1989) worked on the interaction of burrowing nematode, *Radopholus similis* with *Glomus fasciculatum* in banana in pot experiments. The length and fresh and dry weights of the roots were increased by mycorrhizal inoculation, while nematode inoculation retarded root growth. The plants which received mycorrhiza 7 days prior to nematodes were able to offset the effect caused by the latter and had more vigorous roots even compared with uninoculated plants. Although root colonization by mycorrhiza was reduced in plants inoculated with nematodes, spore production was not adversely affected. Nematode numbers both in roots and soil were significantly lower when the mycorrhiza was applied simultaneously or 7 days prior to nematode inoculation. Plants inoculated with mycorrhiza alone or with the nematode had higher contents of N,P,K, Ca and Mg, reducing and total sugars, phenols and total amino acids in their roots.

Thomas *et al.* (1989) investigated the individual and interactive effects of VA mycorrhizal fungi and root-knot nematode *Meloidogyne incognita* on cardamom. They inoculated *Elettaria cardamomum* with 6 species of VAM fungi individually and in different combinations with *M. incognita*, simultaneously and subsequently and recorded its growth 6 and 12 months after inoculation. Compared with the control, the growth of *E. cardamomum* plants was improved by inoculation with VAM fungi, whereas nematode inoculation reduced it. The growth response induced by VAM fungi was similar when inoculation was made simultaneously and after nematode inoculation. VAM fungi significantly reduced the nematode population in the roots, when nematodes were inoculated simultaneously and prior to inoculation by VAM fungi. *Gigaspora margarita* and *Glomus fasciculatum* were the most effective VAM fungi, able to promote maximum growth response in the absence as well as in the presence of root-knot nematodes.

Subhashini (1990) studied the role of VA mycorrhiza in controlling certain root diseases of tobacco. It was observed inoculation with *Glomus fasciculatum* reduced the intensity of infection by root knot nematode but had no effect on *Orobanche* infestation. Attacks of *Pythium* in tobacco nurseries were also reduced by *G. fasciculatum* which was most promising of 4 vesicular-arbuscular mycorrhizal fungi tested.

Baghel *et al.* (1990) worked on the citrus nematode *Tylenchus semipenetrans* and its interaction with VA mycorrhizal fungus *Glomus mosseae*. VAM inoculation stimulated growth of *Citrus jambhiri* seedlings, while the nematode decreased the growth. When the 2 organisms were inoculated simultaneously the adverse effect of the nematode was partly nullified by the fungus and limited the development of the nematode.

Sivaprasad *et al.* (1990) working on root knot nematode - *Meloidogyne incognita* of *Piper nigrum* recorded reduced root-knot index by 32.4 and 36% when pre-inoculated cuttings of the cv. Panniyur - 1 were grown with *Glomus fasciculatum* or *G. etunicatum* respectively. The presence of mycorrhizal plants also reduced populations of the nematode in both roots and surrounding soil. Mycorrhizal plants showed significantly improved growth even in the presence of the nematode.

Singh *et al.* (1990) reported that pre-inoculation of the roots with mycorrhizal inoculum coupled with biochemical changes such as increase in lignins and phenols, make Pusa Ruby cv. of tomato, resistant to root-knot nematode.

Krishna Prasad (1991) reported that the germination and stand of the flue cured tobacco (*Nicotiana tobacum*) was not affected by the incorporation of endomycorrhizal fungus, *Glomus fasciculatum* in root-knot (*Meloidogyne incognita*) infested sandy soils. In nonmycorrhizal nursery plots, the percentage root-knot infestation in seedlings was 67.5% at 50 days after sowing which increased to 95% at 75 days. In mycorrhizal inoculated plots, the percentage of infestation of seedlings ranged from 48-52% at 50 days and 73-75 at 75 days after sowing. The number of galls, endoparasites and egg masses per infested seedling was reduced by 61 to 89 percent as a result of mycorrhizal inoculation, even with an additional nematode inoculation to nursery soil. Although the fecundity of the root-knot females was not affected by the fungus, yet the reproductive potential of the nematode was very much reduced due to fewer egg masses produced per infested seedling. Transplanting of mycorrhizal tobacco seedlings into root-knot nematode infested soil indicated that the growth and yield of tobacco was better than non-mycorrhizal seedlings both quantitatively and qualitatively. In addition, the plants had lower root-knot indices and required half the dose of phosphorus fertilization compared to non-mycorrhizal seedlings. These studies suggested a possible use of mycorrhizal fungus for the biocontrol of root-knot nematode under field conditions in transplanted crops such as tobacco, tomato, brinjal etc.

Mittal *et al.* (1991) showed that tomato root tissues forming galls following inoculation with *Meloidogyne incognita* had no VAM fungi while roots lacking nematode galls had vesicles and arbuscules of *Glomus fasciculatum*, presence of which probably inhibits the formation of nematode galls.

Sundarababu *et al.* (1993) also advocated the same. According to them the prior establishment of mycorrhiza nullified the damaging effects of the nematode and improved the plant growth.

Sharma *et al.* (1994) studied the interaction between *Glomus fasciculatum* and *Meloidogyne incognita* in tomato in two phases under greenhouse conditions. In the first phase, two locally adopted tomato varieties Pusa Ruby and Pusa Early Dwarf were evaluated for VAM susceptibility. Pusa Ruby variety was found to be more susceptible to *Glomus fasciculatum*. VAM colonization was found to reduce the root-knot infestation. Application of VA-mycorrhiza first followed by nematodes resulted greater reduction in nematode infestation than the simultaneous inoculation or nematode first followed by mycorrhiza.

Sankaranarayanan and Sundarababu (1994) reported that the interaction between *Glomus fasciculatum* and *Meloidogyne incognita* inoculated at different timings on blackgram controlled the nematode population and also increased biomass production. But when *G. fasciculatum* was applied 5 days earlier than the nematode, VA mycorrhizal colonization and spore production were reduced.

Santhi and Sundarababu (1995b) studied the effect of three species of VAM *viz*., *Glomus fasciculatum, G. versiforme, G. etunicatum* and root knot nematode *Meloidogyne incognita* on cowpea growing in different types of soil. The interaction of *G. fasciculatum, G. versiforme* and *G. etunicatum* with *Meloidogyne incognita* on cowpeas was tested in different types of soil (clay soil, black loam, red loam and sandy soil) and pot soil mixtures (red soil/sand/ FYM in a ratio of 2:1:1). The effect of different soil types on *G. fasciculatum* and *G. etunicatum* was the same and both species preferred the pot soil mixture followed by sandy soil, clay loam, black loam and red loam, whereas *G. versiforme* preferred the pot soil mixture followed by red loam, clay soil, sandy soil and black loam. *G. fasciculatum* was superior at reducing the nematode population in soil. The pot soil mixture increased growth as well as spore production and colonization, reducing the nematode population.

Sundaram and Arangarasan (1995b) made a study to identify suitable VAM fungus for controlling nematode infection in the tomato growing soil of South Arcot, and exploit the same in practice. To obtain mycorrhizal plants of tomato (*Lycopersicon esculentum* M.) cv. Co 3, a thin layer of mycorrhizal inoculum (at two levels, 5.0 g and 10.0 g) was placed 2 cm below the surface of soil before sowing seeds. The inoculum consisted of chlamydospores (approximately 15 spores/g), roots, and soil from a pot culture of onion (*Allium cepa*) which was colonized individually by *Glomus fasciculatum, G. mosseae, Acaulospora laevis* and *Gigaspora margarita*. Non-mycorrhizal seedlings were transplanted to sterilized soil in 15 cm diameter (8kg) pots-one seedlings per pot. *Meloidogyne incognita* was obtained from stock cultures maintained in tomato. This was inoculated into the rhizosphere of tomato plants at the rate of 5 ml (approximately 8000 larvae) per pot. Pots were maintained in a greenhouse. The interactive effects of VAM fungi and root knot nematode (*Meloidogyne incognita*) were found to be variable. At 5.0 g inoculum level, all four VAM fungal species failed to perform significantly in improving the growth characteristics and the reduction in the number of galls formed. Interestingly, mycorrhizal plants which received 10.0 g inoculum levels were more resistant to root knot nematode infection. The growth and yield benefits were also generally greater than the control. The number of galls were also significantly reduced in mycorrhizal treatments at 10.0g inoculum level.

Sharma *et al.* (1995c) studied the effect of VAM fungus (*Glomus fasciculatum*) on survival and development of root-knot nematode *Meloidogyne incognita* in tomato cv. Pusa Ruby. On the basis of their observations they concluded that VAM fungi reduced the reproduction and development of *M. incognita*. The reduction of galling indices in the biology and development of *M. incognita* suggested the use of VAM as a bio-control agent against the root-knot disease in tomato.

Rajeshwari and Sankaranarayanan (1995) incorporated *Glomus fasciculatum* into nursery beds to help the fungus to colonize tomato roots before transplantation to the main field, thereby preventing the penetration and development of nematodes in colonized plants. Thus, the VAM was able to offset the adverse effects of nematodes and increased the tomato yield by 91.3% compared with the untreated control. Carbofuran 3 G and Phorate 10 G were equally effective in decreasing nematode numbers and increased yield, followed by the mycorrhizal plants which were equivalent to Phorate 10 G.

Santhi and Sundarababu (1995a) investigated the effect of phosphorus on the interaction of vesicular-arbuscular mycorrhizal fungi with *Meloidogyne incognita* on cowpea. A pot experiment was conducted to study the effect of different levels of phosphorus (0, 50 and 100 µg/g soil) against *Meloidogyne incognita*, VAM fungi and the nematode and fungal interaction. They noted all the differences among the growth parameters of cowpea at different phosphorus levels which were significant. Plants with VAM were more resistant to *M. incognita* than those without. Plants treated with phosphorus resulted in increased total phosphorus content. Positive correlation was observed between the phosphorus levels and the nematode population, whereas negative correlation occurred between phosphorus levels and the VAM spore population and colonization. They also studied (1995b) the effect of three species of VAM *viz*., *Glomus fasciculatum, G. versiforme* and *G. etunicatum* and root knot nematode *Meloidogyne incognita* on cowpea growing in different types of soils (clay soil, black loam, red loam and sandy soil) and pot mixtures soil (red soil/ sand/ FYM in ratio of 2:1:1). The effect of different soil types on *G. fasciculatum* and *G. etunicatum* was the same and both species preferred the pot soil mixture followed by sandy soil, clay loam, black loam and red loam, whereas *G. versiforme* preferred the pot soil mixture followed by red loam, clay soil, sandy soil and black loam. *G. fasciculatum* was superior at reducing the nematode population in soil. The pot soil mixture increased growth as well as spore production and colonization, reducing the nematode population.

Butool and Haseeb (1996) studied the effects of VAM fungus *Glomus aggregatum* on the root-knot development and multiplication of *Meloidogyne incognita* in Egyption henbane *Hyoscyamus muticus* in glasshouse conditions. Plant growth was significantly reduced in plants inoculated with nematode alone, but when simultaneously inoculated with *G. aggregatum*, plant growth improved although not as much as when inoculated with the fungus alone, and the root-knot development and nematode multiplication were reduced. However, the beneficial role of VAM fungus was not found to be very effective in suppressing the nematode population and in improving growth.

Reddy *et al.* (1995) evaluated an integrated management method for the control of citrus nematode - *Tylenchus semipenetrans* using VA mycorrhiza and oil cakes. Castor (*Ricinus communis*), karanj (*Pongamia pinnata*) and neem (*Azadirachta indica*) oil cakes at 20 g per plant and vesicular-arbuscular mycorrhiza, *Glomus fasciculatum* at 4000 spores per plant were evaluated either individually or in combination (at half the dosages) for the management of *Tylenchus semipenetrans*, infecting acid lime under greenhouse conditions. The neem cake in combination with VA mycorrhiza gave maximum shoot length, shoot weight, root length and root weight. The highest reduction in nematode population in both roots and soil was obtained with neem cake + *G. fasciculatum* which was at par with karanj cake + *G. fasciculatum*, neem cake and karanj cake. Neem cake gave maximum colonization of *G. fasciculatum* in roots which was at par with karanj cake.

Rao *et al.* (1995a) conducted experiments to determine the effect of integration of neem (*Azadirachta indica*) cake and vesicular-arbuscular mycorrhiza (*Glomus mosseae*) on the control of *Meloidogyne incognita* on tomato. Results indicated that mycorrhizal seedlings of tomato, when transplanted in neem-cake-amended soil, were the least infected by *M. incognita*. An increase in plant growth parameters and a decrease in root-knot index and final population of *M. incognita* were observed in the treatments where mycorrhizal seedlings were transplanted in neem-cake amended soil. In addition, increase in colonization of *G. mosseae* on roots of tomato and in chlamydospore densities of *G. mosseae* in these treatments indicated favourable effects of neem-cake amendment on the growth of *G. mosseae*. It was concluded that the desirably rhizosphere changes caused by addition of neem-cake to the soil could be utilized in the effective utilization of *G. mosseae* for the control of *M. incognita* on tomato.

Rao *et al.* (1995b) investigated the impact of VA mycorrhizal colonization on growth of chickpea infected with *Meloidogyne incognita* and *Fusarium oxysporum* f sp. *ciceri* disease complex. In a field experiment, seeds of the susceptible cv. *annigeri* were sown in soil infested with both the pathogens. The soil was solarized for 6 weeks before sowing the seeds. The seeds were treated with carbendazim and / or carbosulfan and inoculated with *Glomus fasciculatum*. The percentage colonization with indigenous or introduced mycorrhizas was reduced by seed treatment with carbendazim but was higher with carbosulfan alone or in combination with solarization. Solarization followed by carbosulfal treatment and mycorrhizal application resulted in the highest colonization (60.6%). Mycorrhizal inoculation increased shoot length by 16.7%, shoot dry weight by 31.8% and root dry weight by 56.8% over the control. Wilt incidence (*F. oxysporum* f. sp. *ciceri*) was reduced by mycorrhizal inoculation, alone or combined with other treatments.

Siddiqui and Mahmood (1995) used *Bacillus subtilis, Bradyrhizobium japonicum* and *Glomus fasciculatum* alone and in combination for the management of a wilt disease complex of pigeonpea caused by the nematode *Heterodera cajani* and the fungus *Fusarium udum*. Application of all the three management agents alone or in combination to plants inoculated with the pathogens increased shoot dry weight, number of nodules, phosphorus content, and reduced nematode multiplication and wilting index. Application of *B. subtilis* alone to plants inoculated with either of the pathogens caused an increase in shoot dry weight equal to that caused by *G. fasciculatum*. However, use of *B. subtilis* on plants inoculated with both the pathogens resulted in greater shoot dry weight than caused by *G. fasciculatum* or *B. japonicum*. Increase in shoot dry weight was greater when plants inoculated with pathogens were treated with *G. fasciculatum* plus *B. subtilis* or these two combined with *B. japonicum*. Application of all the three management agents against pathogens resulted in the greatest nodulation and the greatest reduction in nematode multiplication. Combined application of *G. fasciculatum* and *B. japonicum* increased root infection by *G. fasciculatum* while combined use with *B. subtilis* reduced mycorrhizal colonisation.

Siddiqui and Mahmood (1996) studied the biological control of *Heterodera cajani* and *Fusarium udum* in pigeonpea by *Glomus mosseae, Trichoderma harzianum* and *Verticillium chlamydosporium*. They were used alone and in combination for management of wilt disease complex of pigeonpea caused by the nematode *Heterodera cajani* and the fungus *Fusarium udum*. Treatment of plants inoculated with pathogens increased plant length, shoot dry weight, number of nodules, and reduced nematode multiplication and wilting index. Simultaneous use of biocontrol agents against pathogens gave better control than their individual application. *T. harzianum* had an adverse effect on root colonization by *G. mosseae*. Parasitism of nematodes by *V. chlamydosporium* was also reduced in the presence of *T. harzianum*. The highest reduction in nematode multiplication was observed when all the three biocontrol agents were used together.

Mishra and Shukla (1996) studied the interrelationship of *Meloidogyne incognita* and *G. fasciculatum* with 3 commonly used fungicides (carbendazim, aluminium-tris and chlorothalonil) and suggested that VAM inoculation improved growth of tomato cv. Pusa Ruby. Pre-establishment of *G. fasciculatum*, increased plant growth, decreased the size and number of galls and improved NPK uptake compared with plants inoculated with nematode alone or pre-inoculated with nematode and VAM. Fungicidal combinations with VAM and nematode produced an overall decrease in plant growth. However, simultaneous applications of carbandazim + *G. fasciculatum* improved plant growth. Reduced uptake of NPK was recorded in all the treatment combinations of VAM and nematode when combined with fungicides.

Rao *et al.* (1997) conducted experiment to study the effect of integration of *Glomus fasciculatum* and castor *Ricinus communis* for the control of *Meloidogyne incognita* on tomato (*Lycopersicon esculentum*). Mycorrhizal seedlings (nursery seedlings colonized with *G. fasciculatum*) were transplanted in castor-cake-amended soil (sterilized or unsterilized). Significant reduction in root-knot index and final population of nematodes, and a significant increase in tomato plant growth were observed in mycorrhizal seedlings transplanted in castor-cake-amended soil. Significant increase in the colonization of *G. fasciculatum* in tomato roots and chlamydospore density (in soil) in the treatments where mycorrhizal seedlings were transplanted in castor-cake-amended soil, indicated that castor cake amendment in the soil affected favourably the growth of *G. fasciculatum* which in turn adversely affected the development of the nematodes. Statistically significant differences were not observed in effects of integration of *G. fasciculatum* and castor cake on tomato growth and control of nematode in sterilized and unsterilized soils.

Nagesh and Reddy (1997) used the oil cakes (at 15g/plant) with *G. mosseae* (1000 chlamydospores/plant) in pot experiments with *Crossandra undulaefolia* to control the *Meloidogyne incognita*. The combination promoted better plant growth, reduced nematode multiplication rate and increased root colonization by the mycorrhiza and the number of mycorrhizal propagules. The treatments were effective in the order: *G. mosseae* + *Azadirachta indica* > + *Pongamia pinnata* > + *Ricinus communis* > *G. mosseae* alone.

Sankaranarayanan and Sundarababu (1997b) studied the effect of oil cakes and nematicides on the growth of blackgram (*Vigna mungo*) inoculated with VAM fungus (*Glomus fasciculatum*) and root-knot nematode (*Meloidogyne incognita*). Pot soil was either treated with *Azadirachta indica, Madhuca longifolia, Ricinus communis* or groundnut oilcakes, 15 days prior to sowing of *V. mungo* cv. Co5, or inoculated with mycorrhizal fungi at 10 g inoculum/plant, 22 days prior to sowing. 37 days later, nematodes were inoculated at one J_2/cc soil. Seven days, following nematode application phorate 10 G was applied. All the treatments were effective but the greatest reduction in nematode populations was found in the oilcake treatments. P content of plants was higher in mycorrhizal fungi treatments.

Sankaranarayanan and Sundarababu (1997a) studied the effect of leaf extracts on growth of blackgram cv. Co_5, inoculated with *Glomus fasciculatum* and *Meloidogyne incognita. Calotropis procera, Tagetus erecta, Catharanthus rosea* and *Bougainvillea spectabilis* were used for the leaf extracts alongwith carbofuron 3G and Phorate 10G. They recorded a significant increase in all the growth parameters due to addition of leaf extracts. Leaf extracts recorded higher spore population and mycorrhizal colonization than the nematicides but the latter recorded less gall-index and nematode population.

Mishra and Shukla (1997) studied the interactions between *Glomus fasciculatum* and *Meloidogyne incognita* on tomato. Interactions between *G. fasciculatum* (VAM) and *M. incognita* were studied on tomato cv. Pusa Ruby. Simultaneous inoculations of *G. fasciculatum* with *M. incognita* caused the greatest reduction in number and size of the nematode - induced root galls. Application of *G. fasciculatum*, 15 days prior to the nematode significantly increased plant growth compared with the nematode alone and inoculation with the nematode 15 days prior to the VAM. Prior application of VAM also exhibited significant decrease in number, size of the galls as compared with the nematode alone or nematode application prior to VAM. NPK content of tomato plants was significantly higher with prior or simultaneous application.

Jothi and Sundarbabu (1997b) studied the management of root lesion nematode, *Pratylenchus zeae* with the endomycorrhizal fungus, *Glomus fasciculatum* on maize. Plant growth was maximum in pots inoculated with vesicular arbuscular mycorrhizal fungi, and least in nematode-alone pots.

Sharma and Trivedi (1997) studied the concomitant effect of *Paecilomyces lilacinus* and vesicular arbuscular mycorrhizal fungi on root-knot nematode infected okra. Combined application of *Glomus fasciculatum* and *G. mosseae*, 2 weeks prior to inoculation with *Meloidogyne incognita* resulted in increased fresh and dry weight of root, shoot and okra fruit as compared to only nematode treated plants, in pot experiments. The combined effect of the mycorrhizal fungi and *Paecilomyces lilacinus* gave maximum reduction of nematode galls.

Siddiqui and Mahmood (1997) studied the effects of a plant growth promoting bacterium *Pseudomonas fluorescens*, an VAM fungus *Glomus mosseae* and different soil types (having a mixture of clay:sand in the ratio of 4:0, 3:1, 2:2, 1:3 and 0:4) on the growth of tomato plants and on the reproduction of *Meloidogyne juvanica* and morphometrics of nematode females. Combination of clay and sand in the ratio of 3:1 gave best plant growth response followed by combination of 4:0 and 2:2 clay:sand. Use of *G. mosseae* or *P. fluorescens* in different soil types caused significant increase in plant growth except in plants grown in pure sand. Use of *G. mosseae* or *P. fluorescens* both in uninoculated and nematode inoculated plants caused almost similar increase in plant growth but the combined use of both the organisms caused greater increase in plant growth than when used singly. Plants grown in pure sand had the largest galling and nematode reproduction. Moreover, plants grown in a mixture of clay and sand of 1:3 supported highest galling and nematode reproduction, followed by 2:2 and 3:1 clay:sand ratio. Applications of *G. mosseae* or *P. fluorescens* caused reduction in galling, nematode reproduction and morphometric parameters of females. Combined use of both organisms caused higher reduction in galling, nematode reproduction and morphometrics than their individual application. All the parameters of morphometrics observed showed reduction in their size except stylet length and stylet knob width. VAM colonization was best in plant roots grown in 3:1 clay:sand ratio and least when grown in pure sand.

Babu *et al.* (1998) studied the management of *Pratylenchus zeae* on maize by biofertilizers and VAM. In a field trial on maize, mycorrhizal fungi, *Glomus fasciculatum* and *Azospirillum* and Phosphobacterium when applied alone or in different combination in soil, reduced the population of *P. zeae* and recorded the highest cob yield.

Jain *et al.* (1998b) investigated the influence of the endomycorrhizal fungus, *Glomus fasciculatum* on *Meloidogyne incognita* and *Tylenchorhynchus vulgaris* infecting berseem (*Trifolium alexandrium*). Gall formation by *M. incognita*, and multiplication of both the nematodes was reduced by 28-50%. Conversely, the prior presence of nematodes reduced the percent root colonization by VAM fungi.

Rao *et al.* (1998b) conducted experiments to test the feasibility of integration of *Glomus fasciculatum* and *Ricinus communis* cake for the management of *Meloidogyne incognita* on egg plant. Seedlings colonized with mycorrhiza were least infected by *M. incognita* when transplanted in soil which was amended with castor cake. Further significant increase in colonization of *G. fasciculatum* on roots of egg plant and chlamydospore densities of mycorrhiza in this treatment indicated favourable effects of castor cake amendment on the growth of *G. fasciculatum*.

Jain *et al.* (1998c) reported that pre-inoculation of lucerne (cv. IGFRI-112) plants with *Glomus fasciculatum* two weeks before introduction of either *Meloidogyne incognita* or *Tylenchorhynchus vulgaris* significantly reduced the adverse effects of nematodes on total biomass of the plant. Similarly, gall formation and also the multiplication of both nematode species were considerably reduced. In contrast, prior presence of the nematodes reduced mycorrhizal colonization.

Sankaranarayanan and Sundarababu (1998) studied the effect of *Rhizobium* on the interaction of vesicular-arbuscular mycorrhizae and root-knot nematode on blackgram (*Vigna mungo*). The interaction of *Rhizobium* with VAM and the nematode resulted in a significant increase in the growth parameters of black gram. The highest root nodulation was recorded in plants inoculated with *Rhizobium* and *G. mosseae*. Inoculation of *M. incognita* reduced the number of nodules per plant whereas the combination of *Rhizobium* + *G. mosseae* + *M. incognita* resulted in increased nodules and was similar to the *Rhizobium* + *G. mosseae* treatment. The combination of *Rhizobium* and VAM resulted in the least nematode population density and in the maximum spore, mycorrhizal colonization, total nitrogen and phosphorus content of black gram.

Siddiqui *et al.* (1998) used *Glomus mosseae, Paecilomyces lilacinus* and *Pseudomonas fluorescens* alone or in combination to control the wilt disease complex of pigeonpea caused by the nematode *Heterodera cajani* and the fungus *Fusarium udum*. All three antagonists alone or in combination increased plant growth, nodulation, phosphorus content and reduced nematode multiplication and wilting in infected plants. Treatment with *G. mosseae* or *P. lilacinus* caused similar increases in plant growth and nodulation. The use of *P. fluorescens* caused a greater increase in plant growth and nodulation, maximum increase was observed when all the three antagonists were used together. The use of antagonists on plants inoculated with both pathogens caused a greater increase in plant growth and nodulation compared with plants inoculated with single pathogen. Treatment with *G. mosseae* increased phosphorus contents both in pathogen inoculated and uninoculated plants whereas other antagonists increased the phosphorus content only in pathogen inoculated plants.

Rao *et al.* (1998a) conducted experiments in a screen house to integrate the use of the vesicular arbuscular mycorrhiza. *Glomus mosseae* with a nematophagous fungus *Paecilomyces lilacinus* for the management of *Meloidogyne incognita* on egg plant [aubergine]. Both biological control agents did not affect each others colonization on the roots resulting in an additive effect of both on the management of *M. incognita*. The final nematode population was significantly less in the treatment where *G. mosseae* and *P. lilacinus* were integrated.

Sosamma *et al.* (1998) studied the effect of VAM, *Glomus mosseae,* on nematodes infesting banana under coconut based high density multispecies cropping system. *Glomus mosseae* was effective in reducing the nematode populations, especially that the *Meloidogyne incognita* and *Radopholus similis* infesting banana, as well as enhancing banana growth and bunch weight in pot experiments. There was 18% increase in height, 2.5% increase in the number of leaves, 30.7% increase in leaf area, 35.7% increase in bunch weight, 30% increase in the number of hands, 39.6% increase in number of fingers and 30.1% increase in shoot weight in the mycorrhizal colonized plants when compared to that of control banana plants in the field.

Sivaprasad and Sheela (1998) conducted an extensive survey in three major pepper [*Capsicum annuum]* growing districts of Kerala (Kannur, Wyanad and Idukki) revealed high VAM colonization and spore count, coupled with low population levels of *Meloidogyne incognita* and *Radopholus similes* in different pepper cultivars. Plant height and dry weight of shoot and root of pepper plants were increased due to *Glomus etunicatum* inoculation (82.60 and 104%, respectively over *M. incognita* infested plants). *G. mosseae* was more effective in suppressing the nematode population in root (60%) and soil (33%).

Koshy *et al.* (1998)studied an interactive effect of VAM with *Radopholus similis* on coconut seedlings. A significant increase in lamina width and leaf area was recorded in plants that received mycorrhizae alone. Plants inoculated with nematodes alone showed maximum reduction in growth, fresh shoot and root weights indicating the pathogenic effect of *R. similis* on coconut seedlings. Increase in shoot and root weight, and decrease in lesion indices was noticed in plants inoculated with mycorrhizae prior to the nematode. Prior inoculation of VAM was found to be more effective against *R. similis* than the simultaneous inoculation of VAM and nematode. Even through root infection and spore production of VAM was slightly reduced in nematode inoculated plants, the lesion index and root nematode population could be successfully lowered with mycorrhizal infection. A mixture of VAM consisting of multiple endophytes (*Acaulospora bireticulata, Glomus fasciculatum, G. macrocarpum, G. mosseae, G. versiforme, Sclerocystis rubiformis* and *Scutellospora nigra*) was found effective in improving the plant growth and reducing the *R. similis* population infesting coconut seedlings.

Sadasivan and Babu (1998) studied interaction of *Glomus fasciculatum* (Thaxter sensu Gerd) Gerd and Trppe and *Heteroderma cajani* Koshy, 1967 on cowpea (*Vigna unguiculata* (I.) Walp.) The yield and biomass production was higher in *G. fasciculatum* inoculated plants compared to uninoculate and *H. cajani* inoculated plants in pot experiments. Cowpea pod yield per plant was high (8.9 g) in *G. fasciculatum* inoculated plants and low (5.9 g) in *H. cajani* inoculated plants. The uninoculated control gave a yield of 6.9 g. The cyst count per 100 cc of soil was 196 in the combined inoculation of G. *fasciculatum* and *H. cajani*, while in *H. cajani* inoculated soil it was 326 indicating that G. *fasciculatum* effectively reduced the cyst population. The colonization of cowpea roots by G. *fasciculatum* hampered the root invasion by *H. cajani* leading to reduction in cyst population. *G. fasciculatum* inoculated plants recorded good biomass production and yield due to the suppression of the cyst population and high uptake of phosphorous. The phosphorous content in *G. fasciculatum* inoculated plants was 0.87% compared to 0.81% in uninoculated control. Phosphorus content was low (0.76%) in *H. cajani* inoculated plants. This clearly indicates that *G. fasciculatum* can be effectively used as a biocontrol agent for *H. cajani*.

Ray and Dalei (1998) investigated the effects of the VAM fungus, *Glomus fasciculatum*, the legume bacterium *Rhizobium phaseoli* (M-14 strain) and *Meloidogyne incognita* on green gram (*Vigna radiata*), alone and in various combinations, in pot experiments. It was observed that all plant growth parameters including pod yield, leaf chlorophyll content, bacterial nodulation, leghaemoglobin content of nodules and NPK contents of the plant were reduced in the nematode inoculated plants compared with the mycorrhizal and rhizobial inoculations either alone or in combination. The mycorrhizal and rhizobial inoculations alone were superior over the control in the absence of nematode. Combined inoculations gave a better result than their single inoculations, showing a compatible and synergistic relationship between the two organisms with common antagonism to the nematode, which in turn suppressed them also. Increase in pod yields in mycorrhiza + nematode, and *Rhizobium* + nematode treatments were 44% and 22%, respectively over that of nematode inoculations alone, and 13%, 26% and 66% over control in nematode free soil due to mycorrhiza, *Rhizobium* and mycorrhiza + *Rhizobium* combinations, respectively.

Manian *et al.* (1999) conducted experiments to evaluate the biocontrol efficiency of VA mycorrhiza (*Glomus fasciculatum*) against root-knot galls in tomato under greenhouse conditions. In the interactions with nematode, mycorrhiza significantly increased plant growth and reduced the number and size of root - knot galls. Early establishment of VAM enhanced shoot and root length in tomato and also reduced severity of *Fusarium* wilt by 40 per cent. It also decreased the percentage disease index of *Alternaria* leaf spot significantly.

Iqbal and Mahmood (1999) used arbuscular mycorrhizal fungus *Glomus mosseae* for the management of the root-knot disease caused by the nematode *Meloidogyne incognita.* Prior or simultaneous inoculation of *G. mosseae* to plants inoculated with *M. incognita* increases plant length, dry weight, P content, fruit yield and reduced nematode population density and root galling. *M. incognita* had an adverse effect on the percentage root colonization and sporulation by AM fungi. Prior inoculation of *G. mosseae* to *M. incognita* caused greater reduction in galling (88%) and nematode multiplication and better plant growth and yield than simultaneous or subsequential inoculation.

Malathy *et al.* (1999) studied the root-knot nematode infestation and growth of *Kaempferia galanga* as influenced by inoculation with vesicular arbuscular mycorrhiza in the green house. VAMF colonization remarkably reduced the nematode population in the rhizosphere soil and root. VAMF inoculated plants even in the presence of nematode registered a significant increase in yield over non-mycorrhizal control (59.5%) and nematode alone treatment (70.8%). The mycorrhizal effect on plant growth characteristics and rhizome yield in the absence of nematode was highly remarkable. The increase in yield over control was 81.2 per cent.

Pandey *et al.* (1999a) studied the influence of vesicular-arbuscular mycorrhizal fungi alone or in combination with *Meloidogyne incognita* on *Hyoscyamus niger* L. Three vesicular-arbuscular mycorrhizal (VAM) fungi, *Glomus aggregatum, G. fasciculatum* and *G. mosseae,* were evaluated for protecting against the reproduction and infection potential of *Meloidogyne incognita* on black henbane, *Hyoscyamus niger*. Different VAM fungi enhanced the growth and biomass of the plant. The results also indicated that treatments of plants with VAM fungi increased fresh and dry root and shoot weight and decreased *M. incognita* population in both soil and roots in pot experiments. A significant reduction in *M. incognita* reproduction potential (Rf=4.4) was observed when *G. aggregatum* was used. Treatment with *G. aggregatum* alone yielded the highest mycorrhizal infection (88%) followed by *G. aggregatum* + *M. incognita* (75%), *G. mosseae* (69%), *G. fasciculatum* (57%), *G. mosseae* + *M. incognita* (33%) and *G. fasciculatum* + *M. incognita* (45%). A similar trend was also observed in the multiplication of VAM spores in soil.

Nagesh *et al.* (1999b) carried out *in vitro* (pot conditions) and *in vivo* (nursery beds) experiments to correlate *G. fasciculatum* spore density (1,2,4,8 and 16 spores/g soil), root colonization of tomato and root infection of *Meloidogyne incognita. In vitro* experiments revealed that mycorrhizal root colonization and formation of chlamydospores were positively correlated with spore density, although the rate of increase in root colonization and subsequent chlamydospore formation declined at 8-spores/g soil and above. The nematode populations, egg masses/plant and root gall index were negatively correlated with spore density. Further, at least 2-spores/g soil were required for 50% root colonization under pot conditions. However, under nursery conditions, 4-spores/g soil were required to obtain 50% root colonization which reduced nematode populations in root (in terms of egg masses/plant) by 59 and 61%, respectively. It was observed that new flushes of roots from precolonized roots after chlamydospore formation (45-60 DAS) were infected by root knot nematode juveniles.

Sosmma *et al.* (1999) conducted an experiment under greenhouse conditions to find out the most effective mycorrhizal fungi suppressing *Radopholus similis* multiplication on coconut. The mycorrhizal isolates used were *Acaulospora bireticulata, Scutellospora coralloidea, Glomus macrocarpum* and *Sclerocystis rubiformis*. The minimum growth characters and maximum multiplication of nematodes were recorded in plants that were inoculated with *Radopholus similis* alone. On combined inoculation of mycorrhizas and *R. simili*, the maximum plant growth and lowest nematode populations were recorded in *A. bireticulata* inoculated plants.

Pandey *et al.* (1999b) studied the efficacy of VAM fungi, green manure, organic materials and pesticides in the management of root-knot nematode in menthol mint. Maximum enhancement in various growth parameters such as fresh and dry root/shoot weight and total oil yield of *Mentha arvensis* cv. Himalaya were obtained in soil treated with VAM fungi (*Glomus fasciculatum* and *G. mosseae*) followed by plant extracts, green manures and nematicides (carbofuran at 0.0020 g and aldicarb at 0.0015 g a.i./kg). A reduction in soil and root populations of *Meloidogyne incognita* as well as root knot indices followed a similar trend in these glasshouse experiments.

Sankaranarayanan and Sundarbabu (1999) studied the interaction of vesicular-arbuscular mycorrhiza (*Glomus mosseae*) and root knot nematode (*Meloidogyne incognita*) on black gram. Black gram (*Vigna mungo*) plants inoculated with phosphorus showed significantly enhanced shoot and root lengths when compared to the uninoculated plants, and nematode alone treatments. Nematode alone treated plants recorded the lowest shoot and root length. Among the combined inoculation of mycorrhiza (*Glomus mosseae*) and nematode and phosphorus; the 50% (0.48g/plant) recommended level of phosphorus registered higher shoot and root lengths and weights, and pod yield, and was at par with the 100% (0.962 g/plant) phosphate treatment.

Nagesh *et al.* (1999a) concluded that in general, mycorrhizal inoculation (especially in combination with neem cake) not only reduced nematode infection but also significantly enhanced plant growth parameters and flower yield in pot experiments with *Crossandra undulaefolia* [*C. infundibulformis*]. Of the 4 fungi, *Glomus fasciculatum* recorded the highest plant growth rates and least nematode damage, followed by *G. mosseae, G. intraradices* and *Aucalospora laevis*. Both the first 2 species performed better then carbofuran treatments.

Bhat and Mahmood (2000) studied the role of *Glomus mosseae* and *Paecilomyces lilacinus* in the management of root knot nematode on tomato. A greenhouse experiment showed that the root knot nematode *Meloidogyne incognita* was controlled more effectively when *P. lilacinus* and *G. mosseae* were applied together in a pot experiment than when either was applied alone. Inoculation of tomato plants with *G. mosseae* did not markedly increase the growth of plants infected with M. incognita. Inoculation of plants with *G. mosseae* and *P. lilacenus* together or alone resulted in a similar shoot and plant height. The highest root development was achieved when mycorrhizal plants were inoculated with *P. lilacinus* to combat root knot nematode inoculation of tomato plant with *P. lilacinus* suppressed galls/root system and eggs/egg masses, compared to seedlings inoculated with *M. incognita* alone. Mycorrhizal colonization was not affected by inoculation of *P. lilacinus*.

Ratti *et al.* (2000) determined the effect of the vesicular arbuscular mycorrhiza (VAM) fungus, *Glomus mosseae*, on the reproduction of root-knot nematode, *Meloidogyne incognita* and on the growth and yield of menthol mint (*Mentha arvensis* subsp. *haplocalyx.*) The treatments were: the control: inoculation with VAM alone: inoculation with *Meloidogyne incognita* alone; simultaneous inoculation with VAM and *M. incognita*; inoculation with *M. incognita* 15 days prior to VAM inoculation; and inoculation with VAM 15 days earlier than M. incognita inoculation. Shoot height, shoot and root fresh and dry weight, and oil yield were significantly higher in plants inoculated with VAM compared to the control. The suppressive effect of *M. incognita* on shoot and root (fresh and dry) weight, and oil yield was highly significant, irrespective of the presence or absence of VAM, compared to the control. Simultaneous inoculation of *M. incognita* and VAM improved plant fresh weight by 26.4% compared to the plants inoculated with nematodes alone. However, VAM had a non-significant effect on shoot fresh weight and oil yield when it was inoculated 15 days after nematode inoculation. Maximum suppression of nematode reproduction was achieved with prior inoculation of VAM fungi. VAM was effective in increasing growth and oil yield of menthol mint irrespective of the presence or absence of *M. incognita*.

Borah and Phukan (2000) studied the effect of VAM fungus, *Glomus fasciculatum* and root knot nematode, *and Meloidogyne incognita* on brinjal, *cv. Pusa Purple Long* in a pot experiment. Mycorrhizal plants had longer shoots than non-mycorrhizal plants, with the longest shoots observed in plants inoculated with VAM at 300 spores per plant, followed by those treated with 150 spores per plant and the control. Plants inoculated with nematodes had the highest number of galls and egg masses, while plants treated with VAM prior to nematode inoculation had the lowest number of galls and egg masses. The application of VAM fungi prior to nematode inoculation was more effective in reducing nematode infestation than the simultaneous application of VAM and nematode or nematode inoculation before VAM treatment.

Mahanta and Phookan (2000) conducted a pot experiment to determine the effects of the vesicular arbuscular mycorrhizal (VAM) *G. fasciculatum* on the penetration and development of the parasitic nematode *M. incognita* on black gram *cv.* T-9. Treatments comprised the simultaneous inoculation with nematode and VAM; VAM inoculation 3, 6 or 9 days before nematode inoculation; and nematode inoculation alone (control). Nematode penetration commenced within 24 h from inoculation on mycorrhizal and non-mycorrhizal plants. The greatest penetration was recorded for plants inoculated with nematodes alone, while the lowest penetration level was observed in plants inoculated with VAM 3, 6 or 9 days before nematode inoculation. VAM inoculation significantly reduced the number of galls, in infested plants, with the lowest gall incidence exhibited by plants inoculated with VAM 9 days before nematode inoculation.

Rao *et al.* (2000) conducted glasshouse and field experiments during August-November 1999 and February-June 2000 in Bangalore, Karnataka to evaluate the efficacy of *Pasteuria penetrans* and *Glomus mosseae* in controlling the root-knot nematode *Meloidogyne incognita* in tomato. Treatment with both organisms significantly reduced root galling, nematode population, egg mass production and fecundity of the nematode. Incorporation of *G. mosseae* in nursery beds resulted in a significant increase in the growth of tomato seedlings. The combination of *P. penetrans* and *G. mosseae* resulted in higher parasitism of female nematodes compared with *P. penetrarns* alone.

Siddiqui and Mahmood (2000) studied the effect of *Bacillus subtilis, Glomus mosseae* and ammonium sulphate on the development of *Meloidogyne javanica* and on growth of tomato, alone and in combination. Treatment with ammonium sulfate both in nematode inoculated and uninoculated plants increased plant growth more than *B. subtilis* or *G. mosseae*. Plant growth promotion was greater for nematode inoculated than for uninoculated plants. Greatest growth of tomato plants occurred when ammonium sulfate was used in combination with *B. subtilis* plus *G. mosseae*. Treatments with ammonium sulfate, *B. subtilis* and *G. mosseae* reduced all the morphometric parameters except stylet length and stylet knob width. Treatment with ammonium sulfate alone caused largest reduction in galling and nematode multiplication and affected nematode females form. Largest reduction in galling. Nematode multiplication and morphometrics were observed when ammonium sulfate was used in combination with *G. mosseae* plus *B. subtilis*. Arbuscular mycorrhizal colonization was greater in the absence of nematodes than in their presence. AM colonization was also reduced by *B. subtilis*.

Sundarababu *et al.* (2001) conducted an experiment to study the management of *Meloidogyne incognita* in chilli nursery with *Glomus mosseae*. Field experiments were conducted at TNAU, Coimbatore and Bhavanisagar during 1998-99 to determine the effectiveness of the vesicular arbuscular mycorrhizal fungus *Glomus mosseae* against *Meloidogyne incognita* in chilli (*Capsicum annuum*) cv. K2. *G. mosseae* was compared to carbofuran 2% at 1.0 kg/ha. The results showed that although carbofuran caused a sudden and large reduction in the soil nematode population immediately after application, it did not give sustained suppression of nematodes, as did *G. mosseae*. Seedlings treated with the mycorrhiza produced significantly higher yield than carbofuran and untreated control.

Rajendran *et al.* (2001) carried out studies on the control of plant parasitic nematodes, *Meloidogyne incognita* and *Tylenchuaus semipenetrans* in horticultural crops, such as citrus, tomato, potato, chilli and aubergine, using *Psuedomonas fluorescens, Pasteuria penetrans, Baasillus subtills, B. cereus* and vesicular arbuscular mycorrhizal fungi, *Glomus fasciculatum* and *G. mosseae*, are presented. Results showed that these organisms could be used as successful biological control agents for the management of plant parasitic nematodes.

Sankaranarayanan and Sundarababu (2001) investigated the effects of different levels of moisture (30, 40, 50, 60, 70, 80, 90 and 100%) and pH (4,5,6,7, 8 and 9) on the biological control efficiency of *Glomus mosseae* against *Meloidogyne incognita* on black gram (*Vigna mungo* cv. Co5). Mycorrhizal colonization increased as moisture level increased up to 70%. Moisture levels of 60 and 70% were favorable for the establishment of *G. mosseae*. Mycorrhizal colonization declined at moisture levels above 70% resulting in increased nematode population. The maximum spore population and maximum mycorrhizal colonization were obtained in the pH range of 6-7. Mycorrhizal development was noticed above pH 7, but this declined rapidly below pH 7. Mycorrhizal growth was negligible in pH 5, and pH 4, the mycorrhizal fungi failed to develop.

Jothi and Sundarababu (2001a) studied the management of root-knot nematode in brinjal by using VAM and crop rotation with green gram and pearl millet. The integrated effect of vesicular arbuscular mycorrhizal fungi (VAM; *Glomus mosseae* at 10 g/kg soil) and brinjal [aubergine]-pearl millet (*Pennisetum glaucum*)- green gram rotation against the root-knot nematode *Meloidogyne incognita* was investigated in a glasshouse. Brinjal growth parameters increased and nematode population decreased in the presence of VAM. VAM colonization increased and nematode population decreased when pearl millet, a non-host for root-knot nematode and a host for VAM, was sown in rotation after brinjal. Nematode population decreased from 143 nematodes per 200 ml soil in the first brinjal crop to 105 nematodes per 200 ml soil in the second brinjal crop (sown after green gram in the crop rotation system).

Rao and Reddy (2001) conducted an experiment to standardize a strategy of integrated management of the root-knot nematode, *Meloidogyne incognita*, on aubergine *cv. Pusa Round* under field conditions using different combinations of the endomycorrhizal fungi *G. mosseae*, the fungus *P. lilacinus* isolate IIHR-2 and neem cake (NC, *Azadirachta indi ca*) in Bangalore, Karnataka. The different treatments were *G. mosseae* alone, NC + *G.mosseae*, NC +*P. lilacinus, G. mosseae* + *P. lilacinus, P. lilacinus* alone, NC + *G. mosseae* + *P. lilacinus* and NC alone. Nursery beds infested with *M. incognita* were amended with neem cake two weeks before the incorporation of *G. mosseae* or *P. lilacinus* isolate IIHR-2, or both. Healthy and vigorous seedlings colonized with G. *mosseae* and *P. lilacinus* were obtained for transplanting. Transplants obtained from nursery beds treated with neem cake + *G. mosseae* + *P. lilacinus* were least infected in the field. The parasitization of eggs of root-knot nematode was significantly increased by *P. lilacinus* and the transplants yielded significantly more fruits. Neem cake amendment in the nursery beds played a positive role in increasing the colonization of *G. mosseae* and *P. lilacinus* on the root of transplants before and after transplanting. The combined effect of these three components facilitated the sustainable management of *M. incognita* on aubergine *cv. Pusa Round* under field conditions.

Jothi and Sundarababu (2001b) studied the interaction between four endomycorrhizal fungi *(Glomus* spp.) and the root-knot nematode, *Meloidogyae incognita* in aubergine (*Solanum melongena*). Mycorrhizal inoculation reduced root-knot nematode infestation. Mycorrhizal plants had higher total soluble protein, phenol, total and reducing sugars. The possible role of some of these chemical constituents in suppressing the development of root-knot nematode is discussed.

Sharma *et al.* (2001a) investigated the effect of 4 soil treatments on the control of root knot disease caused by *Meloidogyne incognita* in 7 mulberry cultivars (K-2, Local, MR-2, S-36, S-34, S-13 and V-1) in a field experiment conducted in Mysore, Karnataka, India [date not given]. The treatments comprised a control, bionema (*Verticillium chlamydosporium)* + neem oil cake at 2 mt/ha, Furadan [carbofuran] at 40 kg/ha, neem oil cake, and vesicular arbuscular mycorrhizal fungi (*Glomus mosseae*) at 40-50 g. K-2, Local and MR-2 were highly susceptible, S-36 was susceptive, whereas S-34, S-13 and V-1 were moderately resistant to the disease. The moderately resistant cultivars had significantly less number of root knots per plant, egg mass per plant and nematode population in soil as well as high leaf yield compared to highly susceptible and susceptible cultivars among the soil treatment Bionema along with neem oil cake showed significantly less disease incidence and maximum leaf yield. Combined soil application of Bionema and neem oil cake on the moderately resistant cultivars reduced the root knot per plant by 85.01, 97.08%, egg masses by 89, 13-98.03%, nematode population by 92.51-99.22% and improved the leaf yield by 26.34-36.12% over the control.

Labeena *et al.* (2002) evaluated five arbuscular mycorrhizal (AM) fungi (*Glomus fasciculatum, Glomus macrocarpum, Gigaspora margarita, Acaulospora laevis,* and *Sclerocystis dussii*) for their ability to mitigate damage caused by *M. incognita* on tomato *cv. Pusa Ruby. Glomus fasisculatum* was most efficient in promoting plant growth in the presence of nematodes. However, *Glomus macrocarpum* and *Gigaspora margarita* were also efficient. Plant biomass and phosphorus uptake were highest in these treatments. The development of nematode stages in the roots and nematode density in soil were suppressed by the AM fungi, with the most pronounced effect exhibited by *Glomus fascicilatum,* followed by *Glomus macrocarpum* and *Gigaspora margarita. A. leavis* and *S. dussii* did not show any effect.

Anusuya and Vadivoo (2003) conducted a study to investigate the effect of vesicular arbuscular mycorrhiza (VAM) and *Trichoderma* spp. on micropropagated carnation (*Dianthus caryophyllus*) plants. Treatment comprised: uninoculated (control); *Glomus mosseae* (Gm); *G. mosseae+Trichoderma harzianum* (Gm + Th); *G. mosseae+T. viride* (Gm + Tv); *G. fasciculatum* (Gf); *G. fasciculatum+T. harzianum* (Gf+Th); and *G. fisciculatum+T. viride* (Gf+Tv). Sixteen weeks after inoculation, plants treated with Gm+Th had higher percentage of mycorrhizal root colonization than other combined inoculation treatments. Combined inoculation with Gm+Th significantly increased the shoot number, shoot length, root length, flower number, flower diameter and plant biomass of micropropagated carnations. The combined inoculation of Gm+Th recorded the highest VAM colonization and VAM spore number; Synergistic interactions between VAM fungi and *Trichoderma* sp. were recorded. Results show that acclimatization of micropropagated carnation plantlets improved physiological competence through the use of *T. harzianum* and *G. mosseae.*

Diseases due to Bacterial Pathogens

Kumar and Sood (2002) conducted pot experiments on seedlings of tomato cultivars EC191536 (moderately resistant) and Solan Gola (susceptible) to determine the efficacy of bacterial antagonists *Pseudomonas fluorescens* and *Bacillus cereus* and vesicular arbuscular mycorrhiza *Glomus mosseae* against tomato bacterial wilt (*Ralstonia solanacearum* race 1 biovar III). The combined use of *P. fluorescens* and *B. cereus* was more effective than the individual applications. No wilt incidence was observed in EC191536 when the antagonists were simultaneously applied compared with the control (11.1% in 1998 and 22.2% in 1999). Wilt incidence in Solan Gola was reduced from 83.3 to 33.3% during 1998 and 88.8-27.7% during 1999. Population studies during 1998 and 1999 revealed that *R. solanacearum* population was least in treatments where the antagonists were combined with vesicular arbuscular mycorrhiza.

CONCLUDING REMARKS

A large number of Indian workers have tried to explore the possibility of controlling the nematode diseases of plants through the use of VAM fungi as biocontrol agents. Their main concentration had been on sedantary nematodes especially *Meloidogyne* and *Heterodera*. Few studies are related also to *Tylenchus*. Diseases due to migratory nematodes (*Pratylenchus* and *Radopholus*) have been attempted rarely, while no published report is available on diseases due to ectoparasitic nematodes. Important diseases of a variety of crops including field crops such as sorghum, maize, cowpea, blackgram, chickpea, pigeonpea, tomato, tobacco and brinjal, fruit crops like banana, citrus, and mulberry have been investigated and suppression of disease due to application of VAM fungi has been reported. Work of few workers clearly shows that VAM fungi have a potential to curtailing the damage caused not only due to nematodes but also soil-borne fungal pathogens like *Fusasarium, Pythium* or *Phytophthora* along with them. Considering that integrated management may provide better results, a number of workers tried to explore the potentiality of VAM technology in combination with nematicides, other biocontrol agents, fungicides (for nematode + fungal disease complex), various types of organic amendments like oil cakes, leaf extracts or other beneficial microbes like nitrogen fixers, PGPR etc. As expected, in majority of cases the integration gave better results with reference to suppression of disease severity as well as growth and yield of the crops.

Since most of the experiments were conducted under pot conditions, the recommendations have limited value unless they are repeatedly evaluated under field conditions. Nematode diseases of many other crops cultivated in the country need investigation because presently the nematode diseases have become serious threat to productivity.

8

VA MYCORRHIZA IN WASTELAND DEVELOPMENT

During the last thirty years, there has been a growing appreciation of the impor tance of VAM fungi in improving the plant growth and yield. They have been extensively studied also in India. Considering the potential importance of these organisms in the recovery of disturbed sites and their possible use in land reclamation, researches regarding their relationship to ecosystem recovery have now been initiated also in India.

Large areas of land in the country called wastelands are degraded and lying unutilized due to various constraints. Part of it called culturable, has the potential for the development of vegetation cover. On the other hand, part of it called unculturable, has no potential for the development of vegetation cover. In addition to these natural wastelands, there are large tracts of wastelands created due to human activities such as mining, deforestation etc. Such wastelands are also culturable and a vegetation cover can be developed on them after removing the constraints with some efforts. In view of the increasing shortage of plant resources due to population explosion it has become imperative that all the wastelands are put to use by developing vegetation cover. A variety of plantation programmes are being executed in the country utilizing known scientific inputs. Attempts are being made world over to utilize VAM technology in plantation programmes. Indian mycorrhizasts have also made important contributions in this direction. Following description gives an account of their contributions:

Thatoi *et al.* (1993) investigated the comparative effect of VAM inoculations with *Glomus mosseae* or *G. fasciculatum* on growth, nodulation and *Rhizobium* population of subabul (*Leucaena leucocephala*) grown in iron mine waste soil (from Orissa). Seedlings were grown in poly bags in 3 media: garden soil: mine soil: or mine waste soil mixed in equal proportions with sand. In general, inoculation with *G. mosseae* improved growth, nodulation and the rhizosphere rhizobial population, while inoculation with *G. fasciculatum* had negative effects. However, in pure mine soil inoculation with *G. mosseae* reduced nodulation and rhizosphere rhizobial population.

Pattanaik *et al.* (1995) working in the some iron-mine area investigated the effect of inoculation of VA mycorrhiza *G. fasciculatum* or *G. mosseae* on horsegram (*Macrotyloma uniflorum*) and reported increased plant growth and nodulation over the uninoculated control. *G. fasciculatum* was superior to *G. mosseae* and after 60 days growth, it increased shoot height by 36.03%, plant dry weight by 94.44%, nodule dry weight by 100% and individual seed dry weight by 27.45% compared with uninoculated control.

Jamaluddin and Chandra (1995) studied the growth response of *Eucalyptus* hybrid for VAM fungi in the coal mine spoils of Korba, M.P. Both ectomycorrhizal and VAM fungi were associated with *Eucalyptus* (*Scleroderma verrucoseum, S. cepa. Glomus mosseae, G. intraradices, G. fasciculatum, G. invermaium, Glomus* spp., *Gigaspora* spp., *Acaulospora scrobiculata, A. delicata, Acaulospora* spp., *Scutellospora* spp., and *Sclerocystis* spp.) The initial colonizers were the VAM fungi while ectomycorrhizal fungi were found associated with three years old plants. *Eucalyptus* proved to be a mycotrophic plant with exhibiting association with both ecto- and endomycorrhizal fungi.

Thatoi *et al.* (1995) studied the effect of *Rhizobium* and VAM fungi inoculation on the growth and rhizosphere microbial population of *Acacia nilotica* in some iron mine waste soils. They raised *A. nilotica* in pure iron mine soil as well as mine soil amended with sand and leaf litter. They inoculated the seedlings with *Rhizobium* and VAM fungi in order to study the potential for revegetation of mine waste areas. Growth responses in terms of shoot height and dry weight and rhizosphere microbial populations (general bacteria, ammonifiers, total fungi, *Rhizobium, Actinomycetes* and *Nitrobacter*) were studied from six-month old nursery seedlings. There was a general increase in growth and rhizosphere microbial populations with the sand and leaf litter amendments. Inoculation of *Rhizobium* and *Rhizobium* along with VAM fungi increased growth in all soil conditions. The rhizosphere microbial populations were selectively stimulated under different soil conditions. While microbial inoculations produced best growth under leaf litter amended conditions, the maximum stimulation in rhizosphere microbial population was observed under sand amended mine soils.

Bisen *et al.* (1996) studied the VAM colonization in tree species planted in Cu, Al and Coal mines of Madhya Pradesh with special reference to *Glomus mosseae*. Soils from the mine were acidic (pH 5.3 - 6.8) and low in available N, P and K, while soil from a non-mining area (Bhopal) was alkaline (pH 8.2). Most of the planted tree spp. showed colonization by VAM fungi, with *G. mosseae* predominating. Colonization was highest on *Dalbergia sissoo* planted on a copper mine site (73%) and *Eucalyptus* spp. in a coal mine area (68%). Mining activities increased the abundance of VAM fungi by lowering the pH.

Mehrotra (1996) examined the potential use of revegetated coal mine spoil as source of arbuscular mycorrhizal inoculum for nursery inoculations. Rhizosphere soils of five tree species were used as sources of mycorrhizal inoculum. Soils contained seven spore-forming species of VAM fungi. The substrate used in the pot experiment was a mixture of unsterilized coal mine spoil (wihout any mycorrhizal propagule) and autoclaved sandy loam soil. *Cassia siamea* and *Derris indica* were used as the test plants. Measurements were made of shoot and root biomass, P uptake, per cent mycorrhizal infection and spore population of AM fungi. Growth measured as shoot and root dry weight was significantly higher in seedlings inoculated with soil inoculum from *Dalbergia sissoo, C. siamea, D. indica* and *A. indica. Acaulospora scrobiculata* was found to be the best fungus in terms of root colonization ability and effectiveness to promote P uptake and growth in plants. The results of the study justify the use of revegetated coal mine spoil as an effective and economical source of endomycorrhizal inoculum for inoculating nursery seedlings.

Maiti (1997) studied the importance of VAM fungi in coal mine overburden reclamation and factors effecting their establishment on overburden dumps. The importance of mycorrhizal establishment for the reclamation of coal mine spoil dumps is reviewed. VAM fungi enhanced nutrient absorption and water transport to plant roots under stress conditions. The successful colonization of VAM fungi on coal mine spoil dumps depended on pH, organic matter, water contents, nutrients, temperature, topsoil cover and age of dumps.

Routray *et al.* (1997) reported the response of cowpea (*Vigna ungiculata*) to inoculation with selected VAM fungi and *Rhizobium* strains in iron mine waste soil of Orissa, under nursery conditions. Dual inoculation of *Glomus fasciculatum Rhizobium* strain increased growth, nodulation and rhizosphere microbial population of cowpea. Sand amendment (25%) to mine soil gave better plant growth responses. They recommended inoculation of VAM fungi and *Rhizobium* in iron mine soil for better plant yield.

Chandra *et al.* (1997) made a comparative study of VA-mycorrhizal root infection and spore density in *Grevilia pteridifolia* in undisturbed and disturbed areas. Comparative studies of VAM in two localities in undisturbed (nursery -site plantation) and disturbed (coal mine overburden) plantation of *G. pteridifolia* were undertaken. Mining operations considerably disturbed the VAM population in coal mine overburden spoil. Highest percentage of root infection (19.36%) with spore density 412 spores/100g soil was recorded in 1991 plantation of undisturbed soil in TFRI, Jabalpur. However, the infection of VAM was 13.88% in 1991 plantation in coal mine overburden (with spore density 397 spores/100g soil). The infection of VAM in roots of *G. pteridifolia* and spores in soil increased with the age of the plants. The *Glomus* spp. were dominant in undisturbed plantations while *Acaulospora* spp. were more in mine site. The population of *Scutellospora* and *Sclerocystis* spp. were scanty in coal mine spoils. There was no definite trend in occurrence of VAM fungi in undisturbed and disturbed plantations. This study confirmed that *G. pteridifolia* is VA-mycorrhizal and its derives benefit from obligate symbiotic system thereby showing better performance in disturbed sites.

Rao *et al.* (1999a) studied the role of VAM fungi in rehabilitation of mine spoils. Surface mining is one of the important activities in arid region of Western Rajasthan which destroy vegetation. *Glomus fasciculatum* isolated from these areas was found to be the most effective than other strains. Out of thirty one plant species occurring naturally on these spoils, 10 species were selected based on their economic value to establish in mined areas after inoculation with the isolated *Glomus fasciculatum* species @ 10 spores /g soil. Per cent root infection increased by 2-9 folds in inoculated plants. Shoot biomass, N, P, K, Zn and Cu concentrations were significantly improved than normal in all the cases. The results demonstrated the possibilities of employing identified VAM-fungi for rehabilitating mine spoils.

Thatoi *et al.* (1999) conducted experiment for the selection of efficient VAM fungi for improvement of growth and nodulation of *Acacia nilotica* in chromite mine waste soil. Comparative assessment on the effect of 4 VAM species *viz, G. mosseae, G. fasciculatum, Gigaspora gilmorei* and *Gigaspora margarita* on growth, nodulation and total N content of *Acacia nilotica* was done with a view to select out most efficient VAM fungi for mine area soil. Results indicated a general increase in shoot height, plant dry weight and total N content of the plant due to all types of VAM inoculation but varied in their degree of stimulating capacity. Among four VAM species tried, *G. mosseae* showed highest stimulatory effect in terms of shoot height, plant dry weight and nodule numbers while *Gigaspora margarita* treatment showed least impact on improving biomass whereas *G. gilmorei* on nodulation of the test plant. Considering the overall effect on growth, nodulation and total nitrogen content of the plant. *G. mosseae* was found to be best species for inoculating *A. nilotica* in chromite soil.

Mehrotra and Prakash (1999) suggested the use of soil inoculum from under mycorrhizal tree species grown on reclaimed coal mine spoil. An experiment using soil inoculum from under *D. sissoo, C. siamea, D.indica* and *A. indica* and *Eucalyptus* sp. growing on reclaimed coal mine spoil was conducted. The study revealed that soils from under *D. sissoo, D. indica* and *A. indica* can be utilized as an effective and economical source of vesicular arbuscular mycorrhizal inoculum for inoculating nursery seedlings of *C. siamea* and *D. indica*. It was also observed that the VAM fungus, *Acaulospora scrobiculata,* commonly found in reclaimed coal mine spoil not only colonize roots extensively, but also consistently increases P uptake and shoot dry matter production. The study also provided evidence that *Scutellospora calospora* also persists in the form of propagules other than spores in the soil and its effect on the growth of seedlings was more conspicuous when used in combination with *A. scrobiculata*. A demonstration on the use of soil inoculum from under tree species growing on reclaimed coal mine spoil for inoculating seedlings of *D. sissoo* and *A. indica* was made at Jayant nursery, Northern Coal Fields Ltd., Singrauli.

Chandra *et al.* (2000b) made an effort to manage the wastelands via petroleum plantations and VAM technology. The study with soils of five different wastelands and four promising petrocrops clearly showed the promise of VAM technology in improving the quality and establishment of petrocrops in wastelands and problematic soils. Out of 26 petrocrops surveyed *viz., Calotropis procera, Catharanthus roseus, Euphorbia tirucalli, Pedilanthus tithymaloides* var. *aureus, P. tithymaloides* var. *cuculatus, P. tithymaloides* var. *variegatus, Pergularia daemia, Sonchus asper,* showed highest potential to develop mycorrhization in their roots. The conclusions drawn from the study may prove appropriate guidelines for future work. The petrocrops have potentiality to form mycorrhizal association with VAM fungi under natural conditions. The magnitude of mycorrhizal status varies with the petrocrops as well as soil and environment related factors. Introduction of VAM inoculants improves not only the mycorrhizal status of the petrocrops but also their biomass and biocrude content. The magnitude of improvement varies with the petrocrops and VAM inoculants. Appropriate VAM inoculants have the potentiality to help the petrocrops to overcome various types of constraints in wastelands and give better performance with reference to biomass and biocrude content. Cultivation of selected petrocrops may be undertaken in wastelands with VAM technology.

Behl (1990) investigated the role of endomycorrhizae in fuelwood plantation nurseries for alkaline soil sites. Initial results suggested that vesicular arbuscular mycorrhizal fungi including *Glomus fasciculatum, G. intraradices, G. dimorphicum, Scutellospora calospora* and *S. gigantea* improved the growth of hardwood tree seedlings particularly on degraded alkaline soils.

Sengupta and Chaudhuri (1995) studied the effect of dual inoculation of *Rhizobium* and mycorrhiza on growth response of *Sesbania grandiflora* L. in coastal saline and sand dune soil. They tested the response of single and dual inoculations with *Glomus fasciculatum* and a salinity tolerant *Rhizobium* stain for *Sesbania grandiflora* growing in pots in coastal saline sand dune soil collected from the Sundarbans of West Bengal. After 120 days, both single and dual inoculations gave significantly higher root and shoot dry matter yields over control, the non-inoculated plants. Growth differences between single inoculations of either *Rhizobium* or mycorrhizas was not significant but dual inoculations of both gave significantly higher yields than single inoculations of either.

Uniyal and Thapar (1995) studied the effectiveness of VAM fungi and rhizobia in promoting growth and biomass of *Prosopis juliflora* seedlings in a pot experiment using nutrient deficient sodic soil. Inoculation with *Glomus fasciculatum* and / or *Rhizobium* (R-1115 or R-1121) markedly improved shoot height and collar diameter. Best results were obtained with dual inoculation treatments, which showed increased nodulation also.

Kumar and Mangal (1995a) studied the response of potato to VAM inoculation under different saline soils with and without phosphorus nutrition. They used *Glomus mosseae* as VAM inoculum and five salinity levels i.e. 0.75 (control), 4,6,8 and 10 d Sm^{-1} and two phosphorus levels. Significant improvement in tuber yield was observed for all treatment combinations over non-VAM-P-deficient control treatments. VAM plants at each salinity level produced significant and maximum tuber yield under P-deficient conditions. At salinity level 10.0 d Sm^{-1}, 42.6% increase in yield over treatments with P alone was recorded which was higher than that recorded at any other salinity level. Under the same treatment combination, the increase in yield over their respective control counter parts were higher with increasing salinities.

Madan *et al.* (1995) studied the effect of vesicular-arbuscular mycorrhizal fungi on different plant species in highly alkaline saline soils. Mycorrhizal inoculum (250 g), consisting of 200 spores/100g soil besides fungal matrices, was inoculated into three-and-half-month old seedlings of *Ailanthus excelsa, Pongamia glabra*, and *Cassia siamea*. No treatment was given to control. A standard amount of farmyard manure (10 tons/ha) was applied in all the cases. After four months, five plants from each plot were removed at random. Individual plants were taken out with care and their roots were washed gently under a fine jet of water, without damaging the root system. At the end of experiment, the soil was analyzed for VAM spore population. The data on the effect of VAM fungi on the mortality and height of plants, the weights of stem and roots, and the percentage of VAM colonization of roots showed that mycorrhizal inoculation resulted in 100% survival in all the three types of plants while in non-inoculated plants. 10, 20 and 12% mortality was observed in the case of *A. excelsa, P.glabra,* and *C.siamea*, respectively. Mycorrhiza-inoculated plants attained better shoot heights and shoot and root weights as compared to non-inoculated ones. The shoot dry weight of *A. excelsa, P. glabera* and *C. siamea* was enhanced by 28.4, 15.6, and 26.9%, respectively. A similar increase in root weight was observed in plants inoculated with VAM fungi. The interaction studies revealed that the infection in roots was maximum in *A. excelsa* followed by *P. glabera* and *C. siamea*.

Sidhu and Behl (1995) investigated the efficiency of VA-mycorrhizae on the growth of *Prosopis juliflora* in salt-affected soils. They transplanted 20 days old seedlings of *P. juliflora* in soils maintained at four pH levels (a) pH 7.8 (b) pH 8.5 (c) pH 9.5 and (d) pH 10.5 and used three VAM fungi as inoculants which were *Glomus fasciculatum, G. macrocarpum* and *G. mosseae*. The results revealed that the soil pH had a significant effect on growth of the seedlings. The shoot biomass decreased to less than one-third with increase in the pH of the soil medium from 7.8 to 10.5. Root biomass of non-mycorrhizal plants decreased by nearly 75% in plants growing in soil with pH of 10.5 as compared to the same with a pH of 7.8. However, VAM fungi increased plants in all the four treatments as compared to that of non-mycorrhizal plants. Total biomass of VAM inoculated plants increased by 67% at pH 7.8 and 151.4% at pH 10.5 as compared to non-mycorrhizal plants. All the three VAM fungi were found to be effective at various pH levels studied. Differences in the efficiency among the three VAM fungi were not statistically significant yet *G. fasciculatum* performed better than the other two. They concluded that since *G. fasciculatum* was originally isolated from high salt conditions the isolate may be a physiological type adapted to similar edaphology.

Sharma (1995) investigated the effect of phosphorus fertilization on VA-mycorrhiza in reclaimed sodic soils. On the basis of a number of experiments, he concluded that judicious use of P-fertilizers in reclaimed sodic soils is very important for proper colonization of plant roots with VAM, and thereby improving the nutrient uptake capacity of plants. It was emphasized to isolate the VAM spp. adapted to high soil-P-levels for their use in sodic soils with naturally high P-levels. Green manuring was found to be favourable to increase the population of VAM types in soil. He also suggested to determine the beneficial effects of VAM spp. adapted to high soil-P-level before their use, because of the fact that fertilization is known to affect the species composition of soils in such a way that the less beneficial VAM fungi become dominant.

Sharma *et al.* (1995a) performed a study to identify efficient species/ strains of mycorrhiza in enhancing survival, establishment and growth of *Acacia auricularis, Casuarina equisetifolia* and *Pterocarpus marsupium* seedlings growing in degraded acid soils of Kerala. In all, they used 36 VAM cultures, 18 monospore cultures collected from the rhizosphere soils of the above trees and 18 obtained from other sources such as ICRISAT, BAIF, TNAU, UAS, TERI and JNU. Out of these only 26 cultures were selected for glasshouse studies. High colonization of roots of *Acacia, Casuarina* and *Pterocarpus* by VAM fungi indicated that these species are mycorrhiza-dependent. However, for each tree species, the efficient VAM species were different. Increase in growth, shoots and root dry weight and P content of inoculated plants was recorded which showed that the VAM fungi may help the plants in their successful establishment in degraded soil.

Janardhanan and Khaliq (1995) studied the influence of VAM fungi on growth and productivity of chamomile in alkaline *Usar* soil collected from Anwari, Lucknow. They used pot cultures of *Glomus mosseae, G. fasciculatum* and *G. aggregatum* to inoculate the chamomile seedlings. Mycorrhizal plants showed higher biomass and a greater number of flowers than non-mycorrhizal plants. The increase in production due to *G. aggregatum, G. fasciculatum* and *G. mosseae* was 76, 230 and 155% respectively. The results indicated the potential utility of VAM fungi for improving crop productivity in alkaline *Usar* land.

Vijaya *et al.* (1996) studied the sodic soil tolerance of teak (*Tectona grandis*) seedlings produced after gibberellic acid seed treatment. They planted seedlings in soil of pH 9.6 in pots, some of the pots had 5 g of the inoculum of *Glomus macrocarpus* mixed in the soil and seedlings survival, growth and leaf chlorophyll were recorded after 3 months. The mycorrhiza treated seedlings showed more tolerance to the sodic soil, better survival and exhibited significantly greater growth and chlorophyll content than controls.

Sharma *et al.* (1996a) searched out the possibility of using mycorrhizal and nitrogen fixing symbionts in reforestation of degraded acid soils of Kerala, Peechi. Several monosporal VAM cultures raised from spores isolated from the rhizosphere soils under *Acacia auriculiformis, Casuarina equisetifolia*, and *Pterocarpus marsupium* and other VAM cultures obtained from various sources, were screened under glasshouse conditions to select the most efficient ones for enhancing the growth of each of the three test tree species. However, response varied with different isolates of VAM fungi. A significant increase in total P content was also observed in the inoculated plants. Based on overall performance and a cluster analysis of the data, the most efficient VAM selected for the field studies were *Glomus caledonium* and *G. mosseae* for *Acacia, G. fasciculatum* for *Casuarina* and *G. intraradices* and *G. mosseae* for *Pterocarpus*. Forty cultures of *Rhizobium* and *Frankia* isolated from the nodules of *A. auriculiformis, P. marsupium* and *C. equisetifolia* were screened for efficiency in increasing growth and nodulating the 3 test tree species in the glasshouse. The two best strains were selected for each species. In glasshouse trials the effect of fertilizers on seedling growth with and without symbionts, were also observed between various treatments (VAM, N_2 fixing symbionts and fertilizer) which showed significant difference in height and diameter of field planted 1 year old *A. auriculiformis*. All the treatments responded favourably to nitrogen and / or phosphorus fertilizer application. The percentage increase in growth with fertilizer application was more in symbiont-inoculated plants than non-inoculated controls. Maximum percentage increase in height was 57.4 (*G. caledonium* + fertilizer) and 56.2 (*G. caledonium* + *Rhizobium* + fertilizer) whereas for diameter growth the maximum increase was 84.5% (*G. caledonium* + *Rhizobium* + fertilizer) followed by 70% (*G. caledonium* + fertilizer). In control treatments the increase in height and diameter growth was 11.5 and 41.4 per cent respectively.

Giri and Mukerji (1999a) conducted a study to determine the effect of VAM fungus *Glomus macrocarpum* on the growth and dependency of *Sesbania grandiflora* in saline soil. Inoculation of *S. grandiflora* with *Glomus macrocarpum* significantly increased fresh and dry weights of root and shoot, percent VAM colonization in roots and production of VAM fungal spores in the rhizosphere soil. There was an increase in number of nodules. VA mycorrhizal dependency of *S. grandiflora* was maximum at 30 days after germination (DAG) which decreased with the increasing age of the plants. VAM fungus *G. macrocarpum* increased the capacity of *S. grandiflora* for uptake of nutrients, establishment, survival and protection against salinity stress.

Reddy and Bagyaraj (1988) attempted to grow *Leucaena* and *Trifolium* with VAM fungi on semi-arid soils and wastelands and reported that VAM inoculations subsided the effects of adverse conditions like low fertility and drought. They screened seven VAM fungi for symbiotic responses with two cultivars (K-8 and K-28) of *Leucaena*. They reported that preinoculation of *Leucaena*, with mycorrhizal fungus *Glomus mosseae* gave a vigorous growth in oxisols and vertisols. Inoculation also improved N fixation and P uptake.

Rano Jagpal and Mukerji (1988) reported that plants of *Leucaena leucocephala* inoculated with *Glomus macrocarpum* or *G. fasciculatum*, resulted in fast and increased growth compared with non-inoculated plants. They also reported a greater growth of *L. leucocephala* and *Sesbania sesban* inoculated with *G. macrocarpum* than non-inoculated ones (1991). Phosphatase activity (used to assess uptake of phosphorus) was greater in inoculated plants. Better performance of *Acacia melinifera, A. nilotica* and *Leucaena leucocephala* was obtained when they were inoculated with *Rhizobium* and selected VAM fungi (Rangarajan *et al.* 1988).

Chandra and Kehri (1994) studied the microbial spectrum with special reference to VAM fungi in brick fields during natural reclamation. An analysis of physico-chemical and microbiological characteristics of the soils of brick fields during their natural reclamation showed that colour, texture and pH of soils, status of Nitrate-N, available-P, organic matter and Potassium as well as their microbial population including that of some beneficial microbes *viz., Rhizobium, Azotobacter* and *Azotobacter* and VAM fungi undergo a change to acquire a status of fertile soil after a long period of many years. Response of selected tree species (*Acacia nilotica, A. catechu* and *Leucaena leucocephala*) with microbial inoculants in unreclaimed soil with specific amendments indicated that the method may be exploited for hastening the enrichment of soil with organic matter and microbes.

Mishra (1994) explored the possibility of using VAM fungi in reclamation of wastelands of Eastern ghat regions of Orissa. Among the tree and crop legumes tested, *Sesbania grandis, Acacia grandis, A. nilotica* and *Leucaena leucocephala*, local crop legume *Cajanus cajan* cv. *kandule* and *Vigna ungiculata* sub species *cylindrica* gave better performance when pre-inoculated with VAM fungi.

Mishra *et al.* (1995a) studied the mycorrhizal development and plant growth in amended flyash of thermal power plants. The population status of VAM fungi in fly ash rehabilitated by planting with 6 multipurpose tree species (*Acacia auriculiformis, A. nilotica, Cassia siamea, Eucalyptus* hybrid (*E. tereticornis*), *Peltophorum ferrugineum* and *Pongamia pinnata* established in pits filled with soil (transported in or *in situ* and various organic amendments) at Chachai, Shahdol District, Madhya Pradesh, was compared with that under natural vegetation and in barren fly ash. A pot culture experiment was also performed using fly ash/soil/compost (1:1:1) as growing medium and 20 nitrogen fixing and non-nitrogen fixing tree species. Varying number of mycorrhizal spores were found in the 3 site types (none in barren fly ash, 50/100g soil under natural annual vegetation and up to 57/100g soil in the rehabilitated areas treated with transported soil and 52/100g soil in those with *in situ* soil) and in the pot experiment. There was a significant and positive correlation between percentage mycorrhizal infection and biomass production by the trees in the pot experiment. Mycorrhizal colonization of the trees in the rehabilitated area was 30-80% in transported soil, and 10-16% in *in situ* soil. It has been suggested that selection and introduction of the mycorrhizal strains identified may assist future afforestation programmes, and help in the establishment of indigenous plants in rehabilitating areas.

Saralabai and Vivekanandan (1995) studied the ability of endomycorrhizae to withstand extreme stress. *Vigna radiata, V. catjung, V. mungo, Cajanus cajan* and *Glycine max* were raised in a soil with and without electrostatic precipitator (ESP) dust at different concentration. On the basis of experimental results, they reported that application of ESP dust up to 200g/kg soil did not affect VAM spore population, spore germination, root colonization and mycorrhizal effectiveness. The results also revealed that VAM fungi have the potential to survive extremities of stress like high pH and Ec values of the soil.

Williams *et al.* (1995) studied the population status of VAM fungi in rehabilitated ash disposal area. The abundance of VAM (vesicular arbuscular mycorrhizas) in a reclaimed fly ash treated soil was compared to that under natural vegetation and barren fly ash in Madhya Pradesh. Different spore number were identified at these sites. The introduction of such strains may assist revegetation by indigenous plants.

Sharma *et al.* (1996b) evaluated the responses of *Acacia nilotica* var. *cupriciformis* to phosphorus application and inoculation with the indigenous consortium of VAM fungi in a nursery experiment using soil from a marginal wasteland in Haryana. A positive growth response to mycorrhizal inoculation was observed at an olsen-P level of 20 ppm in the presence of the natural population of VAM fungi. There was stimulation by either inoculation or additional P at the highest soil P but P content of both shoot and root were positively correlated. Inoculation with the indigenous VAM consortium significantly increased the uptake of P at all levels of applied P. *Acacia* was moderately dependent upon the VAM symbiosis and exhibited a maximal mycorrhizal dependency (MD) of 18.25% at 20 ppm Olsen-P-level under soil conditions studied. A sharp and considerably significant reduction in MD and dry matter yield observed at 40 ppm P suggested that the external P requirement for maximal production of biomass was met at 20 ppm olsen-P.

Sidhu and Behl (1997) studied the relative tolerance of *Glomus mosseae, G. fasciculatum* and *G. macrocarpum* to graded pH levels (7.8 to 10.5) and their influence on productivity and P uptake in *Prosopis juliflora* seedlings. Increase in pH adversely affected seedling growth, their biomass and P concentration. Chlamydospore formation in the rhizosphere soil by all the three *Glomus* spp. decreased with increase in pH. Application of VAM fungi resulted in significant increase in root-shoot length, collar diameter and seedling biomass even at high pH levels. Mycorrhizal plants grown at pH 10.5 reached a level of biomass equivalent to that of non-inoculated ones grown at pH of 7.8. Phosphorus concentration in both root and shoot tissues increased significantly in VAM inoculated seedlings. Root P concentration increased by 68% in VAM inoculated seedlings at pH 7.8, however, the increase was only 40% at 10.5 pH. Increase in growth of VAM inoculated seedlings was attributed to enhance P uptake. Variations between the 3 *Glomus* spp. were insignificant, yet *G. fasciculatum* originally isolated from high pH soil site (pH 9.2) had a relatively high tolerance to pH 8.5 to 10.5 compared to other species, as exhibited by higher degree of chlamydospore formation and root colonization. Their study suggested that VAM isolates can tolerate high pH and assist in alleviating stress of high pH by promoting P uptake and seedling growth.

Dixon *et al.* (1997) studied the VA mycorrhizal symbiosis in relation to forestation in arid land. VA mycorrhizal fungal inoculation of *Prosopis juliflora* seedlings using *Glomus macrocarpum* in the nursery of the University of Delhi, significantly increased the juvenile growth response over 3 months. Three-month old-nursery-raised-mycorrhizal seedlings gave enhanced establishment (survival and growth) 3 months after outplanting (i.e. at 6 months old) in 30x30x30 cm pits in nutrient deficient alkaline soil in Asala wildlife sanctuary (about 45 km from Delhi), a semiarid area with rhizosphere soil temperature of 52°C. The VAM symbiont also increased drought tolerance of *Prosopis* plants in terms of maintaining and enhancing growth under water stress conditions. The results indicated the potential of VA mycorrhizae to partly reduce/replace the fertilizer requirements of trees in degraded, semi-arid sites, where they can be described as acting as biofertilizers.

Kothari *et al.* (1997) studied the mycorrhizal biodiversity of petroeffluent-irrigated fields. Petroeffluent from the Indian Petrochemicals corporation (IPCL) has been tested for irrigation at the Ecoform at IPCL, Baroda, Gujrat, for recycling millions of tonnes of water. The rhizosphere of 6 crops raised in these fields were evaluated for mycorrhizal biodiversity combined with the mycorrhizal status of these crops. All the crops investigated showed efficient mycorrhization with VA mycorrhiza. The rhizosphere was rich in VAM species. *Acaulospora appendicula, A. birelicola, A. foveata, A. sporocarpia, Glomus aggregatum, G. deserticola, G. heterosporum, G. microcarpum* and 2 unidentified species were recorded.

Gaur *et al.* (1998b) studied the variation in the spore density and percentage of root length of tree species colonized by vesicular arbuscular mycorrhizal fungi at a rehabilitated waterlogged site in Haryana, where waterlogging was a usual phenomenon that extended for more than nine months annually. Vesicular arbuscular mycorrhizal (AM) distribution in all the plots represented by various tree species, *viz.*, *Terminalia arjuna, Syzygium cuminii, Populus euphratica* and naturally grown *Typha elephantina* was found to be significantly different. Distribution profile showed the dominance of the two genera, *viz.*, *Glomus* and *Gigaspora, Glomus* was dominant in the *T. arjuna* plot while *Gigaspora* was abundant in the *P. euphratica* plot. *Glomus* showed positive correlation to available soil P while *Gigaspora* showed positive correlation to organic matter content. Both the genera also predominated at the naturally grown *T. elephantina* plot.

Mishra *et al.* (1999) studied the effect of application of VAM fungi in enhancing water stress tolerance capability of pasture legume *Stylosanthes hamata* grown in degraded forest wasteland soil Chandaka, Orissa in presence of Jalshakti (a chemical polymer known for increasing water holding capacity formulated by NCL) under different water regimes (one day and three days gap watering). Application of VAM fungi in presence of *Rhizobium* was seen to be more effective in enhancing water stress tolerance capability in *Stylosanthes*, indicated through enhancement of plant growth, biomass and nodulation capacity along with rhizospheric microbial activity but to a limited extent. Effectiveness of VAM fungi inoculation was seen to vary with water regime and Jalsakti application. On the contrary, there was indication of retardation of growth and nodulation as well as rhizospheric microbial activity in *Stylosanthes*, with Jalsakti application under adequate watering condition (daily watering) but the effectiveness of VAM fungi inoculation stimulating growth and nodulation capacity of *Stylosanthes* was seen to enhance significantly under limited water supply condition (3 days gap) in presence of Jalsakti. There was about 66% of water economy along with induction of drought tolerance capacity due to Jalsakti application and VAM fungi inoculation.

Bhattacharya *et al.* (1999) studied the mycorrhizal dependency, phosphorus utilization efficiency and relevance of mycorrhiza for bamboo cultivation in laterite wasteland. Most commercially important bamboo species have a coarse surface spreading root system. Natural root colonization by arbuscular mycorrhiza and inoculum availability in the rhizosphere of most species of bamboo in wasteland soils are reported to be low. *Dendrocalamus strictus*, an important bamboo species in a phosphorus deficient laterite soil showed above 40% vesicular arbuscular mycorrhizal dependency for shoot dry matter production and phosphorus acquisition at 360 days. Phosphorus Utilization Efficiency (PUE) of mycorrhiza inoculated seedling plants was lower than that of uninoculated plants due to a higher net phosphorus acquisition rate than that could be currently utilized by the inoculated plants. Both inoculation efficiency and PUE declined at high phosphorus level but significant response efficiency was available even upto about 25 ppm soil P. Mycorrhiza inoculation at 4 ppm level of soil P showed response equivalence to 130 ppm soil supply. Based on these evidence, the root morphological traits and the reported mycorrhizal relation, bamboo species, such as *D. strictus* may be considered as suitable for application of VAM technology in wasteland plantations.

Bhatia *et al.* (1999) assessed the growth of *Prosopis juliflora* and its contribution to soil enrichment, following inoculation with three VA mycorrhizal isolates, namely *Glomus caledonium, Gigaspora calospora*, and an indigenous strain; and two *Rhizobium* isolates, namely P-5 and Tal-600. The trees were 6-year old, growing on a semi-arid wasteland. Results showed significant improvement in biomass of closely spaced *P. juliflora* inoculated with *G. caledonium alone*. Growth of *P. juliflora* on a relatively nutrient deficient wasteland significantly restored the soil productivity by ameliorating and enriching the soil underneath. A significant reduction in soil reaction (pH) and considerable improvement in soil organic carbon build up and Olsen's-P at both the depths was observed, in all the treatments, over zero-time. Moreover, although total nitrogen content increased over control, but the increment was not significant when comparisons were made between respective treatments at zero-time and after six years of growth.

Sharma *et al.* (1995b) studied the response of *Acacia nilotica* to indigenous VAMF inoculation and single superphosphate fertilization in degraded wasteland soils. They found the overall growth of mycorrhizal plants higher over non-mycorrhizal plants at all levels of added P (10, 20, 30 and 40 ppm). The plant shoot and root dry weights increased over their respective controls at all levels of P application. However, the values were higher in mycorrhizal *A. nilotica* compared to non-mycorrhizal plants. Maximum plant growth improvement was recorded at 20 ppm of added P. This level of phosphorus indicates the net saving of 50% in P application.

Mathur and Vyas (1995b) evaluated the efficacy of indigenous VAM fungi in biomass production, nutrient uptake and nodulation in *Acacia nilotica* in the Indian Thar Desert. Five different soils (collected from Bikaner, Jodhpur, Nagpur, Osian and Pali) were used in the study and two sets of experiments, one with sterilized soils and other with non-sterilized soils. Though the soils used in the experiment had different mycorrhizal populations, considerable VAM infection occurred in the plants grown in unsterilized soils. (59 to 68% in different soil types). Nodule number, biomass and nutrient uptake especially of N and P was more in mycorrhizal plants.

Mathur and Vyas (1996a) studied the incidence of root infection by indigenous VAM fungi and its effects on biomass production, nodulation and nutrition of *Prosopis cineraria* in pots in 5 samples of sandy soils from different locations in the Indian Thar desert. Soil sterilization completely inhibited root colonization by indigenous VAM fungi. However, considerable root colonization occurred in roots of the plants in unsterilized soils. Root colonization by indigenous VAM fungi resulted in improved nodulation, biomass production, and uptake of N,P,K, Cu, Mn, Fe and Zn in all the 5 soil samples.

Meshram *et al.* (1997) studied the effects of a selected isolate of *Azotobacter chroococcum* (SM_3) and of the VAM fungus *Glomus fasciculatum* on growth and biomass production of *Eucalyptus camaldulensis* on waste/ barren land at Nagpur, Maharashtra. The field treatment included factorial combinations of biofertilizer (*Azotobacter chroococcum* SM_3 + *Glomus fasciculatum*) with farmyard manure and chemical fertilizers (N as urea and P) and a control treatment. Of the 8 treatment combinations, *A. chroococcum* + mycorrhizas amended with FYM and N + P gave maximum height and biomass production and plant nitrogen content after two years.

Prasad *et al.* (1997) studied the influence of VAM, macro - and micro-nutrients on vegetative propagation of *Dendrocalamus strictus*. Seeds of *Dendrocalamus* were collected from the Forest Research Institute, Dehra Dun, Uttar Pradesh and germinated in mother beds in April - May, 1995. Young seedlings with 2-3 culms pots filled with 8 kg degraded laterite soil, given various treatments, NPK micronutrients (Zn+Mn+B+Cu+Mo), inoculation with VAM from a mass culture, inoculation with *Azotobacter* or all possible combinations of these 4 treatments. After 150 days, maximum culm numbers and height were recorded in the NPK + micronutrients + VAM treatment followed by NPK alone, NPK + VAM + *Azotobacter* etc.

Bhatia *et al.* (1998) assessed the growth of *Prosopis juliflora* and its contribution to soil enrichment following inoculation with three vesicular-arbuscular mycorrhizal isolates, *Glomus caledonius, Gigaspora calospora*, and an indigenous strain, and two *Rhizobium* isolates, P-5 and Tal-600. The trees were 6 years old and grew on a semi-arid wasteland. There was a significant increase in the biomass of closely spaced *P. juliflora* inoculated with *G. caledonius* alone. *P. juliflora*, growing on a relatively nutrient-deficient wasteland, significantly restored the soil productivity by ameliorating and enriching the soil. A significant reduction in the soil reaction (pH) and a considerable improvement in soil organic carbon build-up and phosphorus, at both depths, were observed in all the treatments by the end of the experiment. Moreover, although the total nitrogen content increased in comparison to controls, this increase was not statistically significant when comparisons were made between respective treatments at the beginning of the experiment and after 6 years' growth.

Tarafdar and Kumar (1996) studied the role of VAM fungi on crops, trees and grasses grown in an arid environment. They inoculated *Prosopis juliflora, Cenchrus ciliaris* and *Vigna aconitifolia* with VAM fungi on a sandy soil with low fertility in Rajasthan. The percentage root infection in inoculated plants increased by a factor of 2.9 after 8 weeks. Nitrogen, P,K,Zn and Cu contents increased significantly, and Mn decreased significantly at crop maturity or after one year of grass or tree growth. The effect was most evident in *P. juliflora* inoculated with *Glomus fasciculatum.*

Giri *et al.* (1999) studied the growth response and dependency of *Sesbania aegyptiaca* on vesicular arbuscular mycorrhiza in salt stressed soil. Seedlings of *Sesbania aegyptiaca* [*S. sesban*] with and without inoculation by the vesicular arbuscular mycorrhizal (VAM) fungus *Glomus macrocarpum* were grown in salt-stressed soil (pH 8.9). Seedling growth was assessed 30, 45 and 60 days after germination. At each assessment, length, and fresh and dry weight of both roots and shoots were greater in inoculated than uninoculated plants. There was an increase in the number of nodules formed on roots of inoculated plants. The percentage mycorrhizal dependency of *S. sesban* increased with seedling age.

Rodrigues (1999) made an assessment of the VA (Vesicular arbuscular) mycorrhizal association in 21 naturally occurring seedlings of shrub and tree species belonging to 19 genera and distributed among 16 families collected from a 12-year-old iron ore mine reject dump in Goa. Spore densities of the iron ore mine rejects and adjacent undisturbed site were also studied. The seedlings of the species studied showed mycorrhizal infection, which varied from 25 to 98% in the different species. The spore densities of the iron ore mine rejects were very low compared to those of adjacent undisturbed areas.

Rodrigues (2000) isolated and identified the arbuscular mycorrhizal (AM) species in the rhizosphere of 15 herbaceous and woody plant species growing on fairly established iron ore mine waste sites (pH 6.02 with an EC value of 0.051, low in all essential micro- and macro-nutrients) in Goa. A total of 27 AM species were identified, belonging to 4 genera, *viz., Glomus, Gigaspora, Acaulospora and Scutellospora* with 16, 3, 7 and 1 species, respectively. Maximum AM species diversity was recorded in Blumea mollis which showed the presence of seven AM fungal species (*Glomus mosseae, G. fasciculatum, G. australe, G. hoi, Gigaspora candida, A. spinosa* and *A. nicolsonii*), while minimum AM species diversity was recorded in *Lindernia crustacea, Eugenia corymbosa, Smithia sensitiva* and *Tephrosia purpurea* each of which had 3 AM species, Maximum mycorrhizal infection was observed in *Tephrosia purpurea* (99%), while minimum mycorrhizal infection was recoded in *Spermacoce hispida* (37%).

Mathur and Vyas (2000a) studied the mycorrhizal dependency of *Tamarix aphylla* in saline areas of Thar desert. Four vesicular arbuscular mycorrhizal species (VAM) namely

Scutellospora calospora, Gigaspora margarita, Acaulospora morrowiae and *Glomus fasciculatum,* were collected from rhizosphere soils of *T. aphylla* from salty arid and semi-arid regions of the Indian Thar Desert and cultured on *Cenchrus ciliaris*. Pot cultures of these four species were used to inoculate. *T. aphylla* to assess the effect of these VAM species on biomass production and nutrient uptake of this salry tree. The results showed that this tree shows tremendous dependency on these VAM endophytes for biomass production and nutrient uptake. However, among the four VAM species, *Gigaspora margarita* was the most efficient VAM fungi for *T. aphylla,* as it significantly increased the plant biomass and nutrient uptake by effectively colonizing the roots, *Glomus fasciculatum* was the least efficient in increasing plant biomass production and nutrient uptake.

Rao and Tak (2001) studied the effect of soil inoculation with an arbuscular mycorrhizal (AM) fungus, *Glomus fasciculatum*, isolated from a gypsum mine spoil, on dry mater production, nutrient uptake and rhizosphere microbial activity in five tree species (*Acacia ampliceps, A. eriopoda, Albizia lebbeck, Azadirachta indica* and *Calophospermum mopana*) growing in gypsum mine spoil. Dry matter production was significantly enhanced upon inoculation to varying degrees depending on the tree species. The level of root infection increased by 14-36% and spore densities in the rhizosphere increased by 49-81. AM fungal inoculation resulted in enhanced activities of various soil enzymes (dehydrogenase [oxidoreductase], phosphatases [phosphoric monoester hydrolases] and nitrogenase) in the rhizosphere. Total uptake of many nutrients, but not of potassium and sodium, was significantly enhanced upon AM fungal inoculation. It is concluded that arbuscular mycorrhizal fungi are important symbionts for use in the revegetation of gypsum mine spoils.

Rao *et al.* (2002) investigated arbuscular mycorrhizal dependency of two forest tree species, *viz. Enterolobium saman [Samanea saman]* and *Acacia melanoxylon* in eleven coalmine-disturbed soils. The study revealed that the mycorrhizal dependency could be correlated with the degree of disturbance of the soil. Mycorrhizal dependency of *E. saman* in different soils ranged between 23 and 63 percent with minimum dependency in 3 incline soils and maximum in 11A incline soils. Mycorrhizal dependency of *A. melanoxylon* ranged between 30 and 61 with minimum in 3 incline soils and maximum in 7B incline soils. Nodulation of the two tree species seedlings also varied with the type of disturbed soil.

CONCLUDING REMARKS

To develop vegetation cover on the mine spoils is a difficult venture, however a number of workers have explored the possibility of utilizing VAM technology in combination with other scientific inputs in the development of vegetation cover over the mine spoils of iron, copper, aluminium, chromium and coal. In general, some economically important shrubs and trees like *Leucaena, Eucalyptus, Acacia, Cassia, Sesbania, Prosopis, Casuarina, Tectona, Petroplants* etc. have been included in their study. In order to overcome nutritional deficiency of the spoils some workers have used fertilizers, others have used biofertilizers while few others have used different types of organic matter. Although most of the work has been conducted under pot conditions, however, the results clearly indicate that VAM technology can be successfully utilized to assist the programmes related to plantation of mine spoils.

But a thorough understanding of the constraints and extensive location trials will be needed before the technology is put in practice.

Comparatively less effort is needed to develop vegetation on some wastelands like sandy dunes, brickfields, waterlogged areas, fly ash dumps and certain problem soils such as laterite soil, alkali soil, acid soil, saline soil and sodic soil because the constraints associated with them are not so severe. VAM technology in combination with other scientific inputs has been evaluated for its potential to assist the development of plantation over these wastelands and problem soils. A variety of shrubs and trees of economic value as well as diverse crops including fruit crops have been included in the evaluation trials. In most of the trials, VAM technology in combination with other scientific inputs has been proved to be a good package to remove the constraints associated with wastelands and problem soils. However, practical application of these packages to develop vegetation cover on wastelands and problem soils will depend on their success in multilocational field trials.

9

VA MYCORRHIZA INOCULUM PRODUCTION

Since a method to cultivate VAM fungi in pure culture is not available as yet, clean and high quality VAM inoculum production is not possible. However, VAM technology is gradually getting a place in agriculture as well as reclamation and afforestation programmes. The inoculum currently being used includes infested fields soils, sievings from such soils, soil and roots of greenhouse plants, or pot culture inoculum derived from surface disinfected spores of a single VAM fungus on a host plant grown in a sterilized growth medium. Attempts are being made to manipulate VAM fungi, rooting media, hosts and fertilizers in order to secure superior quality of inoculum by these methods on a large scale. In addition, other techniques like tissue culture, hydroponic system etc. are also being attempted to cultivate VAM fungi in pure culture. The Indian mycorrhizasts have also attempted to devise and develop improved methods for production and storage of commercial VAM inoculum. The following description gives an account of Indian work related to production and storage of consistently high quality disease free VAM inoculum.

Potty (1988a) studied the multiplication and inoculation of *Glomus microcarpum* var. *microcarpum* in vegetatively propagated tuber crops. Mycorrhizal association in *Coleus parviflorus, Impomeae batatas* and *Manihot esculenta* was investigated. Artificial inoculation of spores and root masses through the inert material (lignite) was effective. Spores in lignite slurry were viable for more than 6 months. The roots of coleus, sweet potato and cassava registered VAM infection within 10-12, 15-29 and 18-25 days of the root formation, respectively. Colonization of roots by *Glomus* sp. in coleus, sweet potato and cassava was to the tune of 80-90, 80-91 and 80-91 and 80-85%. Higher soil pH and P retarded the growth and establishment of VAM fungi in respective hosts.

Sreenivasa and Bagyaraj (1988a) compared perlite, vermiculite, soilrite and soil and their combinations in various proportions to select a suitable substrate for mass multiplication of *Glomus fasciculatum* inoculum. Perlite-soilrite mixture (1:1 by volume) was found to be the best substrate on the basis of root colonization and spore production and of the infective propagules.

Sreenivasa and Bagyaraj (1988b) studied the effect of culture, age and pruning on mass production of the VA mycorrhizal fungi. Rhodes grass (*Chloris gayana*) seeds were sown in *G. fasciculatum* pot cultures. Perlite-soilrite mixture (1:1 by volume) was used in 16 cm pots holding 700g substrate. The maximum number of infective propagules was obtained 75 days after sowing. Pruning reduced the number of propagules for 4 weeks, after which the number increased.

Sreenivasa and Bagyaraj (1988c) studied the suitable level of zinc, copper and manganese for mass production of VAM fungus, *Glomus fasciculatum*. According to them, the percentage root colonization of *Chloris gayana* by *G. fasciculatum* in a perlite soilrite mixture, extramatrical chlamydospore number and number of infective propagules/unit weight of inoculum were maximum at half the recommended level of Ruakura plant nutrient solution *i.e.* 0.18, 0.04 and 0.73 nM of Zn, Cu and Mn respectively.

Sreenivasa and Bagyaraj (1988d) studied the optimum frequency of addition of nutrient solution for mass production of *Glomus fasciculatum* inoculum. Rhode's grass seeds were sown in 3 kg pots inoculated with 30 g/pot (210 propagules/g) *G. fasciculatum*. Pots were maintained at 28-32°C in greenhouse and the plant population was kept uniform in all 4 replications in each treatment. Ruakura plant nutrient solution was added (50 ml/pot with 2.5 litre substrate mix) to pots once in 2,4,6,8 or 10 days (50 ml water added to control pots.) After 60 days plants were harvested and the shoot and root dry weight recorded. Percentage mycorrhizal root colonization, extramatrical chlamydospore number and the number of infective propagules/ unit weight were highest when the solution was added after 8 days. Shoot and root dry weight were also high in the 8 day treatment. The control gave similar results to addition of solution at 2-days intervals.

Secilia and Bagyaraj (1988) studied the fungi associated with VA mycorrhizal pot cultures of *Glomus fasciculatum, Gigaspora margarita, Acaulospora laevis* and *Sclerocystis dussii*. They made quantitative as well as qualitative analysis of the fungi. The fungal population in the root zone of the different mycorrhizal pot cultures was not altered except with *S. dussiii*, where a significant decrease was recorded. *Cladosporium herbarum* was the predominant species associated with all the mycorrhizal pot cultures.

Sreenivasa and Bagyaraj (1989) studied the use of pesticides for mass production of VA mycorrhizal inoculum. Nine fungicides, 3 nematicides and 5 insecticides and acaricides were incorporated into the pot cultures of *Glomus fasciculatum* in order to study their effect on the mycorrhizal association and the contaminant organisms. All the nematicides were found to be deleterious to VAM at the recommended level. Captan and carbofuran, at half the recommended level (125 mg and 144.5 mg/ 2.5 litre of the substrate mix) significantly increased the percentage root colonization, extramatrical chlamydospore number and inoculum potential. Captan and carbofuran at this level also significantly suppressed the contaminant fungi and nematodes in the pot cultures of *G. fasciculatum*. The insecticides/ acaricides for mothion and melathion at half the recommended level, had no deleterious effect on *G. fasciculatum* but suppressed mites and collumbola in pot cultures.

Sreenivasa and Bagyaraj (1990) studied the suitable source and level of nitrogen for mass production of the VA mycorrhizal fungus *Glomus fasciculatum*. Mass production of these spores was best achieved using *Chloris gayana* in a perlite:soilrite (1:1 v/v) mixture using ammonium nitrate fertilizer at 80 ppm N.

Janardhanan *et al.* (1990b) tried for an axenic culture of a VA mycorrhizal fungus. *G. aggregatum* associated with *Cymbopogon martinii* var. *motia* was cultured and maintained on a synthetic medium. Mycorrhizal association of the isolate was successfully established in callus culture, axenic plants and potted plants. The authors claimed it to be the first report of culturing this obligate symbiont.

Adholeya and Khanna (1990) examined the requirements for the sporulation of *Glomus mosseae* under *in vitro* condition. They observed the germination with different combinations of agar, sucrose and soil extract, and reported very marginal differences. The addition of root extracts either from maize or mungbean, soil extract or sucrose (2% w/v) respectively did not enhance hyphal proliferation and spore formation as compared to agar (0.8% Himedia w/v) alone. They also studied the effect of light, moisture and aeration on spore germination. The spores incubated in dark, germinated and showed hyphal elongation while in the presence of light they failed to do so. Likewise, Petridishes wrapped with nescofilm just after inoculation, resulted in extended hyphal proliferation (up to three weeks) whereas the proliferation of hyphae in Pteriplates without nescofilm stopped after two weeks. They observed during experiment that freshly prepared Petriplates showed improved spore germination and proliferation of hyphae as compared to plates which were more than three days old at room temperature. Incubated spores on the simple water agar medium formed germ tubes after 48 hours. After 6 days, on proliferating hyphae, small globular structures were visualized which became double walled dark globular structures. These secondary spores did not germinate even after 6 weeks of incubation, but when tested for their viability and infectivity, positive infectivity was observed.

Singh (1990) conducted preliminary trials for inoculum production with *Glomus* spp. on various plants *viz.*, maize, sorghum, bajra, bahia grass, rhodes grass, angan grass and varieties of guinea grass, raised in different substrates such as sand, soil, soilrite, perlite, peat and vermiculite in different proportions e.g. soil:sand (1:3 by volume), soilrite:sand (1:3 by volume), vermiculite:peat (4:3:1 by weight), perlite:soilrite (1:1 by volume) and devised methods of processing of the resulting VAM inoculum for distribution. He also devised method of lypholization of infected root segments.

Tilak *et al.* (1990) reported a stimulation of spore germination of *Glomus fasciculatum* by cell free extracts of non- symbiotic nitrogen fixers (*Azotobacter chroococcum, Azospirillum brasilense* and *A. lipoferum*). Ca free extracts of non-nitrogen fixing bacteria did not stimulate significant increase in spore germination, except the extracts of *Pseudomonas putida* and *P. fluorescens*.

Singh (1992) screened eight forage grasses *viz., Cenchrus ciliaris, Cenchrus* sp., *Pennisetum parviflorum, P. pedicellatum, Stylosanthes guianensis, S. scarbaeoides, S. humilis* and *S. hamata* for potential use in mass culturing of VA mycorrhizal inoculum. He also studied the effect of *Azospirillum brasilense* on the formation of vesicles and arbuscules as well as spore production. The root dry weight and root volume increased in all species following combined inoculation with *A. brasilense* and *Glomus macrocarpum*, with the greatest increases in *Pennisetum* spp. No significant differences in VAM root infection were observed between species with or without *A. brasilense*, although the maximum infection occurred in *S. hamata* and *S. guianensis*. Spore production was not related to infection intensity. Vesicle formation was greatest in *P. pedicellatum, P. parviflorum, S. guianesis, S. humilis* and *S. hamata*, particularly when *A. brasilense* was present and arbuscule formation and sporulation were also enhanced in some species by the presence of *A. brasilense*.

Mallesha *et al.* (1992) showed that dual inoculation with *Rhizobium* and VAM fungi resulted in superior growth in *Leucaena leucocephala*. They worked further for the simple method to provide the inoculum. A mixture of perlite and soilrite (a non-soil medium) has been shown to be suitable for VAM multiplication. Their results showed survival of *Rhizobium* and VAM fungi for more than 11 months when inoculated either separately or together into a mixture of perlites/soilrite. Population levels of both rhizobia and VAM were successfully maintained for 6 months in the mixed preparation, but decreased rapidly in the single inoculated treatment.

Singh and Jha (1994) studied the effect of various bacteriological media and fertilizer solutions on the mass inoculum production of VA mycorrhizae. According to them, the incorporation of bacteriological media (Okon's, Jensen's and Pikovaskaya's) and NPI fertilizer solution enhanced the root volume of *Pennisetum* spp. and the formation of spores in VAM inoculum (soil and sand mixture). The NPI fertilizer solution was more effective in increasing root volume and spore number than the bacteriological media and as effective as *Azospirillum* inoculation. Differences in spore number between 2 *Glomus* species (*G. macrocarpum* and *G. fasciculatum*) were also recorded under the influence of NPI fertilizer solution.

Raja and Mahadevan (1995) experimented with several plant systems for mass multiplication of vesicular arbuscular mycorrhizae under controlled conditions. According to them *Kalanchoe pinnata* is an ideal plant for mass multiplication of *Glomus aggregatum, G. mosseae* and *Gigaspora margarita* employing different supporting bases. The percent infection ranged from 15 to 75 in vermiculite, 15 to 70 in soil and 10 to 60 in vermiculite and soil base. Total number of spores were 35/10 g vermiculite, 30/10 g soil and 21/10 g vermiculite and soil base after 90th day of inoculation.

Rao *et al.* (1995c) carried out a study to standardize techniques of VAM inoculum production in large containers using non-organic substrate and in modified hydroponic system, and to study the effectiveness of the inocula on selected crop plants. They established a facility for production of 2000 kg of inoculum per annum. The inoculum was tested on selected multipurpose trees and crop plants for assessing mycorrhizal potency e.g. *G. fasciculatum,* single and combined inoculations with *Azotobacter chroococcum* on *Allium cepa* crop showed significant effect in germination of seeds and percentage of VAM root colonization as compared to non-inoculated control.

Singh (1995) studied the impact of N_2 fixing and phosphate solubilizing bacteria on root colonization and spore production of VAM fungi. He investigated the effect of various fertilizer solutions on the production of VAM inoculum. VAM fungus *Glomus macrocarpum* was multiplied under potted conditions on *Trifolium alexandrium* and 400 chlamydospores per 100 g of carrier based inoculum was obtained. Using *Pennisetum padicillatum* as host, *Azospirillum brasilense, A. lipferum* and *Azotobacter chroococcum* as N_2 fixers and *Bacillus polymyxa, P. striata* as P solubilizers, a superior inoculum of *G. macrocarpum* was produced.

Varma and Adholeya (1995) made efforts to identify a suitable host for *in vitro* inoculation by VAM fungus *Gigaspora margarita*. For inducing transformation

Agrobacterium rhizogenes strain was used. Root cultures of different mother plant species namely carrot (*Daucus carota*), tomato (*Lycopersicon esculentum*) cucumber (*Cucumis sativus*) and chilli (*Capsicum annum*) were grown in modified Whites medium and minimal medium. During the process of transformation Ri-T DNA gets incorporated at different sites, developing number of clones of individual species on the basis of their growth pattern. In the compatibility study, all the clones tested were found to be mycorrhizal, and there was not much of variation in the time taken for incidence. Carrot was found to be the best (GP-IA strain) as it was one of the fast growing clones and was compatible for the VAM fungus.

Varma and Adholeya (1996) studied the cost-economics of existing methodologies for inoculum production of vesicular-arbuscular mycorrhizal fungi. Nutrient film technique, aeroponics, and root organ culture are the three different systems which are proposed to be used for mass inoculum production of vesicular-arbuscular mycorrhizal fungi. A comparative cost analysis of these three systems considering the state-of-art can help in identifying the varied components and their resultant costs if applied on a commercial scale. The inoculum produced in these systems differs in terms of quality and quantity and consequently, reflects on the net cost incurred. To apply these systems on a pilot scale, a comparative cost study would, therefore, be of immense help.

Mehrotra (1996a) demonstrated that *Paulownia* is an ideal species for multiplication of VAM fungi, since its seedlings in polypots develop massive and dense root systems with many branching lateral roots. The plants are propagated from seed or root and shoot cuttings. Composition of the potting mixture is important (soil/ sand at 1:1 + 2% FYM treated with methyl bromide to eliminate contaminant VAMF and competitive microflora). A nominal dose of phosphate fertilizer stimulates growth and it is applied at the base, at the time of pricking of seedlings when making a hole in the soil deep enough to accommodate the roots and the requisite VAMF inoculation. Application of Furadan 3G or deoiled neem is recommended to control nematode infestation. According to him mass multiplication of *Glomus mosseae* (and allied *Glomus* spp.) using *Paulownia* as host is likely to strengthen seedling production in forest nurseries.

Gadkar *et al.* (1997) made efforts to grow VAM fungi in pure culture. They used randomly amplified polymorphism, if any, in the spores of *Gigaspora margarita* produced both *in vitro* and *in situ* (soil). The fingerprint patterns obtained from *in vitro* and *in situ* spores were similar. Extramatrical structures, such as auxilary cells, were also examined by DNA fingerprinting. Their amplification pattern did not vary from the mother or daughter spores. A few interesting observations were made. For instance, the mother spore, which seemed hollow and inactive after germination, nevertheless contained nuclei after 4 months under *in vitro* conditions and generated a fingerprint pattern. The fingerprint pattern for *Gigaspora margarita* was different from that of *Gigaspora gigantea*, indicating that the minisatellite sequence could be exploited for identifying VAM fungi. ROC appears to be a truly representative system, in the sense that it mimics the essential features of the complex rhizosphere, allowing the fungi to complete their life cycle without any induced genetic changes per se.

Adholeya *et al.* (1997) carried out an *in vitro* study with RiT-DNA transformed roots of carrot (*Daucus carota*) to evaluate the role of macro-elements contributed as impurities in the gelling agent (phytagel) over and above those present in the minimal (M) medium. Production of root biomass was taken as a measure to quantify the influence of macro-elements added to the minimal medium. The levels of phosphorus when adjusted to 1.19 mg/1 and 1.09 mg/1, lead to dry root biomass production at par with the control. Attempts made to lower the amount of impurities in phytagel by de-ionization using different alkalies, proved NaOH to give the best results in terms of relatively high amount of root biomass. In an *in vitro* dual culture system with carrot as host and *Gigaspora margarita* as the vesicular-arbuscular mycorrhizal fungus, phytagel impurities helped to produce maximum number of infection units and auxillary cells when phytagel was added to the minimal medium.

Gaur *et al.* (1998a) compared different types of nursery inocula formulations, namely mixed indigenous cultures and *Glomus intraradices* Schenck and Smith, with commercially available inoculants of VAM fungi in a pot experiment using two horticultural crops, *Capsicum* and *Polianthes*. Soil-based inocula and soil beads produced the highest response in both crops. *Glomus intraradices* resulted in the highest yield in both *Polianthes* (45% increase in spike length) and *Capsicum* (112% increase in fruit yield). Among the commercial inocula tested, only Mycorise enhanced spike length (33%) and fruit yield (11%) in the two hosts. Overall VAM colonization was higher in *Polianthes* than in *Capsicum*. Sheared root inocula of *G. intraradices* resulted in high colonization (upto 68%) but the yield enhancement was lower than with soil-based formulations. The mixed indigenous culture produced the highest number of spores and propagules and commercial inocula the lowest.

Kumar and Bagyaraj (1999) used two fungicides (rilon and captan), two insecticides (tafaban and sumicidin), a herbicide (atrataf) and an acaricide (tafethion) to study their effects on the vesicular arbuscular mycorrhizal (VAM) fungus, *Glomus mosseae*, at two levels (*viz.*, full and half the recommended level). Contaminant fauna was considerably reduced by the rilon and sumicidin treatments at half the recommended level. Rilon and sumicidin, both applied at half the recommended level resulted in a higher mycorrhizal infective propagules compared to other treatments. It was therefore concluded that rilon and sumicidin can be used for suppressing the contaminants during mass production of the VAM fungus, *Glomus mosseae*.

Rao and Tarafdar (1999) conducted experiment employing soil solarization for mass scale production of vesicular arbuscular mycorrhizal fungal inoculum in arid zone of the country. Summer irrigation and soil solarization combined with mustard cake, eradicated about 91% native arbuscular mycorrhizal fungal spores during an experiment conducted 1995. This treated soil then allowed the large scale multiplication of mycorrhizal inoculum.

Gadkar *et al.* (1999) studied *Gigaspora margarita* during its colonization with RiT-DNA transformed carrot roots under *in vitro* conditions. The RiT-DNA transformed root organ culture technique is a useful tool for studying VAM fungal events *in situ*. Dual culture of the vesicular arbuscular mycorrhizal fungus *Gigaspora margarita* and RiT-DNA transformed carrot roots (*Daucus carota* L.) was established which allowed a closer examination of the extraradical phase of the fungus than possible using the soil based cultivation systems. In their study of the colonization of *G. margarita* with RiT-DNA transformed carrot roots they observed series of structures formed by the fungus which gave an appearance of an injury. Wounds occurred near the growing tips of the hyphae or well behind the tips. Anastomosis to bypass the wounded sections ranged in complexity from simple clamp-type connections to figure eight. These injuries were a regular feature of the symbiosis suggesting that such an event were part of the fungal life cycle. They tried to document these injuries and the associated healing mechanisms. Another question which they tried to answer was that whether these injuries if induced *in vitro* lead to similar mechanism of healing? Some of these structures have been reported for this fungi under soil based cultivation systems. Comparative analysis of such structures and its possible significance in the fungal life cycle have been discussed. They postulated that these wound might be better classified as "malfunctions" since they occurred in the absence of biotic and abiotic stresses known to cause injury.

Mehrotra and Mehrotra (1999) developed a suitable potting mixture for use in root trainers, which ensures adequate mycorrhization and abundant sporulation of VAM (vesicular arbuscular mycorrhizal fungi in the rhizosphere. The mixture consisted of soil/sand/FYM (1:2:2) with added deoiled neem seed cake, superphosphate, phorate and Dithane M-45 (mancozeb], and enabled the production of very high numbers of spores per 100 g of soil by volume. The seeds of 7 tree species (*Dalbergia sissoo, Dendrocalamus strictus, Bauhinia purpurea, Chukrasia velutina [C. tabularis], Taxodium mucronatum, Entolobium timbouva [Enterolobium contortisiliquum*], and *Tectona grandis*] and one medicinal plant (*Pogostemon heyneanus*] were treated with Bavistin prior to sowing in the medium in root trainers. At about 8 months old they examined for VAM infection and soil VAM spore content. Maximum VAM infection on roots was recorded in *Dendrocalamus strictus* followed by *Tectona grandis*. In general, the spore population in the rhizosphere was very high; exceptions were *Taxodium mucronatum* and *Tectona grandis*. The maximum number of spores in the rhizosphere was recorded for *Bauhinia purpurea* (3565-9047 with an average of 6761 per 100 g of soil in the root trainer); these included 3.5 *Glomus* spp. and *Acaulospora scrobiculata*. Because of this high spore content of the medium in *B. purpurea*, and the profuse root development and high level of VAM infection in *D. strictus*, these 2 species are recommended to be raised for establishment of mycorrhizal banks for the purpose of supply or starter cultures of VAM fungi and use of VAM inoculum for containerized seedlings in forest nurseries. No correlation could be drawn between percentage of root infection and spore population of VAM fungi in the rhizosphere.

Gadkar and Adholeya (2000) used the ROC technique with *Gigaspora margarita* and RiT-DNA transformed carrot roots to examine the fungal growth and physiology under long-term axenic symbiosis. The fungus formed spores inside the host roots under *in vitro* conditions. Sporulation was a temporal phenomenon found in dual cultures more than 18-20 month old. The spores were formed singly or in rare cases cluster of two or three. No preferential zone of formation was found. The spores formed intraradically were 10-15% of the total spores formed in a single culture. These intraradical spores were analyzed for their morphology and DNA polymorphism pattern with spores, which had formed conventionally in the medium. Both analyses showed no detectable variation. This is the first report of spore formation in *Gigaspora* inside the roots of a host.

Khaliq and Bagyaraj (2000) examined the colonization of arbuscular-mycorrhizal fungi on RiT-DNA transformed roots in synthetic medium. Hairy root culture of tomato (*Lycopersicon esculentum*) was induced using 3 strains of *Agrobacterium rhizogenes* (A4, ATCC T5834 and LB 9402). The greatest hairy root growth was obtained with *A. rhizogenes* strains A4 and LB9402. Hairy roots were maintained on Murashige and Skoog medium. Spores of *Gigaspora margarita* were isolated and a method for surface sterilization of spores is described. Surface-sterilized spores of *G. margarita* were used for inoculation of transformed roots on minimal medium. Mycorrhizal spore germination and penetration into the root cortex were observed. Inter- and intracelular mycelial spread and formation of arbuscules also occurred in the cortical region of transformed roots.

Raman and Sambandan (2000) examined the axenic germination of *Scutellospora erythropa* and *S. nigra* in various media. Positive results were obtained with media containing root exadates, soil extract, inositol etc. Spore germination was increased in a combined medium. There was significant increase in hyphal growth when roots were introduced into the medium 3 days after spore germination.

Babu *et al.* (2001) conducted pot experiments for mass production of vesicular-arbuscular mycorrhizae using different hosts. Tomato, cumbu (pearl millet, (*Pennisetum typhoides [Pennisetum glaucum)*, cotton (*Gossypium* spp.), bhendi [okra], brinjal (aubergine)), cowpea, green gram, maize, red gram [*Cajamus cajan*] and black gram were grown in pots and cultured with *Glomus fasciculatum* (10 -/kg of soil) and after one and 2 months, number of spores in the soil and root colonization were recorded to find the most suitable host for endomycorrhizal multiplication. The spore count was highest for pearl millet, followed by bhendi (450 and 268.8 spores/200 ml of soil, respectively after one month). In cotton, spore count reached 312.9/200 ml of soil after 2 months. The spore count was more than100 in 200 ml of soil in cowpea, maize and red gram.

Tiwari and Adholeya (2002) studied the *in vitro* co-culture of two AMF isolates *Gigaspora margarita* and *Glomus intraradices* on RiT -DNA transformed carrot roots as the host bearing simultaneously different vesicular arbuscular mycorrhizal (VAM) fungi belonging to two genera *Glomus intraradices* and *Gigaspora margarita* was developed. Coexistence appeared healthy and harmonious, as both the generic species showed extensive hyphal proliferation and sporulation. The co-culture model under the in vitro system appears especially appropriate for further investigations on the competition and on the interaction mechanism involved in such types of associations occurring in nature and also as a model approach towards mass production of multiple mycorrhizal fungal isolates. This is a first report of successful co-culture of two genera of VAM fungi under in vitro conditions.

Harikumar and Potty (2002b) conducted a study to develop a technology for the mass multiplication and mass inoculation of arbuscular mycorrhizal fungi (*Glomus microcarpum*) to field-grown sweet potato cv. Kanjangad. Sweet potato vine cuttings were collected, stacked and kept for defoliation. Mycorrhizal inoculation was done batch-wise by planting the stacks in pots containing inoculum at two fertility phosphorus levels: high (140 ppm,) and low (69.75 ppm). Root samples were collected at 10-day intervals up to 40 days after inoculation (DAI). Co-inoculation of *G. macrocarpum* in the roots of sweet potato vine cuttings was observed from 10 DAI on onwards. At low P level, the percentage of colonization and the intensity of infection were greater than at high P level. From 10 DAI onwards, both the percentage of colonization and the intensity of infection showed an upward trend. Even at 40 DAI, the percentage of colonization approximated 100.00 and 84.61 in roots at low and high P levels, respectively. The intensity of infection also showed the same trend. Spore density in the soil increased after continuous cultivation for 3-4 seasons.

CONCLUDING REMARKS

Soil culture method is the most common and dependable method for producing VAM inoculum. Success of this method depends on an appropriate host which should be rapidly growing and should have an extensive root system. Many Indian mycorrhizasts have tried a variety of hosts including crops like maize, sorghum, pearlmillet and a number of cultivated and wild grasses. Some of the hosts have been found to be very suitable for the purpose, however, it has been emphasized that health and vigour of host is more important than the plant species. A number of workers have tried different substrates and their modifications with reference to their physico-chemical characteristics in order to ensure better health and vigour of the host plants. Likewise, few other workers have tried comparatively lighter growth media and their combinations (like vermiculite, soilrite, perlite, peat etc.) in order to ensure an easy transfer of inoculum. VAM inoculum contaminated with unwanted microbes is unfit for use. Pesticides application may solve the problem but selection of pesticides compatible with VAM production is an important aspect. Even Indian mycorrhizasts have done commendable work in this direction and have tried to identify appropriate pesticides including fungicides and insecticides. Hydroponic culture technique is a good method to produce high quality inoculum under controlled conditions but the inoculum consists of mainly mycelium and has a limitation because of its restricted survival ability. Indian work on this technique, its utility and its limitations is meagre and includes only few reports.

Growth of VAM fungi on a synthetic medium may produce a consistently high quality, disease-free VAM inoculum. Being obligate symbionts, VAM fungi are not expected to grow on synthetic media, however, attempts are being made to grow them on such media. A number of media with modifications at various levels have been tried in order to explore the possibility of growing VAM fungi in culture. Some information related to this has been collected also by Indian mycorrhizasts.

In order to obtain a pure inoculum of VAM fungi attempts are being made to grow VAM fungi in tissue culture with different hosts and VAM combinations. Success has been achieved in developing dual culture but the VAM culture has been found to show limitation with reference to its potentiality as an inoculum. Few Indian mycorrhizasts have also worked on this aspect but complete success has not been achieved.

10

MISCELLANY

The understanding of VAM symbiont and VAM symbiosis is of prime importance in manipulating VAM associations for the welfare of plants. Many contributions have been made worldover in the areas of taxonomy, morphology, anatomy, histology, cytology and genetics of VAM fungi and/or their associations with the plants. However, the share of Indian mycorrhizasts in such contributions is negligible. The following description gives an account of work done by the mycorrhizasts of the country in these areas.

Taxonomy

Sharma *et al.* (1986a) has given the descriptions of *Acaulospora laevis* and *A. scrobiculata* collected from the rhizosphere soils and *Alnus nepalensis* respectively. Roots of both tree species were heavily mycorrhizal.

Sharma *et al.* (1986b) collected eight species of *Glomus, Gigaspora* and *Sclerocystis* from the rhizosphere of trees in sub-tropical evergreen forest of NE India. *S. microcarpus* and *Gigaspora gregaria* constitute the first report from India.

Rani and Mukerji (1988) studied the taxonomy of Indian vesicular-arbuscular mycorrhizal fungi. With a view to study the taxonomy of Endogonaceae, soil samples were collected from different places in India. Microscopic examination was done by mounting the spores in Hoyer's medium which preserved the features of the spores in original form. The spores were identified with the help of different keys provided by different workers (Hall & Fish 1979). They have isolated 17 spp. of *Glomus*, 1 of *Acaulospora*, 4 of *Gigaspora*, 3 of *Sclerocystis* and 1 of *Endogone*.

Khan and Sarbhoy (1990) recovered 20 species of VAM fungi belonging to 6 genera from 3 different soil types in the IARI fields. Out of these 12 were described as new records from India.

Mehrotra and Baijal (1992a) isolated *Glomus sterilum*, a new species of VA-mycorrhiza from potato roots, which had pale yellow to reddish brown spores with walls composed of 1-2 layers. Both fertile sporocarps and sterile structures resembling sporocarps but without chlamydospores were produced. The fungus formed vesicular-arbuscular mycorrhizas with maize, sunflower and bahia grass (*Cenchrus ciliaris*) but no vesicles were produced in potato roots.

Neeraj *et al.* (1993) isolated a new species of endogonaceous fungus *Gigaspora* from desert soils of Rajasthan. The soil samples collected from Madhogarh in the Alwar district in Rajasthan were studied. The fungus, named *G. tuberculata*, produced rusty-brown azygospores with septate subtending hyphae. The shape, size and general appearance of these spores resembled those of *Scutellospora persica*.

Shankar and Varma (1993) recovered spores of *Glomus deserticola, G. fasciculatum, Scutellospora gilmorei* and *S. calospora* from rhizosphere soil of plants growing in arid and semiarid regions of western India. They were characterized by subjecting spore extract to polyacrylamide gel electrophoresis and selective enzyme staining. The banding patterns of malate dehydrogenase were used to distinguish the isolates at intra-generic and intra-specific levels. Results indicated that isoenzymic studies are useful in the differentiation of genera and species of VAM fungi. Species growing under environmentally different conditions gave almost identical banding patterns. It has been proposed that isoenzymic studies should be included in the descriptions of new taxa, to show, how the banding pattern of different enzymes of newly identified species is related to that of existing genera and other known species.

Mehrotra (1997) described a new species of the arbuscular mycorrhizal fungus, *Glomus*, isolated from contaminated pot cultures of other glomalean fungi. The fungus, named *G. bagyarajii*, produced abundant bright yellow to yellow brown spores, 36-100.8 um, with four wall layers in two groups. Group A consisted of an outer, hyaline evanescent wall layer, 0.5-3.0 um thick and an inner yellow to yellow brown laminated layer, 1.6-3.6 um thick. Group B consisted of an outer hyaline unit wall layer, 0.25-0.5 um thick. Reaction of wall layers to Melzers's reagent was negative. Subtending hypha hyaline to pale yellow, cylindrical or funnel-shaped, 0.7-10.8 um wide at the spore base. Germination was by regrowth of the subtending hyphae.

Mukerji *et al.* (1983) described two new species of vesicular-arbuscular mycorrhizal fungi *viz., Glomus delhiense* and *G. multisubstensum* from Indian soils. In pot tests the former formed VAM with *Trigonella foenum-graecum* and the later with maize.

Anatomy/ morphology

Nair and Mahabale (1975) studied morphology of the endophytes *in vivo* in *Ophioglossum reticulatum* L., *O. pedunculosum* Prantle, *O. vulgatum* L., *O. fibrosum* Schum., *O. aitchisoni* d'Alm., *O. nudicaule* L., *O. gramineum* Wild, *Botrychium virginianum* var. *Daucifolium* Bedd. and *Helminthostachys zeylanica* L. It was observed that in the material of *Helminthostachys zeylanica* collected, no endophytes were present. In all other plants they were present. The mode of entry of fungus and three stages of endophytes development *viz.*, mycelial, vesicular and arbuscular were found in succession in all of them. The authors discussed the mode of living of mycorrhiza in the host plant.

Chandel and Srivastava (1988) studied development of VAM in *Cenchrus ciliaris* and *C. setigerous* . *Cenchrus ciliaris* and *C. setigerous* plants alongwith rhizosphere soils were collected periodically from different localities around Jodhpur for VAM fungi associated with the roots. In all, 4 VAM fungal species namely *Glomus constrictum, G. fasciculatum, G. aggregatum* and *Gigaspora calospora* were collected from the fields of *C. ciliaris* and *C. setigerous*. The young feeder roots of 15-25 days plants started developing VAM and the process continued upto maturity of the plants. The infection in the roots was always caused by soil-borne fungal hyphae. All the stages in root colonization, i.e., formation of appressoria, hyphal penetration, hyphal branching, hyphal coiling, vesicles and arbuscules formation were observed.

Sivaprasad *et al.* (1988) studied the ultrastructures of nodules of *Cajanus cajan* Millsp. inoculated with vesicular-arbuscular mycorrhizal (VAM) fungus. Ultrastructural observations of pigeon pea nodules revealed that the bacteroid zone present in transformed cells is surrounded by a membrane envelop. Association with *Glomus fasciculatum* had no effect on the number of bacteroids formed per membrane nevelop, which was always 1. Both the transformed cell size and number of bacteroid/ *Rhizobium* present in transformed cells were higher in mycorrhizal associations.

Vietti and Staden. (1990) studied the histology of VA mycorrhizal development in guayule seedlings. Guayule (*Parthenium argentatum*) seed was germinated and grown in acid-washed quartz sand seeded with inoculum of the vesicular arbuscular mycorrhizal (VAM) fungus *Glomus intraradices*. Examination of seedlings at weekly intervals indicated no fungal colonization until the third week when hyphae spread over the root surface. Following the development of appressoria, root penetration and intercellular colonization took place between the forth and sixth week of inoculation. Mature fungal infections exhibiting vesicles became established after the sixth week.

Punj *et al.* (1992) carried out histopathological studies on mycorrhiza infected roots of *Leucaena leucocephala*. The changes which occur in roots of *L. leucocephala* infected by *Glomus fasciculatum* are reported. Vesicles, arbuscules (hyphae) and spores were observed in infected roots. The arbuscules developed intercellularly and the spores intracellularly. The arbuscules did not infect the root nodules suggesting that mycorrhizal effects on nitrogen-fixing ability may be phosphorus mediated.

Doss and Bagyaraj (1990) studied the anatomical observations on roots of finger millet colonized by VA mycorrhiza. *Glomus fasciculatum* was shown to penetrate lignified sclerenchymatous exodermal cells in roots of *Eleusine coracana* plants grown in inoculated soil.

Nagarajan *et al.* (1995) made light and electron microscopic observations of the spores of *Glomus deserticola*. *Glomus deserticola* was cultured using *Kalanchoe pinnata* as the trap plant. The spores were collected and stored in Ringer's solution and treated with 0.1 N HCl for an hr to remove outer peridial layer. Semithin sections (1 um) were stained with 1% toluidine blue and examined under a Nikon light microscope. Ultrathin sections were stained for 30 min in 2% (w/v) aqueous uranyl acetate followed by 10 min in lead citrate and examined in a 80 KV on Philips CM 10 transmission electron microscope. Mature chlamydospores were white to slightly yellowish, predominantly globose to infrequently subglobose, opaque and 45-80 um in diameter. The spores generally had a single wall layer and thick inner layer. The inner wall layer appeared lamellated. Semithin sections (1 um) were stained with 1% toluidine blue; they exhibited darkly stained outer large cell and relatively less intensely stained thick inner layer. TEM of chlamydospores revealed in some instances a wall of more or less uniform composition and relatively very low preservation of spore cytoplasm. At higher magnification, single lamellated wall layer of uniform thickness was discernible.

Pant *et al.* (1995) studied nature of endomycorrhizal associations in the prothalli of six species of *Ophioglossum*, four species of *Botrychium* and the monotypic *Helminthostachys*

(H. zeylanica). The structure, anatomy, ecology and taxonomy of the endophytes and the etymology of the term mycorrhiza were discussed in detail.

Ramesh *et al.* (1995) made light and electron microscopic observations of vesicular-arbuscular mycorrhizal symbiosis between *Pennisetum pedicellatum* and *Gigaspora margarita*. *P. pedicellatum* is common fodder grass of the tropics containing 4-6% leaf protein. The seeds were surface sterilized with 0.5% w/v mercuric chloride, repeatedly washed with sterile distilled water and grown in 30 cm earthen pots containing sterilized sand and soil mixture (1:1). The plants were inoculated with 50 sterilized spores of *Gigaspora margarita*. The roots were harvested at regular intervals of 15th, 30th, 45th and 60th day to assess infection. The spore population was assessed between 60th and 90th days. The highly infected roots were used for electron microscopic observation. Infection was detected in cleared and stained roots at 15th day. The infection was limited to a few longitudinally running hyphae in the outer and middle cortex and more towards the entry points. At the 30th day, the infection was much more extensive containing new arbuscules and senescing arbuscules. Electron microscopic observation indicated that the hyphae were present both inter- and intracellularly. The hyphae in cross section appeared spherical, ovoid or elliptical in outline. Inter- and intracellular hyphae were highly cytoplasmic containing numerous organelles, tonoplast bound electron dense bodies. The cytoplasm contained numerous small Golgi bodies. The intracellular hyphae contained primarily irregular shaped vacuoles. Nuclei were apparent in the hyphae. Intermediate stages between highly cytoplasmic and vacuolated hyphae were also present. Highest number of spores were seen on 90th and 120th days. Further work is being carried out to localize enzymes such as acid, alkaline and neutral phosphatases to ascertain the physiological significance of this important symbiotic association.

Selveraj and Subramanian (1988) carried light and scanning electron microscopic studies of VAM in *Sesamum indicum* roots. Occurrence of VAM fungi in the roots and rhizosphere of *Sesamum indicum* cv. Co-1, growing in P-deficient (4 mg/kg) sandy loam soil (pH 7.2) was reported. Six different VAM fungal species were noticed in roots and rhizosphere soil of the host plant. Of them, 4 belonged to the genus *Glomus, i.e. G. convolutus, G. macrocarpum, G. fasciculatum, G. fasciculatum* strain E3 and the remaining were *Sclerocystis sinuosa* and *Gigaspora gigantea*. The spores and sporocarps of *G. macrocarpum, G. fasciculatum* and *S. sinuosa* were very prevalent in the rhizosphere soil. The roots of the test plant harboured arbuscules, vesicles and young spores characteristic of *G. fasciculatum* and *G. fasciculatum* strain E3. The appressoria, arbuscules and vesicles of *G. fasciculatum* from inoculated roots of *S. indicum* were examined under SEM.

Mehrotra (1993) critically investigated the nature of vesicles formed inside the host. According to him in vesicular-arbuscular mycorrhizal literature there is considerable confusion regarding the true nature of vesicles. In order to clarify the identity of these structures, a study of their formation of different species of vesicular-arbuscular mycorrhizal fungi in association with various field crops, under different edaphic conditions, was made. Wide variation in the morphology of vesicles, which differed with fungi, host and soil was observed. The study revealed that the formation, development and structure of the so-called vesicles is similar to chlamydospores.

Zeuske *et al.* (1999) discussed possible series related to mycotrophy in the Araliaceae.Roots of selected species within the Araliaceae (*Schefflera arboricola, Hedera helix* var. conglomerata, *H. canariensis* [H. *helix* subsp *canariensis, H. colchica* cv. *Schulze, and Fastsia japonica*, x *Fatshedera lizei, Panax ginseng* and *P. quinquenfolius*) were inoculated with spores of *Glomus albidum, G. mosseae* and *G. intraradices*. After successful establishment of the endophytes, the structural characteristics of the mycorrhizas were investigated. Three groups with different morphotypes could be seen, depending on the plant species. In addition, data from roots collected from natural sites were used to support the results.

Sailo *et al.* (2002) studied anatomical and histochemical changes in the roots and leaves of *Simarouba glauca* [*Quassia simarouba*] because of inoculation with the arbuscular mycorrhizal (AM) fungus *Glomus mosseae* alone and together with the mycorrhizal helper bacterium (MHB) *Bacillus coagulans* in comparison with the uninoculated plants. The root cells of plants inoculated with AM + MHB had thicker cell walls compared to plants inoculated with either AM fungus alone or uninoculated plants. The size and/or number of metaxylem, protoxylem, sclerenchyma cells, stele region and pith cells were also higher in plants inoculated with AM fungus + MHB compared to either mycorrhiza alone inoculated or uninoculated plants. In the leaves there was a striking increase in the size of upper and lower epidermal cells, palisade cells, spongy cells, number and size of plastids when plants were inoculated with *G. mosseae+B. coagulans* compared to plants inoculated with *G. mosseae* alone, the least being in uninoculated plants. The same was true for the concentration of RNA and protein, and sclereid frequency occurring in leaf tissue.

Muthukumar and Udaiyan (2002b) described the morphology of *Glomus decipens* spores, host plants and the wide distribution of this species isolated from Western Ghats of Southern India.

Genetics

Gunasekaran *et al.* (1988) carried out extraction of DNA from *Glomus etunicatum*. A rapid procedure to isolate DNA from the external hyphae is described. Hyphae were isolated from the soil inoculum of *Glomus etunicatum* and homogenized in a micropestle and mortar. About 5 to 10 ug of DNA was isolated from the hyphae recovered from 100 ml soil water suspension having a hyphal density of 43-m/g soils. However, there was a considerable loss of hyphae during isolation from the soil.

METHODOLOGY

Khan (1999) compared the plate, wet sieving and decanting; flotation adhesion and density gradient centrifugation methods for the isolation of spores of mycorrhizal fungi from the soil of the rhizosphere of maize plants in New Delhi. Results showed that the plate method gave the best rapid isolation, while density gradient centrifugation was the best for harvesting maximum numbers of spores with the fewest impurities.

Backiyavathy and Perumal (1999) studied the effect of different levels of nitrogen and phosphorus on atmospheric nitrogen fixation and computation by various methods in four species of *Acacia.*^{15}N urea (at 50,75 and 100% concentration) and single super phosphate (at 50, 75 or 100% concentration) were supplied alone or in combination with *Rhizobium leguminosarum, Glomus fasciculatum* (VAM) or both to 4 species of *Acacia (A. nilotica, A. leucophloea, A. holosericea and A. planifrons*). The accuracy of various methods for estimating N-fixation (% Ndfs, isotope dilution technique and 'A' value method) was evaluated. The treatment 75% urea + *R. leguminosarum* + VAM resulted in the highest N-fixing potential, while higher N rates decreased this potential. The percent N derived from the atmosphere (% Ndfa) was highest when 50% single super phosphate was applied and a further increase in P concentration decreased the % Ndfa. The percent N derived from fertilizers (%Ndff) ranged from 0.91 to 2.85 regardless of species. %Ndff and nitrogen use efficiency was low in the treatment 50% urea + *R. leguminosarum* + VAM. The highest amount of N was fixed from the atmosphere by *A. nilotica*, and the values obtained using the %Ndfs, isotype dilution and 'a' value methods were 167, 177 and 176 mg/seedling.

Singh and Tiwari (2001) studied the modified wet sieving and decanting technique for enhanced recovery of vesicular-arbuscular mycorrhizal fungi from forest soils. A modified wet sieving and decanting technique is presented for extraction and recovery of vesicular-arbuscular mycorrhiza spores from loamy sand soils. The modified techniques gave 21% greater recovery of spores than the standard procedure.

CONCLUDING REMARKS

The work on taxonomy of VAM fungi in the country was initiated by Mukerji and his colleagues at Botany Department, Delhi University, Delhi nearly 25 years ago. They published a number of reports on the occurrence of different taxa of VAM fungi in soils of the country collected from diverse agroclimatic regions. These reports were followed by many other reports by Bakshi and his colleagues at Forest Research Institute, Dehra Dun. A number of new taxa were reported by both the schools (e.g. *Glomus multicaule, G. reticulatum, G. multisubtensum, G. delhiense, Sclerocystis sinuosa, S. indicus* etc.) Some new records from the country were also reported *viz., Sclerocystis dussi, S. rubriformis, Gigaspora corolloidea, Modicella reniformis* etc. With increasing knowledge of the potentiality of VAM fungi in providing benefit to the plants, the attention of the Indian mycorrhizasts shifted to VAM associations instead of VAM fungi. Consequently they started collecting informations related to distribution of mycorrhizal associations in different plant communities from different habitats and regions. However, work on taxonomy of VAM fungi was not completely neglected. As a result, not only the list of VAM fungi occurring in the country has increased but many new taxa have been created. According to the estimate of Manoharachary, of Osmania University, Hyderabad, a well known expert on VAM taxonomy, more than hundred species of VAM fungi have been recorded from the country till now.

During its complex life cycle in contact with the roots, the fungal symbiont gives rise to different structures, e.g. extramatrical hyphae, spores, intracellular hyphae, vesicles and arbuscules. A lot of information has been collected through the efforts of mycorrhizasts worldover regarding morphology, anatomy and cytology of these structures through light microscopy as well as electron microscopy. These informations have provided not only a clear insight about host response to fungal colonization and host-fungus relationship but also methodology to identify VAM more accurately. The Indian mycorrhizasts have made few but important contributions regarding histopathological details of mycorrhizal associations of a number of plants including *Leucaena leucocephala, Eleusine coracana, Parthenium argentatum, Kalanchoe pinnata, Pennisetum pedicellatum* and *Sesamum indicum* using light as well as electron microscopy. They have collected valuable information related to not only morphology, anatomy and cytology of extramatrical and intramatrical structures of the symbiont but also mode of its penetration, infection and spread within the host. Since wide variations are expected in different phases of the life cycle of the symbiont as well as its interaction with the host, many more host-symbiont combinations have to be studied to reach definite generalizations.

A clear understanding of the functioning of genomes of the symbiont and host is a prerequisite for explaining the compatibility or incompatibility between them. Furthermore, a knowledge of the characteristics of the genome of symbionts will go a long way to help the efforts to manipulate them for improving their specific desired properties. However, it is unfortunate that the VAM fungi, inspite of their proven worth, have not been studied with this reference even at the global level. The mycorrhizasts of the country have also paid no attention to this important aspect of study of VAM fungi.

11

FUTURE PROSPECTS AND STRATEGIES

Current status of researches on VA mycorrhiza and their benefits to the plants carried out by the mycorrhizasts of the country has been described in detail in chapters II to X.

Prior to the establishment of any specific forum, the pace of research on VA mycorrhiza in the country was not of desired level. However, the activities related to research on VA mycorrhiza gained unexpected momemtum when Mycorrhiza network came into exixtence in the year 1989. The network provided not only assistance and guidance of various types to the scientists working on VA mycorrhiza in the country but also offered them periodical platforms to exchange their informations and thoughts on status of researches on VA mycorrhiza and to propose future strategies of research. During the National conferences held in different parts of the country the panel of experts made certain observations regarding future prospects and strategies for research on VA mycorrhiza. They have been mentioned in the present chapter.

A number of senior soil microbiologists of the country are actively engaged in exploring the verstality of VA mycorrhiza as a biotool for plant health and care. They have made significant contributions and provided prodigious information regarding application of VA mycorrhiza and its dimensions of utility. Personal observations of some of them on the future prospects and strategies of research on VA mycorrhiza in the country were invited. They have also been mentioned in the present chapter.

On the basis of the observations of the panel of experts made during national conferences and personal observations of some senior mycorrhizasts of the country it, has been possible to frame certain priorities for research on VA mycorrhiza in the country. They have also been mentioned in the present chapter.

OBSERVATIONS MADE DURING NATIONAL CONFERENCES

First National Conference on Mycorrhiza held at H.A.U., Hisar 1990

1. Floristic survey of ectomycorrhizal and VAM fungi should be conducted in different parts of the country including soils supporting mangroves, oilseed crops, millets, medicinal plants and arid zone plants.
2. Studies should be initiated on taxonomy and life cycle of VAM fungi. A key for identification of Indian VAM fungi should be prepared.
3. Emphasis should be laid on host susceptibility, culivar response, mycorrhizal effectiveness, viable spores, spore germination, establishment of indigenous strains under field conditions for biomass/dry matter production/enhancement.

4. Attempts should be made for manipulating nutrients and other parameters for mass multiplication of VAM fungi.
5. The following regional centres are recommended for deposition of mycorrhizal materials: Haryana Agricultural University, Hisar, Indian Agricultural Research Institute, New Delhi; University of Delhi, Delhi; University of Agricultural Sciences, Bangalore; Bharatiya Agro Industries Foundation, Pune; Himachal Pradesh University, Shimla; Centre of Advanced Study in Botany, Madras; and Botany Department, Osmania University, Hyderabad. Mycorrhizal isolates/ cultures should be given accession number and deposited in any one of the above centres.
6. Most recent techniques such as ELISA, finger printing, isozyme patterns, and protein profiles should be employed, along with morphological characters and ultrastructural details for identification of VAM fungi.
7. Critical ecological factors that help mycorrhizal effectivity and mycorrhizal symbiosis in relation to plant growth should be analysed.
8. Emphasis should be laid on the role of associated mycoflora in the development of mycorrhizal association.
9. In-depth ecological data should be obtained on seasonal variation, distribution, qualitative composition, rhizosphere interaction and mycorrhizal infection.
10. Studies should be initiated on development and cytology of ecto- and endomycorrhizal fungi.
11. Genes responsible for mycorrhizal symbiosis in each isolate should be isolated, studied and manipulated.

Physiology/ Biochemistry

1. The area of research in relation to physiology and biochemistry of VAM with plants, particularly at the root VAM interface (endorhizosphere) needs to be pursued more vigorously.
2. There is a need to precisely show that there is effective increase in photosynthetic rates in VAM plants besides the increase in net photosynthetic area.
3. The approach to multiply VAM inocula under defined axenic cultural conditions or in organ culture should receive greater emphasis.

Biological Interactions

1. Studies should be initiated on mechanisms of synergism between VAM and other beneficial soil micro-organisms and also on mechanisms of biocontrol by VAM.
2. Appropriate controls should be maintained in planning experiments on VA mycorrhiza.
3. Research should be intensified on bacteria like organelles (BLOs), bacterial associates, hyperparasites and microfauna associated with VAM fungi as these play an important role in the mass production of VAM.

Soil-Plant Relationships

1. All field trials employing mycorrhizal fungi both at laboratory and farmer's levels need a standard methodology for inoculation under different agroclimatic conditions.
2. Research is needed to understand factors influencing mycorrhiza formation.
3. Studies should be conducted on specific metabolites, of either plant or fungus origin, and their interactions which govern fruiting response to VAM fungi.
4. Studies should be conducted to understand the processes occurring in the host-fungus interface, mechanisms controlling them and the factors affecting them.
5. Studies should be conducted to understand more about plant-microbe interaction and to develop new and more efficient fungi, either by selection processes or by genetically manipulating mycorrhizal fungi.
6. Normally, in nature, more than one fungus form mycorrhizae at the same time in every plant. Studies on interactions among various mycorrhizal fungi, especially between introduced and indigenous ones, would help in evaluating net beneficial responses of mycorrhiza under field conditions.

Biocides

1. As the results obtained in fungicide-VAM interactions studies are contradictory and inconsistent, standard methods of VAM inoculation need to be evolved. Use of pesticides should be judicious.
2. Efforts should be made to evaluate, on the basis of colonization in the roots, the population in the soil and chlamydospores contents.
3. Recommendation of fungicides should be based on microplot experiments rather than on pot experiments.
4. Interaction studies should be made with initial inoculation level and should be quantified.
5. If possible, multilocation trials should be made with the same biocide and the same quantity of the inoculum in one lot, keeping all other conditions similar, as far as possible.

Second National Conference on Mycorrhiza held at Bangalore Agriculture University, Bangalore, 1990

1. There is a need to go in for field trials whereever possible, for which standardization of inoculum requirement is essential; and
2. Selection of efficient strain(s) for various crops grown under different soil conditions need to be intensified.

Mycorrhizal Inoculum Production

1. Dr. Janardhan, who had isolated *Glomus aggregatum* in pure culture on synthetic medium, was advised to ascertain the authenticity of his culture by looking into the

DNA replication of the fungus and also by employing other necessary methods of molecular biology; and

2. Intensive efforts should also be continued to culture VAM fungi on nutrients media.

Ecology of Mycorrhizal Fungi

1. Work on dissemination and persistence of mycorrhizal fungi should be intensified;
2. There is need for more work on agronomic practices with native VAM fungi; and
3. Role of mycorrhizal fungi in wastelands, problem/disturbed areas should be studied.

Taxonomy, Occurrence and Distribution of Mycorrhizal Fungi

1. Taxonomical work for identification of mycorrhizal fungi should be intensified;
2. Centres should be recognized which can accord identification services to mycorrhiza workers;
3. Mycorrhizal species/strains which are tolerant of saline/alkaline soil conditions should be identified; and
4. Research on ectomycorrhizae should be expanded.

Biological Interactions with Mycorrhizal Fungi

1. Work should be carried out on selection and development of tree genotype specific strains of *Rhizobia* and efficient strains of mycorrhizae.
2. Combinations which will promote higher dry matter accumulation in woody biomass need to be found out for different soil types.
3. The benefits of *Azospirilla* and mycorrhizae together can possibly be best harnessed by combinations or relay cropping as in agroforestry systems, by suitable selection of physiologically dependent and responsive crop types. Phosphorus response of plants, which is highly variable, should be one of the criteria for selecting plant species for such purposes.
4. On fundamental aspects, there is a need for partitioning the physiological basis of improved plant response upon such multiple interactions which could only allow selection of suitable combinations of plants and bio-inoculants.

Mycorrhizal Fungi and Biological Control

1. Concerted efforts should be made to generate meaningful information on the biocontrol efficiency of mycorrhizal symbionts and mechanisms involved in such controls;
2. A practical approach should be attempted to suppress the progression of soil-borne plant diseases caused by fungi and nematodes through an integrated management system;

3. The information from the points raised above should be integrated with the role and management of abiotic factors such as drought, salinity, and stress, and effective correlative models should be developed; and

4. Mycorrhizal endophytes having biocontrol potential may be catalogued for future research investigations.

Third National Conference on Mycorrhiza held at New Delhi, 1995

A special session was held at the end of the conference where the present trends in mycorrhiza research were were reviewed and recommendations were discussed on future thrust areas in mycorrhiza research by a panel of experts drawn from different fields of mycorrhiza. Six thrust areas were identified. A brief summary of recommendations is given below:

Production of mass inoculum of mycorrhizal fungi

Ultimate objective of any mycorrhizal research is to make mycorrhizal biofertilizers available to the actual users in easily transportable form in which the mycorrhizal propagules are in concentrated forms, free from contaminants. To achieve this objective, mycorrhizal inoculum would be needed to develop on a mass scale to cut down the cost of production. But to reach that stage of inoculum development, different fungi/strains of fungi suitable for a wide range of plants under a particular set of edapho-climatic conditions will have to be identified. It was, therefore, considered that

Increasing reseach efforts should be oriented to bulk production of contamination-free inoculum of mycorrhizal fungi for use by a common man.

Mycorrhizal fungi should be evaluated for their performances under different edapho-climatic conditions. Ultimately mycorrhizal fungi found efficient under broad spectrum of site conditions should be selected for mass multiplication.

There should be a centre for quality and purity assurance of inoculum produced on a mass scale. Centre for Mycorrhizal Culture Collection (CMCC) established at Tata Energy Research Institute may serve as such a centre.

Standardization of technologies for determination of mycorrhizal dependency of plants

A lot of work has been done on beneficial role of mycorrhiza in increasing plant growth, affording protection to plants against soil borne diseases, insect pests, drought, heavy metal toxicity, rehabilitation of disturbed lands, difficult and poor sites, and sand dunes. Results are, however, scanty on the extent of dependency of a plant species on VA mycorrhizal infection in its roots. It was, therefore, considered that

Methodologies should be standardized to obtain uniform results on different species of agricultural crops and economically important forest tree species for determining their dependency on mycorrhiza.

Dependency of different plant species and trees of economic importance should be determined under various site conditions such as degraded lands, saline, alkaline and acid soils, polluted soils, water-logged soils and sand dunes.

Phosphorus response curves with and with out mycorrhiza should be obtained through extensive research experiments for crop plants and economically important forest tree species.

Technology transfer from the laboratory to the field

Significant achievements have been made on mycorrhiza research under laboratory/nursery conditions. A large number of mycorrhizal fungi have been identified under different edapho-climatic conditions. The performance of these isolates has been evaluated in different laboratory all over the country. In a few cases strains within a mycorrhizal fungal species have also been identified for their differential performance. It was, therefore, considered that

Laboratory/nursery experiments should be extended to field conditions and benefits from mycorrhizal inoculation should be evaluated.

Uniform parameters for different crops i.e., agricultural crops, horticultural crops, forest trees, fodder crops etc., should be adopted for evaluation of the field performance results.

For forest trees, evaluation should be based on survival percentage one year after planting in the field and stem diameter and plant height three years after planting in case of fast growing species, and five years after planting for other species should be the main parameters in addition to the data on other growth characters.

Use of molecular techniques for taxonomy of mycorrhizal fungi

Vesicular-arbuscular mycorrhizal (VAM) fungi are difficult to identify in the absence of knowledge on their complete life cycle. These fungi have also not been cultured on artificial media. The identification of VAM fungi is, therefore, done on the basis of morphology of azygospores/chlamydospores/sporocarps, and the hyphae attached to these structures. It was, therefore, considered that

In order to delimit mycorrhizal fungal species/strains from one another, research should be now focussed on modern molecular techniques. These include Isozyme techniques, Immunological probes, DNA library and fingerprinting, and PCR techniques combined with RAPID/RELP. Some of these techniques can also prove excellent tools in detection of mycorrhizal fungi in the field conditions.

Research on taxonomy of mycorrhizal fungi should be encouraged.

Combined inocula of mycorrhizal fungi with associated beneficial bacteria

Dual inoculation of plants with nodule forming nitrogen fixing bacteria or non-nodule forming nitrogen fixing bacteria and VAM fungi has often yielded increased biomass production, greater nodule number and increased nitrogenase activity. Inoculation of plants with VAM fungi combined with phosphorus solubilizing bacteria or with plant growth promoting rhizobacteria have also yielded good responses on plant growth. It was, therefore, considered that

Field scale experiments should be conducted to evaluate the performance of different forms of combined inocula on various agricultural crops or forest tree species.

Possibilities should also be explored to develop inocula of different beneficial bacteria in the form of a mixed consortium.

Research should also be conducted to evaluate growth promoting ability of beneficial bacterial bacteria on growth of VAM fungi for their ultimate use along with VAM fungi for mass production of VAM inoculum.

Information Network

The main constraint for mycorrhiza scientists in India has been the non-availability of latest literature on mycorrhiza. It was, therefore, considered that

Database on mycorrhiza literature maintained by Mycorrhiza Network at TERI should be continuously enriched and made available to the mycorrhiza scientists.

An information network should be established so that the literature entered in TERI data base is made available on-line to mycorrhiza scientists in the country on National Network on the pattern of Educational Research Network (ERNET).

Fourth National Conference on Mycorrhiza held at Bhopal, 1999

Continued efforts to develop mass production technology for VAM fungi.

Attempts to exploit VAMF in nutrient-deficient areas.

VAM dependency of plant species should be intensified in horticultural, medicinal, and forestry plants particularly where low productivity is obtained due to low input system of production and management.

Initiate intensive research programmes for the integration of mycorrhizal research with other micro-organisms-based biofertilizers, biocontrol agents, etc. for holistic plant nutrition.

Involve expert group in the development of the taxonomy of VAM employing molecular study tools. All described cultures should be maintained live and deposited with the Centre for Mycorrhizal Culture Collection at TERI.

Determine maximum phosphorus level at which the benefits from VAM inoculation can accrue to plants for each plant species which is dependent on mycorrhiza under different edaphoclimatic conditions.

OBSEVATIONS MADE BY SOME ACTIVE MYCORRHIZASTS

Professor D.J. Bagyaraj, Bangalore Agriculture University, Bangalore

The role of VA mycorrhizal fungi in improving plant growth is well documented. The enhanced plant growth is because of their role in increased uptake of diffusion limited nutrients (like P, Zn, Cu etc.,) and water, hormone production, and biological control of root pathogens, This has attracted scientists from different disciplines to work on VAM fungi. Thus, interest in mycorrhiza research has reached a pinnacle in recent years. Inspite of these advances, there is lot more to be done for manipulating VAM symbiosis to our advantage in the future. Some areas worth investigating in future are suggested below :

Anatomy and Morphology

1. Anatomical changes in the host because of VAM fungal infection needs more study. Recently we found that VAM inoculated *Simarouba* plants had more number of sclereids in the palisade and spongy layers of the leaf tissue, which are indicators of drought resistance.
2. Is there a correlation between root architecture and mycorrhizal dependency? This needs investigation.

Ecology

1. Influence of land use change and agricultural practices on VAM fungal diversity.
2. Is there a succession of VAM fungi in perennial plants?
3. Experiments to track down the introduced VAM fungi to study their competitive ability with indigenous VAM fungi, using molecular techniques in general, and PCR based techniqus, in particular.

Taxonomy/Biodiversity

1. Use of gene cloning technology as powerful tool to study diversity and distribution of VAM fungi, and also VAM symbiosis in natural and cultivated ecosystems.

Physiology and Molecular Biology

1. Mechanisms of recognition between host plant and VAM fungi.
2. What makes some plants non-mycorrhizal?
3. Production of growth promoting substances by VAM fungi.

Interactions with other soil organisms

1. Synergistic interaction between VAM fungi and rhizobia is well known. The mechanisms by which VAM fungi help nodulation and nitrogen fixation is fairly well understood. But the information as to how rhizobia enhance colonization and sporulation by VAM fungi is scanty. This needs investigation.
2. Research on parasites and predators of VAM fungi should be intensified as they will play an important role in the success or failure of VAM inoculation trials.
3. Investigations on mycorrhiza helper bacteria.

Mass Production

1. Innovative research on mass production technology should receive top priority.
2. Possibility of culturing VAM fungi on culture media in the laboratory.

Field experiments

1. It is high time to switch over from green house pot culture studies to field experiments with VAM fungi.
2. Studies on bioremediation on stressed environments with VAM inoculation.
3. Possibility of using VAM fungi along with less costly low grade rock phosphates, perhaps together with phosphate solubilising microorganisms.

Practical applications

1. Since VAM fungi are obligate symbionts and difficult to mass-produce, we can concentrate our research on transplanted crops raised through nursery beds/ containerized bags/root trainers/micropropagation so that this tehnology would be relevant to practical agriculture, horticulture and forestry.
2. In a recent study we found that perennial plants already established in the field respond to VAM inoculation. More studies are needed in this area.
3. Restoration of plants under stressed environments with VAM inoculation.
4. Preparation of suitable microbial consortia consisting of VAM fungi + N fixers + PGPR etc, for different crops.

I strongly believe that eco-friendly organisms like VAM fungi and other beneficial soil organisms play an important role in not only improving crop production but in reducing the use of hazardous pesticides and fertilizers in agriculture, which are deleterious to soil, animal and human health. I believe that these organisms would provide a cleaner environment for man to live.

Dr. B.L. Jalali, Hisar Agriculture University, Hisar

Over the last two decades, researches in India have given qualitative momentum in initiating and thereby generating useful research information on various dimensions of mycorrhizal association. By and large, most studies on the mycorrhizal impact on plant growth and development have been assessed under *in-vitro* conditions, which necessarily may not be translated under natural field conditions. Therefore, mycorrhizal model system must be considered in the framework of the sustainable agricultural systems. The probiotic influence exerted by mycorrhizae must be seen as inseparable component of agroecosystems which may exhibit both positive and negative influences on the overall performance of the crop, depending on the prevailing macro- and microenvironment.

For the past several years, many research centres in the country have evinced great interest in mycorrhizal research. Several initiatives have accordingly been launched to encourage basic and applied works in various areas of mycorrhizae. The Indian Council of Agricultural Research (ICAR), Department of Biotechnology (DBT), Department of Science and Technology (DST), and other such central agencies have sponsored several projects to encourage in-depth mycorrhizal investigations. The ICAR has also initiated networking of mycorrhizal pursuits.

The areas of research studies which, however, need to gain qualitative momentum in the years ahead should include practical applications of mycorrhizal technology. Inconsistencies frequently occur in field trials. The degree to which cultivars are colonized and benefited by mycorrhizae is a heritable trait; therefore, crop-breeding programmes need to select genotypes/lines which are responsive to mycorrhizal inoculations. Certain crops preferentially form mycorrhizae with only certain mycorrhizal fungal species, and so cropping sequences may influence species composition of fungal communities. No clear-cut generalizations can be made about the impact of pesticidal applications of mycorrhizae. A well-defined protocol need to be devised for uniform adoption; this will facilitate to monitor appropriate use of pesticides under natural farming systems, and ultimately make the IPM technology more meaningful.

In-depth studies on non-nutritional influences of mycorrhizal endophytes on plant pathogens may yield a more conclusive picture. The selection of efficient strains of mycorrhizal fungi is critical for harnessing better benefits from such symbiotic associations. In the face of major biotic and abiotic challenges in the zone of root influence, establishment of inoculating mycorrhizal symbiont should be assessed. Above all, workable technology for appropriate delivery system of the mycorrhizal inoculum need to be perfected for harnessing fuller benefits.

Professor Ajeet Verma, Jawahar Lal Nehru University, New Delhi

There are many fungi, conservative estimates are in the order of 100000 species and there are many ways to study them, from descriptive accounts of organisms found in nature to laboratory experimentation at the cellular and molecular level. All such studies expand our knowledge of fungi and the fungal processes and improve our ability to utilize and to control fungi for the benefit of humankind. Indian mycologists have played a key role in the development of new subject areas shows how versatile and productive they have been during the twentieth century. The future is, however, uncertain. Concentration of funding on "molecular" aspects of biology, and of research into a few organisms, has had a debilitating effect in Universities, research stations, and even in museums and herbaria on whole organisms biology, on taxonomy and ecology. There is alarming trend not only in India but all over the world, not to replace traditional mycologists with similarly qualified staff as they retire. The ability to recognize and name common fungi and their hosts is becoming increasingly rare. It may be true to say that many amateur mycologists are better qualified to recognize what is known than professionals. This should matter of concern and make possible every effort to reverse this unfortunate trend.

There is converging evidence that the genomes of glomalean fungi are unusual in both their complexity and structure, and that these organisms cannot be classed as either sexual or clonal sensu stricto. There could be a primitive type form of sexuality (or parasexuality) in glomalean fungi which does not imply nuclear fusion but involves nuclear exchange through anastomosis between hyphae from different spores of a species so that rare sequences persists in populations. If the Glomales are clonal organisms, then the nuclear lineages could be considered as being clones, and multinucleate hyphae or spores will be populated by a mixture of several nuclear lineages. Research into the genomics of the Glomales is still very much at its beginning. There remains a lot to learn in order not only not to progress in an academic pursuit of the genetics of these fungi but also to more efficiently exploit them in plant productive system.

The molecular biology of VAM fungi is very complicated and time-consuming. The particular characteristics of these biotrophic symbionts and their impact on plant communities and on plant production systems implicate, however, the meaning of this field of research. The first task to achieve will be to understand in more detail the molecular cross-talk of the AM fungi with plant roots and with other soil microorganisms. In the long term, it will then be necessary to use the molecular probes obtained from pot cultures in field experiments in order to investigate their expression and relevance in terrestrial ecosystems. The results from experiments will help to reach the final goal of applied AM research: the design of tools for formulating inocula for their use in horti- and agriculture, forestry, revegetation of degraded soils, and in the acclimatization of micropropagated plants.

Inoculum production of arbuscular mycorrhizal fungi presents a very different problem. These fungi will not grow like any other fungi apart from their hosts (obligate symbionts). Although genetic engineering of these fungi may, in future, provide the possibility of their production in the absence of the host plants, the classical technique will still be in use for considerable time. Obligate symbiotic mode of growth, non-availability of pure culture and expensive means of production and their unreliability for the beneficial effects have greatly jeopardized/undermined the mycorrhizal science, and may be appropriately called a biological enigma. Because of the absence of an authentic pure cultue, commercial production is the greatest bottleneck in the use and application of mycorrhizal biotechnology. However, there is silver lining for the scientists dealing with the mycorrhizal species, as Varma and co-workers have discovered a new root endophyte designated as *Piriformospora indica* belonging to the hymenomycetes (Basidiomycota).

PRIORITIES FOR FUTURE RESEARCH

A Indian mycorrhizasts have collected a huge bulk of information on the distribution of VAM fungi in soils having different physicochemical characters representing variety of habitats and geographical regions of the country. However, for better utilization of the benefits which the plants acquire from their association, the mycorrhizasts of the country have to make more extensive surveys to understand the ecological adaptations of these specific microbes. They have to identify the ecological factors which govern the dispersal, germination, colonization, sporulation, survival, persistence of VAM fungi which is possible only by undertaking more purposeful surveys.

A variety of plant communities growing in diverse habitats have been surveyed for their VA mycorrhizal association, however, there is still much to be learned. The mycorrhizasts of the country have to pay attention to many other plant communities and habitats. It will provide definite information about VAM species best adapted to certain soil conditions or hosts. It will provide information also about the conditions which make the symbiont a pathogen. More elaborate studies about host-symbiont-habitat combinations will provide knowledge about the specificity of VAM fungi with reference to plants and their habitats. Wihout a clear understanding about the edaphic conditions and host to which a symbiont is best adapted, our ability to manipulate the mycorrhizal symbiosis to the benefit of plant communities will be severely limited.

B Indian mycorrhizasts have investigated the effects of VAM on the growth and productivity of a variety of plant communities including agricultural, horticultural, medicinal, ornamental and plantation crops as well as forest trees in order to exploit the VAM symbiosis in the welfare of plants. However, in many cases because of defective experimental designs the interpretations of data can not be regarded as authentic and the conclusions of the experiments can not be safely made a basis for practical application of the symbiotic association in improving the growth and productivity of the plants under field conditions. It is suggested that the following points should be given due consideration in experiments designed to test the effect of VAM on plant growth:

- Net beneficial response of VA-mycorrhizae under field condition can be evaluated when interactions between indigenous and introduced VAM fungi are well understood. In view of this, pot experiments should always be conducted with sterilized and unsterilized field soils. However, the sterilization may not remove only the indigenous VAM fungi from the soil but may also cause a change in the microbial and chemical characteristics of the soil. Addition of soil filterate to the sterilized soil may partially overcome this problem. However, this fact should be given due consideration while interpreting the data of the experiment.
- Plant species differ in the amount of nutrients they require in the media to achieve their maximum growth. Furthermore, the amount of nutrients required for maximum growth of plants varies greatly with the soil. Effects of inoculation with VAM fungi in increasing plant growth can generally be overcome by increasing nutrient supply to a level where it it not limiting for the growth of uninoculated plants. Therefore, the magnitude of the effect of VAM on plant growth will depend on the nutrition status of soil for plant growth. In view of this, in all the experiments which are designed to test the effect of VAM on plant growth, due consideration should be given to plant growth response curves to added phosphorus and other nutrients while making final conclusions.
- There are several procedures which are employed for applying VAM inoculum in experiments related to evaluation of the potentiality of VAM in improving growth and productivity of plants. However, the inoculum in all of them consists of either only spores or a mixture of spores, infected root pieces and soil. For ensuring authentic

data standarization of the inoculum is a prerequisite. Most probable number (MPN) should be determined and given due attention while designing the experiment. In addition, purity of inoculum and viability of spores should be tested so that the trial may provide dependable information.

- The degree to which cultivars are colonized and benefitted by mycorrhizae is a heritable tract, therefore, emphasis should be laid on testing the response of different cultivars of host to mycorrhizal inoculation. Raising of genotypes/lines responsive to mycorrhizal inoculation will ensure better production because of their superior potentiality to receive the benefit of mycorrhization. Cross breeding programmes should also pay due attention to this trait of cultivars while developing and releasing new genotypes/ lines for cultivation.
- When VAM inoculum is applied to the pot soil, a little amount of organic matter (present in spores, root pieces or soil) gets introduced in the soil. Presence of this additional organic matter in the soils of treatment series and its absence in the soils of control series may not allow to reach to a fully correct conclusion from the experiments designed to test VAM effect on plant growth. This difficulty may be partially solved by adding VAM inoculum (same amount as in treatment series) also in the pot soil of control series but only after sterilizing it completely.
- On one hand, the nutritional requirements of a plant species vary both qualitatively and quantitatively with stages of its growth. On the other hand, the effectiveness of a VAM fungus is related to its ability to associate itself to a plant at a time most appropriate for increased uptake of a nutrient which is deficient for the growth of that plant. In view of this, an association of more than one VAM symbionts with a plant is expected to be more beneficial for its growth than an association of a single VAM symbiont. Thus in order to provide maximum benefit from mycorrhizal association to a plant species for its performance, a number of VAM fungi should be evaluated for their potentiality to improve the performance of the plant species individually as well as in different combinations. Most potential combination should be selected and employed to inoculate that plant species so that it may receive maximum benefit from the mycorrhizal inoculation for its performance.
- In order to have a holistic picture of the effect of VAM inoculation on plant performance, data on all the possible parameters related to growth, productivity and biochemical constituents of the candidate plants should be collected. In addition, instead of single and final harvest, sequential harvesting should be included in the experimental design. Furthermore, the data on stages in the formation and development of mycorrhiza should also be collected sequentially to have a clear understanding about the relationship between the intensity of mycorrhization and growth phases of the host plant. A comparison of the biochemical constituents of plants of control and treatment series may provide an information regarding the extent to which plants have received nutritional benefits due to mycorrhizal inoculation. A comparison of the nutritional status of the soils of the control and treatment series may assist in gaining this information which is of prime importance in deciding the contribution of VAM in improving the performance of the plants.

- The work of Indian mycorrhizasts on a variety of soil types, host species and endophytes has clearly established that VAM fungi improve the performance of plants in sterilized and unsterilized soils. However, most of the conclusions are based on pot trials and have not been confirmed under field conditions. There are only few reports of successful field inoculation with VAM symbiont. In order to ensure utility and applicability of VAM technology for the welfare of plants, the mycorrhizasts of the country should pay more attention to the field trials.
- The mycorrhizasts of the country have been successful in collecting informations to show that a variety of plant communities of the country are able to acquire benefits from the mycorrhizal inoculation under different habitats and climates. However, practical application of their informations is not feasible since inexpensive and convenient method for mass production of pure inoculum of VAM symbionts is still not available. Nevertheless, in spite of this limitation, the VAM technology is relevant for transplanted crops which are raised through nursery beds/ contianerized bags/root trainers/ micropropagation where requirement of VAM inoculum for establishing infection is much less. In view of this, at present the mycorrhizast of the country should concentrate their attention to such crops.

C The Indian mycorhizasts have made some important contributions related to mineral nutrition, water relations, as well as carbon and nitrogen metabolism of mycorrhizal and non-mycorrhizal plants in order to understand the physiology of mycorrhizal associations. However, there are several fundamental problems which are still unsolved and research should be directed towards solving them.

- Since VAM fungi are obligate symbionts, they do not have their separate identity. Consequently, it is difficult to characterize their physiological properties. However, physiological characteristics of their extramatrical and intramatrical phases may be elucidated separately after their recovery through suitable techniques like dual membered solution culture and enzymatic degradation of the host tissues. Such elucidation is expected to help in determining the activity of symbiont in mycorrhizal association in a better way.
- The extramatrical mycelium is responsible for absorbing nutrients from soil and transfering it to the roots. However, metabolically active hyphae might also be making some contribution to the host nutrition. Attempts should be made employing suitable techniques to distinguish the metabolically active proportion of the extramatrical mycelium and its role in host nutrition.
- Mycorrhizal association induces many changes in the host plants. Some of these e.g. increased uptake of certain nutrients, may be easily assigned to the activity of the symbiont. Many others may be linked directly with improved host nutrition. However, reasons of many other changes induced in the host plants due to mycorrhizal infection are still unknow. Attempts should be made to identify such changes and explore the factors responsible for them. It will be worthwhile to known that whether the changes are directly related to the activity of the symbiont or indirectly to some modification in the physiology of the host due to improved nutrition.

- VA endophytes show little specificity for infection of their hosts. However, they show great variation in providing benefits to the host. These variations may be due to the differences in their ability to infect roots, to compete with other microbes in the root zone or to transfer nutrients to the host. They may be also due to the difference in their ability to adapt to various nutritional environment and soil conditions. Attempts should be made to identify the physiological basis of different behaviour of various endophytes. The information will be helpful in selecting most efficient VAM symbiont for inoculation.

- Although bulk of information has been collected regarding physiology and biochemistry of mycorrhizal plants, however, researches in relation to the physiology and biochemistry of root-VAM interface (endorhizosphere) need to be pursued more vigrously. Studies should be conducted to understand the process occurring in the host-fungus interface, mechanisms controlling them and the factors affecting them. Concerted efforts should be made to explore the mechanisms of recognition between host plant and VAM fungi. It is also worthwhile to identify the factors which made some plants non-mycorrhizal.

D The results obtained so far with interaction studies between VAM fungi and other soil organisms carried out by Indian mycorrhizasts are encouraging and indicates the need for strengthening research in this area. In general, a synergistic effect on plant growth and productivity has been reported when VAM is employed as coinoculant with other beneficial soil organisms like symbiotic, non-symbiotic and associative nitrogen fixers, plant growth promoting rhizobium, phosphate solubilizing microbes, and beneficial actinomycetes. Following points need due attention by the mycorrhizasts of the country in relation to this area of research:

- Most of the interaction studies have been conducted in sterilized and unsterilized field soil under pot condition. In view of the success achieved with mixed bioinoculants field scale experiments should be conducted to evaluate the performance of different forms of combined inocula in various crops or forest tree species.

- Development of mycorrhizal association depends on other microbes present in the root zone. Role of these microbes in stimulating or suppressing the mycorrhizal infection should be identified. Research on parasites and predators of VAM fungi should be intensified as they may play an important role in the success or failure of VAM inoculation. Interaction studies should be made with initial inoculum level and should be quantified.

- In view of the heavy cost of P fertilizers, concerted efforts should be made to explore the possibility of using VAM fungi alongwith less costly low grade rock phosphate together with phosphate solubilizing microorganisms.

- A number of fungi and bacteria have been found to be closely associated with VAM fungi (companion fungi, bacterial associates, bacterium like organelles, helper bacteria). Research should be conducted to evaluate their ability to promote growth of the plants alone or in combination with VAM fungi.

E The extent to which a VAM symbiont may colonize and provide benefit to its host is expected to be influenced by the factors which affect its life cycle. the Indian mycorrhizasts have collected some information on the effect of certain factors related to soil and climate on the survival, germination, root colonization and sporulation of VAM endophytes. However, for application of VAM technology in the field, a more thorough understanding of this will be critical.

A variety of agronomic practices are employed to ensure better health and productivity of plants. These practices are expected to affect the diversity and life cycle of VAM symbionts in the field. In view of this, a knowledge of the interactions between such practices and VAM symbionts appears to be an important prerequisite for application of VAM technology at the field level successfully. Some attempts have been made by the mycorrhizasts of the country to investigate the effect of agronomic practices such as application of fertilizers, plant protection chemicals, cropping schedules etc. on the mycorrhizal associations and their benefits but the information is far from complete. For field application of VAM technology, due attention to the following is urgently required.

- It is not enough to know that the VAM fungal species differ in their distribution, their benefit to the host plants and their ability to sporulate or colonize roots but it is essential to know specifically how and why each fungal species differs from others and to which edaphic conditions and host each VAM fungus is best adapted. On the basis of this information it will be possible to identify most appropriate VAM inoculant for particular plant community under particular set of soil and climate conditions. In view of this, attempts should to made in this direction.

- Without a clear understanding of the interactions between VAM technology and agronomic practices, the ability to manipulate the mycorrhizal symbiosis to the benefits of agriculture will be severely limited. In order to ensure optimum benefit from VAM technology as well as agronomic practices, integrated package of management should be developed for every host-symbiont partnership.

- Recommendation of integrated package should be based on microplot experiments rather than on pot experiments. It should be finally tested under field conditions before releasing it for large scale application.

F Bulk of information has been collected by the mycorrhizasts of the country on the interaction between VAM endophytes and soil borne plant pathogens. In general, VAM decrease the severity of root diseases, however, few cases have been reported where they had no effect on the disease or they supported the severity of disease. Neverthless, through numerous studies it has been established that VAM endophytes have potential to suppress the activity of soil-borne pathogens and diseases caused by them. Thus they can be safely exploited as biocontrol agents since their use will be inexpensive and ecofriendly method for disease control. In order to utilize VAM as biocontrol agent for soil-borne disease on a field scale studies should be initiated on the following lines:

- Most of the studies on the interaction between VAM and soil-borne pathogens have been conducted under pot conditions. For practical application of the results of these studies, microplot and field trials should be carried out. If possible, multilocational trials should also be made with same biocide and same quantity of the inoculum in one lot, keeping all other conditions similar, as far as possible.
- Only fragmentary information is available on the mechanisms involved in the suppression of the severity of soil borne diseases by VAM endophytes. Concerted efforts should be made to generate meaningful information on the biocontrol efficiency of mycorrhizal symbionts and mechanisms involved in such controls. In depth studies on non-nutritional influences of mycorrhizal endophytes on the pathogens may yield a more conclusive picture.
- Abiotic factors such as drought, salinity and stress are expected to affect the interactions between VAM symbiont and pathogens. Efforts should be made to collect information with reference to this and effective correlation models should be developed integrating management of abiotic factors and pathogens through VAM symbiont.
- Attempts should be made to develop integrated management systems including effective VAM for suppressing the soil-borne diseases. This appears to be a practical approach especially for disease complexes e.g. fungal and nematode disease complex.

G In order to meet the scarcity of plant resources because of population explosion attempts are going on in the country to bring cultivable wastelands under green cover. In addition, a number of revegetation programmes are in progress for the lands which have been degraded due to human activities, urbanization and industrialization. Because of various types of stresses associated with wastelands or degraded lands, establishment of plantation on them requires suitable scientific inputs and technologies. Because of its verstality, VAM technology has been proved to be a very fruitful technology in the plantation programmes. Mycorrhizasts of the country have made extensive studies on some specific wastelands and degraded lands and have developed suitable host-VAM models for plantation programmes. However, most of their studies are of limited value because they have been conducted under pot conditions. Moreover, many other types of wastelands and degraded lands have not been tackled. Due attention to the following points is required for a better success of plantation programmes:

- VAM technology simply improves the efficiency of the roots to absorb certain nutrients available in the soil but has negligible role in providing any additional nutrient itself to the host. In view of this, the nutritional stress, a common limitation of wastelands or degraded lands cannot be overcome only through VAM technology. It should be integrated with other technologies and inputs like use of organic matter, biofertilizers, fertilizers etc., enriching the soil with nutrients. The constraints associated with different wastelands and degraded lands should be properly identified and an appropriate package of management with VAM technology as a component should be developed to tackle them.
- Natural vegetation consists of three types of plant communities viz. herbs, shrubs and trees. To provide sustainability to the raised plantation, all the three types of plant

communities, preferably selected from the indigenous flora of the region should be included in the programmes for the development of wastelands or degraded lands. Appropriate package of managment including VAM technology should be developed for these plant communities for a better success of the programmes.

- Mycorrhizal species/strains tolerant to specific stresses will be most appropriate VAM inoculants for wastelands or degraded lands. Such species/strains should be recovered from the specific wastelands or degraded land, which is to be developed, and the most potent one should be identified and selected for use as VAM inoculant.
- A variety of wastelands or degraded lands are occupying a large area in different parts of the country. Technological package for their management have not been developed. Due attention should be given to them and attempt should be made to develop VAM technology for their development. Results of pot experiments should be tested at site before making any recommendations for large-scale application.
- Attempts are being made for bioremediation of stressed envionments by raising selected plant species tolerant to specific stress conditions. VAM technology may go a long way to assist such attempts; therefore, research should be directed also to this area.

H Inoculum production of VAM fungi presents a very different problem. These fungi fail to grow apart from their hosts. Obligate symbiotic mode of growth, non-availability of pure culture and expensive means of inoculum production have greatly jeopardized their practical exploitation. Although genetic engineering of these fungi may, infuture, provide the possibility of their production in the absence of the host plants, classical techniques for the inoculum production will still be in use for considerable time. A major step towards mass production of pure inoculum of these fungi is the success in growing them in root organs cultures or T-DNA transformed roots. However, this success is of limited value because subculturing from mother culture has not given positive results. Since the inoculum produced by this method has been shown to be effective for plants raised under natural conditions, it has been suggested that possibly any residual host material is necessary for persistence of infectivity or development of symbiosis. Nevertheless, attempts are going on to explore the possibility of practical application of the inoculum produced through this technology and making it available to the actual users in easily transportable form. Indian mycorrhizasts have made attempts to improve the techniques related to mass production of VAM inoculum. Few of them have tried to perfect the tissue culture technique for the purpose. Following points should receive due attention by the mycorrhizasts of the country interested in mass production of VAM inoculum.

- The approach to multiply VAM inocula under defined axenic cultural conditions or in organ culture should receive greater emphasis.
- Intensive efforts should also be continued to culture VAM fungi on synthetic media. Attempts should be made for manipulating nutrients and other parameters for their mass multiplication.

- Research should be intensified on bacteria like organelles (BLOs), bacterial associates, hyperparasites and microfauna associated with VAM fungi as these play an important role in the mass production of VAM.
- Mycorrhizal fungi should be evaluated for their performanceunder different edapho-climatic conditions. Ultimately, mycorrhizal fungi found efficient under broad spectrum of site conditions should be selected for mass multiplication.
- Attempts to improve the performance of classical techniques of mass production of VAM inoculum should be continued.

I VAM fungi are difficult to identify in the absence of a complete knowledge of their life cycle. They have also not been cultured on artificial media because of being obligate symbionts. More information on taxonomy, morphology, anatomy, cytology and genetics of these forms may go a long way to overcome these difficulties. In view of this, Indian mycorrhizasts are required to direct their efforts towards these aspects of study on VAM fungi with due attention to the following:

- Studies on taxonomy and life cycle of VAM fungi should be intensified. Most recent techniques such as ELISA, finger printing, isozyme patterns, and protein profiles should be employed alongwith morphological characters and untrastructural details for identification of fungi. A key for the identification of Indian VAM fungi should be prepared.
- Expert groups should be involved in the development of taxonomic studies on Indian VAM fungi employing molecular study tools. All described cultures should be maintained live and deposited with authorized and established centres for mycorrhizal collection in the country and / or abroad.
- Parameters related to growth and productivity in mycorrhizal plants are not sufficient to provide a clear understanding of the mechanisms through which the symbionts provide benefits to their hosts. Additional parameters such as anatomical and cytological changes in the host because of mycorrhizal infection should be included and given due consideration for better interpretation of results. Correlation between root architecture and mycorrhizal dependency needs thorough investigations.
- Only fragmentary information is available on the genomes of the Glomales. A knowledge of genetics of VAM fungi will enable the mycorrhizasts to exploit these biotools in plant productive systems more efficiently. Attempts should be made to identify genes of the symbiont governing difficult phases of its life cycle and development of mycorrhizal symbiosis. This venture will go a long way to overcome many hurdles related to multiplication of VAM inocula under defined axenic cultural conditions or in organ culture through a favourable genome manipulation.

REFERENCES

Abbott, L.K. 1982. Comparative anatomy of vesicular-arbuscular mycorrhizas formed on subterranean clover. *Australian Journal Botany*. **30** : 485-500.

Abbott, L.K. and Robson, A.D. (1984). Colonization of the root system of subterranean clover by three species of vesicular-arbuscular mycorrhizal fungi. *New Phytologist*. **96**: 275-281.

Abbott, L.K. and Robson, A.D. (1991). Field management of VA mycorrhizal fungi. *Belsville Symposia in Agricultural Research*. **14**: 355-362.

Abdel-Fatiah, G. M. and Mohamedin, A. H. (2000). Interactions between a vesicular-arbuscular mycorrhizal fungus (*Glomus intraradices*) and *Streptomyces coelicolor* and their effects on sorghum plants grown in soil amended with chitin of brawn scales. *Biology and Fertility of Soils* **32** (5): 401-409.

Abhijeet, K., Halbe, S. and Nair, L.N. (1997). VAM of *Casuarina* from Konkan. In: *Microbial Biotechnology*. (Eds. S.M. Reddy, H.P. Srivastava, D.K. Purohit and S. Ram Reddy) Jodhpur University, Jodhpur : 64-66.

Addy, H.D., Schaffer, G.F., Miller, M.H. and Peterson, R.L. (1994). Survival of the external mycelium of a VAM fungus in frozen soil overwinter. *Mycorrhiza*. **5**(1): 1-5.

Adholeya, A. and Cheema, G.S. (1990). Evaluation of VA mycorrhizal inoculation in micropropagated *Populus deltoides* Marsh clones. *Current Science*. **59**(23):1244-1247.

Adholeya, A. and Khanna, S. (1990). Sporulation in *Glomus mosseae* under *in vitro* conditions. In: *Current Trends in Mycorrhizal Research* (Eds. B.L.Jalali and H. Chand), H.A.U. Hissar: 53-54

Adholeya, A., Johri, B.N. and Chauhan, R.K.S. (1988a). Effect of VAM-*Rhizobium* interaction on productivity, N-P uptake, dry weight and infection per cent in Indian mung bean (*Vigna radiata*). In: *Mycorrhizae for Green Asia*. (Eds. A. Mahadevan, N. Raman and K. Natrajan), Madras University, Madras :195-197.

Adholeya, A., Johri, B.N. and Chauhan, R.K.S. (1988b). VAM - *Rhizobium* interaction and its effect on colonization and nitrogen fixation in mung bean (*Vigna radiata*) under field trials. In: *Mycorrhizae for Green Asia*. (Eds. A. Mahadevan, N. Raman and K. Natrajan), Madras University, Madras :198-200.

Adholeya, A., Verma, A. and Bhatia, N. P. (1997). Influence of media gelling agents on root biomass and *in vitro* VA-mycorrhizal symbiosis of carrot with *Gigaspora margarita*. *Biotropia*.**10**:63-74.

Agrawal, J. and Chauhan, U.K. (1995). Growth response of vesicular-arbuscular mycorrhizae on *Dalbergia sissoo*. In: *Mycorrhizae: Biofertilizers for the Future*. (Eds. Alok Adholeya and Sujan Singh), TERI, New Delhi: 438-440.

Ahiabor, B.D. and Hirata, H. (1994). Characteristic responses of three tropical legumes to the inoculation of two species of VAM fungi in Andosol soils with different fertility. *Mycorrhiza*. **5**(1): 63-70.

Aldwell, F.E.B. (1983). *Immunological techniques to identify VAM fungi.* Ph.D. thesis, Otago University.

Aldwell, F.E.B., Hall, I.R. and Smith, J.M.B. (1983). *Soil Biology and Biochemistry.* **15** : 377-378. cf. A review of serological techniques for the identification of mycorrhizal fungi. (Aldwell and Hall). In : *Mycorrhiza News* **3** (3) 1991. TERI, New Delhi.

Aldwell, F.E.B., Hall, I.R. and Smith, J.M. (1985). Enzyme-linked immunosorbent assay as an aid to taxonomy of the Endogonaceae. *Transactions British Mycological Society.* **84**: 399-402.

Allen, F.B. and Allen, M.F. (1988). Facilitation of succession by the nonmycotrophic colonizer *Sasola kali* (Chenopodiaceae) on a harsh site: Effect on mycorrhizal fungi. *American Journal of Botany.* **75**: 257-266.

Allen, F.B. and Cunningham, G.L. (1983). Effect of vesicular arbuscular mycorrhizae on *Distichlis spicata* under three salinity levels. *New Phytologist.* **93**: 227-236.

Allen, F.B., Chambers, J.C., Connor, K.F., Allen, M.F. and Brown, R.W. (1987). Natural reestablishment of mycorrhizae in disturbed alpine ecosystems. *Arctic Alpine Research.* **19**: 11-20.

Allen, M.F. (1982). Influence of vesicular arbuscular mycorrhizae on water movement through *Bouteloua gracilis* (H.B.J.) Lag. ex. Stend. *New Phytologist.* **91**: 191-196.

Allen, M.F. (1984). Physiology of mycorrhizae: a key to understanding successful plant establishment. In: *VA-mycorrhizae and reclamation of arid lands.* (Eds. S.E. Williams and M.F. Allen), Wyoming University, Wyoming, U.S.A.: 69-80.

Allen, M.F., Moore, T.S. and Christensen, M. (1980). Phytohormone changes in *Bouteloua gracilis* infected by vesicular arbuscular mycorrhiza. I. Cytokinin increases in the host plant. *Canadian Journal of Botany.* **58**: 371-374.

Allen, M.F., Moore, T.S. and Christensen, M. (1982). Phytohormone changes in *Bouteloua gracilis* infected by vesicular arbuscular mycorrhiza. II. Altered levels of gibberellin-like substances and abscisic acid in the host plant. *Canadian Journal of Botany.* **60** :468.

Allen, M.F., Smith, W.K., Moore., T.S. and Christensen, M. (1981). Comparative water relations and photosynthesis of mycorrhizal and non-mycorrhizal *Bouteloua gracilis* (HBK) Lag ex. Stend. *New Phytologist.* **88**: 683-693.

Ambika, P.K., Das, P.K., Katiyar, R.S. and Choudhury, P.C.(1994) The influence of vesicular arbuscular mycorrhizal association on growth, yield and nutrient uptake in some mulberry genotypes (*Morus* spp.). *Indian Journal of Sericulture.* **33**(2):166-169.

Ames, R.N., Mihara, K.L. and Bethlenfalvay, G.J. (1987). The establishment of microorganisms in vesicular-arbuscular mycorrhizal and control treatments. *Biological Fertility of Soils.* **3**: 217-223.

Ames, R.N., Reid, C.P.P. and Ingham, E.R. (1984). Rhizosphere bacterial populations responses to root colonization by a vesicular-arbuscular mycorrhizal fungus. *New Phytologist.* **96**: 555-563.

Ammaiyappan Salvam and Ayyamperumal Mahadevan (2002). Distribution of mycorrhizas in an abandoned fly ash pond and mind sites of Neyveli Lignite Corporation, Tamil Nadu, India. *Basic and Applied Ecology* **2** (3): 277-284.

Ammani, K. and Rao, A.S. (1995). Effect of RD 9 and quinolphos on vesicular-arbuscular mycorrhizal colonization in cotton. In: *Mycorrhizae: Biofertilizers for the Future*. (Eds. Alok Adholeya and Sujan Singh), TERI, New Delhi: 131-132.

Ammani, K. and Rao, A.S. (1996). Effect of two arbuscular mycorrhizal fungi *Acaulospora spinosa* and *A. scrobiculata* on upland rice varieties. *Microbiological Research*. **151**(3): 235-237.

Ammani, K., Kosaiah, G., Rao, A.S., Satyanarayana and Seenaiah, P. (1989). Increased mycorrhizal colonization in gamma ray-induced green gram micromutants. *Current Science*. **58**(11): 630-631.

Ammani, K., Rao, V.U. and Rao, A.S. (1988). Incidence of vesicular-arbuscular mycorrhizae in sugarcane. In: *Mycorrhizae for Green Asia*. (Eds. A. Mahadevan, N. Raman and K. Natrajan), Madras University, Madras : 29-31.

Ammani, K., Venkateswarlu, K. and Rao, A.S. (1985) Development of vesicular-arbuscular mycorrhizal fungi on upland rice variety. *Current Science*. **56**(21): 1120-1122.

Ammani, K., Venkateswarlu, K. and Rao, A.S. (1986). Vesicular-arbuscular mycorrhizal fungal sporocarps associated with *Pennisetum pedicillatum* Proc. *Indian Academy of Sciences*. (Plant Sciences) **96**(2): 153-158.

Anandraj, M. and Sharma, Y.R. (1994). Effect of vesicular arbuscular mycorrhiza on rooting of black pepper (*Piper nigrum* L.). *Journal of Spices and Aromatic Crops*. **3**:39-42.

Anandraj, M., Kandiannan, K., Sivaraman, K. and Sarma, Y.R. (1999). Identification of efficient strains of vesicular arbuscular mycorrhizae for black pepper (*Piper nigrum* L.). Abstracts: *National Conference on Mycorrhiza*, Barkatulla University, Bhopal: 71.

Ananthapadmanabha, H.S., Nagaveni, H.C., Vijayalakshmi, G. and Somashekar, P.V. (1998). Comparative effect of inorganic and biofertilizer on growth of teak plants. *Indian Journal of Forestry*. **21**(1):1-3.

Antibus, R.K. and Lesica, P.C. (1985). Effect of soil parent material on mycorrhizal infection of alpine plants. In: *Proc. of 6th North American Conference on Mycorrhiza* (NACOM). (Ed. R. Malina), Forest Research laboratory, Cornvallis, OR, pp. 313.

Anusuya, D. (1995a). Interaction of carbendzim and VAM fungi on growth and yield of chilli (*Capsicum annuum* L.). In: *Mycorrhizae: Biofertilizers for the Future*. (Eds. Alok Adholeya and Sujan Singh), TERI, New Delhi: 144-146.

Anusuya, D. (1995b). Interaction of carbendzim and VAM fungi on growth and yield of tomato. *Indian Journal of Mycology and Plant Pathology*. **25**(3): 307.

Anusuya, D. and Dhaneshwari, D. (1995). Effect of fungicides on nodulation and vesicular-arbuscular mycorrhizal development in groundnut. In: *Mycorrhizae: Biofertilizers for the Future*. (Eds. Alok

Anusuya, D. and Shobha, B.A. (1995). Effect of glyphosate on vesicular-arbuscular mycorrhizal association in green gram (*Phaseolus aureus* L.). In: *Mycorrhizae: Biofertilizers for the Future*. (Eds. Alok Adholeya and Sujan Singh), TERI, New Delhi: 336-337.

Anusuya, D. and Vadivoo, K.S. (2003). Multualistic symbiosis of VAM fungi and *Trichoderma* on micropropagated *Dianthus caryophyllus* L. *Mycorrhiza News* **14** (4): 13-15.

Ardey, J. and Mukerji, K.G. (1986). Fungi associated with the roots of herbaceous plants, *Current Science*. **55**(18): 934-935.

Arya, A. (1999). Application of AM fungi to increase biomass of neem seedlings. Abstracts: *National Conference on Mycorrhiza*, Barkatulla University, Bhopal: 71.

Awasti, R.P., Godara, R.K. and Kaith, N.S. (1999). Correlation study between VA-mycorrhizae spore number, root colonization, *Azotobacter* population and fruit yield of July Elberta peach. *Journal of Hill Research* **12** (1): 1-4.

Azcon-Aguilar, C. and Barea, J.M. (1975). Synthesis of auxins, gibberellins and cytokinins by *Azotobacter vinelandii* and *Azotobacter beijerinckii* related to the effects produced on tomato plants. *Plant and Soil*. **43**: 609-619.

Azcon-Aguilar, C. and Barea, J.M. (1978). Effects of interactions between different culture practices of Phosphobacteria and *Rhizobium* on mycorrhizal infection, growth and nodulation of *Medicago sativa. Canadian Journal of Microbiology*. **24**:

Azcon-Aguilar, C. and Barea, J.M. (1981). Field inoculation of *Medicago* with VA mycorrhiza and *Rhizobium* in phosphate-fixing agricultural soil. *Soil Biology and Biochemistry*. **13**: 19-22.

Azcon-Aguilar, C., Gianinazzi-Pearson, V., Fardeau, J.C.and Gianinazzi, S. (1986). Effect of vesicular-arbuscular mycorrhizal fungi and phosphate-solubilizing bacteria on growth and nutrition of soybean in a neutral-calcareous soil amended with ^{32}P-^{45}Ca-tricalcium phosphate. *Plant and Soil*. **96**:3-15.

Azcon, R., Azcon-Aguilar, C. and Barea, J.M. (1978). Effects of plant hormones present in bacterial cultures on the formation and responses to VA endomycorrhiza. *New Phytologist*. **80**: 359-364.

Azcon, R., El-Atrach, F. and Barea, J.M. (1988). Influence of mycorrhiza vs. soluble phosphate on growth nodulation and N_2 fixation ^{15}N in alfa-alfa under different levels of water potential. *Biological Fertility of Soils*. **7**: 28-31.

Babu, K.S. and Manoharachary, C. (2003). Occurrence of arbuscular mycorrhizal fungi in rhizosphere soils of some medicinal plants. *Indian Phytopathology* **56** (2): 223-227.

Babu, R. S., Poornima, K. and Suguna, N. (2001). Mass production of vesicular-arbuscular mycorrhizae using different hosts. *Mycorrhiza News* **13** (1): 20-21.

Babu, R.S., Sankaranarayanan, C. and Jothi, G. (1998). Management of *Pratylenchus zeae* on maize by biofertilizers and VAM. *Indian Journal of Nematology*. **28**(1):77-80.

Babu, R.S., Saralabai, V.C. and Vivekanandan, M. (1995). Vesicular-arbuscular mycorrhizal colonization and biomass production in economically important crop plants grown directly over the sludge. In: *Mycorrhizae: Biofertilizers for the Future*. (Eds. Alok Adholeya and Sujan Singh), TERI, New Delhi: 159-163.

Babu, R.S.H., Lokeshwar, D., Rao, N.S. and Rao, B.R. (1988). The response of chilli (*Capsicum annuum* L.) plants to early inoculation with mycorrhizal fungi at different levels of phosphates. *Journal of Horticultural Science*. **63**(92): 315-32..

Baby, U.I. and Premkumar, R. (2001). Incidence of vesicular arbuscular mycorrhizal fungal spores in tea soils. *Newsletter-UPASI Tea Research Foundation* **11**(2): 5.

Backiyavathy, M.R. and Rani Perumal (1999). Effect of different levels of nitrogen and phosphorus on atmospheric nitrogen fixation and computation by various methods in four species of *Acacia*. *Madras Agricultural Journal* **86** (7/9): 419-422.

Baghel, P.P.S., Bhatti, D.S. and Jalali, B.L. (1990). Action of VA mycorrhizal fungus and *Tylenchulus semipenetrans* on citrus. In: *Current Trends in mycorrhizal research. Proceedings of the National Conference on Mycorrhiza.* (Eds. B.L. Jalali and H. Chand) H.A.U., Hisar: 118-119.

Bagyaraj, D.J. and Machado, C. (1996). Phosphorus concentration in soil solution optimum for VA mycorrhizal symbiosis in *Leucaena leucocephala. Annals of Forestry.* **4**(2): 123-128.

Bagyaraj, D.J. and Manjunath, A. (1980). Response of crop plants to VA-mycorrhizal inoculation in an unsterile Indian soil. *New Phytologist.* **87**: 33-36.

Bagyaraj, D.J. and Menge, J.A. (1978). Interaction between a VA mycorrhiza and *Azotobacter* and the effect on the rhizosphere microflora and plant growth. *New Phytologist.* **80**: 567-573.

Bagyaraj, D.J. and Shreeramulu, K.R. (1982). Preinoculation with VA mycorrhiza improves growth and yield of chilli transplanted in the field and saves phosphatic fertilizer. *Plant and Soil.* **69**(3): 375-381.

Bagyaraj, D.J. and Varma, A.(1987). State of art of mycorrhizae. In: *Proceedings of a workshop on mycorrhiza* J.N.U., New Delhi, pp. 25-32.

Bagyaraj, D.J., Manjunath, A. and Patil, R.B. (1979a). Interaction between a vesicular-arbuscular mycorrhiza and *Rhizobium* and their effects on soybean in the field. *New Phytologist.* **82**(1):141-145.

Bagyaraj, D.J., Manjunath, A. and Patil, R.B. (1979b). Occurrence of vesicular-arbuscular mycorrhizas in some tropical aquatic plants. *Transactions of the British Mycological Society.* **72**(1):164-167.

Bagyaraj, D.J., Reddy, M.S.B. and Nalini, P.A. (1989). Selection of an efficient inoculant of VA mycorrhizal fungus for *Leucaena. Forest Ecology and Management.* **27**(2): 81-85.

Balachandar, D. and Nagarajan, P. (1999). Dual inoculation of vesicular arbuscular mycorrhiza and *Rhizobium* in green gram. *Legume Research* **22** (3): 177-180.

Balasubramanian, A. and Srinivasan, P.S. (1995). Response of certain tree species to vesicular-arbuscular mycorrhizae inoculation. In: *Mycorrhizae: Biofertilizers for the Future.* (Eds. Alok Adholeya and Sujan Singh), TERI, New Delhi: 345-347.

Balasubramanian, A., Srinivasan, P.S. and Gurumurthi, K. (1995). Interactive effect of vesicular-arbuscular mycorrhizae and *Frankia* on *Casuarina* seedling growth. In: *Mycorrhizae: Biofertilizers for the Future.* (Eds. Alok Adholeya and Sujan Singh), TERI, New Delhi: 261-263.

Bali, M. and Mukerji, K.G. (1988). Effect of VAM fungi on *Fusarium* wilt of cotton and jute. In: *Mycorrhizae for Green Asia.* (Eds. A. Mahadevan, N. Raman and K. Natrajan), Madras University, Madras : 233-234.

Bansal, M. and Mukerji, K.G. (1994). Positive correlation between VAM induced changes in root exudation and mycorrhizosphere mycoflora. *Mycorrhiza.* **5**(1): 39-44.

Baon, J.B., Smith, S.E. and Alston, A.M. (1994). Phosphorus uptake and growth of barley as affected by soil temperature and mycorrhizal infection. *Journal of Plant Nutrition.* **17**(2-3): 479-492.

Barea, J.M. (1988). Endomycorrhiza biotechnology to improve plant productivity in intercropping systems. In : *2nd European Symposium on Mycorrhizae*, Prague: 9-10.

Barea, J.M. and Azcon-Aguilar, C. (1982). Production of plant growth regulating substances by the vesicular-arbuscular mycorrhizal fungus *Glomus mosseae. Applied Environmental Microbiology*. **43**: 810-813.

Barea, J.M. and Azcon-Aguilar, C. (1983). Interactions between phosphate solubilizing bacteria and VA mycorrhiza to improve the utilization of rock phosphate by plants in non-acidic soils. In: *3rd International Congress on Phosphate Compounds*. Brussels.

Barea, J.M. and Azcon-Aguilar, C. (1985). Mycorrhizas and their significance in nodulating nitrogen fixing plants. In: *Advances in Agronomy*. (Ed. N.C. Brandy), Academic Press, New York **36** : 1-54.

Barea, J.M., Azcon, R. and Hayman, D. (1975). Possible synergistic interactions between *Endogone* and phosphate-solublizing bacteria in low-phosphate soil. In : *Endomycorrhiza*. (Eds. B. Mosse, F.E. Sanders and P.B. Tinker), Academic Press, New York: 409-417.

Barea, J.M., Azcon-Aguilar, C. and Azcon, R. (1987). Vesicular-arbuscular mycorrhiza improves both symbiotic N_2 fixation and uptake from soil as assessed with a ^{15}N technique under field condition. *New Phytologist*. **106**: 717: 725.

Barea, J.M., Azcon-Aguilar, C. and Azcon, R. (1988). The role of mycorrhiza in improving the establishment and function of the *Rhizobium*-legume system under field conditions. In: *Nitrogen fixation by legumes in mediterranean agriculture.* (Eds. D.P. Beck and L.A. Materon), Martinus Nijhoff Publishers. Netherlands,: 153-160.

Barea, J.M., Bonis, A.F. and Olivares, J. (1983). Interactions between *Azospirillum* and VA mycorrhiza and their effects on growth and nutrition of maize and ryegrass. *Soil Biology and Biochemistry*. **15**: 705-709.

Barea, J.M., Navarro, E. and Montoya, E. (1976). Production of plant growth regulators by rhizosphere phosphate solubilizing bacteria. *Journal of Applied Bacteriology*. **40**.

Barthakur, M.P. and Bordoloi, D.N. (1990). Occurrence of vesicular-arbuscular mycorrhizal fungi in some *Cymbopogon* species of north-east India. *Current Science*. **59**(9): 470-471.

Basu, M. and Srivastava, N.K. (1998). Root Endophytes in medicinal plants: Their population and effects. *Abstract : 7th International Congress of Plant Pathology*, Edinburgh, Scotland : 98.

Baya, A.M., Boethling, R.S. and Ramos-Cormenzana (1981). Vitamin production in relation to phosphate solubilization by soil bacteria. *Soil Biology and Biochemistry*. **13** : 527-531.

Bayan, N. and Talukdar, N.C. (1997). Effect of coinoculation with *Gigaspora and Bacillus* in green gram (*Vigna radiata). Indian Journal of Microbiology*. **37**(2): 103-104.

Beamer, E.E.J. (1992). Plant life span and response to inoculation with vesicular arbuscular mycorrhizal fungi. III. Responsiveness and residual soil P levels. *Mycorrhiza*. **1**(4): 169-174.

Becard, G. and Fortin, J.A. (1988). *New Phytologist*. **108** : 211-218.

Becker, W.N. (1976). *Quantification of onion Vesicular Arbuscular Mycorrhizae and their Resistance to Pyrenochaeta terrestris*, Ph.D. thesis, Univ. of Illinois, Urbana.

Beena, K. R.; Raviraja, N. S.; Arun, A. B.;and Sridhar, K. R. (2000). Diversity of arbuscular mycorrhizal fungi on the coastal sand dunes of the west coast of India. *Current Science* **79** (10): 1459-1466

Beena, K.R., Arun, A.B., Raviraja, N.S. and Sridhar, K.R. (2001). Association of arbuscular mycorrhizal fungi with plants of coastal sand dunes of west coast of India. *Tropical Ecology* **42**(2): 213-222.

Beena, K.R., Arun, A.B., Raviraja, N.S. and Sridhar, K.R. (1999). Geographical diversity of arbuscular mycorrhizal fungi of coastal sand dunes of west coast of India. Abstracts: *National Conference on Mycorrhiza*, Barkatulla University, Bhopal: 38.

Behl, H.M. (1990). Role of endomycorrhizae in fuelwood plantation nurseries for alkaline soil sites. In: *Trends in mycorrhizal research. Proceedings of the National Conference on Mycorrhiza.* (Eds. B.L. Jalali and H. Chand), H.A.U., Hisar: 137-138.

Behl, R.K., Sharma, H., Kumar, V. and Narula, N. (2003). Interactions amongst mycorrhiza, *Azotobacter chroococcum* and root characteristics of wheat varieties. *Journal of Agronomy and Crop Science* **189** (3): 151-155.

Bellgard, S.E. (1993). The top soil as the major store of the propagules of vesicular arbuscular fungi in south east Australian sand stone soils. *Mycorrhiza.* **3**(1): 19-24.

Bellgard, S.E., Whelan, R.J. and Muston, R.M. (1994). Impact of wild fire on vesicular-arbuscular mycorrhizal fungi and their potential to influence re-establishment of post fire plant communities. *Mycorrhiza.* **4**: 139-146.

Beniwal, R.S., Toky, O.P. and Sharma, P.K. (1992). Effect of VA mycorrhizal fungi and phosphorus on growth and nodulation of *Acacia nilotica* (L.) Willd ex Del. *Crop Research* (Hisar). **5** (Supplement): 172-176.

Benjaminn, J.R. (1979). Zygomycetes and their spores. In: *The Whole Fungus.* (Ed. B. Kendrick), National Museums of Canada, Ottawa, pp. 573-616.

Berch, S.M. (1986). Endogonaceae : Taxonomy, specificity, fossil record, phylogeny. In: *Frontiers in Applied Microbiology*. Vol.2 (Eds. K.G. Mukerji, N.C. Pathak and V.P.Singh), Indian Print House, Lucknow: 161-186.

Berch, S.M., Gamiet, S. and Deom, E. (1988). Mycorrhizal status of some plants of South Western British Columbia. *Canadian Journal of Botany*. **68**(10): 1924-1928.

Bethlenfalvay, G.J. and Yoder, J.F. (1981). The *Glycine-Glomus-Rhizobium* symbiosis. I. Phosphorus effect on nitrogen fixation and mycorrhizal infection. *Physiology of Plant.* **52**: 141-145.

Bethlenfalvay, G.J., Brown, M.S., Mihara, K.L. and Stafford, A.E. (1987). *Glycine-Glomus-Rhizobium* symbiosis. V. Effects of mycorrhiza on nodule activity and transpiration in soybeans under drought stress. *Plant Physiology*. **85**: 115-119.

Bhadauria, S. and Yadava, R. (2000). Endomycorrhizal associations of *Acacia nilotica* in nurseries and plantations from alkaline region of U.P. *Indian Journal of Soil Conservation* **28**(1): 48-53.

Bhadraiah, B., Kanakadurga, V.N., Ramarao, P., and Manoharachary, C. (2002). Effect of AM fungi and rock phosphate on phosphatase activities in *Tectona grandis* L. In: *Frontiers in Microbial Biotechnology and Plant Pathology* [Eds. C. Manoharachary, D.K. Purohit, S. Ram Reddy, M.A. Singara Charya, S. Girisham.]: 259-262 ISBN 81-7233-291-2.

Bhadraiah, B., Kankadurga, V.V., Ramarao, P. and Manoharachary, C. (1999). Effect of VAM fungi and rock phosphate on phosphatase acivities in *Terminalia arjuna*. Abstracts: *National Conference on Mycorrhiza*, Barkatulla University, Bhopal: 58.

Bhandari, S.C. and Rawat, U.S. (1995). Nutrient use efficiency in mycorrhizal legume-soybean. In: *Mycorrhizae: Biofertilizers for the Future*. (Eds. Alok Adholeya and Sujan Singh), TERI, New Delhi: 365-368.

Bharat, N. K. and Bhardwaj, L. N. (2002). Association of vesicular-arbuscular mycorrhizal fungi with apple in Himachal Pradesh. *Plant Disease Research* **17** (2): 326-328.

Bhaskaran, C. and Selvaraj, T. (1997). Seasonal incidence and distribution of VA-mycorrhizal fungi in native saline soils. *Journal of Environmental Biology*. **18**(3): 209-212.

Bhat, M.N., Jeyarajan, R. and Ramaraj, B. (1993). Response of subabul to vesicular-arbuscular mycorrhizal inoculation in sterile and unsterile soil. *Indian Journal of Forestry*. **16**(4):309-312.

Bhat, M.N., Jeyarajan, R. and Ramaraj, B. (1994a). Field response of three forest tree seedlings to inoculation with VAMF. *International Tree Crops Journal*. **8**(1):37-41.

Bhat, M.N., Jeyarajan, R. and Ramaraj, B. (1994b). Occurrence of vesicular-arbuscular mycorrhizae in forest trees at Nilgiris. *Indian Journal of Forestry*. **17**(2): 175-177.

Bhat, M.S. and Mahmood, I. (2000). Role of *Glomus mosseae* and *Paecilomyces lilacinus* in the management of root knot nematode on tomato.*Archieves of Phytopathology and Plant Protection* **33** (2): 131-140.

Bhat, P.R. and Kaveriappa, K.M. (1997). Occurrence of vesicular-arbuscular mycorrhizal fungi in the roots of *Colocasia esculenta* (L.) Schott. *Mycorrhiza News*. **9**(2): 12-13.

Bhatia, N.P., Adholeya, A. and Sharma, A. (1998). Biomass production and changes in soil productivity during long-term cultivation of *Prosopis juliflora* (Swartz) DC inoculated with VA mycorrhiza and *Rhizobium* spp. in a semi-arid wasteland. *Biological Fertility of Soils*. **26**: 208-214.

Bhatia, N.P., Adholeya, A. and Sharma, A. (1999). Inoculation effects of AM fungi and *Rhizobium* spp. on growth performance of *Prosopis juliflora* and consequent improvement in productivity of semi-arid soil. Abstracts: *National Conference on Mycorrhiza*, Barkatulla University, Bhopal: 63.

Bhatt, S., Johri, B.N. and Sharma, A.K. (1999). Influence of *Peudomonas*-arbuscular mycorrhizal fungal interaction on growth of maize (*Zea mays* L.). Abstracts: *National Conference on Mycorrhiza*, Barkatulla University, Bhopal: 21.

Bhattacharjee, M. and Mukerji, K.G. (1980). *Acta Botanica Indica*. **8**: 99-102.

Bhattacharjee, M., Mukerji, K.G. and Mishra, S.(1980). *Sydowia*. **33** : 14-17.

Bhattacharjee, M., Mukerji, K.G., Tiwari, J.P. and Skoropad, W.P. (1982). Structure and hyperparasitism of new species of *Gigaspora*. *Transactions British Mycological Society*. **78**: 184-188.

Bhattacharya, P.M., Misra, D., Saha, F. and Chaudhuri, S. (2000). Arbuscular mycorrhizal dependency of *Eucalyptus tereticornis* Sm: how real is it? *Mycorrhiza News* **12**(3): 11-15.

Bhattacharya, P.M., Misra, D., Saha, J. and Chaudhuri, S. (1999). Mycorrhizal dependency, phosphorus utilization efficiency and relevance of mycorrhiza for bamboo cultivation in laterite wasteland. Abstracts: *National Conference on Mycorrhiza*, Barkatulla University, Bhopal: 47.

Bhattacharya, P.M., Paul, A.K., Saha, J. and Chaudhuri, S. (2002). Change in the root development pattern in bamboo and sweet orange plants upon arbuscular mycorrhization. *Mycorrhiza News* **14** (1): 15-18.

Bhattacharya, P.M., Saba, J., Bannerjee, K., Ali, Md. M. and Chaudhuri, S. (1995). Mycorrhizal relations of bamboo species in different soil situations in West Bengal. In: *Mycorrhizae: Biofertilizers for the Future.* (Eds. Alok Adholeya and Sujan Singh), TERI, New Delhi: 17-21.

Bhattacharya, S. and Bagyaraj, D. J. (2002). Effectiveness of arbuscular mycorrhizal fungal isolates on arabica coffee (*Coffea arabica* L.). *Biological Agriculture & Horticulture* **20**: 125-131.

Bianciotto, V. and Bonfante, P. (1992). Quantification of the nuclear DNA content of two arbuscular mycorrhizal fungi. *Mycorrhizal Research.* **96**(12): 1071-1076.

Bisen, P.S., Dev, A., Gour, R.K., Jain, R.K. and Sengupta, L.K. (1995). Study of vesicular-arbuscular mycorrhizal fungus *Glomus mosseae* in soil samples of Bhopal. In: *Mycorrhizae: Biofertilizers for the Future.* (Eds. Alok Adholeya and Sujan Singh), TERI, New Delhi: 73-76.

Bisen, P.S., Gour, R.K., Jain, R.K., Dev, A. and Sengupta, L.K. (1996). VAM colonization in tree species planted in Cu, Al and coal mines of Madhya Pradesh with special reference to *Glomus mosseae. Mycorrhiza News.* **8**(1): 9-11.

Bisht, V.S., Krishna, K.R. and Nene, Y.L. (1985). Interaction between vesicular-arbuscular mycorrizal fungi and *Phytophthora drechsleri* f. sp. *cajani. International Pigeonpea Newsletter.* **4**:63-64.

Bloss, H.E. (1985). Studies of symbiotic microflora and their role in the ecology of desert plants. *Desert Plants.* **7**(3):119-127.

Bojan, C.R., Ramesh, C. and Mahadevan, A. (1995). Outplanting performance of VAM inoculated sugarcane (*Saccharum officinarum* L.) seedlings raised through tissue culture. *Biology and Biotechnology of Mycorrhizae.* **56**:57-64.

Bolan, G. and Wheelar, C.T. (1980). Non-legume nodule systems. In: *Methods for Evaluating Biological Nitrogen Fixation.* (Ed. F.J. Bergersen), John Wiley and Sons. New York: 185.

Bolan, N.S. and Robson, A.D. (1987). Effects of vesicular arbuscular mycorrhiza on the availability of iron phosphates to plants. *Plant and Soil.* **99**: 401-410.

Bolan, N.S., Robson, A.D. and Barrow, N.W. (1984). Increasing phosphorus supply can increase the infection of plant roots by vesicular-arbuscular mycorrhizal fungi. *Soil Biology and Biochemistry.* **16**: 419-420.

Bond, G. and Wheeler, C.T. (1980). Non-legume nodule systems In: *Methods for Evaluating Biological Nitrogen Fixation* (Eds. F.J. Bergersen). John Wiley & Sons, New York: 185

Bonfante, P. and Perotto, S. (1995). Strategies of arbuscular mycorrhizal fungi when infecting host plants. *New Phytologist.* **130**(1): 3-21.

Borah, A. and Phukan, P.N. (2000). Effect of VAM fungus, *Glomus fasciculatum* and root knot nematode, *Meloidogyne incognita* on brinjal. *Journal of the Agricultural Sciences Society of North-East India* **13**(2): 212-214.

Borges, R.G. and Chaney, W.R. (1989). Root temperature affects mycorrhizal efficacy in *Fraxinus pennsylvanica* Marsh. *New Phytologist.* **112**(3): 411-417.

Brockwell, J. (1980). Experiments with crop and pasture legumes-principles and practice. In: *Methods for Evaluating Biological Nitrogen fixation.* (Ed. F.J. Bergersen), John Wiley and Sons, New York: 417.

Brown, M.E. and Carr, G.H. (1980). Effects on plant growth of mixed inocula of VA endophytes and root-microrganisms. *Rothamsted Experimental Station Annual Report.* **1** : 189.

Burr, T.J. and Caesar, A. (1984). Beneficial plants bacteria. *Critical Reviews in Plant Sciences.* **2**: 1-20.

Butler, E.J. (1939). The occurrences and systematic position of the vesicular arbuscular type of mycorrhizal fungi. *Transactions of British Mycological Society.* **22**: 274.

Butool, F. and Haseeb, A. (1996) Influence of *Glomus aggregatum* on root-knot development caused by *Meloidogyne incognita in Hycoscyamus muticus. Afro-Asian Journal of Nematology.* **6**(2): 188-190.

Buwalda, J.G., Ross, G.J.S., Stribley, D.P., and Tinker, P.B. (1982). The development of endomycorrhizal root system. III. Mathematical representation of the spread of vesicular arbuscular mycorrhizal infection in root system. *New Phytologist.* **9**: 669.

Callow, J.A., Capaccio, L.C.M., Parish, G. and Tinker, P.B. (1978). Detection and estimation of polyphosphate in vesicular-arbuscular mycorrhizas, *New Phytologist.* **80** : 125.

Caron, M., Richard, C. and Fortin, J.A. (1986). Effect of preinfestation of the soil by a vesicular-arbuscular mycorrhizal fungus, *Glomus intraradices* on *Fusarium* crown and root rot of tomatoes. *Phytoprotection.* **67**(1): 15-19.

Chakravarty, P. and Mishra, R.R. (1986). Influence of endotrophic mycorrhizae on the fusarial wilt of *Cassia tora* L. *Journal of Phytopathology.* **115**(2): 130-133.

Champawat, R.S. (1988). Influence of three VAM fungi on nutrient uptake and growth in groundnut (*Arachis hypogaea*). In: *Mycorrhizae for Green Asia* . (Eds. A. Mahadevan, N. Raman and K. Natrajan), Madras University, Madras : 132-133.

Champawat, R.S. (1989). Effect of vesicular-arbuscular mycorrhizal fungi on growth and nutrition of chickpea. *Madras Agricultural Journal.* **76**(6)L 310.

Champawat, R.S. (1990a). Effect of different VAM fungi under varying levels of phosphorus on growth and nutrition uptake of pigeon pea (*Cajanus cajan*). In: *Current trends in mycorrhizal research. Proceedings of the National Conference on Mycorrhiza.* (Eds. B.L. Jalali and H. Chand), H.A.U., Hisar :141-145.

Champawat, R.S. (1990b). Effect of dual inoculation of *Rhizobium* and vesicular arbuscular mycorrhizal fungi on *Pisum sativum. Folia Microbiologica.* **35**(3): 236-239.

Champawat, R.S. (1990c). Response of chickpea (*Cicer arietinum*) to *Rhizobium* and vesicular arbuscular mycorrhiza dual inoculation. *Acta Microbiologica Polonica.* **39**(3/4):163-169.

Champawat, R.S. (1992). Effect of vesicular-arbuscular mycorrhizal fungi on growth and nutrition of chickpea. *Madras Agricultural Journal.* **79**(2): 91-93.

Champawat, R.S. and Pathak, V.N. (1993). Effect of vesicular-arbuscular mycorrhizal fungi on growth and nutrition uptake of pearl millet. *Indian Journal of Mycology and Plant Pathology.* **23**(1): 30-34.

Champawat, R.S., Pathak, V.N. and Krishna, K.R. (1987). Influence of vesicular-arbuscular mycorrhizal fungi on growth and nutrition uptake in barley (*Hordeum vulgare*). *Current Science*, India. **56**(24): 1294-1295.

Chandel, S. and Srivastava, H.P. (1988) Development of VAM in *Cenchrus ciliaris* and *C. setigerous*. In: *Mycorrhizae for Green Asia.* (Eds. A. Mahadevan, N. Raman and K. Natrajan), Madras University, Madras : 35-36.

Chandra, K.K. and Jamaluddin (1999a). Dynamics of VAM infectivity and their population under different seasons in teak and bamboo plantations of Chandrapur (M.S.). *Vaniki Sandesh* **23** (1) 16-20

Chandra, K.K. and Jamaluddin (1999b). Distribution of vesicular arbuscular mycorrhizal fungi in coalmine overburden dumps. *Indian Phytopathology* **52** (3): 254-258.

Chandra, K.K. and Ujjaini, M.M. (2002). Interaction of AMF with three levels of soil organic matter and their influence on seedling biomass and root infection of six forest species. *Myforest* **38** (2): 155-161.

Chandra, K.K., Chaturvedi, P. and Jamaluddin (1997). Comparative study of VA-mycorrhizal root infection and spore density in *Grevilia pteridifolia* in undisturbed and disturbed areas. *Mycorrhiza News.* **9**(2): 9-12.

Chandra, S. and Kehri, H.K. (1994). Microbial spectrum with special reference to VAM fungi in brick fields during natural reclamation. In: *Crop Epidemics, Microbes and Ecosystem Conservation.* (Eds. G.P. Agarwal and S.K. Hasija), Narendra Publishing House, Delhi:185-191.

Chandra, S. and Kehri, H.K. (1995). Microbial technology for wasteland development. *Proceedings of National Academy of Sciences India.* **66**(B), II: 123-145.

Chandra, S. and Zaidi, S.M.H. (1987). Mycorrhizal association in certain crops as affected by growth stages. In : *Proceedings of National Symposium on Fungal Ecology*, Osmania University, Hyderabad : 57-61.

Chandra, S. Srivastava, N. and Kehri, H.K. (2000a). A low input technology for commercial cultivation of *Chrysanthemum* and *Gladiolus*. *Abstracts: Proceedings Symposium "Emerging Trends in Plant Biotechnology*, Botany Section, ISCA, Pune : 8.

Chandra, S., Dubey, R. and Kehri, H.K. (2000b). Wasteland management via petroleum plantation and VAM technology. In : *Innovative Approaches in Microbiology*. (Eds. D.K. Maheshwari and R.C. Dubey), Bishen Singh Mahendra Pal Singh, Dehra Dun : 75-97.

Chandra, S., Kehri, H.K. and Dwivedi, R. (2000c). Managing soil-borne diseases of horticultural crops through VAM fungi. In : *Glimpses in Plant Sciences.* (Eds. K.R. Aneja, M.U. Charaya, A. Aggarwal, D.K. Hans and S.A. Khan), Pragati Prakashan, Meerut : 82-93.

Chandra, S., Kehri, H.K. and Maheshwari, S. (1995a). Macrophomina dry root-rot of Leguminous crops and its management through VAM Biopesticides. In: *Detection of Plant Pathogens and Their Management.* (Eds. J.P. Verma, Anupam Varma and Dinesh Kumar), Angkor Publishers (P) Ltd. New Delhi: 357-364.

Chandra, S., Kehri, S.S. and Kehri, H.K. (1996). Synergism of biofertilizers and their integration for improved crop productivity. In: *Perspectives in biological sciences.* (Eds. V.Rai, M.L. Naik and C. Manoharacharry), Raipur University, Raipur :141-151:

Chandra, S., Srivastava, N and Kehri, H.K. (2004). Selection of VA Mycorrhizal Fungi for *Gladiolus*. *Environmental Biology and Conservation* **9**: 9-13.

Chandra. S., Kehri, H.K. and Chatterjee, T. (1995b). Improved performance of medicinal plants through VAM technology. In: *Recent Advances in Phytopathological Researches*. (Eds. A.K. Roy and K.K. Sinha), M.D. Publications Pvt. Ltd., New Delhi : 7-24.

Chandraghatgi, P.S. and Sreenivasa, M.N. (1995). Possible synergistic interactions between *Glomus macrocarpum* and *Bacillus polymyxa* in chilli. In : *Mycorrhizae: Biofertilizers for the Future*. (Eds. Alok Adholeya and Sujan Singh), TERI, New Delhi: 180-183.

Chandrashekara, C.P. and Patil, V.C.(1997). Influence of VAM inoculation and phosphorus on dry matter production and seed yield of two sunflower (*Helianthus annuus* L.) genotypes. *Journal of Oilseeds Research* . **14**(2): 210-215.

Chandrashekara, C.P., Patil, V.C. and Sreenivasa, M.N. (1995). VA-mycorrhiza mediated P. Effect on growth and yield of sunflower (*Helianthus annuus* L.) at different P levels. *Plant and Soil*. **176**(2): 325-328.

Chang, D.C.W. (1994). What is the potential for management of vesicular-arbuscular mycorrhizae in horticulture? In: *Management of mycorrhizae in Agriculture, Horticulture and Forestry*. (Eds. A.D. Robson, L.K. Abbott, and N. Malajczuk), Kluwer Academic Publishers, The Netherlands. pp. 187-190.

Chapin, F.S. (1980). The mineral nutrition of wild plants. *Annual Review Ecological System* **11**: 233-260.

Chaterjee, T., Arya, A. and Chandra, S. (1994). Application of VAM to increase the alkaloid contents in *Catharanthus roseus* L. In : *Advances in Plant Sciences Research. Volume II.* (Ed. K.C. Sahni), New Delhi : 183-196.

Chaturvedi, C. and Kumar, A. (1991) Nodulation and nitrogenase activity in gram (*Cicer arietinum* L.) as influenced by *Rhizobium* and VA mycorrhiza interaction. *National Academy Science Letters*. **14**(7):289-292.

Chaturvedi, C. and Sharma, A.K. (1988). Interaction of *Glomus caledonium* and *Rhizobium* in relation to uptake and translocation of Labelled phosphatic fertilizers in chickpea (*Cicer arietinum*). In : *Mycorrhizae for Green Asia*. (Eds. A. Mahadevan, N. Raman and K. Natrajan), Madras University, Madras : 134-135.

Chaturvedi, C. and Singh, R. (1990). ^{32}P uptake and translocation in chickpea (*Cicer arientinum* L.) inoculated with vesicular-arbuscular mycorrhiza. *National Academy Science Letters*. **13**(10):371-372.

Chaturvedi, C., Singh, R. and Sharma, A.K. (1987). Occurrence of vesicular-arbuscular (VA) mycorrhiza in *Ascochyta* blight resistant and susceptible cultivars of chickpea. *Current Science*. **56**(16):831-831.

Chaubal, R., Sharma, G.D. and Mishra, R.R. (1982). Vesicular arbuscular mycorrhiza in subtropical aquatic and marshy plant communities. *Proceedings, Indian Academy of Sciences* (Plant Sciences). **91**(1): 69-77.

Chaudhuri, S. and Sengupta, A. (1990). Mycorrhizal relations of successional stages of mangrove vegetation at the Ganges of estuary in India. In: *Current trends in mycorrhizal research*.

Proceedings of the National Conference on Mycorrhiza. (Eds. B.L. Jalali and H. Chand), H.A.U., Hisar: 7-9.

Chauhan, Y. S. and Pokhriyal, T. C. (2002). Nodulation and nitrogen fixation as influenced by *Rhizobium* and mycorrhizal inoculation in *Albizia lebbek* (L.), Benth. *Annals of Forestry* **10**(2): 243-251.

Chaurasia, B. and Khare, P.K. (1999). VAM fungi associations in a protected grassland. *Mycorrhiza News.* **11**(1): 13-14.

Chezhiyan, N., Balasubramani, P., Harris, C.V. and Ananthan, M. (1999). Effect of inorganic and biofertilizers on growth and yield of hill banana var. Virupakshi. *South Indian Horticulture* **47** (1/6): 161.

Chhabra, M.L. and Jalali, B.L. (1995). Effect of vesicular-arbuscular mycorrhiza on utilization of rock phosphate in pigeonpea. In: *Mycorrhizae : Biofertilizers for the Future.* (Eds. Alok Adholeya and Sujan Singh), TERI, New Delhi: 261-263.

Chhabra, M.L. and Jalali, B.L. (1997). Impact of VA-mycorrhizal endophyte *Glomus mosseae* inoculation on utilization of rock-phosphate in wheat. *Plant Disease Research.* **12**(2): 413-415.

Chhabra, M.L., Singh, R.P. and Jalali, B.L. (1990). Studies on vesicular-arbuscular (VA) mycorrhizal impact on growth and development of cowpea (*Vigna unguiculata* (L.) (Walp.). In: *Trends in mycorrhizal research. Proceedings of the National Conference on Mycorrhiza.* (Eds. B.L. Jalali and H. Chand), H.A.U., Hisar : 150-151.

Chinnamuthu, C.R. and Venkatakrishnan, A.S. (2001). Effect of integrating inorganic fertilizer with vermicompost and vesicular arbuscular Mycorrhizae (VAM) on the productivity of sunflower. *Madras Agricultural Journal* **88** (7/9): 424-427.

Chokraborty, P. and Sharma, B.B. (1993). Leaf nutrient composition of sweet orange as affected by combined use of bio and chemical fertilizers. *South Indian Horticulture* **41**(3): 131-134.

Chu-Chou, M. and Grace, L.G. (1979). *Endogone flammicorona* as a mycorrhiza symbiont of Douglas fir in New Zealand *New Zealand Journal for Science.* **9**: 344-347.

Claassen, V.P., Zasoski, R.J. and Tyler, B.M. (1996). A method for direct soil extraction and PCR amplification of endomycorrhizal fungal DNA. *Mycorrhiza.* **6**(5): 447-450.

Clapp, J.P., Young, J.P.W., Merryweather, J.W. and Fitter, A.H. (1994). Diversity of fungal symbionts in arbuscular mycorrhizas from a natural community. *New Phytologist* :

Clark, M.F. (1981a). *Ann. Review of Phytopathology.* **19** : 83-106. cf. A review of serological techniques for the identification of mycorrhizal fungi. (Aldwell and Hall). *Mycorrhiza News* **3** (3) 1991. TERI, New Delhi.

Clark, M.F. (1981b). In : *Immuno Assays for the Eighties.* (Ed. A. Voller), Lancaster, M.Y.P. cf. A review of serological techniques for the identification of mycorrhizal fungi. (Aldwell and Hall). *Mycorrhiza News* **3** (3) 1991. TERI, New Delhi.

Clarkson, D.T. (1985). Factors affecting mineral nutrients acquisition by plants. *Annual Review of Plant Physiology.* **36**: 77-115.

Cluett, H.C. and Boucher, D.H. (1983). Indirect mutualism in the legume-Rhizobium-mycorrhizal fungus interaction. *Ecologia.* **59**: 405-408.

Cooper, K.M. (1984). Physiology of VA mycorrhizal association. In: *VA mycorrhizae*. (Eds. C.LL. Powell and D.J. Bagyaraj), CRC Publication : 155-203.

Cooper, K.M. and Grandison, G.S. (1987). Effect of vesicular arbuscular mycorrhizal fungi on infection of tamarillo (*Cyphomandra betacea*) by *Meloidogyne incognita* in fumigated soil. *Plant Disease*. **71**(12): 1101-1106.

Cox, G., Moran, K.J., Sanders, F., Nockolds, C. and Tinker, P.B. (1980). Translocation and transfer of nutrients in vesicular-arbuscular mycorrhizas. III. polyphosphate granules and phosphorus translocation. *New Phytologist*. **84**: 649.

Cui-Muyi and Nobel Park, S. (1992). Nutrient status, water uptake and gas exchange for three desert succulents infected with mycorrhizal fungi. *New Phytologist*. **122**: 643-649.

Daft, J. and Hacskaylo, E. (1976). Arbuscular mycorrhiza in the anthracite and bituminous coal waste of Pennsylvania. *Journal of Applied Ecology*. **13**: 523-531.

Daft, M.J. and Hacskaylo, E. (1977). Growth of endomycorrhizal and non-mycorrhizal red maple seedlings in sand and anthracite soil. *Forest Science*. **23**: 207-216.

Dakessian, M.S., Brown, M.S. and Bethlenfalvay, G.J. (1986). Relationship of mycorrhizal growth enhancement and plant growth with soil, water and texture. *Plant and Soil*. **94**:439-443.

Dalal, S. and Hippalgaonkar, K.V. (1995). The occurrence of vesicular-arbuscular mycorrhizal fungi in arable soils of Konkan and Solapur. In : *Mycorrhizae: Biofertilizers for the Future*. (Eds. Alok Adholeya and Sujan Singh), TERI, New Delhi: 3-7.

Dangeard, P.A. (1900). Le *"Rhizophagus populinus"*, Dangeard, *Botaniste*. **7**: 285.

Danso, S.K.A. (1988). The use of ^{15}N enriched fertilizers for estimating nitrogen fixation in grain and pasture legumes. In: *Nitrogen fixation by legumes in Mediterranean agriculture*. (Eds. D.P. Beck and L.A. Materon), ICARDA and Martinus Nijhoff. Dordrecht: 245-358.

Das, P. K., Sethi, A. K., Jena, M. K. and Patra, R.K. (1999). Effect of P sources and dual inoculation of VA-mycorrhiza and *Rhizobium* on dry matter yield and nutrient uptake by green gram *(Vigna radiate* L.*) Journal of the Indian Society of Soil Science* **47** (3): 466-470.

Das, P.K., Katiyar, R.S., Gowda, M.H., Fathima, P.S. and Chodhury, P.C. (1995a). Effect of vesicular arbuscular mycorrhizal inoculation on growth and development of mulberry (*Morus* spp.) saplings. *Indian Journal of Sericulture*. **34**(1): 15-17.

Das, U., Bayan, N., Talukdar, N.C. and Deka, P.C. (1995b). Effect of coinoculation with vesicular-arbuscular mycorrhizal fungus and phosphate solubilizers on tomato and green gram grown in sterilized soils. In : *Mycorrhizae : Biofertilizers for the Future*. (Eds. Alok Adholeya and Sujan Singh), TERI, New Delhi: 264-269.

Davis, E.A. and Young, J.L. (1985). Endomycorrhizal colonization of greenhouse grown wheat as influenced by fertilizer salt when banded on soil mixes. *Canadian Journal of Botany*. **63**: 1196-1203.

Davis, R.M., Menge, J.A. and Zentmeyer, G.A. (1978). Influence of vesicular arbuscular mycorrhizae on *Phytophthora* root rot of three crop plants. *Phytopathology*. **86:** 1614.

Dehne, B. and Schuepp, H. (1990). Influence of VA mycorrhizae on the uptake and distribution of heavy metals in plants. *Agriculture, Ecosystem and Environment*. **29**(1-4): 70-84.

Dehne, H.W. and Backhaus, G.F. (1986). *Zt. Pfl. Krankh.* **93** : 415-424.

Dehne, H.W. and Schonbeck, F. (1979). Investigations on the influence of endotrophic mycorrhiza on plant diseases. II. Phenol metabolism and lignification. *Phytopathology*. **95**: 210.

Deka, H.K., Mishra, R.R. and Sharma, G.D. (1990). Effect of fuel burning on VA mycorrhizal fungi and their influence on the growth of early plant colonizing species. *Acta Botanica Indica.* **18**(2): 184-189.

Delorenzini, G., Barea, J.M. and Olivares, J. (1979). Fertilizacion biologica (micorrhiza + *Rhizobium* - phosphobacterias) de *Trifolium pratense* en diferentes condicions de cultivo. *Rev. Lat-amer, Microbiol.,* **21**: 129-134.

Devagiri, G.M., Dhiman, R.C., Singh, Y.P. and Patil, S.K. (2001). Seed source dependent variation in mycorrhizal colonization and nutrient uptake in *Dalbergia sissoo* seedlings. *Journal of Tropical Forest Sciences* **13** (2):246-257.

Devi M. C. and Reddy, M.N. (2002). Phenolic acid metabolism of groundnut (*Arachis hypogaea* L.) plants inoculated with VAM fungus and *Rhizobium. Plant Growth Regulation* **37** (2): 151-156.

Devi, G. and Phookan, A.K. (1999). Relative mycorrhizal dependency of french bean (cv. *Contender*) to different *Glomus* species. Abstract: *National Conference on Mycorrhiza*, Barkatulla University Bhopal: 48.

Devi, G.U. and Sitaramaiah, K. (1990). Effect of superphosphate, rock phosphate as sources of phosphorus in combination with *Glomus fasciculatum* on root colonization, growth and chemical composition of blackgram. *In: Trends in mycorrhizal research. Proceedings of the National Conference on Mycorrhiza.* (Eds. B.L. Jalali and H. Chand), H.A.U., Hisar : 144-145.

Devi, G.U. and Sitaramaiah, K. (2001). Response of endomycorrhizal fungi at different levels of inoculum in blackgram. *Indian Phytopathology* **54**(3): 377-379.

Devi, M.C. and Reddy, M.N. (2001). Growth response of groundnut to VAM fungus and *Rhizobium* inoculation. *Plant Pathology Bulletin* **10** (2): 71-78.

Dhinakaran, R. and Savithri, P. (1997). Phosphorus use efficiency of tomato as influenced by phosphorus and vesicular arbuscular mycorrhizal (VAM) fungi inoculation. *Journal of Nuclear Agriculture and Biology* **26**(3): 142-146.

Dhingra, K.K., Gupta, R.P., Jagrup Singh and Harpal Kaur (1994). Response of lentil to vesicular-arbuscular mycorrhiza (*Glomus fasciculatum*) and *Rhizobium leguminosarum* inoculation in relation to phosphorus levels. *Lens Newsletter*. **21**(1): 26-38.

Diaz, G. and Honrubia, M. (1993). Infectivity of mines soils from south-east Spain. II. Mycorrhizal population levels in spoil sites. *Mycorrhiza* **4**: 85-88.

Dixon, R.K. (1990). Cytokinin activity in *Citrus jambhiri* seedlings colonized by mycorrhizal fungi. *Agriculture Ecosystem and Environment.* **29** (1-4): 103-106.

Dixon, R.K., Garg, V.K. and Rao, M.V. (1993). Inoculation of *Leucaena* and *Prosopis* seedlings with *Glomus* and *Rhizobium* species in saline soil: rhizosphere relations and seedlings growth. *Arid Soil Research and Rehabilitation.* **7**:133-144.

Dixon, R.K., Mukerji, K.G., Chamola, B.P. and Kaushik, A. (1997). Vesicular arbuscular mycorrhizal symbiosis in relation to forestation in arid land. *Annals of Forestry.* **5**(1): 1-9.

Doss, D.D. and Bagyaraj, D.J. (1990). Anatomical observations on roots of finger millet colonized by VA mycorrhiza. *Current Science*. **59**(8): 423-424.

Doss, D.D., Bagyaraj, D.J. and Syamsunder, J. (1988). Anatomical and histochemical changes in the roots and leaves of finger millet colonized by VA mycorrhiza. *Indian Journal of Microbiology*. **28**(3-4): 276-280.

Dubey, S.K. (1993). Effect of mycorrhiza, *Bradyrhizobium* inoculation and phosphorus level on nodulation, yield and yield attributes in soybean (*Glycine max). Indian Journal of Agricultural Sciences* .**63**(11): 737-739.

Dubey, S.K. (1998). Response of soybean *(Glycine max)* to biofertilizers with and without nitrogen, phosphorus and potassium on swell-shrink soil. *Indian Journal of Agronomy* **43** (3): 546-549.

Dueck, T.A., Visser, P., Ernst, W.H.O. and Schat, H. (1986). Vesicular-arbuscular mycorrhizae decrease zinc-toxicity to grasses growing in zinc-polluted soil. *Soil Biology and Biochemistry*. **18**(3): 331-333.

Durga, V.V.K. and Gupta, S. (1995). Effect of vesicular arbuscular mycorrhizae on the growth and mineral nutrition of teak (*Tectona grandis* Linn. F.). In *Special issue: focus on teak. Indian Forester*. **121**(6): 518-527.

Dwivedi, O.P., Yadav, R.K., Vyas, D. and Vyas, K.M. (2003). Distribution of AM fungi in the rhizosphere soils of betel vine. *Indian Phytopathology* **56** (2): 228-229.

Earanna, N., Mallikarjuna, R.R., Bagyaraj, D.J. and Suresh, C.K. (2001). Response of *Coleus aromaticus* to *Glomus fasciculate* and other beneficial soil microflora. *Journal of Spices and Aromatic Crops* **10** (2): 141-143.

Earanna, N., Tanuja, B.P., Bagyaraj, D.J. and Suresh, C.K. (2002). Vesicular arbuscular mycorrhizal selection for increasing the growth of *Rauvolfia tetraphylla. Journal of Medicinal and Aromatic Plant Science* **24** (3): 695-697.

Edathil, T.T., Manian, S. and Udaiyan, K. (1994a). Early establishment of native vesicular arbuscular mycorrhizas in three vegetable crops of South India - a comparative study. *Pertanika Journal of Tropical Agricultural Science* **17**(3): 157-161.

Edathil, T.T., Manian, S. and Udaiyan, K. (1994b). The effect of vesicular-arbuscular mycorrhizal exposure period on their colonization and spore production in tomato seedlings (*Lycopersicon esculentum* Mill.), and on host biomass. *Agriculture, Ecosystems & Environment*. **51**(3): 287-292.

Edathil, T.T., Manian, S. and Udaiyan, K. (1996). Interaction of multiple VAM fungal species on root colonization, plant growth and nutrient status of tomato seedlings (*Lycopersicon esculentum* Mill.). *Agriculture, Ecosystem & Environment*. **59**(1/2):63-68.

Elmes, R.P. and Mosse, B. (1984). Vesicular-arbuscular mycorrhizal inoculum production. II. Experiments with maize (*Zea mays*) and other hosts in nutrient flow culture. *Canadian Journal of Botany*. **62** : 1531-1536.

Eranna, A. and Paranna, V.R.R. (1994). Effect of liming, mycorrhizal inoculation and rock phosphate on phosphorus uptake and growth of soybean in an acid soil. *Mysore Journal of Agricultural Sciences*. **28**(4): 292-296.

Ernst, W.H.O., Dium, W.E. and Van Oolbekking, G.T. (1984). Vesicular-arbuscular mycorrhiza in dune vegetation. *Acta Botanica Neerlandica*. **33**(2): 151-160.

Eswarappa, H., Mohandas S. and Gowda, K.N.C (2002). Effect of VAM fungi on banana. *Current Research – University of Agricultural Sciences (Bangalore)* **31** (5/6): 69-70.

Fahey, T.J. (1992). Mycorrhizae and forest ecosystems. *Mycorrhiza* **1**(2):83-90.

Fathima, P.S., Das, P.K. and Katiyar, R.S. (2000). Effect of different levels and sources of phosphorus on VA-mycorrhizal root colonization and spore load in mulberry (*Morus alba* L.). *Crop Research* (Hisar) **20** (3): 504-508.

Fathima, P.S., Katiyar, R.S., Das, P.K., Gowda, H.M. and Choudhury, P.C. (1995). Influence of vesicular-arbuscular mycorrhizal inoculation on mulberry (*Morus* sp.) under different sources and levels of phosphorus. In: *Mycorrhizae: Biofertilizers for the Future.* (Eds. Alok Adholeya and Sujan Singh), TERI, New Delhi: 369-373.

Ferrera-Cerrato, R. and Perez-Moreno, J. (1994). Mycorrhizal interactions with plants and soil organisms in sustainable agroecosystems. In: *15th World Congress of Soil Science, Acapulo, Mexico, Transactions.* **4** : 53-54.

Firdaus-E-Bareen (1990). Vesicular arbuscular mycorrhiza in aquatics. In: *Current trends in mycorrhizal research. Proceedings of the National Conference on Mycorrhiza,* (Eds. B.L. Jalali and H. Chand), H.A.U., Hisar: 1-3.

Forster, S.M. and Nicolson, T.H. (1981). Aggregation of sand from a maritime embryo sand dune by microorganisms and higher plants. *Soil Biology Biochemistry.* **13**: 199-203.

Frank, A.B. (1885). *Uber die aufwurzelsymbiose beruhende Ernahrung gewisser Baume durch unterirdische pilze. Ber. Dtsch Bot. Ges.*,**3**: 128.

Frank, A.B. (1887). Ueber neue Mykorrhiza-foremen, *Ber. Bot. Gescll.* **5**:395.

Gadkar V. and Adholeya, A. (2000). Intraradical sporulation of AM *Gigaspora margarita* in long-term axenic cultivation in RiT-DNA carrot root colonization. *Mycological Research* **104** (6): 716.

Gadkar, V., Adholeya, A. and Satyanarayana, T. (1997). Randomly amplified polymorphic DNA using the M-13 core sequence of the VAM fungi *Gigaspora margarita* and *G. gigantea. Canadian Journal of Microbiology.* **43**(8): 795-798.

Gadkar, V., Adholeya, A. and Satyanarayana, T. (1997). Randomly amplified polymorphic DNA using the M 13 core sequence of the vesicular-arbuscular mycorrhizal fungi *Gigaspora margarita* and *Gigaspora gigantea. Canadian Journal of Microbiology.* **43**: 795-798.

Gadkar, V., Satyanarayana, T. and Adholeya, A. (1999). Mycelial injuries in *Gigaspora margarita* during the colonization with Ri-T-DNA transformed carrot roots under *in vitro* conditions. Abstracts: *National Conference on Mycorrhiza*, Barkatulla University, Bhopal: 29.

Gallaud, I. (1905). *Etude sur les mycorhize endotrophes. Review of General Botany.* **7**: 5-48.

Galli, U., Meier, M. and Brunold, C. (1993). Effect of cadmium on non-mycorrhizal and mycorrhizal Norway spruce seedlings (*Picea abies* L.) and its ectomycorrhizal fungus *Laccaria laccata* (Scop. ex. Fr.) Bk. and Br.: sulphate reduction, thiols and distribution of the heavy metal. *New Phytologist.* **125**(4): 837-843.

Ganny, F., Diem, H.G. and Dommergues, Y.R. (1982). Effect of inoculation with *Glomus mosseae* on nitrogen fixation by field grown soybeans. *Plant and Soil.* **68**: 321-329.

Gaonker, S.B.N. and Sreenivasa, M.N. (1994). Effects of inoculation with *Glomus fasciculatum* in conjunction with different organic amendments on growth and yield of wheat (*Triticum aestivum* L.). *Microbiological Research.* **149** (4): 419-423.

Gaonker, S.B.N., Sreenivasa, S.B. and Alagawadi, A.R. (1993). Rhizosphere microflora, growth and yield of wheat as influenced by inoculation of *Glomus fasciculatum* in conjunction with organic amendments. *Karnataka Journal of Agriculrural Science.* **6**(4): 371-377.

Gardes, M., Mueller, G.M., Fortin, J.A. and Kropp, B.R. (1991). Mitochondrial DNA polymorphisms in *Laccaria bicolor, L. laccata, L. proxima* and *L. amythystina. Mycological Research.* **95** : 206-216.

Gardner, I.C. (1986). Mycorrhizae of actinorrhizal plants. *MIRCEN Journal of Applied Microbiology and Biotechnology.* **2**(1): 147-160.

Gardner, I.C., Clelland, D.M. and Scott, A. (1984). Mycorrhizal improvement in non-leguminous nitrogen fixing associations with particular reference to *Hippophae rhamnoides. Plant and Soil.* **78**: 189-199.

Garg, R.K., Pandey, A.K., Goyal, D. and Sharma R.K. (1999). Effect of mycorrhizae on the growth of *Eucalyptus tereticornis* seedlings in the nursery in different geo-climatic conditions. *Indian Forester* **125** (7): 684-689.

Gaur, A. and Adholeya, A. (1999). Mycorrhizal effects on the acclimatization, survival, growth and chlorophyll of micro propagated *Syngonium* and *Draceana* inoculated at weaning and hardening stages. *Mycorrhiza* **3** (4): 21-27.

Gaur, A. and Adholeya, A. (2000a). Responses of three vegetable crops to VAM fungal inoculation in nutrient deficient soils amended with organic matter. *Symbiosis (Rehovot)* **29**(1): 19-31.

Gaur, A. and Adholeya, A. (2000b). Effects of the particle size of soil-less substrates upon AM fungus inoculum production. *Mycorrhiza* **10** (1): 43-45.

Gaur, A. and Adholeya, A. (2002). Arbuscular-mycorrhizal inoculation of five tropical fodder crops and inoculum production in marginal soil amended with organic matter. *Biology and Fertility of Soils* **35**(3): 214-218.

Gaur, A., Adholeya, A. and Mukerji, K.G. (1998a). A comparison of AM fungi inoculants using *Capsicum* and *Polianthes* in marginal soil amended with organic matter. *Mycorrhiza.* **7**:307-312.

Gaur, A., Adholeya, A. and Mukerji, K.G. (2000). On-farm production of VAM inoculum and vegetable crops in marginal soil amended with organic matter. *Tropical Agriculture* **77** (1): 21-26.

Gaur, A., Sharma, M.P. and Adholeya, A. (1998b). Variation in the spore density and percentage of root length of tree species colonized by arbuscular mycorrhizal fungi at a rehabilitated waterlogged site. *Journal of Tropical Forest Science.* **10**(4): 542-551.

Gaur, A.C. and Rana, J.P.S. (1990). Role of VA mycorrhizae, phosphate solubilizing bacteria and their interactions on growth and uptake of nutrients by wheat crops. In: *Current trends of mycorrhizal research. Proceedings of the National Conference on Mycorrhiza.* (Eds. B.L. Jalali and H. Chand), H.A.U., Hisar : 105-106.

Gautam, A. and Irshad, M. (2002). A survey of some cultivated legume crops of Aligarh district to determine the occurrence of AM fungi. *Mycorrhiza News.* **14**(1): 13-15

Gautam, A. and Mahmood, I. (2002). A survey of some cultivated legume crops of Aligarh district to ditermine the occurrence of AM fungi. *Mycorrhiza News* **14** (1): 13-15.

Gautam, S.P and Maitra, A. (1995). Impact of vesicular-arbuscular mycorrhizal fungi on growth of *Dendrocalamus strictus*. In: *Mycorrhizae: Biofertilizers for the Future*. (Eds. Alok Adholeya and Sujan Singh), TERI, New Delhi: 400-402.

Gautam, S.P. and Prasad, K. (2000). Effect of *Glomus macrocarpum* VAM on the fiber strength properties and production of holocellulose, hemicellulose, and cellulose by *Dendrocalamus strictus* under field conditions. *Journal of Phytological Research* **13**(1): 35-40.

Gautam, S.P. and Prasad, K. (1999). Impact of *Glomus macrocarpum* VAM fungus on proximate chemical composition of *Dendrocalamus strictus* under field conditions. Abstract: *National Conference on Mycorrhiza*, Barkatulla University, Bhopal: 57.

Gerdemann, J.W. and Bakshi, B.K. (1976). *Transactions of British Mycological Society*. **66**: 340-343.

Gerschefske, S.K., Daniels-Hetrick, B.A. and Wilson, G.W.T. (1988). Relationship of soil fertility to suppression of the growth response of mycorrhizal big bluestem in non-sterile soil. *New Phytologist*. **109**: 473-481.

Ghazala Nasim (1990). Vesicular arbuscular mycorrhiza in portions other than root. In: *Current trends in mycorrhizal research. Proceedings of the National Conference on Mycorrhiza*. (Eds. B.L. Jalali and H. Chand), H.A.U., Hisar: 4-6.

Gianinazzi, S., Trouvelot, A., Lovato, P., Tuinen, D.Van., Franken, P. and Gianinazzi-Pearson, V. (1995). Arbuscular mycorrhizal fungi in plant production of temperate agroecosystems. In *Special issue on Arbuscular Mycorrhizae: Biotechnological applications: an environmental sustainable biological agent. Critical Reviews in Biotechnology*. **15**(3/4): 305-311.

Gianinazzi-Pearson, V., Smith, S.E., Gianinazzi, S. and Smith, F.A. (1991). Enzymatic studies on the metabolism of vesicular-arbuscular mycorrhizas. V. Is H^+- ATPase a component of ATP-hydrolysing enzyme activities in plant-fungus interfaces? *New Phytologist*. **117**(1): 61-74.

Gildon, A. and Tinker, P.B. (1983). Interactions of vesicular-arbuscular mycorrhizal infection and heavy metal in plants. I. The effect of heavy metals on the development of vesicular-arbuscular mycorrhizas. *New Phytologist*. **95**(2): 247-261.

Gill, T.S., Singh, R.S. and Kaur J. (2002). Comparison of four arbuscular mycorrhizal fungi for root colonization, spore population and plant growth responses in chickpea. *Indian Phytopathology* **55** (2): 210-212.

Ginwal, H.S., Verma, R.K. and Srivastava, R.L. (1996 Publ. 1997). Natural occurrence of arbuscular mycorrhizae in different provenances of *Acacia nilotica* Willd. ex Del. seedlings. *Indian Journal of Forestry*. **19**(4): 311-314.

Giovannetti, M. and Gianinazzi-Pearson, V. (1994). Biodiversity in arbuscular mycorrhizal fungi. *Mycological Research*. **98**(7): 705-715.

Giri, B. and Mukerji, K.G. (1999a). Dependency of *Sesbania grandiflora* on vesicular-arbuscular mycorrhiza in saline soil. Abstract: *National Conference on mycorrhiza*, Barkatulla University Bhopal: 11.

Giri, B. and Mukerji, K.G. (1999b). Improved growth and productivity of *Sesbania aegyptiaca* (Pers) under salinity stress through mycorrhizal technology, *Sesbania grandiflora. Journal of Phytological Research* **12**(1/2): 35-38.

Giri, B., Kapoor, R. and Mukerji, K.G. (2000). *Sesbania aegyptiaca* pers. seedling response to VA mycorrhization in two types of soil. *Phytomorphology* **50** (3/4): 327-332.

Giri, B., Mandeep Kaur and Mukerji, K. G. (1999). Growth responses and dependency of *Sesbania aegyptiaca* on vesicular arbuscular mycorrhiza in salt stressed soil, *Annals of Agricultural Research* **20** (1): 109-112.

Girija, V.K. and Nair, S.K. (1988). Incidence of VAM in banana varieties. *Indian Journal of Microbiology*. **28**(4): 294-295.

Gnanadevi, G. and Haripriya, K. (1999). Screening of efficient VAM fungi for chrysanthemum. *South Indian Horticulture* **47** (1/6): 325-326.

Godse, D.B., Madhusudan, T., Bagyaraj, D.J. and Patil, R.B. (1976). Occurrence of vesicular-arbuscular mycorrhizas on crop plants in Karnataka. *Transactions of the British Mycological Society*. **67**(1): 169-171.

Godse, D.B., Wani, S.P., Patil, R.B. and Bagyaraj, D.J. (1978). Response of cowpea (*Vigna unguiculata* (L.) Walp.) to *Rhizobium* VA mycorrhizal dual inoculation. *Current Science*. **87**(20): 784-785.

Gokuldas Pal, Bagyaraj, D.J. and Prasad, T.G. (1990). Mycorrhizal reproduction as influenced by moisture stress. In: *Current trends in mycorrhizal research. Proceedings of the National Conference on Mycorrhiza*. (Eds. B.L. Jalali and H. Chand), H.A.U., Hisar : 25-26.

Gopalkrishna, M.N., Bagyaraj, D. J. and Vasanthakrishna, M. (1990). Interaction between *Glomus fasciculatum* and two phosphate solubilizing fungi in finger millet. In: *Trends in mycorrhizal research. Proceedings of the National Conference on Mycorrhiza*. (Eds. B.L. Jalali and H. Chand), H.A.U., Hisar :113-114.

Gosal, S.K., Gupta, R.P. and Gosal, S.S. (2001). *In vitro* establishment of vesicular-arbuscular mycorrhiza in roots of micropropaged sugarcane. *Annals of Agri Bio Research* **6** (1): 19-22.

Goswani, M. G. and Jamaluddin (2001). A note on vesicular-arbuscular mycorrhiza development in *Albizia procera* in root trainers. *Mycorrhiza News* **13** (1): 21-22

Gowda, M.H. and Rao, C.G.P. (1994). Rhizosphere ecology of *Morus alva* var. Mysore local, inoculated with VA mycorrhizal fungus *Glomus intraradices*. *Journal of Ecobiology*. **6**(3): 161-167.

Gowda, M.H., Das, P.K., Katiyar, R.S., Fathima, P.S. Choudhury, P.C. and R.K. Datta (1995). The role of synergistic effect of vesicular-arbuscular mycorrhiza and *Azotobacter* in Mulberry. In: *Mycorrhizae: Biofertilizers for the Future*. (Eds. Alok Adholeya and Sujan Singh), TERI, New Delhi: 197-201.

Graf, F. and Gerber, W. (1997). (The influence of mycorrhizal fungi on soil structure and their importance for bio-engineering) Der Einfluss von mycorrhizapilzen auf..die Bodenstruktur und deren Bedeutung fur den Lebendverbau. *Schweizerische Zeitschrift fur Forstwesen*. **148**(11): 863-886.

Greep, S., Muthukumar, T., Udaiyan, K. and Bai, V.N. (1997). Response of siratro (*Macroptilium atropurpureum* Urb. Fabaceae) to vesicular-arbuscular mycorrhizal fungi and *Rhizobium* sp. in sterilized soil. *Penanika Journal of Tropical Agricultural Science*. **20**(1): 19-20.

Griffioen, W.A.J. (1994). Characterization of a heavy metal tolerant endomycorrhizal fungus from the surroundings of a zinc refinery. *Mycorrhiza.* **5**: 197-200.

Gunasekaran, P., Raja, N. Ubalthoose, Sundaresan, P. and Lakshmanan, M. (1988). Extraction of DNA from *Glomus etunicatum.* In: *Mycorrhizae for Green Asia.* (Eds. A. Mahadevan, N. Raman and K. Natarajan), Madras University, Madras : 309-310.

Gunasekaran, P., Sundaresan, P., Raja, N.U. and Lakshmanan, M. (1987). Effect of pH, temperature and nutrients on the germination of a vesicular-arbuscular mycorrhizal fungus, *Glomus fasciculatum in vitro. Proceedings of the Indian Academy of Sciences (Plant Sciences).* **97**(3): 231-234.

Gupta, M. and Lakhanpal, T.N. (1993). Identification of mycorrhizal types, mycorrhizosphere mycoflora and vesicular-arbuscular mycorrhizal (VAM) fungi in apple. *Annals of Biology* (Ludhiana). **9**(1): 87-93.

Gupta, M. L, Khalio, A., Pandey, R., Shukla, R. S., Singh, H. N. and Kumar, S. (2000). Vesicular-arbuscular mycorrhizal fungi associated with *Ocimum* spp. *Journal of Herbs, Spices and Medicinal plants.*7 (2): 57-63.

Gupta, M.L., Janardhanan, K.K., Chattopadhyay, and Jain, A. (1994). Association of *Glomus* with palmarosa and its effect on growth and biomass production. *Mycological Research.* :561-563.

Gupta, M.L., Prasad, A., Muni Ram and Kumar, S. (2002). Effect of the vesicular-arbuscular mycorrhizal (VAM) fungus *Glomus fasciculatum* on the essential oil yield related characters and nutrient acquisition in the crops of different cultivar of menthol mint (*Mentha arvensis*) under field conditions. *Bioresource Technology* **81** (1): 77-79.

Gupta, N. and Ali, S.S. (1990). Incidence of vesicular arbuscular mycorrhizal fungi in Raipur soils. In: *Current trends in mycorrhizal research. Procedings of the National Conference in Mycorrhiza.* (Eds. B.L. Jalali and H. Chand), H.A.U., Hisar: 24.

Gupta, N. and Rahangdale, R. (1999). Response of *Albizia lebbek* and *Dalbergia sissoo* towards dual inoculation of *Rhizobium* and arbuscular mycorrhizal fungi. *Indian Journal of Experimental Biology* **37** (10): 1005-1011.

Gupta, N., Basak, U.C. and Das, P. (2001). Arbuscular mycorrhizal association of mangroves in saline and non-saline soils. *Mycorrhiza News***13** (4): 14-19.

Gupta, N., Routaray, S., Basak, U. C. and Das, P. (2003). Distribution of AM fungi in the rhizosphere of different tea clones. *Mycorrhiza News* **15** (1): 13-18.

Gupta, N., Shukla, A. and Ali, S.S. (1990a). Possible role of weeds in maintaining mycorrhizal infectivity of rice soils. In: *Mycorrhizal Symbiosis and Plant Growth.* (Eds. D.J. Bagyaraj and A. Manjunath), U.A.S., Bangalore :30.

Gupta, P.P., Chhabra, M.L., Jalali, B.L. and Kumar, P.R. (1990b). Influence of VA-mycorrhizal inoculation on growth and development of rapeseed. In: *Trends in mycorrhizal research. Proceedings of the National Conference on Mycorrhiza.* (Eds. B.L. Jalali and H. Chand), H.A.U., Hisar: 135-136.

Gupta, P.P., Kumar, R. and Jalali, B. L. (1999). Response of different vesicular-arbuscular mycorrhizal fungi on cowpea. *Plant Disease Research* **14** (1): 25-30.

Gupta, R. and Krishnamurthy, K.V. (1994). Receptivity of various wheat cultivars to infection by vesicular-arbuscular mycorrhizal fungi. *Journal of Phytological Research.* 7(2): 183-187.

Gupta, R. and Mukerji, K.G. (1996). Studies on the vesicular arbuscular mycorrhizal associations in the root of common Delhi grasses. *Phytomorphology*. **46**(2): 117-121.

Gupta, R. and Mukerji, K.G. (1999). Effect of VAM fungus on the growth and flowering of *Helianthus annuus* L. var. Peredovik EC- 68414. *Phytomorphology* **49** (1): 37-42.

Gurumurthy, S. B., Sreenivasa, M. N., Kulkarni, J.H. and Nadagoudar, B.S. (1999a). Screening and selection of an efficient VAM fungus for shisham *(Dalbergia sissoo). Karnataka Journal of Agricultural Sciences* **12** (1/4): 88-92.

Gurumurthy, S.B and Sreenivasa, M.N. (1998). Screening and selection of efficient VA mycorrhizal fungus for teak (*Tectona grandis* L.). *Karnataka Journal of Agricultural Sciences* **11** (4): 956-960.

Gurumurthy, S.B. and Sreenivasa, M.N. (1995). Influence of different levels of *Glomus macrocarpum* on growth and yield of chilli. In: *Mycorrhizae: Biofertilizers for the Future*. (Eds. Alok Adholeya and Sujan Singh) , TERI, New Delhi: 429-431.

Gurumurthy, S.B. and Sreenivasa, M.N. (1999). Effect of inoculation of efficient VAM fungus at different levels of P on nutrition and growth of silveroak (*Grevillea robusta*). *Advances in Forestry Research in India.* **21**: 262-271.

Gurumurthy, S.B. and Sreenivasa, M.N. (2000). Occurrence and distribution of VAM fungi in the rhizosphere of five tree species grown under agro forestry system. *Environment and Ecology* **18** (2): 500-502.

Gurumurthy, S.B., Sreenivasa, M.N., Kulkarni, H. and Nadagoundar, B.S. (1999b). Effect of inoculation of efficient VAM fungus and P. solubilizer on growth and nutrition of teak *Tactona grandis*. Abstract: *National Conference on Mycorrhiza*, Barkatulla University Bhopal: 22.

Habte, M., Fox, R.L., Aziz, T. and EL-Swaify, S.A. (1988). Interaction of vesicular-arbuscular mycorrhizal fungi with erosion in an oxisol. *Applied Environmental Microbiology*. **45**(4): 945-950.

Hall, I.R. and Abbott, L.K. (1983). Photographic slide collection illustrating features of Endogonaceae. *Bulletin of British Mycological Society*. **17**: 54-55.

Hall, I.R.(1984). Taxonomy of VA mycorrhizal fungi. In: *VA mycorrhizae.* (Eds. C.LL. Powell and D.J. Bagyaraj), C.R.C. Press. Boca Ratoon, FL.: 57-94.

Hardie, K. (1985). The effect of removal of extraradical hyphae on water uptake by vesicular-arbuscular mycorrhizal plants. *New Phytologist*. **101**: 677-684.

Harikumar, V. S. and Potty, V. P. (2002a). Genotype dependent variation in arbuscular mycorrhizal colonization and its response on the growth of sweet potato. *Bulgarian Journal of Agricultural Science* **8** (2/3): 161-166.

Harikumar, V. S. and Thomas, G. V. (2003). Occurrence of arbuscular mycorrhizal (AM) species under different management in coconut plantations. *Plant Archives* **3** (1): 105-108.

Harikumar, V.S. (1995). Vesicular-arbuscular mycorrhizal association in *Kaempferia galanga* L. in relation to soil type. In: *Mycorrhizae: Biofertilizers for the Future*. (Eds. Alok Adholeya and Sujan Singh), TERI, New Delhi: 70-72.

Harikumar, V.S. (2001). Arbuscular mycorrhizal colonization in some wetland plants in Kerala. *Mycorrhiza News* **12** (4): 14-15.

Harikumar, V.S. and Potty, V.P. (1998). Performance of two geographical isolates of *Glomus microcarpum* var. *microcarpum* on infectability, yield and nutrient uptake in sweet potato. *Journal of Root Crops* **24** (1): 49-53.

Harikumar, V.S. and Potty, V.P. (1999). Effect of inoculation of arbuscular mycorrhizal fungi *Glomus microcarpum* on phosphorus kinetics in sweet potato soil. In: *Proceedings of National Conference on Mycorrhiza*. Barkatulla University, Bhopal: 18

Harikumar, V.S. and Potty, V.P. (2002b). Technology for mass multiplication of arbuscular mycorrhizal fungi for field inoculation to sweet potato. *Mycorrhiza News* **14** (1): 11-12.

Harinikumar, K.M. and Bagyaraj, D.J. (1988a). Effect of crop rotation on native vesicular arbuscular mycorrhizal propagules in soil. *Plant and Soil*. **110**(1): 77-80.

Harinikumar, K.M. and Bagyaraj, D.J. (1988b). Effect of intercropping and organic soil amendments on native VAM. In: *Mycorrhizae for Green Asia*. (Eds. A. Mahadevan, N. Raman and K. Natrajan), Madras University, Madras : 207-208.

Harinikumar, K.M. and Bagyaraj, D.J. (1988c). The effect of season on VA mycorrhizae of leucaena and mango in a semi-arid tropic. *Arid Soil Research and Rehabilitation*. **2**(2): 139-143.

Harinikumar, K.M. and Bagyaraj, D.J. (1989). Effect of cropping sequence, fertilizers and farmyard manure on vesicular-arbuscular mycorrhizal fungi in different crops over three consecutive seasons. *Biology and Fertility of Soil*. **7**(2): 173-175.

Harinikumar, K.M. and Bagyaraj, D.J. (1994). Potential of earth-worms, ants, millipedes, and termites for dissemination of vesicular-arbuscular mycorrhizal fungi in soil. *Biology and Fertility of Soils*. **18**(2): 115-118.

Harinikumar, K.M. and Bagyaraj, D.J. (1995). Spread of vesicular arbuscular mycorrhizal fungal hyphae in soil. *Microbiological Research*. **15**(1): 77-80.

Harinikumar, K.M. and Bagyaraj, D.J. (1996). Persistence of introduced *Glomus intraradices* in the field as influenced by repeated inoculation and cropping system. *Biology and Fertility of Soils*. **21**(3): 184-188.

Harinikumar, K.M. and D.J. Bagyaraj (1990). Do wasps disperse vesicular arbuscular mycorrhizal fungi? In: *Mycorrhizal symbiosis and plant growth*. (Eds. D.J. Bagyaraj and A. Manjunath), U.A.S., Bangalore :14-15.

Harinikumar, K.M., Bagyaraj, D.J. and Kale, R.D. (1988). Vesicular arbuscular mycorrhizal propagules in earth worm cast. In: *Advances in management and conservation of soil fauna. Proceedings of the 10th International Soil Zoology Colloquium*. (Eds. G.K. Veeresh, D. Rajgopal and C.A. Viraktamath), U.A.S., Bangalore: 605-610.

Harinikumar, K.M., Bagyaraj, D.J. and Mallesha, B.C. (1990a). Effect of intercropping and organic soil amendments on native VA mycorrhizal fungi in an Oxisol. *Arid Soil Research and Rehabilitation*. **4**(3): 193-197.

Harinikumar, K.M., Savalgi, V.P., Bagyaraj, D.J. and Math, K.K. (1990b). Selection of an efficient vesicular arbuscular mycorrhizal fungus for sunflower. In: *Mycorrhizal symbiosis and plant growth*. (Eds. D.J. Bagyaraj and A. Manjunath), U.A.S., Bangalore :74-75.

Haripriya, K. and Sekharan, M. V. S. (2002). Effect of VA mycorrhizal inoculation on the growth and yield of onion. *Advances in Plant Sciences* **15** (1): 319-321.

Haripriya, K. and Sriramachandrasekharan, M.V. (2002). Effect of VAM inoculation on the growth and yield of chrysanthemum. *Journal of Ecobiology* **14**(1): 39-42.

Hariram, N. and Santhaguru, K. (1995). Vesicular-arbuscular mycorrhiza in rhizosphere soils of some plant species from Madurai. *Journal of Ecobiology*. 7(4): 243-248.

Harley, J.L. (1969). *The Biology of Mycorrhiza*. Leonard Hill, London: 334.

Harley, J.L. (1989). Significance of mycorrhiza. *Mycological Research*. **92**(2): 129-139.

Harley, J.L. and Smith, S.E. (1983). *Mycorrhizal symbiosis*. Academic Press, New York.: 483.

Hartmond, U., Schaesberg, N.V., Graham, J.H. and Syvertsen, J.P. (1987). Salinity and flooding stress effect on mycorrhizal and non-mycorrhizal citrus rootstock seedlings. *Plant and Soil*. **104**: 37-43.

Hasan A. (2002). Arbuscular mycorrhizal root colonization of some pulses in 7 districts of eastern Uttar Pradesh – a preliminary report. *Mycorrhiza News* **14** (3): 15-16.

Hasan, A., Verma, A.C. and Khan, M.N. (2001). Occurrence of arbuscular mycorrhizae on wasteland weeds in eastern Uttar Pradesh: a preliminary survey report. *Mycorrhiza News* **13** (2): 17-21.

Hasan, N. and Jain, R.K. (1987). Parasitic nematodes and vesicular-arbuscular mycorrhizal (VAM) fungi associated with berseem (*Trifolium alexandrinum* L.) in Bundelkhand region. *Indian Journal of Nematology*. **17**(2): 184-188.

Hasselwandter, K. (1995). Mycorrhizal fungi and siderophore production. In: *Special issue on Arbuscular Mycorrhizae: Biotechnological applications: an environmentally sustainable biological agent. Critical Reviews in Biotechnology*. **15**(3/4): 287-291.

Hayman, D.S. (1975). The occurrence of mycorrhiza in crops as affected by soil fertility. In: *Endomycorrhizas*. (Eds. F.E. Sanders, B. Mosse and P.B. Tinker), Academic Press, London: 495-509.

Hayman, D.S. (1982). Influence of soils and fertility on activity and survival of vesicular-arbuscular mycorrhizal fungi. *Phytopathology*. **72**: 1119-1125.

Hayman, D.S. (1984). The physiology of vesicular-arbuscular mycorrhiza. In: *Forage legumes, British Grass land Society Occasional Symposium*. (Ed. D.J. Thompson), Hurley. **16**: 44-47.

Hayman, D.S. and Mosse, B. (1978). Clover in grassland. *Roth Amsted Experimental Station Annual Report*. **1**: 241.

Hayman, D.S. and Mosse, B. (1979). Improved growth of white clover in hill grasslands by mycorrhizal inoculum. *Annals of Applied Biology*. **93** : 141.

Hayman, D.S., Johnson, A.M. and Ruddlesdin, I. (1975). The influence of phosphate and crop species on *Endogone* spores and vesicular-arbuscular mycorrhizae under field condition. *Plant and Soil*. **43**: 489-495.

Hazarika, D. K., Das, K. K. and Dubey, L. N. (1999). Effect of vesicular arbuscular mycorrhizal fungi inoculation on growth and nutrient uptake of black gram. *Journal of Mycology and Plant pathology* **29** (2): 201-204.

Hazarika, D.K., Das, K.K., Dubey, L.N. and Phookan, A. (2000). Effect of vesicular arbuscular mycorrhiza (VAM) fungi and *Rhizobium* on growth and yield of green gram [*Vigna radiata* (L.) Wilczek]. *Journal of Mycology and Plant Pathology* **30** (3): 424-426.

Hazarika, R.L. and Phookan, A.K. (1995). Growth response of chilli (*Capsicum annum* L.) to vesicular-arbuscular mycorrhizal infection with different levels of applied phosphorus in acid soil. In: *Mycorrhizae: Biofertilizers for the Future*. (Eds. Alok Adholeya and Sujan Singh), TERI, New Delhi: 421-423.

Hegde, S.V. and Rai, P.V. (1984). Influence of *Glomus fasciculatum* on damping-off of tomato. *Current Science*. **53**(11): 588-589.

Helgason, T., Merryweather, J.W., Deneison, J., Wilson, P., Young, J.P.W., and Fitter, A.H. (2002). Selectivity and functional diversity in arbuscular mycorrhizas of co-occuring fungi and plants from temperate deciduous woodland. *J. Ecol.* **90**: 371-384

Helm, D.J. and Carling, D.E. (1993). Use of soils transfer for reforestation on abandoned mined lands in Alaska. II. Effect of soil transfers from different successional stages on growth and mycorrhiza formation by *Populus balsamifera* and *Alnus crispa*. *Mycorrhiza*. **3**(3): 107-114.

Hemalatha, M. and Selvaraj, T. (2003). Association of AM fungi with Indian borage (*Plectranthus amboinicus* (Lour.) Spreng) and its influence on growth and biomass production. *Mycorrhiza News* **15** (1): 18-21.

Henrison, B., Le Tacon F. and Martin, F. (1992). *New Phytologist*. **122** : 289-298.

Herrera, M.A., Salamanca, C.P. and Barea, J.M. (1993). Inoculation of woody legumes with selected arbuscular mycorrhizal fungi and Rhizobia to recover desertified mediterranean ecosystems. *Applied Environmental Microbiology*. **59**: 129-133.

Hetrick, B.A.D., Wilson, G.W.T., Gills, B.S. and Cox, T.S. (1995). Chromosome location of mycorrhizal responsive genes in wheat. *Canadian Journal of Botany*. **73**(6): 891-897.

Hewitt, E.J. (1966). Rev. 2nd Ed. *Commonw. Bur. Hort. Plant Crops* (GB) Rech. Commun. 22.

Hicks, P.M. and Loynachan, T.E. (1987). Phosphorus fertilization reduces vesicular-arbuscular mycorrhizal infection and changes nodules occupancy of field-grown soybean. *Agronomy Journal*. **79**: 841-844.

Hirrel, M.C. and Gerdemann, J.W. (1980). Improved growth of onion and bell pepper in saline soils by two vesicular-arbuscular mycorrhizal fungi. *Proc. Soil Science Society of America*. **44**: 654-655.

Ho, I. (1987). Phytohormonal control of seedling growth by ectomycorrhizal fungi. In : *7th North American Conference on Mycorrhizae*. University of Florida, Gaineswille : 228.

Ho, I. and Trappe, J.M. (1979). Interaction of VA mycorrhizal fungus and a free living nitrogen fixing bacterium on growth of tall fecue. *4th North American Conference. Mycorrhizae*, Fortcollins, Colorado.

Homkar, U., Chandra, K. K., Srivastava, J. L. and Tiwari, K.P. (1998). Natural colonization of VA-mycorrhizae in roots of medicinal plants growing under gene bank. *Vaniki Sandesh* **22** (3): 7-10.

Howeler, R.H., Sieverding, E. and Saif, S.R. (1987). Practical aspects of mycorrhizal technology in some tropical crops and pastures. *Plant and Soil*. **100**: 249-283.

Hulkoti, I.D. and Sreenivasa, M.N. (2002). Least number of AM infective propagules required to initiate root colonization in onion. *Journal of Maharashtra Agricultural Universities* **26** (3): 332-333.

Ibrahim, K.K., Arunachalam, V., Rao, P.S.K. and Tilak, K.V.B.R. (1995). Seasonal response of groundnut genotypes to arbuscular mycorrhiza-*Bradyrhizobium* inoculation. *Microbiological Research.* **150**(2): 218-224.

Indi, D.V., Konde, B.K. and Sonar, K.R. (1990a) Response of brinjal genotypes in terms of dry weight and phosphorus uptake as influenced by VAM inoculation. In: *Current trends in mycorrhizal research. Proceedings of the National Conference on Mycorrhiza.* (Eds. B.L. Jalali and H. Chand), H.A.U., Hisar: 146-147.

Indi, D.V., Konde, B.K., Wani, P.V. and Kale, P.N. (1990b). Yield and nutrient uptake by brinjal as influenced by *Azospirillum brasilense* and/or *Glomus fasciculatum* inoculations under graded phosphorus levels. In :*Trends in mycorrhizal research. Proceedings of the National Conference on Mycorrhiza.* (Eds. B.L. Jalali and H. Chand) H.A.U., Hisar: 157-159.

India, Centre for Mycorrhizal Culture Collection (1997). Lyopholization: a method for long-term preservation of mycorrhizal fungi. *Mycorrhiza News.* **9**(1): 18-19.

India, Centre for Mycorrhizal Culture Collection (CMCC) (1996). Effect of storage conditions on infectivity level of VA mycorrhizal fungi. *Mycorrhiza News.* **8**(1): 15.

Iqbal, J. and Gautam, A. (1996). Status of vesicular-arbuscular mycorrhizal fungi of Aligarh District - a survey report. *Mycorrhiza News.* **8**(3): 12-14.

Iqbal, J. and Mahmood, I. (1998). Studies on seasonal fluctuations of VAM on some commonly cultivated crops in Aligarh district. *Mycorrhiza News.* **9**(4): 12-15.

Iqbal, J. and Mahmood, I. (1999). Effect of single and multiple VAM inoculants on the growth parameters of tomato. *Mycorrhiza News.* **10**(1): 13-15.

Ismail, S.M., Venkatesan, T., Bose, M.S. and Balasubramanian, S. (1999). Evaluation of different biofertilizers on biomass production of two tree seedlings growth in ordinary and sludge amended soil with sewage water. Abstract: *National Conference on Mycorrhiza*, Barkatulla University, Bhopal: 67.

Iyer, R., Sastry, R.K. and Korikanthimath, V.S. (1989). Effect of fertilizer levels on vesicular arbuscular mycorrhiza colonization in cardamom. *Journal of Plantation Crops.* **16**(Supplement): 429-432.

Iyer, R. and Bhat M.N. (1995). Prevalence of vesicular-arbuscular mycorrhizae in palm-based cropping systems under varying agroclimatic conditions. In: *Mycorrhizae: Biofertilizers for the Future.* (Eds. Alok Adholeya and Sujan Singh), TERI, New Delhi:294-296.

Jadhav, D.N. and Patil, M.S. (1996). Effect of VAM inoculation on growth and NPK uptake of groundnut (*Arachis hypogaea* L. cv. JL-24). *Acta Agronomica Hungarica.* **44**(2): 153-159.

Jagpal, R. and Mukerji, K.G. (1988). VAM fungi in reforestation. In: *Plant roots and their environment. Proceedings of an ISRR symposium.* (Eds. B.L. McMichael and H. Persson), Uppsala, Sweden :309-313.

Jagpal, R., Panigrahi, A. and Mukerji, K. G. (1988a). Effect of automobile pollution on VAM spore population of soil. In: *Mycorrhizae for green Asia.* (Eds. A. Mahadevan, N. Raman and K. Natarajan), Madras University, Madras : 46-47.

Jagpal, R., Sharma, P.D. and Mukerji, K.G. (1988b). Distribution of VAM fungi in relation to pollution. In: *Mycorrhizae for green Asia.* (Eds. A. Mahadevan, N. Raman and K. Natarajan), Madras University, Madras : 48-49.

Jahan, K. and Khan, W. (1996). Impact of sulphur dioxide on root nodulation and VAM colonization of roots of mungban. *Current Researches in Plant Sciences*. **II**:229-238.

Jain, R. K., Hassan, N., Pandey, P.N. and Singh, R.K. (1998). Response of some forage legumes to inoculation with VA-mycorrhiza at different P levels. *Range Management & Agroforestry* **19** (2): 121-125.

Jain, R.K. and Hasan, N. (1986). Association of vesicular arbuscular mycorrhizal (VAM) fungi and plant parasitic nematodes with forage sorghum (*Sorghum bicolor* L.). *Sorghum Newsletter*. **29**: 84

Jain, R.K. and Sethi, C.L. (1987). Pathogenicity of *Heterodera cajani* on cowpea as influenced by the presence of VAM fungi, *Glomus fasciculatum* or *G. epigaeus*. *Indian Journal of Nematology*. **17**(2):165-170.

Jain, R.K. and Sethi, C.L. (1988). Interaction between vesicular-arbuscular mycorrhiza, *Meloidogyne incognita* and *Heterodera cajani* on cowpea as influenced by time of inoculation. *Indian Journal of Nematology*. **18**(2): 263-268.

Jain, R.K. and Sethi, C.L. (1989). Influence of endomycorrhizal fungi *Glomus fasciculatum* and *G. epigaeus* on penetration and development of *Heterodera cajani* on cowpea. *Indian Journal of Nematology*. **18**(1):89-93.

Jain, R.K., Hasan, N., Singh, R.K. and Pandey, P.N. (1997). Vesicular-arbuscular mycorrhizal (VAM) association in some weeds of forage legumes. *Mycorrhiza News*. **9**(1): 10-12.

Jain, R.K., Hasan, N., Singh, R.K. and Pandey, P.N. (1998a). Influence of the endomycorrhizal fungus, *Glomus fasciculatum* on *Meloidogyne incognita* and *Tylenchorhynchus vulgaris* infecting berseem. *Indian Journal of Nematology*. **28**(1): 48-51.

Jain, R.K., Hasan, N., Singh, R.K. and Pandey, P.N. (1998b). Interaction between plant parasitic nematodes, *Meloidogyne incognita, Tylenchorhynchus vulgaris* and VAM fungi, *Glomus fasciculatum* on lucerne. *Annals of Plant Protection Sciences*. **6**(1): 37-40.

Jaiswal, V. and Rodrigues, B. F. (2001). Occurrence and distribution of arbuscular mycorrhizal fungi in coastal sand dune vegetation of Goa. *Current Science* **80** (7): 826-827.

Jaiswal, V., Bukhari, M. J., Khade, S. W. and Gaonkar, U. C. (2003). Preliminary survey of arbuscular mycorrhizal association in mangrove vegetation of Goa. *Plant Archives* **3** (1): 73-76.

Jalali, B.L. (1986). Vesicular-arbuscular mycorrhiza: Currence status. In: *Vistas in Plant Pathology*. (Eds. A. Varma and J.P. Verma), Malhotra Publishing House, New Delhi: 437-450.

Jalali, B.L. and Chhabra, M.L. (1991). Effect of pesticides on vesicular-arbuscular mycorrhizal development on growth of pearl millet. *Indian Phytopathology*. **44**(3): 314-318.

Jalali, B.L. and Thareja, M.L. (1981). Suppression of *Fusarium* wilt of chickpea in vesicular-arbuscular mycorrhizal inoculated soils. *International Chickpea Newsletter*. **4** : 21-22.

Jalali, B.L. and Thareja, M.L. (1985). Plant growth response to vesicular-arbuscular mycorrhizal inoculation in soils incorporated with rock phosphate. *Indian Phytopathology*. **38**(2): 306-310.

Jalali, B.L., Chhabra, M.L. and Singh, R.P. (1990). Interaction between vesicular-arbuscular mycorrhizal endophyte and *Macrophomina phaseolina* in mungbean. *Indian Phytopathology*. **43**: 527-530.

Jamaluddin and Chandra, K.K. (1995). Growth performance of *Eucalyptus* hybrid for vesicular-arbuscular mycorrhizal fungi in the coal mine spoils of Korba. In: *Mycorrhizae: Biofertilizers for the Future.* (Eds. Alok Adholeya and Sujan Singh), TERI, New Delhi:260-264.

Jamaluddin and Chandra, K.K. (1997). Distribution of VAM fungi in bauxite mine overburden plantation of Amarkantak (Madhya Pradesh). *Indian Forester*. **123**(5): 412-418.

Jamaluddin and Chaturvedi, P. (1996). Colonization of VAM in *Dendrocalamus asper. Mycorrhiza News*. **8**(3): 10-12.

Jamaluddin, Chandra, K.K. and Goswani, G. H. (2001a). Studies on VAM fungi associated with tree species planted in a sludge garden at the Ballarampur paper mills,Maharashtra, *Mycorrhiza News* **13** (4): 10-12.

Jamaluddin, Chandra, K.K. and Kumar, R. (1998). Effect of fungicides on colonization and development of VAM in *Dalbergia sissoo. Indian Forester*. **124**(8): 658-662.

Jamaluddin, Chandra, K.K. and Malakar, M.G. (2001b). Effectiveness of various typed of VAM inocula on growth and biomass of *Bambusa nutans*. *Mycorrhiza News* **13** (3): 15-22.

Janakirani, S. and Manoharachary, C. (1990). Occurrence and distribution of VAM fungi in safflower. In: *Mycorrhizal symbiosis and plant growth.* (Eds. D.J. Bagyaraj and A. Manjunath), U.A.S., Bangalore : 16.

Janardhanan, K.K. and Khaliq, A. (1995). Influence of vesicular-arbuscular mycorrhizal fungi on growth and productivity of German chamomile in alkaline *Usar* soil. In: *Mycorrhizae: Biofertilizers for the Future.* (Eds. Alok Adholeya and Sujan Singh), TERI, New Delhi: 410-412.

Janardhanan, K.K., Ahmad, A. and Gupta, M. (1990a). Infection by a fungal endophyte, *Balansia sclerotica* enhances vesicular-arbuscular mycorrhizal (VAM) association in lemon grass. In: *Current trends in mycorrhizal research. Proceedings of National Conference on Mycorrhiza.* (Eds. B.L. Jalali and H. Chand), H.A.U., Hisar: 115-116.

Janardhanan, K.K., Gupta, M.L. and Husain, A. (1990b). Axenic culture of a vesicular-arbuscular mycorrhizal fungus. *Current Science*. **50**(10): 509-513.

Janse, J.M. (1896-1897). Les endophytes radicaux de quelques plantes Javanaises. *Ann. Jard. Bot. Buitenzorg*. **14**:53.

Jasper, D.A., Abbott, L.K. and Robson, A.D. (1989). *Acacias* respond to addition of phosphorus and to inoculation with VA mycorrhizal fungi in soils stockpiled during mineral sand mining. *Plant and Soil*. **115**: 99-108.

Jasper, D.A., Abbott, L.K. and Robson, A.D. (1990). *Acacias* respond to additions of phosphorus and to inoculation with VA mycorrhizal fungi in soils stockpiled during mineral sand mining. *Plant and Soil*. **115**(1): 99-108.

Jasper, D.A., Robson, A.D. and Abbott, L.K. (1987). The effect of surface mining on the infectivity of vesicular-arbuscular mycorrhizal fungi. *Australian Journal of Botany*. **36**(6): 641-652.

Jayaraman, J. and Kumar, D. (1995). VAM fungi-pathogen-fungicide interactions in gram. *Indian Phytopathology*. **48**:294-299.

Jha, D.K., Sharma, G.D. and Mishra, R.R. (1988). Status of VAM in degraded forest ecosystem. In: *Mycorrhizae for Green Asia*. (Eds. A. Mahadevan, N. Raman and K. Natarajan), Madras University, Madras : 50-52.

Jha, D.K., Sharma, G.D. and Mishra, R.R. (1993). Mineral nutrition in the tripartite interaction between *Frankia, Glomus* and *Alnus* at different soil phosphorus regimes. *New Phytologist*. **123**(2): 307-311.

Jha, M. N., Dayaram, Prasad, A. N., Mishra, S. K. (2002). Observed a new arbuscular mycorrhizal association in *Typha angustata* under waterlogged condition. *Indian Phytopathology* **55**(1): 79-80.

Jindal, V., Atwal, A., Sekhon, B.S. and Singh, R. (1993). Effect of vesicular arbuscular mycorrhizae on metabolism of moong plants under NaCl salinity, *Plant Physiology and Biochemistry (Paris)*. **31**(4): 475-881.

Johnsen, A., Jakobsen, I. and Jensen, E.S. (1992). Hyphal transport of ^{15}N-labelled nitrogen by a vesicular-arbuscular mycorrhizal fungus and its effect on the depletion of inorganic soil N. *New Phytologist*. **122**: 281-288.

Johnston, P.J. (1979). Ethanol-an alternative to its use as a fuel. Paper presented in *Workshop on Fermentation Alcohol in developing countries*, Vienna, Austria.

Jones, N.P. and Sreenivasa, M.N. (1993). Effect of inoculation of VA mycorrhiza and/or phosphate solubilizing bacteria on rhizosphere microflora of sunflower. II. *Azotobacter* and phosphate solubilizing bacteria. *Journal of Ecotoxicology & Environmental Monitoring*. **3**(1):55-58.

Joseph, K., Kothandaraman, R. and Mathew, J. (1988). VAM infection in different leguminous cover crops of rubber. In: *Mycorrhizae for Green Asia*. (Eds. A. Mahadevan, N. Raman and K. Natarajan), Madras University, Madras :216-218.

Joseph, K., Kothandaraman, R. and Methew, J. (1999). A comparative study of different isolates of VAM on growth, nutrient uptake and biochemical constituents of *Pueraria phaseoloides*. Abstracts: *National Conference on Mycorrhiza*, Barkatulla University, Bhopal: 18.

Joseph, P.J., Sivaprasad, P., Lekha, P.S. and Vijayan, M. (1995a). Vesicular arbuscular mycorrhizal colonization in Ginger and its influence on the natural incidence of rhizome rot. In: *Mycorrhizae: Biofertilizers for the Future* (Eds. Alok Adholeya and Sujan Singh), TERI, New Delhi: 77-80

Joseph, U., Kothari, I.L. and Patel, N. (1995b). Effect of mycorrhization on two crops raised on land disturbed by petro-effluents. In: *Mycorrhizae: Biofertilizers for the Future*. (Eds. Alok Adholeya and Sujan Singh), TERI, New Delhi: 164-167.

Joshaline, C.M. and Balasubramanian, A. (1995). Growth response of *Casuarina equisetifolia* under tripartite interaction between *Frankia, Glomus* and *Casuarina* at different phosphorus regimes. In: *Mycorrhizae: Biofertilizers for the Future*. (Eds. Alok Adholeya and Sujan Singh), TERI, New Delhi: 258-260.

Joshi, K.C. and Singh, H.P. (1995). Interrelationships among vesicular arbuscular mycorrhizae population, soil properties and root colonization capacity of soil. *Journal of the Indian Society of Soil Science*. **43**(2): 204-207.

Joshi, P.K. (1995). Prospects of groundnut improvement by VAM fungi in medium-black calcareous soils of Gujrat. In: *Mycorrhizae: Biofertilizers for the Future*. (Eds. Alok Adholeya and Sujan Singh), TERI, New Delhi: 374-377.

Joshi, S.R., Chauhan, M. Sharma, G.D. and Mishra, R.R. (1991). Effect of deforestation on microbes, VAM fungi and their enzymatic activity in eastern Himalaya. In: *Advances in Himalayan Ecology* . (Eds. G.S. Rajwar) **6**: 141-152.

Jothi, G. and Sundarababu, R. (2001a). Management of root-knot nematode in brinjal by using VAM and crop rotation with green gram and pearl millet. *Journal of Biological Control* **15**(1): 77-80.

Jothi, G. and Sundarababu, R. (2001b). Effect of endomycorrhizal and root knot nematode on the growth and chemical composition in brinjal. *Pest Management in Horticultural ecosystems* **7** (2): 162-167.

Jothi, G. and Sundarbabu, R. (1997). Studies on the management of rot lesion nematode, *Pratylenchus zeae* with the endomycorrhizal fungus, *Glomus fasciculatum* on maize. *Indian Journal of Nematology*. **27**(2): 264-266.

Jothi, G., Rajeshwari, and Sundarbabu, R. (1997). Vesicular arbuscular mycorrhizae penetration in maize roots. *Indian Journal of Nematology*. **27**(1): 125-126.

Juniper, S. and Abbott, L.K. (1993). Vesicular-arbuscular mycorrhizae and soil salinity. *Mycorrhiza*. **4**: 45-57.

Jyrwa, S. and Sharma, G.D. (1988). VAM, phosphatase, dehydrogenase and nitrification in root region of paddy under different agricultural practices. In: *Mycorrhizae for Green Asia*. (Eds. A. Mahadevan, N. Raman and K. Natarajan), Madras University, Madras : 142-144.

Kalavathi, B.P., Santthanakrishna, P. and Divya, M.P. (2000). Effect of VA-mycorrhizal fungus and phosphorus solubilizing bacterium in neem. *Indian Forester* **126**(1): 67-70.

Kale, R.D., Kubra Bano, Sreenivasa, M.N. and Bagyaraj, D.J. (1987). Influence of worm cast (Vee.comp.E. UAS'83) on the growth and mycorrhizal colonization of two ornamental plants. *South Indian Horticulture*. **35**(6): 433-437.

Kamalakannan, S. and Manivannan, K. (2002). Influence of *Azospirillum*, phosphobacteria and vesicular arbuscular mycorrhizae on growth parameters of radish (*Raphanus sativus* L.) *Research on Crops* **3** (1) 138-141.

Kanakadurga, V.V. and Rao, P.R. (1995). Effect of vesicular-arbuscular mycorrhizae on root and shoot phenol concentrations of *Terminalia arjuna*. In: *Mycorrhizae: Biofertilizers for the Future*. (Eds. Alok Adholeya and Sujan Singh), TERI, New Delhi: 478-481.

Kanakdurga, V.V., Manoharachary, D. and Rama Rao, P. (1990). Occurrence of endomycorrhizal fungi teak. In: *Mycorrhizal symbiosis and plant growth*. (Eds. D.J. Bagyaraj and A. Manjunath), U.A.S., Bangalore :17.

Kandasamy, D., Oblisami. G., Mohandass, S. and Santhanakrishnan, P. (1986a). Influence of VA-mycorrhizal inoculation on the growth of pyrethrum in the nursery. *Pyrethrum Post* **16**(3): 81-83.

Kandasamy, D., Palanisamy, D. and Oblisami, G. (1988). Screening of germplasm of sweet potato for VA-mycorrhizal fungal occurrence and response of crop to the inoculation of VAM fungi and *Azospirillum*. *Journal of Root Crops*. **14**)(1): 37-42.

Kandasamy, D., Radha, N.V. and Oblisami, G. (1986b). Response of different mulberry varieties to the inoculation of VA-mycorrhizal fungi. *Indian Journal of Sericulture*. **25**(1): 36-39.

Kandiannan, K.,Sivaraman, K., Anandaraj, M. and Krisnamurthy, K.S. (2000). Growth and nutrient content of black pepper (*Pipper nigrum* L.) cuttings as influenced by inoculation with biofertilizers. *Journal of Spices and Aromatic Crop* **9**(2): 145-147.

Kannan, K. and Lakshminarasimhan, C. (1988). Survey of VAM of maritime strand plants of point calimere. In: *Mycorrhizae for green Asia.* (Eds. A. Mahadevan, N. Raman and K. Natarajan), Madras University, Madras :53-55.

Kannan, K. and Lakshminarasimhan, D. (1990). Development of endomycorrhizae in *Phyla nodiflora.* In: *Mycorrhizal symbiosis and plant growth.* (Eds. D.J. Bagyaraj and A. Manjunath), U.A.S., Bangalore :18-20.

Kapoor, A. and Mukerji, K.G. (1990). A strategy for selection and application of VAM fungi for *Glycine max.* In: *Trends in mycorrhizal research. Proceedings of the National Conference on Mycorrhiza.* (Eds. B.L. Jalali, and H. Chand), H.A.U., Hisar:139-140.

Kapoor, A. and Mukerji, K.G. (1998). Microbial interactions in mycorrhizosphere of *Anethum graveolens* L. *Phytomorphology.* **48**(4): 383-389.

Kapoor, A. and Mukerji, K.G. (1999). Mycorrhizal technology to enhance growth and essential oil yield in *Coriandrum sativum. Abstracts: National Conference on Mycorrhiza*, Barkatulla University Bhopal: 17.

Kapoor, A., Pathak, A., Mago,P., Mukerji, K.G. (1996). VAM in two crucifers. *Cruciferae Newletter.* **18**: 116-117.

Kapoor, A., Singh, V.P. and Mukerji, K.G. (1988). Studies on the phosphatases of mycorrhizal and non-mycorrhizal *Trigonella* roots. In: *Mycorrhizae for Green Asia.* (Eds. A. Mahadevan, N. Raman and K. Natarajan), Madras University, Madras : 125-127.

Kapoor, R., Giri, B. and Mukerji, K.G. (2001). Effect of vesicular arbuscular mycorrhiza (VAM) on growth and essential oil yield of *Anethum graveolens* L. *Indian Journal of Plant Physiology* **6**(1): 77-80.

Kapoor, R., Giri, B. and Mukerji, K.G. (2002). *Glomus macrocarpum*:a potential bioinoculant to improve essential oil quality and concentration in dill (*Anethum graveolens* L.) and carum (*Trachyspermum ammi*(linn.) Sprague). *World Journal of Microbiology and Biotechnology* **18** (5): 459-463.

Karthikeyan, B., Pandiyarajan, P. and Santhanakrishnan, P. (1995). Effect of dual inoculation of phosphobacteria and vesicular-arbuscular mycorrhizal fungi on the growth of neem. In: *Mycorrhizae: Biofertilizers for the Future.* (Eds. Alok Adholeya and Sujan Singh), TERI, New Delhi:245-249.

Kashappanavar, S. and Sreenivasa, M. N. (2000). Influence of inoculation of efficient AM fungi on onion at different P-levels. *Karnataka Journal of Agricultural Sciences* **13** (3): 761-764.

Kathiresan, C. and Venkatesha, J. (2002). Effect of biofertilizers with levels of N and P on gladiolus. In: *Floriculture research trend in India.* Proceedings of the National symposium on Indian floriculture in the new millennium, Bangalore, (Eds. R.L Misra and Sanyat Misra): 118-121.

Katiyar, R.S., Das, P.K. Ghosh, A., Choudhury, P.C. and Datta, R.K. (1990). Interaction between vesicular arbuscular mycorrhizal colonization and phosphorus application in mulberry. In: *Mycorrhizal symbiosis and plant growth.* (Eds. D.J. Bagyaraj and A. Manjunath), U.A.S., Bangalore :79-80.

Katiyar, R.S., Das, P.K., Choudhury, P.C., Ghosh, A. and Sengupta, K. (1989). Vesicular-arbuscular mycorrhizal association in mulberry. *Current Science*. **58**(8(: 461-464.

Katiyar, R.S., Das, P.K., Choudhury, P.C., Ghosh, A., Singh C.B. and Datta, R.K. (1995). Response of irrigated mulberry (*Morus alba* L.) to VA-mycorrhizal inoculation under graded doses of phosphorus. *Plant and Soil*. **170**(2): 331-337.

Kaur, M.J., Anand Sagar and Lakhanpal, T.N. (1997). Observations on vesicular arbuscular mycorrhizal association of *Grewia optiva. Journal of Mycology and Plant Pathology*. **27**(3):323-324.

Kaushal, S. (2000). Influence of edaphic factors on VAMF spore population and root colonization in *Acacia niloica* in Rajasthan. *Journal of Mycology and Plant Pathology* **30** (3): 386-388.

Kaushal, S. (2001) studied partial substitution of mycorrhizal fungi for phosphorus fertilization in *Acacia nilotica* seedlings. *Advances in Plant Sciences* **14** (2): 607-609.

Kaushal, S. (2002). Factors affecting spore population and root colonization in *Prosopis cineraria. Advances in Plant Sciences* **15** (1): 327-330.

Kaushal, S. and Srivastava, H.P. (1995). Biology of vesicular-arbuscular mycorrhizal fungi associated with *Prosopis cineraria* and *Acacia nilotica*. In: *Mycorrhizae: Biofertilizers for the Future.* (Eds. Alok Adholeya and Sujan Singh), TERI, New Delhi: 91-94.

Kaushik, A., Dixon, R.K. and Mukerji, K.G. (1992). Vesicular arbuscular mycorrhizal relationships of *Prosopis juliflora* and *Zizyphus jujuba. Phytomorphology*. **42**(1-2): 133-137.

Kaushik, J. C. and Dabas P. (2003). Interaction between VA mycorrhizal fungi, phosphorus fertilizer and water stress application on nodulation in *Acacia nilotica* (L.) Del. and *Dalbergia sissoo* Roxb. *Forage Research* **28** (4): 220-222.

Kaushik, J.C. and Kaushik, N. (1995). Interaction between vesicular-arbuscular mycorrhizal fungi and *Rhizobium* and their influence on *Albizia lebbek* (L.) Benth. In: *Mycorrhizae: Biofertilizers for the Future.* (Eds. Alok Adholeya and Sujan Singh), TERI, New Delhi: 212-215.

Kawani, Y. and Yamamoto, Y. (1986). Increase in the formation and nitrogen fixation of soybean nodules by vesicular-arbuscular mycorrhiza. *Plant Cell Physiology*. **27**: 399-405.

Kehri, H.K. and Chandra, S. (1988). Mycorrhizal infection and its relation to the rhizosphere microflora in urd under water stress conditions. In : *Mycorrhizae for Green Asia*. (Eds. A. Mahadevan, N. Raman and K. Natarajan), Madras University, Madras : 219-221.

Kehri, H.K. and Chandra, S. (1989a). Interaction of *Glomus fasciculatum* and *Macrophomina phaseolina* in chickpea. *Journal of Indian Botanical Society*. **68** : 363-365.

Kehri, H.K. and Chandra, S. (1989b). Mycorrhizal association in ceratian leguminous weeds. *National Academy Science Letters*. **12** : 15-16.

Kehri, H.K. and Chandra, S. (1989c). Mycorrhizal association in crops under sewage farming. *Journal of Indian Botanical Society*. **69**: 267-269.

Kehri, H.K. and Chandra, S. (1990). VAM association in urd as affected by water stress condition in soil and foliar spray. *Acta Botanica Indica*. **18**: 316-318.

Kehri, H.K. and Chandra, S. (1991). Microbial spectrum in the mycorrhizosphere of green gram as affected by non-indigenous VAM fungi under normal and strained water schedules. In : *Botanical Researches in India*. (Eds.N.C. Aery and B.L. Chaudhury), Himanshu Publications, Udaipur : 513-517.

Kehri, H.K. and Chandra, S. (1993a). Effects of Bavistin spray on soil microorganisms and VAM formation in greengram in relation to its yield. *Journal of Indian Botanical Society*. **72**: 55-57.

Kehri, H.K. and Chandra, S. (1993b). Response of mungbean to different VA-mycorrhizal fungi under normal and strained supply of water. *Acta Botanica Indica*. **21**: 196-200.

Kehri, H.K. and Chandra, S. (1994a). Effect of foliar spray on the VA-mycorrhiza formation in chickpea. *Journal of Indian Botanical Society*. **73**: 181-184.

Kehri, H.K. and Chandra, S. (1994b). Interaction between VA-mycorrhizae and *Rhizobium* and their effects on chickpea in unsterilized soil. *Geobios*.**21**(3):177-181.

Kehri, H.K. and Chandra, S. (1994c). Selection of VA-mycorrhizal inoculants for some upland varieties of rice. *Philippine Journal of Science* . **123**(3):229-237.

Kehri, H.K. and Chandra, S. (1994d). VA-mycorrhiza formation in crops in drier tracts of Bundelkhand region. *Proceeding National Academy of Sciences, India* . **64** (B)II: 221-224.

Kehri, H.K. and Chandra, S. (1995a). Interaction between vesicular-arbuscular mycorrhiza and phosphorus application in mungbean. In : *Microbes and Man*. (Eds. Sudhir Chandra, K.K.Khanna and H.K. Kehri), Bishen Singh Mahendra Pal Singh, Dehra Dun : 377-386.

Kehri, H.K. and Chandra, S. (1995b). Relative efficacy of single and multiple inocula of VAM fungi for chickpea. *Mycorrhiza News* . **7:** (1): 9-10.

Kehri, H.K. and Chandra, S. (2001). Performance of blackgram with VAM inoculation and phosphate fertilization. *Philippine Journal of Science* **130** (1) :33-38.

Kehri, H.K., Chandra, S. and Maheshwari, S. (1987). Occurrence and intensity of VAM in weeds, ornamentals and cultivated plants at Allahabad and areas adjoining it. In: *Proceedings of a National Workshop* JNU, New Delhi: 273-283.

Kehri, S.S., Chandra, S. and Kehri, H. K. (1997). A package of microbial inoculants for linseed. *Journal of Indian Botanical Society*. **76**: 21-24.

Kendrick, W.B. and Berch, S.M. (1985). Mycorrhizae : applications in agriculture and forestry. In: *Comprehensive Biotechnology* . (Ed. M. Moo-Young), Pergamon Press, Oxford, **3**:109-152.

Kennedy, R.W. and Chellapillai, K.L. (1999). Synergistic effect of VAM, *Azospirillum*, and phosphobacteria on growth response and nutrient uptake of shoal tree species. *Indian Journal of Forestry* **21**(4): 308-312.

Khade, S. W. and Rodrigues, B. F. (2003). Occurrence of arbuscular mycorrhiza; fungi in tree species from Western Ghats of Goa, India. *Journal of Tropical Forest Science* **15** (2): 320-331.

Khade, S.W. and Rodrigues, B.F. (2002). Arbuscular mycorrhizal fungi associated with some pteridophytes from western ghat region of Goa. *Tropical Ecology* **43** (2): 251-256.

Khaliq, A. and Bagyaraj, D.J. (2000). Colonization of arbuscular-mycorrhizal fungi on Ri T-DNA transformed roots in synthetic medium. *Indian journal of Experimental Biology* **38**(11): 1147-1151

Khaliq, A. and Janardhanan, K.K. (1994). Variation of native VA-mycorrhizal association on cultivated species of mint. *Symbiosis (Rehovot)*. **16**(1): 75-82.

Khaliq, A. and Janardhanan, K.K. (1997). Influence of vesicular-arbuscular mycorrhizal fungi on the productivity of cultivated mints. *Journal of Medicinal and Aromatic Plant Sciences*. **19**(1): 7-10.

Khaliq, A., Gupta, M.L. and Kumar, S. (1999). Influence of *Glomus mosseae* on growth and uptake of mineral nutrients and essential oil contents of menthol mint. Abstracts: *National Conference on Mycorrhiza*, Barkatulla University, Bhopal: 57.

Khan, M.W. and Kulshreshtha, M. (1995). Impact of fly ash on vesicular-arbuscular mycorrhizal fungus and root nodulation in black gram. In: *Mycorrhizae: Biofertilizers for the Future*. (Eds. Alok Adholeya and Sujan Singh), TERI, New Delhi:29-32.

Khan, Q.A. (1999). Evaluation of techniques for isolation of vesicular arbuscular mycorrhizal fungi. *Annals of Plant Protection Sciences* **7** (2): 239-241.

Khan, Q.Z. and Sarabhoy, A.K. (1990). Identification of endogonaceous fungi from Delhi. In: *Current trends in mycorrhizal research. Proceedings of the National Conference on Mycorrhiza.* (Eds.B.L. Jalali and H. Chand), H.A.U., Hisar: 201.

Khan, S.N. and Uniyal, K. (1999a). Growth response of two forest tree species to VAM and *Rhizobium* inoculations. *Indian Forester* **125** (11): 1125-1128.

Khan, S.N. and Uniyal, K. (1999b). Association of arbuscular mycorrhizal fungi with different bamboo species at a reforested site. *Indian Journal of Forestry* **22** (4): 405-408.

Khan, S.N. and Uniyal, K. (1999c). Growth response of some forest trees to VAM and *Rhizobium* inoculations and fertilizer application. Abstracts: *National Conference on Mycorrhiza*, Barkatulla University, Bhopal:66.

Khare, A.K., Rawat, A.K., Dubey, S.B., Patel, K.S., Rathore, G.S., Thompson, J.P. and Takkar, P.N. (1998). Role of native vesicular-arbuscular mycorrhizal fungi in wheat (*Triticum aestivum*)-based cropping sequence for efficient use of phosphorus and zinc in black soils of Madhya Pradesh. *Indian Journal of Agricultural Sciences*. **68**(5): 247-250.

Kichadi, S.N. and Sreenivasa, M.N. (1998). Interaction effects of *Glomus fasciculatum and Trichoderma harzianum on Sclerotium rolfsii* in presence of biogas-spent slurry in tomato. *Karnataka Journal of Agricultural Microbiology* **11** (2): 419-422.

Kiran Bala, Rao, A.V. and Tarafdar, J.C. (1989). Occurrence of VAM association in different plant species of the Indian Desert. *Arid Soil Research and Rehabilitation*. **3**(3): 82-88.

Koide, R.T. and Li, M. (1990). On host regulation of the vesicular-arbuscular mycorrhizal symbiosis. *New Phytologist*. **114**:59-74.

Konde, B.K., Rathi, S.R. and Ruikar, S.K. (1988). Bulb yield and phosphorus uptake in onion as influenced by *Glomus, Gigaspora* and Microphos inoculations under graded phosphorus levels. In: *Mycorrhizae for Green Asia*. (Eds. A. Mahadevan, N. Raman and K. Natarajan), Madras University, Madras :151-152.

Korikanthimath, V. S., Gayathri, A. G., Gowda, S. J. A Hegde, R. and Hosani, M. M. (2000). Vesicular-arbuscular mycorrhizae and phosphate solubilizers in robusta coffee and cardamom mixed cropping system. *Karnataka Journal of Agricultural Sciences* **13** (2): 498-499.

Korikanthimath, V.S. and Gayathri, A. (1999). Vesicular-arbuscular mycorrhiza and phosphate solubilizers in coffee, pepper and cardamom mixed cropping systems. Abstracts: *National Conference on Mycorrhiza*, Barkatulla University, Bhopal:65.

Koshy, P.K., Sosamma, V.K. and Samuel, R. (1998). Interactive effect of vesicular arbuscular mycorrhizae (VAM) with *Radopholus similis* on coconut seedlings. In *Nematology: Challenges and Opportunities in 21 st Century.* Proceedings of the Third International Symposium of Afro-Asian Society of Nematologists (TISAASN), Sugarcane Breeding Institute (ICAR), Combatore, India (ed. U.K. Mehata) Luton UK; Afro-Asian Society of Nematologists 106-11.

Koske, R.E. (1988). Vesicular-arbuscular mycorrhizae of some Hawaiian dune plants. *Pacific Science.* **42**(3-4):217-229.

Kothari, I.L. (1992). Isolation and characterization of VAM fungi from effluent irrigated fields. *33rd Annual Conference of the Association of Microbiologist of India*, Goa University, Goa : 31.

Kothari, I.L., Udhyaya J. and Patel, C.R. (1997). Mycorrhizal biodiversity of petroeffluent-irrigated fields. *Journal of Mycology and Plant Pathology.* **27**(3): 275-278.

Kothari, S.K. and Singh, U.B. (1996). Response of citronella Java (*Cymbopogon winterianus* Jowitt) to VA mycorrhizal fungi and soil compaction in relation to P supply. *Plant and Soil.* **178**(2): 231-237.

Kough, J. Malajczuk, N. and Linderman, R.C. (1983). *New Phytologist.* **94** : 57-62.

Koul, K., Kapoor, R. and Luikham, N. (2001). Influence of lead in soil on mycorrhizal development and plant growth of *Cyamopsis tetragonoloba* (Linn.). *Indian Journal of Experimental Biology* **39** (5): 459-463.

Krishna Prasad, K.S. (1991). Influence of a vesicular arbuscular mycorrhiza on the development and reproduction of root-knot nematode affecting flue cured tobacco. *Afro-Asian Journal of Nematology.* **1**(2):130-134.

Krishna, K.R. and Bagyaraj, D.J. (1982). Influence of vesicular arbuscular mycorrhiza on growth and nutrition of *Arachis hypogaea. Legume Research.* **5**(1): 18-22.

Krishna, K.R. and Bagyaraj, D.J. (1983). Interaction between *Glomus fasciculatum* and *Sclerotium rolfsii* in peanut. *Canadian Journal of Botany.* **67**(9): 2349-2357.

Krishna, K.R. and Bagyaraj, D.J. (1985). Phosphates in the rhizospheres of mycorrhizal and non-mycorrhizal groundnut. *Journal of Soil Biology and Ecology.* **5**(2):81-85.

Krishna, K.R. and Bagyaraj, D.J. (1986). Phenolics of mycorrhizal and uninfected groundnut var. MGS-7. *Current Research.* **15**(5-6):51-52.

Krishna, K.R., Bagyaraj, D.J. and Rai, P.V. (1982). Response of groundnut to VA mycorrhizal inoculation in black clayey soil. *Indian Journal of Microbiology.* **22**(3): 206-208.

Krishna, K.R., Balakrishna, A.N. and Bagyaraj, D.J. (1982). Interaction between VA mycorrhizae and *Streptomyces cinnamomeus* and their effect on finger millet. *New Phytologist.* **92**: 401.

Krishnamurthy, K.V., Nelson, R. and Senthilkumar (1998). VAM association in sandal-a hemi-root parasite. *Mycorrhiza News.* **9**(4): 11-12.

Krishnaraj, P.U. and Sreenivasa, M.N. (1992). Increased root colonization by bacteria due to inoculation of vesicular-arbuscular mycorrhizal fungus in chilli (*Capsicum annuum*). *Zentralblatt fur Mikrobiologie.* **147**(1-2): 131-133.

Krone, W., Bichler, B., Viebrock, E. and Moawad, A.M. (1987). Interactions between VA mycorrhiza and phosphate solubilizing bacteria. In: *7th North American Conference on Mycorrhizae,* University of Florida, Gainesville: 328.

Krukelmann, H.W. (1975). Effect of fertilizers, soils, soil tillage and plant species on the frequency of *Endogone* chlamydospores and mycorrhizal infections in arable soils, In: *Endomycorrhizas* . (Eds. F.E. Sanders, B. Mosse, and P.B. Tinker), Academic Press, London,: 511.

Kshattriya, S., Jha, D.K., Sharma, G.D. and Mishra, R.R. (1990). Nutrient release from litter and the development of mycorrhizae in more disturbed and less disturbed forest communities. In: *Current trends in mycorrhizal research. Proceedings of the National Conference on Mycorrhiza* . (Eds. B.L. Jalali and H. Chand), H.A.U., Hisar: 37-39.

Kulkarni A., Halbe, S. and Nair L.N. (1995). Vesicular-arbuscular mycorrhizae of *Casuarina* from Konkan:II. In: *Mycorrhizae: Biofertilizers for the Future*. (Eds. Alok Adholeya and Sujan Singh), TERI, New Delhi:48-51.

Kulshreshtha, M. and Khan, M. W. (2002). Effect of ozone on root colonization by VAM fungi and root nodulation on black gram. *Indian Phytopathology* **55** (3): 264-268.

Kulshreshtha, M. and Khan, M.W. (1999a). Impact of sulphur dioxide on root colonization by a VAM fungus and nodulation on black gram. *Indian Phytopathology* **52** (2): 182-184.

Kulshreshtha, M. and Khan, M.W. (1999b). Impact of fly ash application in soil on root colonization by a VAM fungus root nodulation by *Rhizobium. Indian Phytopathology* **52** (2): 185-187.

Kulshreshtha, M. and Khan, M.W. (1999c). Effect of ozone on root colonization by VAM fungi and root nodulation by *Rhizobium* on black gram. Abstracts: *National Conference on Mycorrhiza*, Barkatulla University, Bhopal:64.

Kumar R., Gupta, P.P. and Jalali, B.L. (2001). Impact of VA-mycorrhiza, *Azotobacter* and *Rhizobium* on growth and nutrition of cowpea. *Journal of Mycology and Plant Pathology* **31** (1): 38-41.

Kumar, D. and Mehrotra, V.S. (1996). Potentials of microbe-activating technology for plantation health management. *Tropical Forest* : 153-161.

Kumar, D., Gaikwad, S.H. and Singh, C.P. (1995a). Influence of plant growth promoting rhizobacteria on mycorrhizal associations in wheat. In: *Mycorrhizae: Biofertilizers for the Future*. (Eds. Alok Adholeya and Sujan Singh), TERI, New Delhi: 206-208.

Kumar, D., Rai, K. and Jayaraman, J. (1995b). Effect of mushroom-mycorrhizal interactions on yield and quality of pearl millet seeds. In: *Mycorrhizae: Biofertilizers for the Future*. (Eds. Alok Adholeya and Sujan Singh), TERI, New Delhi: 378-380.

Kumar, G.S. (2002). Seasonal variations in the biodiversity of arbuscular mycorrhizal fungi in forest ecosystems. *Journal of Ecobiology* **14**(1): 35-38.

Kumar, G.S. and Murugesh, S. (2002). A study on the VAM fungi improves growth of some medicinal plants. *Advances in Plant Sciences* **15** (1): 43-46.

Kumar, K.A.P. and Bagyaraj, D.J. (1999). Mass production of arbuscular mycorrhiza as influenced by some agrochemicals. *Proceedings of National Academy of Sciences* India. **69**(B),: 62-66.

Kumar, N. and Chandra, R. (1999). Response of mungbean (*Vigna radiata*) to *Rhizobium* and VA mycorrhiza inoculation with varying phosphorus levels. Abstracts: *National Conference on Mycorrhiza*, Barakatulla University Bhopal: 23.

Kumar, P. and Sood, A.K. (2002). Management of bacterial wilt of tomato with VAM and bacterial antagonits. *Indian Phytopathology* **55** (4): 513-515.

Kumar, P.C., Garibova, C.L.V. and Velikanov, L.L. (1997). Alkaline phosphatase levels as indicator for screening VAM fungi for spring wheat. In: *Microbial biotechnology.* (Eds. S.M. Reddy, H.P. Srivastava, D.K. Prohit and S. R. Reddy), Jodhpur University, Jodhpur: 67-69.

Kumar, P.P., Reddy, S.R. and Reddy, S.M. (1999a). Effect of inoculations of VAM and other bioinoculants on the growth of two agro-forestry tree seedlings. Abstracts : *National Conference on Mycorrhiza*, Barkatulla University, Bhopal: 49.

Kumar, P.P., Reddy, S.R. and Reddy, S.M. (2000). Mycorrhizal dependency of some agroforestry tree species. *Indian Forester* **126**(4): 397-402.

Kumar, P.P., Shailaja, K.M. and Reddy, S.R. (1999b). Distribution and occurrence of VAM fungi in polluted soils. Abstracts: *National Conference on Mycorrhiza*, Barkatulla University Bhopal: 5.

Kumar, R., Jalali, B. L. and Chand, H. (2002b). Effect of soil application of fungicides on mycorrhizal colonization and growth of chickpea. *Plant Disease Research* **17** (1) 1-8.

Kumar, R., Jalali, B. L. and Harichand (2002a). Influence of vesicular arbuscular mycorrhizal fungi on growth and nutrient uptake in chickpea. *Journal of Mycology and Plant Pathology* **32** (1): 11-15.

Kumar, R.R. and Potty, V.P. (1998). Biomass production of *Pachyrhizus erosus* (L.) by individual and coinoculation with *Glomus microcarpum* and *Bradyrhizobium*. *Journal of Root Crops* **24** (2): 143-145.

Kumar, S.S. and Arockiasamy, D.I. (1995). Distribution of zinc in various parts of *Sorghum bicolor* (L.) Moench seedlings in the presence of VAM fungi. *Indian Journal of Microbiology.* **35**(3):185-188.

Kumar, V. and Mangal, J.L.(1995a). Response of potato to mycorrhiza at different levels of soil salinity with and without phosphorus nutrition. In: *Mycorrhizae: Biofertilizers for the Future.* (Eds. Alok Adholeya and Sujan Singh), TERI, New Delhi:519-521.

Kumar, V. and Mangal, J.L.(1995b). Vesicular-arbuscular mycorrhiza improved the performance of potted onions and saved phosphorus input. In: *Mycorrhizae: Biofertilizers for the Future.* (Eds. Alok Adholeya and Sujan Singh), TERI, New Delhi:209-211.

Kumaran, K. and Santhanakrishnan, P. (1995). Vesicular-arbuscular mycorrhizal fungi in tea (*Camellia sinensis* (L.) O. Kuntz) soil. In: *Mycorrhizae: Biofertilizers for the Future.* (Eds. Alok Adholeya and Sujan Singh), TERI, New Delhi: 33-37.

Kumari, K., Mishra, D.P. and Johri, B.N. (1990). Characterization of mycorrhiza-specific alkaline phosphatase from french bean. In : *Current trends in mycorrhizal research. Proceedings of National Conference on Mycorrhiza.* (Eds. B.L. Jalali and H. Chand), H.A.U. Hisar : 57-58.

Kumari, K.S.M. and Nair, S.K. (1989). Influence of host variety on the natural incidence of VA-mycorrhiza in cowpea. *Agricultural Research Journal of Kerala.* **27**(1-2): 59-60.

Kumari, S.M.P. and Balasubramanian, A. (1993). Effect of combined inoculation of VAM and *Azospirillum* on the growth and nutrient uptake by coffee seedlings. *Indian Coffee.* **57**(12) 5-11.

Kumutha, K. and Santhanakrishnan, P. (1995). Effect of insoluble phosphate and dual inoculation on soybean. In: *Mycorrhizae: Biofertilizers for the Future.* (Eds. Alok Adholeya and Sujan Singh), TERI, New Delhi:202-205.

Kunwar, L.K., Reddy, P.J.M. and Manoharachary, C. (1999). Occurrence and distribution of AMF associated with garlic rhizosphere soil. *Mycorrhiza News*. **11**(2): 4-6.

Labeena, P., Sreenivasa, M. N. and Lingaraju, S. (2002). Interaction effect between arbuscular mycorrhizal fungi and root knot nematode *Meloidogyne incognita* on tomato. *Indian Journal of Nematology* **32** (2) 118-120.

Lakshman, H. C., Mulla, F. I. And Inchal, R.F.(1999). Studies on the seasonal fluctuations of arbuscular mycorrhizal fungi on some commonly cultivated crops of Dharwad district in Karnataka. In: *Proceedings of National Conference on Mycorrhiza*. Barkatulla University, Bhopal: 40

Lakshman, H.C. (1992). Development and response of vesicular arbuscular mycorrhizal fungi in *Terminalia bellerica* Roxb. *Journal of Tropical Forestry*. **8**(2): 179-182.

Lakshman, H.C. and Raghavendra, S. (1990a). Occurrence of vesicular arbuscular mycorrhizal fungi in medicinal plants. In: *Mycorrhizal Symbiosis and Plant Growth*. (Eds. D.J. Bagyaraj and A. Manjunath), U.A.S., Bangalore : 21-23.

Lakshman, H.C. and Raghavendra, S. (1990b). Spore dispersal of Endogonaceae by worms, wasps, dung rollers and Indian domestic fowls. In *Current trends in mycorrhizal research. Proceedings of the National Conference on Mycorrhiza*. (Eds. B.L. Jalali and H. Chand), H.A.U., Hisar: 40-41.

Lakshman, H.C. and Raghavendra, S. (1995). Effect of leaf litter and soil in different proportions on growth and biomass production in six timber tree species colonized with vesicular-arbuscular mycorrhizae. In: *Mycorrhizae: Biofertilizers for the Future*. (Eds. Alok Adholeya and Sujan Singh), TERI, New Delhi: 445-450.

Lakshman, H.S. (1996). Possible role of weeds in maintaining VA-mycorrhizal colonization in rice soils. *Worlds Weeds*. **3**(1/2): 53-56.

Lakshminarasimaham, A.V. and Vijaykumar, B.S. (1994). VAM association in *Arachis hypogaea* in Andhra Pradesh. *Indian Journal of Mycology and Plant Pathology*. **24**(3): 232-233.

Lakshmipathy, R., Balakrishana, A.N. and Bagyaraj, D.J. (2000). Studied the symbiotic response of cashew rootstocks to different VA mycorrhizal fungi. *Cashew* **14** (3): 20-24.

Lambert, D.H. and Cole (Jr), H. (1980). Effect of mycorrhizae on establishment and performance of forage species in mine spoil. *Agronomy Journal*. **72**: 257-260.

Land, S. and Schonbeck, F. (1991). Influence of different soil types on abundance and seasonal dynamics of vesicular arbuscular mycorrhizal fungi in arable soils of North Germany. *Mycorrhiza*. **1**: 39-44.

Lanfranco, L., Wyss, P., Marzachi, C. and Bonfante, P. (1995). Generation of RAPD-PCR primer for the identification of isolates of *Glomus mosseae*, an arbuscular mycorrhizal fungus. *Molecular Ecology*. **4** : 61-68.

Largent, D.L., Sugihara, N. and Brinitzer, A. (1980). *Amanita gemmata*, a non-host specific mycorrhizal fungus of *Actostaphylos manzanita. Mycologia*. **72**: 435-439.

Lee, A. and Bagyaraj, D.J. (1986). Effect of soil inoculation with vesicular arbuscular mycorrhizal fungi and other phosphate rock dissolving bacteria or thiobacilli on dry matter production and uptake of phosphorus by tomato plants. *New Zealand Journal Agricultural Research*. **29**: 525-531.

Lee, K.K., Reddy, M.V., Wani, S.P. and Trimurtulu. N. (1996). Vesicular-arbuscular mycorrhizal fungi in earthworm casts and surrounding soil in relation to soil management of a semi-arid tropical Alfisol. *Applied Soil Ecology*. **3**: 177-181.

Lekha, K.S., Sivaprasad, P., Joseph, P.J. and Vijaran, M. (1995). *Glomus fasciculatum* - a predominant vesicular-arbuscular mycorrhizal fungus associated with black pepper in forest soils of Kerala. In: *Mycorrhizae: Biofertilizers for the Future*. (Eds. Alok Adholeya and Sujan Singh), TERI, New Delhi : 81-85.

Letham, D.S., Goodwin, P.B. and Higgins, T.J.W. (1978). *Phytohormone and related compounds: A comprehensive treatise.* Vol. 1 and 2, Elsevier, Amsterdam.

Levy, Y., Syvertsen, J.P. and Nemec, S. (1983). Effect of drought stress and vesicular-arbuscular mycorrhiza on citrus transplantation and hydraulic conductivity of roots. *New Phytologist*. **93**(1): 61-66.

Li, C.Y. and Castellano, M.A. (1984). Nitrogen-fixing bacteria isolated from within sporocarps of three ectomycorrhizal fungi. In: *6th North American Conference on Mycorrhizae*, Bend, OR: 264.

Li, C.Y. and Castellano, M.A. (1987). *Azospirillum* isolated from within sporocarps of the mycorrhizal fungi *Hebeloma crustuliniforme, Laccaria laccata* and *Rhizopogon vinicolor. Transactions of British Mycological Society*. **88** : 563-565.

Li, X.L., George, E. and Marshher, H. (1991). Phosphorus depletion and pH decrease at the root-soil and hyphae-soil interfaces of VA mycorrhizal white clover, fertilized by ammonium. *New Phytologist*. **119**: 397.

Lilly, S.S. and Santhanakrishnan, P (2000). Interaction effect of *Glomus fasciculatum* and *Trichoderma viride* on the growth of sunflower. *Madras Agricultural Journal* **86** (1/3): 54-57.

Linderman, R.G. (1988). Mycorrhizal interactions with the rhizosphere microflora: The mycorrhizosphere effect. *Phytopathology*. **78**: 366-371.

Logan, V.S., Clarke, P.J. and Allaway, W.G. (1989). Mycorrhizae and root attributes of plants of coastal sand-dunes of New South Wales. *Australian Journal of Plant Physiology*. **16**(1): 141-146.

Lopez-Sanchez, M.E. and Honrobia, M. (1992). Seasonal variation of vesicular-arbuscular mycorrhizae in eroded soils from Southern Spain. *Mycorrhiza*. **2**(1): 33-39.

Machado, C. and Bagyaraj, D.J. (1995). Mycorrhization helper bacteria and their influence on growth of cowpea. In: *Mycorrhizae: Biofertilizers for the Future*. (Eds. Alok Adholeya and Sujan Singh), TERI, New Delhi: 192-196.

Mada, R.J. and Bagyaraj, D.J. (1993). Root exudation from *Leucaena leucocephala* in relation to mycorrhizal colonization. *World Journal of Microbiology and Biotechnology.* **9**(3):342-344.

Madan, M., Sharma, S. and Vasudevan, P. (1995). Role of mycorrhiza on wasteland : effect of vesicular-arbuscular mycorrhizal fungi on different plant species. In: *Mycorrhizae: Biofertilizers for the Future*. (Eds. Alok Adholeya and Sujan Singh), TERI, New Delhi : 545-548.

Mago, P. and Mukerji, K.G. (1994). Vesicular arbuscular mycorrhizae in Lamiaceae. I. Seasonal variation in some members. *Phytomorphology*. **44**(1-2): 83-88.

Mago, P., Pasricha, P. and Mukerji, K.G. (1993). Survey of vesicular-arbuscular mycorrhizal fungi in scale leaves of modified stems. *Phytomorphology* **43**:81-85.

Mahanta, B. and Phukan, P.N. (2000). Effect of *Glomus fasciculatum* on penetration and development of *Meloidogyne incognita* on blackgram. *Journal of the Agricultural Sciences Society of North-East India* **13**(2): 215-217.

Maiti, D. and Variar, M. (2000). Effect of different rice (*Oryza sativa* L.) based cropping systems on population dynamics of native vesicular-arbuscular mycorrhizal fungi under rain fed upland ecosystem. *Mycorrhiza News* **12**(3): 15-18.

Maiti, D. and Variar, M. and Saha, J. (1999). Exploring possibilities of enhancing VAM association for improvement of upland rice (*Oryza sativa* L.) under rainfed Agro-ecosystem. Abstracts: *National Conference on Mycorrhiza*, Barkatulla University Bhopal: 28.

Maiti, D., Rana, S. K., and Prasad, A. N. (2002). Studied the Influence of rice legumes intercropping on native vesicular-arbuscular mycorrhizal association in rice. *Indian Phytopathology* **55**(1): 75-78.

Maiti, D., Variar, M. and Singh, R.K. (1996). Perpetuation of native VAM fungi under monocropped, rainfed upland agro-ecosystem. *Mycorrhiza News*. **8**(1): 7-9.

Maiti, S.K. (1997). Importance of VAM fungi in coal mine overburden reclamation and factors affecting their establishment on overburden dumps. *Environment and Ecology*. **15**(3): 602-608.

Maitra, A. and Gautam, S.P. (1994). Ecological studies of vesicular arbuscular mycorrhizal fungi (VAMF) associated with *Dendrocalamus strictus* in the central Madhya Pradesh. *Journal of Physiological Research*. **7** (1) : 29-31.

Malathy, M.R., Kumar, K.R. and Sivaprasad, P. (1999). Root knot nematode infestation and growth of *Kaemperia galanga* as influenced by inoculation with arbuscular mycorrhiza. Abstracts: *National Conference on Mycorrhiza*, Barkatulla University, Bhopal: 63.

Mallesha, B.C. and Bagyaraj, D.J. (1990). Reaction of leucaena to single inoculation versus dual inoculation with *Glomus mosseae* and /or *Gigaspora margarita. Leucaena Research Reports*. **11**: 56-57.

Mallesha, B.C., Bagyaraj, D.J. and Pai, G. (1992). Perlite Soilrite mixture as a carrier for mycorrhiza and *Rhizobium* to *Leucaena leucocephala. Leucaena Research Reports*. **13**: 32-33.

Malloch, D.W., Pirozynski, K.A. and Raven, P.H. (1980). Ecological and evolutionary significance of mycorrhizal symbiosis in vascular plants. *Proceedings of National Academy of Sciences U.S.A.*. 77: 2113-2118.

Mamatha, G., Bagyaraj, D. J. and Jaganath, S. (2002). Inoculation of field established mulberry and papaya with arbuscular mycorrhizal fungi and a mycorrhizal helper bacterium. *Mycorrhiza* **12** (6): 313-316.

Mandal, B.S. and Kaushik, J.C. (1994). Effect of VAM inoculations on growth and development of *Acacia nilotica* (L.) Willd. ex.del. *Annals of Biology (Ludhiana)*. **10**(2): 264-267.

Mandal, B.S. and Kaushik, J.C. (1995). Interaction between VA mycorrhizal fungi and *Rhizobium* and their effect on the growth parameters of *Acacia nilotica* (L.) Willd ex. Del. *Haryana Agricultural University Journal of Research*. **25**(3): 107-111.

Mandal, B.S., Kaushik, J.C. and Singh, R.R. (1995). Relative efficacy of single and multiple inocula of some VAM fungi for *Acacia nilotica* (L.) Willd ex Del. *Crop Research (Hisar)*. **9**(3):437-440

Mandhare, V.K., Patil, P.L. and Gadekar, D.A. (1998). Phosphorus uptake of onion as influenced by *Glomus fasciculatum, Azotobacter* and phosphorus levels. *Agricultural Science Digest* (Karnal) **18** (4): 228-230.

Mane, S.S., Raut, R.S. and Kothire, O.D. (1993). Comparative performance of vesicular arbuscular mycorrhizal fungus and *Rhizobium* inoculation with groundnut. *Annals of Plant Physiology*. 7(1): 116-118.

Mangal, J.K. and Kumar, V. (1998). Vesicular arbuscular Mycorrhizae (VAM) improved performance of potted onion under saline conditions with and without phosphorous nutrition. *Vegetable Science* **25** (2): 119-123.

Manian, S., Thomson, T.E., and Kumar, A.A. (1999). Management of pests and diseases of tomato seedlings (*Lycopersicon esculentum*) by using vesicular-arbuscular mycorrhiza. Abstracts: *National Conference on Mycorrhiza*, Barkatulla University, Bhopal:62.

Manjunath, A. and Bagyaraj, D.J. (1982). Vesicular arbuscular mycorrhizas in three plantation crops and cultivars of field bean. *Current Science*. **51**(14): 707-708.

Manjunath, A. and Bagyaraj, D.J. (1984). Response of pigeon pea and cowpea to phosphate and dual inoculation with vesicular arbuscular mycorrhiza and *Rhizobium. Tropical Agriculture*. **61**(1): 48-52.

Manjunath, A. and Bagyaraj, D.J. (1986). Response of black-gram, chickpea and mungbean to vesicular-arbuscular mycorrhizal inoculation in an unsterile soil. *Tropical Agriculture*. **63**(1): 33-35.

Manjunath, A., Bagyaraj, D.J., and Gopal Gowda, H.S. (1984). Dual inoculation with VA mycorrhiza and *Rhizobium* is beneficial to *Leucaena. Plant and Soil*. **78**: 445-448.

Manjunath, A., Mohan, R. and Bagyaraj, D.J. (1981). Interactions between *Beijerinckia mobilis, Aspergillus niger and Glomus fasciculatus* and their effects on growth of onion. *New Phytologist*. **87**: 723.

Manjunath, A., Mohan, R. and Bagyaraj, D.J. (1983). Response of citrus to vesicular-arbuscular mycorrhizal inoculation in unsterile soils. *Canadian Journal of Botany*. **67**(10): 2729-2732.

Manjunath, V.G., Patil, C.P., Swamy, G.S.K. and Patil, P.B. (2002a). Effect of different VAM fungi on growth parameters of papaya cv. Sunset Solo. *Journal of Maharashtra Agricultural Universities* **26** (3): 269-271.

Manjunatha, V.G., Patil, C.P., Swamy, G.S.K. and Patil, P.B. (2002b). Effect of different VAM fungi and phosphorus levels on yield and yield components of papaya. *Karnataka Journal of Agricultural Sciences* **15** (2): 336-342.

Manoharachary, C. and Reddy, P.J.M. (1995). Role of vesicular-arbuscular mycorrhizal fungi in forestry. In : *Mycorrhizae: Biofertilizers for the Future*. (Eds. Alok Adholeya and Sujan Singh), TERI, New Delhi:297-302.

Mathew, J. and johri, B. N. (1989). Effect of indigenous and introduced VAM fungi on growth of mungbean. *Mycological Research*.**92**: 491-493.

Mathew, J., Neeraj, Shankar, A., Kaur, R. and Varma, A. (1990). VAM fungi from the rhizosphere of desert cacti. In: *Current trends in mycorrhizal research. Proceedings of the National Conference on Mycorrhiza.* (Eds. B.L. Jalali and H. Chand), H.A.U., Hisar: 44-46.

Mathur, N. and Vyas, A. (2000a). Mycorrhizal dependency of *Tamarix aphylla* in saline areas of Thar Desert. *Naturalia (Sao Paulo)* **25**:105-110.

Mathur, N. and Vyas, A. (1994). Vesicular arbuscular mycorrhizal relationship of *Simmondsia chinensis. Phytomorphology.* **44**(1-2): 11-14.

Mathur, N. and Vyas, A. (1995a). Changes in isozyme patterns of peroxidase and polyphenol oxidase by VAM fungi in roots of *Zizyphus* species. *Journal of Plant Physiology.* **145**(4)-498-500.

Mathur, N. and Vyas, A. (1995b). Efficacy of indigenous vesicular-arbuscular mycorrhizal fungi on biomass production, nutrient uptake, and nodulation in *Acacia nilotica* in the Indian Thar Desert. In : *Mycorrhizae: Biofertilizers for the Future.* (Eds. Alok Adholeya and Sujan Singh), TERI, New Delhi: 256-259.

Mathur, N. and Vyas, A. (1995c). I. Influence of VA mycorrhizae on net photosynthesis and transpiration of *Zizyphus mauritiana. Journal of Plant Physiology.* **147**(3/4): 328-330.

Mathur, N. and Vyas, A. (1995d). II. Changes in enzymes of nitrogen metabolism by VA mycorrhizae in *Zizyphus nummularia. Journal of Plant Physiology.* **147**(3/4):331-333.

Mathur, N. and Vyas, A. (1995e). Physiological changes in *Prosopis cineraria* by vesicular-arbuscular mycorrhizae. In: *Mycorrhizae: Biofertilizers for the Future.* (Eds. Alok Adholeya and Sujan Singh), TERI, New Delhi: 458-461.

Mathur, N. and Vyas, A. (1995f). Production of highest nutritive pods of *Cyamopsis tetragonoloba* by VA mycorrhiza. *Indian Journal of Experimental Biology.* **33**(6): 464-465.

Mathur, N. and Vyas, A. (1996a). Biomass production, nutrient uptake and nodulation in *Prosopis cineraria* by indigenous VAM fungi. *Annals of Arid Zone.* **35**(1):53-56.

Mathur, N. and Vyas, A. (1996b). Physiological changes in *Zizyphus mauritiana* by differnt VAM fungi. *Indian Forester.* **122**(6): 501-506.

Mathur, N. and Vyas, A. (1996c).Relative efficiency of different VAM fungi on growth and nutrient uptake in *Zizyphus mauritiana. Indian Journal of Forestry.* **19**(2): 129-131.

Mathur, N. and Vyas, A. (1999). Improved biomass production, nutrient uptake and establishment of *in vitro* raised *Zizyphus mauritiana* by mycorrhiza. *Journal of Plant Physiology.* **155**(1): 129-132.

Mathur, N. and Vyas, A. (2000b). Influence of arbuscular mycorrhizae on biomass production, nutrient uptake and physiological changes in *Ziziphus mauritiana* Lam. under water stress. *Journal of Arid Environments* **45** (3): 191-195.

Mathur, N. and Vyas, A.(1997). Enzymatic changes in cluster bean by different VAM species. *Agricultural Science Digest (Karnal)* **17**(1):16-18.

McLennan, E.I. (1926). The endophytic fungus of *Lodium*. II. The mycorrhiza on the roots of *Lolium temulentum* L. with a discussion on the physiological relationships of the organism concerned. *Annals of Botany.* **40**: 43.

Medeiros, C.A.B., Clark, R.B. and Ellis, J.R. (1994). Growth and nutrient uptake of sorghum cultivated with vesicular arbuscular mycorrhizae isolates at varying pH. *Mycorrhiza.* **5**: 185-191.

Meenakumari, K.S. and Nair, S.K. (1992). Effect of phosphatic fertilizer application on vesicular-arbuscular mycorrhizal infection in cowpea. *Agricultural Research Journal of Kerala.* **30**(2): 129-131.

Mehotra, M.D. and Mehrotra, A. (1999). Suitability of potting mixture for VAM infection and spore population in root trainer raised seedlings. *Indian Journal of Forestry* **22** (1/2): 49-52.

Mehrotra, A., Mehrotra, M.D. and Sachan, S.N. (1998). Effects of VAM, fertilizer and deoiled neem cake application on the growth of *Paulownia tomentosa* raised from root cuttings. *Indian Journal of Forestry.* **21**(2):111-114.

Mehrotra, M.D. (1996a). Multiplication of VAMF on *Paulownia*- a veritable possibility. *Indian Forester.* **122**(9): 260.

Mehrotra, M.D. (1996b). New VAM technology and its application. *Indian Forester.* **122**(2): 348-350.

Mehrotra, M.D. (1997). Fantastic growth response of *Paulownia* to mycorrhization and fertilization. *Indian Forester.* **123**(9): 873-875.

Mehrotra, V.S. (1993). The so-called vesicles in vesicular-arbuscular mycorrhizae are chlamydospores. *Philippine Journal of Science* . 377-395.

Mehrotra, V.S. (1995). Arbuscular mycorrhizal association in plants colonizing overburdened soil at an opencast coal mine site. In : *Mycorrhizae: Biofertilizers for the Future.* (Eds. Alok Adholeya and Sujan Singh), TERI, New Delhi: 22-27.

Mehrotra, V.S. (1996). Use of revegetated coal mine spoil as source of arbuscular mycorrhizal inoculum for nursery inoculations. *Current Science.* **70**(1): 73-77.

Mehrotra, V.S. (1997). *Glomus bagyarajii* sp. Nov., A new species of Glomaceaa (Glomales, Zygomycetes) from India. *The Philippine Journal of Science.* **126**(3): 233-242.

Mehrotra, V.S. (1998). Arbuscular mycorrhizal association of plants colonizing coal mine spoil in India. *Journal of Agricultural Science.* **130**(2): 125-133.

Mehrotra, V.S. and Baijal, U. (1992a). *Glomus sterilum,* a new species of VA-mycorrhiza isolated from potato roots. *Philippine Journal of Science.* **121**(3): 305-316.

Mehrotra, V.S. and Baijal, U. (1992b). Influence of certain agronomic practices on mycorrhizal associations in field crops. *Proc. National Academy of Sciences India.* **62**(B): 604-613.

Mehrotra, V.S. and Baijal, U. (1993). Vesicular-arbuscular mycorrhizal formation in an ornamental plant, *Codiaeum pictum* L. *New Agriculturist.* **4**(2): 215-218.

Mehrotra, V.S. and Baijal, U. (1994). Effect of vesicular-arbuscular mycorrhiza on the growth and nutrition of sunflower plant at different levels of added N and P fertilizers. *Proceedings of the National Academy of Sciences India. Section B, Biological Sciences.* **64**(3): 299-304.

Mehrotra, V.S. and Prakash, A. (1999). Use of soil inoculum from under mycorrhizal tree species grown on reclaimed coal mine spoil. Abstracts: *National Conference on Mycorrhiza,* Barkatulla University, Bhopal:72.

Mehrotra, V.S., Baijal, U., Mishra, S.D., Pandey, D.P. and Mathews, T. (1995). Movement of ^{32}P in sunflower plants inoculated with single and dual inocula of VAM fungi. *Current Science*. **68**(7): 751-753.

Mengistu Fentahun and Singh, N. (1999). Effects of biofertilizers on growth, yield and nutrient uptake of onion (*Allium cepa* L). *Vegetable Science* **26** (2): 193-195.

Mercy, M.A., Shrivashankar, G. and Bagyaraj, D.J. (1990). Mycorrhizal colonization in cowpea is host dependent and heritable. *Plant and Soil*. **121**(2): 292-294.

Meshram, A.T., Jadhav, A.C., Konde, B.K. and Wani, P.V. (1999, publ. 2000). Effect of VAM fungi and P-sources on nutrient uptake and yield of chickpea. *Journal of Maharashtra Agricultural Universities* **24** (3): 340-341.

Meshram, A.T., Jadhav, A.C., Konde, B.K. and Wani, P.V. (2000). Effect of VAM fungi and P-sources on nodulation, dry matter and yield of chickpea. *Journal of Maharashtra Agricultural Universities* **25** (1): 99-101.

Meshram, S.U., Peshwe, S.A., Joshi, S.N. and Dongre, A.B. (1997). Response of (to) Biofertilizers on (of) biomass production of *Eucalyptus camaldulensis. Annals of Forestry*. **5**(1): 43-49.

Methew, J. and Johri, B.N. (1989). Effect of indigenous and introduced VAM fungi on growth of mungbean. *Mycological Research*. **92**: 491-493.

Meyer, J.R. and Linderman, R.G. (1986). Selective influence on populations of rhizosphere or rhizoplane bacteria and actinomycetes by mycorrhizas formed by *Glomus fasciculatum. Soil Biology and Biochemistry*. **18**: 191-196.

Miranda, J.C.C. DE and Miranda, L.N.De (1997). (Arbuscular Mycorrhizas)-Mycorrhiza arbuscular. In: *Biologia dos solos dos solos dos cerrados*. (Eds. M.A. Teixeira Vargas, M.A. and M. Hungria): 67-123.

Mishra, A. (1999). Pesticide-*Glomus fasciculatum* (VAM) interaction on growth and development of cultivars of tomato. *Journal of Research, Birsa Agricultural University*.Ranchi. **11** (1): 39-44.

Mishra, A. and Shukla, B.N. (1996). Interaction between *Glomus fasciculatum, Meloidogyne incognita* and fungicide in tomato. *Indian Journal of Mycology and Plant Pathology*. **26**(1): 38-44.

Mishra, A. and Shukla, B.N. (1997). Interactions between *Glomus fasciculatum* and *Meloidogyne incognita* on tomato. *Journal of Mycology and Plant Pathology*. **27**(2):199-202.

Mishra, A.K. (1994). Strategies for reclamation of mine waste areas through the help of selected microbes : problems and prospects. In: *Crop Epidemics, Microbes and Ecosystem Conservation*. (Eds. G.P.Agarwal and S.K. Hasija), Narendra Publishing House, Delhi: 177-184.

Mishra, A.K., Thatoi, H.W. and Mallik, D. (1999). Induction of drought tolerance in *Stylosanthes hamata* (L.) Taub. with Am fungi and Jalsakti growth in forest degraded wastes of Chandaka in Orissa. Abstracts : *National Conference on Mycorrhiza*, Barkatulla University, Bhopal : 11.

Mishra, A.K., Williams, A.J. and Banerjee, S.K. (1995a). Mycorrhizal development and plant growth in amended flyash of thermal power plants. *Advances in Forestry Research in India*. **12**: 213-227.

Mishra, O.R., Tomar, U.S., Sharma, R.A. and Rajput, A.M. (1995b). Response of maize to chemicals and biofertilizers. *Crop Research (Hisar)*. **9**(2): 233-237.

Mishra, R.R. and Verma, N.K. (1983). Effect of different mycorrhizal treatments on the growth of onion. *Acta Botanica Indica.* **11**(1):49-52.

Mishra, S., Mishra, S.,Mishra, D.R., Tiwari, G.L. and Shukla, M.(2005). Endomycorrhizal association from the early Devonian of Rhynie Chert. In : Challenges in Indian Palaeobiology-Current status, Recent Developments and Future Directions", BSIP, Lucknow.

Mishra,A.,Patel, C.R., and Kothari, I.L. (2001). Status of mycorrhizal fungi in turmeric fields subjected to soil solarization. *Mycorrhiza News* **13** (4): 12-13.

Mittal, N., Sharma, M., Saxena, G. and Mukerji, K.G. (1991). Effect of VA mycorrhiza on gall formation in tomato roots. *Plant Cell Incompatibility Newsletter.* **23** : 39-43.

Mohan, V. (1999). Effect of VAM inoculation on the growth of *Prosopis* spp. in soils of arid zone. Abstracts : *National Conference on Mycorrhiza*, Barkatulla University, Bhopal: 53.

Mohan, V. (2000). Endomycorrhizal interaction with rhizosphere and rhizoplane mycoflora of forest tree species in Indian arid zone. *Indian Forester* **126**(7): 749-755.

Mohan, V. and Mishra, R.N. (1998). Studies on VAM association in Marwar Teak [*Tecomella undulata* (Smith) Seeman] in desert soils of Rajasthan. *Mycorrhiza News.* **10**(1): 16-18.

Mohan, V. and Natarajan, K. (1988). VAM association in sand dune plants in the coromandel coast. In: *Mycorrhizae for green Asia.* (Eds. A. Mahadevan, N. Raman and K. Natarajan). Madras University, Madras : 73-76.

Mohan, V. and Singh, Y.P. (1997). Endomycorrhizal associations of acacias in nurseries and plantations of (the) Indian arid zone. *Indian Forester.* **123**(4): 323-330.

Mohan, V. and Verma, N. (1995). Studies on vesicular-arbuscular mycorrhizae association in seedlings of forest tree species in arid zones of Rajasthan. In : *Mycorrhizae : Biofertilizers for the Future.* (Eds. Alok Adholeya and Sujan Singh), TERI, New Delhi:52-55.

Mohan, V., Verma, N and Singh, Y.P. (1995). Distribution of VAM fungi in nurseries and plantations of neem (*Azadirachta indica*) in arid zone of Rajasthan. In : *Special issue: neem-gift of the gods. Indian Forester.* **121**(11): 1069-1076.

Mohanan, C. and Sebastian, M. (1999). Mycorrhizal status of nineteen species of bamboos in Kerala, India. Abstracts : *National Conference on Mycorrhiza*, Barkatulla University, Bhopal: 37.

Mohandas, S. (1987). Field response of tomato (*Lycopersicon esculentum* Mill. 'Pusa Ruby') to inoculation with a VA mycorrhizal fungus *Glomus fasciculatum* and with *Azotobacter vinelandii. Plant and Soil.* **98** (2) : 295-297.

Mohandas, S. (1988). Response of papaya (*Carica papaya*) to VAM fungal inoculation. In : *Mycorrhizae for Green Asia.* (Eds. A. Mahadevan, N. Raman and K. Natarajan), Madras University, Madras : 260-261.

Mohandas, S. (1990a). Enhanced phosphatase activity in mycorrhizal papaya (*Carica papaya* cv. Coorg Honey Dew) roots. In: *Mycorrhizal Symbiosis and Plant Growth.* (Eds. D.J. Bagyaraj and A. Manjunath), U.A.S., Bangalore : 55-56.

Mohandas, S. (1990b). Occurrence of vesicular-arbuscular mycorrhizal fungi in root stocks of mango. In : *Mycorrhizal Symbiosis and Plant Growth.* (Eds. D.J. Bagyaraj and A. Manjunath), U.A.S., Bangalore : 37-38.

Mohandas, S. (1995). Utilization of vesicular arbuscular mycorrhizal fungi in banana cultivation.In:Mycorrhizae:*Biofertilizers for the Future* (Eds. Alok Adholeya and Sujan Singh) TERI: 441.

Mohandas, S. Shivananda, T.N. and Iyergar, B.R.V. (1995). Uptake of ^{32}P labelled superphosphate by endomycorrhizal papaya (*Carica papaya* cv. Coorg Honey Dew). *Journal of Nuclear Agriculture and Biology*. **24** (4) : 220-224.

Mohankumar, V. and Mahadevan, A. (1986). Survey of vesicular-arbuscular mycorrhizae in mangrove vegetation. *Current Science*. **55**(18): 936.

Mohankumar, V. and Mahadevan, A. (1987). Effect of litter leachates on vesicular arbuscular mycorrhizae. *Indian Journal of Experimental Biology*. **25**(2): 129-131.

Mohankumar, V. and Mahadevan, A. (1988a). Seasonal changes in spore density and root colonization of VAM in a tropical forest. In: *Mycorrhizae for Green Asia*. (Eds. A. Mahadevan, N. Raman and K. Natarajan), Madras University, Madras : 77-79.

Mohankumar, V. and Mahadevan, A. (1988b). Vesicular-arbuscular (VA) mycorrhizal distribution with respect to organic matter content of the soil in a tropical forest. *Tropical Ecology*. **29**(1) : 55-62.

Mohankumar, V. and Mahadevan, A. (1988c). Viability of VAM spores in a tropical forest soils. In: *Mycorrhizae for Green Asia*. (Eds. A. Mahadevan, N. Raman and K. Natrajan), Madras University, Madras: 80-81.

Mohankumar, V. and Sivaswamy, S.N. (1992). Vesicular arbuscular mycorrhizae. *Advances in Plant Sciences*. **5**(1): 17-28.

Mohankumar, V., Ragupathy, S., Nirmala, C.B. and Mahadevan, A. (1988). Distribution of vesicular arbuscular mycorrhizae (VAM) in the sandy beach soils of Madras University, Madras Coast. *Current Science*. **57**(7): 367-368.

Moore, L.W. and Cooksey, D.A. (1981). Biology of *Agrobacterium tumefaciens*: plant interactions. In: *Biology of the Rhizobiaceae*. (Eds. K.L. Giles and A.G. Atherly), Academic Press, New York: 15.

Morely, C.D and Mosse, B. (1976). Abnormal vesicular-arbuscular mycorrhizal infections in white clover induced by Lupin. *Transactions British Mycological Society*. **67**: 510.

Morton, J.B. (1988). Taxonomy of VA mycorrhizal fungi: classification, nomenclature and identification. *Mycotaxon*. **32**: 267-324.

Morton, J.B. and Benny, G.L. (1990). Revised classification of arbuscular mycorrhizal fungi (Zygomycetes): a new order, Glomales, two new suborders, Glominae and Gigasporinae, and two new families, Acaulosporaceae and Gigasporaceae, with an emendation of Glomaceae. *Mycotaxon*. **37**: 471-491.

Morton, J.B., Bentivenga, S.P. and Bever, J.D. (1994). Discovery, measurement and interpretation of diversity in symbiotic endomycorrhizal fungi (Glomales, Zygomycetes). *Canadian Journal of Botany* :

Mosse, B. (1973). Advances in the study of vesicular-arbuscular mycorrhiza. *Annual Review Phytopathology*. **11**: 171-196.

Mosse, B. (1975). Specificity in VA mycorrhizas. In: *Endomycorrhizas* (Eds. F.E., Sanders, B. Mosse and P.B. Tinker), Academic Press, London: 469.

Mosse, B. (1981). VAM research on tropical agriculture. *Research Bulletin of Hawaii Institute of Tropical Agriculture and Human Resources*. **19**: 5-81.

Mosse, B. (1986). Mycorrhiza in a sustainable agriculture. *Biol. Agricultural Horticulture*. **3**: 191-209.

Mosse, B. (1990). Practical problems of field inoculation with vesicular arbuscular mycorrhiza (VAM): inoculum production. *Mycorrhiza News*. **2**(4):1-4.

Mosse, B. and Hepper, C. (1975). Vesicular-arbuscular mycorrhizal infections in root organ culture. *Physiological Plant Pathology*. **5** : 215-233.

Mosse, B. and Thompson, J.P. (1984). Vesicular-arbuscular endomycorrhizal inoculum production. I. Exploratory experiments with bean (*Phaseolus vulgaris*) in nutrient flow culture. *Canadian Journal of Botany*. **62** : 1523-1530.

Mosse, B., Stribley, D.P. and LeTacon, F. (1981). Ecology of mycorrhizae and mycorrhizal fungi. *Advanced Microbial Ecology* . **5**: 137.

Mudge, K.W. (1987). Hormonal involvement in ectomycorrhizal development. In: *7th North American Conference on Mycorrhizae* University of Florida, Gainesville: 228.

Mugnier, J. and Mosse, B. (1987). Vesicular-arbuscular mycorrhizal infection in transformed root-inducing T-DNA roots grown axenically. *Phytopathology*. **77** : 1045-1050.

Mukerji, K.G. (Editor) (1996). *Concepts in mycorrhizal research*. Dordrecht, Netherlands, Kluwer Academic Publishers, xi + 374 pp.

Mukerji, K.G. and Ardey, J. (1986). Studies on vesicular-arbuscular mycorrhiza. 1. *Kalanchoe spicata* and *Eclipta alba*. *Acta Botanica Indica*. **13**(1): 101-103.

Mukerji, K.G. and Kapoor, A. (1986).Occurrence and importance of vesicular-arbuscular mycorrhizal fungi in semiarid regions of India. *Forest Ecology and Management*. **16**(1-4): 117-126.

Mukerji, K.G. and Kapoor, A. (1990). Taxonomy of VAM fungi with special reference to Indian taxa. In: *Perspectives in mycological research (volume II)*. (Eds. S.K. Hasija, and K.S. Bilgrami), New Delhi:7-10.

Mukerji, K.G., Bhattacharjee, M. and Mohan, M. (1982). *Angewandte Botanik* **56**: 121-132.

Mukerji, K.G., Bhattacharjee, M. and Tewari, J.P. (1983). New species of vesicular-arbuscular mycorrhizal fungi. *Transactions of the British Mycological Society*. **87**(3): 641-643.

Mukerji, K.G., Gopal, R., Ball, M. and Rekha Rani (1991). The importance of mycorrhiza for roots. In: *Plant roots and their environment. Proceedings of an ISRR symposium*. (Eds. B.L. McMichael and H. Persson) : 290-308.

Mukherjee P.K. and Rai, R.K. (1999). Sensitivity of P uptake to change in root growth and soil volume as influenced by vesicular arbuscular mycorrhizae (VAM), phosphate solubilizing bacteria (PSB) and P levels in wheat and chickpea. *Annuals of Agricultural Research* **20** (4): 528-530.

Mukherjee, P.K. and Rai, R.K. (2000). Effect of vesicular arbuscular mycorrhizae and phosphate-solubilizing bacteria on growth, yield and phosphorus uptake by wheat (*Triticum aestivum*) and chickpea (*Cicer arietinum*). *Indian Journal of Agronomy* **45** (3): 602-607.

Muniyamma, M., Bharati, B.K. and Reddy, C.N (2000). Effect of VAM on root induction in *in vitro* sugarcane (*Saccharum officinarum* L.) seedlings - a new technique. *Mycorrhiza News* **12**(1): 13-16.

Munnus, D.N. and Mosse, B. (1980). Mineral nutrition of legume crops. In: *Advances in Legume Sciences*. (Eds. R.J. Summerfield and A.H. Bunting), University of Reading Press, England: 115.

Munyanziza, E., Kehri, H.K. and Bagyaraj, D.J. (1997). Agricultural intensification, Soil biodiversity and agro-ecosystem function in the tropics : the role of mycorrhiza in crops and trees. *Applied Soil Ecology.* Elsevier Science B.V. **6**:77-85.

Murashinge and Skoog (1962). A revised medium for rapid growth and bioassays with tobacco tissue culture. *Physiol. Plant* **15** : 473-479.

Murugan, R. and Selvaraj, T. (2003). Reaction of kashini (*Cichorium intybus* L.) to different native arbuscular mycorrhizal fungi. *Mycorrhiza News* **15** (1): 10-13.

Murumkar, D.R. and Patil, P.L. (1996). VAM-diazotrophs-bell pepper symbiosis. *Journal of Mâharashtra Agricultural University*. **21**(3): 394-397.

Muthukumar, T. and Udaiyan, K. (2000). Influence of organic manures on arbuscular mycorrhizal fungi associated with *Vigna unguiculata* (L.) Walp. in relation to tissue nutrients and soluble carbohydrate in roots under field conditions. *Biology and Fertility of Soils* **31** (2): 114-120.

Muthukumar, T. and Udaiyan, K. (1995). Vesicular-arbuscular mycorrhizae in dicots of nutrient deficient semi-arid grassland. In: *Mycorrhizae: Biofertilizers for the Future*. (Eds. Alok Adholeya and Sujan Singh), TERI, New Delhi:541-544.

Muthukumar, T. and Udaiyan, K. (1999a). Arbuscular mycorrhizal association in medicinal plants of western ghats of Southern India. Abstracts : *National Conference on Mycorrhiza*, Barkatulla University, Bhopal: 36.

Muthukumar, T. and Udaiyan, K. (1999b). Spore-in-spore syndrome in vesicular-arbuscular mycorrhizal fungi and its seasonality in a tropical grassland. *Nova Hedwigia*. **68**(3/4): 339-349.

Muthukumar, T. and Udaiyan, K. (2000a). Arbuscular mycorrhizas of plants growing in the Western Ghats region, Southern India. *Mycorrhiza* **9**(6): 297-313.

Muthukumar, T. and Udaiyan, K. (2000b). The role of seed reserves in arbuscular mycorrhizal formation and growth of *Leucaena leucocephala* (Lam.) de Wit. and *Zea mays L. Mycorrhiza* **9** (6): 323-330.

Muthukumar, T. and Udaiyan, K. (2001). Vesicular arbuscular mycorrhizal association in medicinal plants of Maruthamalai hills, Western Ghats, southern India. *Journal of Mycology and Plant Pathology* **31** (2): 180-184.

Muthukumar, T. and Udaiyan, K. (2002a). Arbuscular mycorrhizal fungal composition in semi arid soils of Western Ghats,Southern India . *Current Science* **82** (6): 624-628.

Muthukumar, T. and Udaiyan, K. (2002b). *Gigaspora decipens* (Glomales) an arbuscular mycorrhizal fungus from Western Ghats of Southern India. *Journal of Mycology and Plant Pathology* **32** (1): 96-99.

Muthukumar, T. and Udaiyan, K. (2002c). Arbuscular mycorrhizas in cycads of Southern India. *Mycorrhiza* **12** (4): 213-217.

Muthukumar, T. and Udaiyan, N. (2002d). Growth and yield of cowpea as influenced by changes in arbuscular mycorrhiza in response to organic manuring. *Journal of Agronomy and Crop Science* **188** (2): 123-132.

Muthukumar, T., Udaiyan, K. and Manian, S. (1994). Vesicular-arbuscular mycorrhizae in certain tropical wild legumes. *Annals of Forestry*. **2**(1): 33-43.

Muthukumar, T., Udaiyan, K. and Manian, S. (1996). Vesicular-arbuscular mycorrhizae in tropical sedges of Southern India. *Biology and Fertility of Soils*. **22**(1/2): 96-100.

Muthukumar, T., Udaiyan, K. and Rajeshkannan, V. (2001). Response of neem (*Azadirachta indica* A. Juss) to indigenous arbuscular mycorrhizal fungi, phosphate solubilizing and asymbiotic nitrogen-fixing bacteria under tropical nursery conditions. *Biology and Fertility of Soils* **34** (6): 417-426.

Muthukumar, T., Udaiyan, K., Karthikeyan, A. and Manian, S. (1997). Influence of native endomycorrhiza, soil flodding and nurse plant on mycorrhizal status and growth of purple nutsedge (*Cyperus rotundus* L.). *Agriculture, Ecosystems and Environment*. **61**(1): 51-58.

Nagabhushanam, P., Reddy, S.M. and Reddy, S.R. (1999). VAM fungi associated with some common legume trees of Godavari belt. *Indian Journal of Forestry* **22** (1/2): 129-131.

Nagabhushanam, P., Reddy, S.M. and Reddy, S.R. (1999). VAM fungi associated with some common legume trees of Godavari Belt. *Indian Journal of Forestry*. **22**(2): 129-131.

Nagarajan, G., Bojan, C.R. and Mahadevan, A. (1995). Light and Electron microscopic observations of the spores of *Glomus deserticola*. *Biology and Biotechnology of Mycorrhizae*. **56**:149-154.

Nagarajan, P., Radha, N.V., Kandasamy, D., Oblisami, G. and Jayaraj, S. (1989). Effect of combined inoculation of *Azospirillum brasilense* and *Glomus fasciculatum* on mulberry. *Madras Agricultural Journal*. **76**(11): 601-605.

Nagaraju, R., Haripriya, K. and Purushotham, K. (2001). Influence of VA-mycorrhizal inoculation under varying phosphorus levels and sources on onion. In *changing scenario in the production system of horticultural crops. Proceedings of a National Seminar, Coimbatore, Tamil Nadu, India. South Indian Horticulture* **49** (special): 204-207.

Nagaraju, R., Haripriya, K., Rajalingam, G.V., Sriramchandrasekaran, V. and Mohideen, M.K. (2000). Effect of VAM on growth and yield of aggregatum onion (*Allium cepa* L. Var. *aggregatum* Don). *South Indian Horticulture* **48** (1/6): 40-45.

Nagaraju, S., Rajalingam, G.V., Kumar, R.S. and Haripriya, K. (1999). Screening of efficient VAM fungi for aggregatum onion. *South Indian Horticulture* **47** (1/6): 255-258.

Nagesh, M. and Reddy, P.P. (1997). Management of *Meloidogyne incognita* on *Crossandra undulaefolia* using vesicular arbuscular mycorrhiza, *Glomus mosseae*, and oil cakes. *Mycorrhiza News*. **9**(1) :12-14.

Nagesh, M., Reddy, P.P. and Rao, M.S. (1999a). Comparative efficacy of VAM fungi is combination with neem cake against *Meloidogyne incognita* on *Crossandra undulaefolia. Mycorrhiza News***11** (3): 11-13.

Nagesh, M., Reddy, P.P., Kumar, M.V.V. and Nagaraju B.M. (1999b). Studies on correlation between *Glomus fasciculatum* spore density, root colonization and *Meloidogyne incognita* infection on *Lycopersicon esculentum. Zeitschrift für Pflanzenkrankheiten und Pflanzenschutz* **106** (1): 82-87.

Naik, B.H., Nalawadi, U.G. and Sreenivasa, M.N. (1999). Efficacy of different vesicular-arbuscular mycorrhizal fungi in *Callistephus chinensis* (L.) Nees. *Karnataka Journal of Agricultural Sciences* **12** (1/4): 154-158.

Naik, B.H., Nalawadi, U.G. and Sreenivasa, M.N. (1998). Role of vesicular-arbuscular mycorrhizal fungi in the production of China aster. *Karnataka Journal of Agricultural Sciences* **11** (3): 737-741.

Naik, B.H., Nalawdi, U.G., Shreenivasa, M.N. and Patil, A.A. (1995). Field responses of China aster (*Callistephus chinensis* (L.) Nees.) cv. 'Ostrich plume' to the inoculation of vesicular arbuscular mycorrhizal fungi at different phosphorus levels. *Scientia Horticulturae.* **62**(1/2): 129-133.

Nair, L.N. and Mahabale, T.S. (1975). Mycorrhiza in Ophioglossaceae: I. Morphology of endophytes *in vivo*. *Geophytology*. **5**(1): 16-23.

Nair, S.K. and Girija, V.K. (1988). Incidence of vesicular arbuscular mycorrhiza in certain tree crops of Kerala. *Journal of Plantation Crops*. **16**(1): 67-68.

Nair, S.K., Girija, V.K. and Nayar, K. (1991). Effect of certain plant protection chemicals on the incidence of VA mycorrhiza in black gram. *Agricultural Research Journal of Kerala*. **19**(1-2): 61-62.

Nair, S.K., Peethambar, C.K., Nayar, G. K. and Wilson, K.I. (1990). Effect of soil solarization on infection by mycorrhizal fungi and yield of cowpea. *Plant and Soil*. **125**(1): 153-154.

Nalini, P. A., Reddy, M.S.B. and Bagyaraj, D.J. (1986). Selection of an efficient VA mycorrhizal fungus for *Leucaena*- a preliminary report. *Leucaena Research Reports*. **7** : 61-62.

Naqvi, N.S. and Mukerji, K.G. (1998). Mycorrhization of micropropagated *Leucaena leucocephala* (Lam.) de Wit. *Symbiosis (Rehovot)*. **24**(1): 103-114.

Narayana, R.C. and Gudige, C.A. (1999). Growth response of carrot (*Daucus carota* L.) to a selective VA mycorrhizal inoculant in an unsterile soil. *Mycorrhiza News*. **11**(1): 11-13.

Narayanan, R., Rangarajan, M. and Kandasamy, D. (1990). Response of a shola plant *Rhododendron nilgricass* to endomycorrhizal inoculation. In: *Current trends in mycorrhiza research. Proceedings of the National Conference on Mycorrhiza*. (Eds. B.L. Jalali and H. Chand), H.A.U., Hisar :11-12.

Neeraj and Verma, A.(1995). *Cyamopsis*, Vesicular-arbuscular mycorrhizae, and *Rhizobium* interaction studies. In: *Mycorrhizae: Biofertilizers for the Future* . (Eds. Alok Adholeya and Sujan Singh), TERI, New Delhi: 220-223.

Neeraj, Mukerji, K.G., Sharma, B.C. and Varma, A.K. (1993). A new species of *Gigaspora* from desert soils of Rajasthan, India. *World Journal of Microbiology and Biotechnology*. **9**(3): 291-294.

Neeraj, Verma,S., Kaur, R. and Varma, A.K. (1990). Effect of inoculation with endomycorrhizal fungi on the growth of *Cymopsis tetragonoloba*. In : *Mycorrhizal Symbiosis and Plant Growth*. (Eds. D.J. Bagyaraj and A. Manjunath), U.A.S., Bangalore : 82-83.

Negi, M. and Tilak, K.V.B.R. (1990). Interaction between *Glomus versiforme* and *Azospirillum brassilense* in barley. In: *Current trends in mycorrhizal research. Proceedings of the National Conference on Mycorrhiza.* (Eds. B.L. Jalali and H. Chand), H.A.U., Hisar :109-110.

Negi, M., Sachdev, M.S. and Tilak, K.V.B.R. (1990). Influence of soluble phosphorus fertilizer on the interaction between a vesicular-arbuscular mycorrhizal fungus and *Azospirillum brasilense* in barley (*Hordeum vulgare* L.). *Biology and Fertility of Soils.* **10**(1): 57-60.

Nema, S., Behari, B. and Jamaluddin. (1998). Effect of VAM on growth and development of bamboos planted under agroforestry system. *Indian Forester.* **124**(7): 516-523.

Nemec, S. (1987). Effect of storage, temperature and moisture on *Glomus* species and their subsequent effect on citrus root-stock seedling growth and mycorrhizae development. *Transactions of the British Mycological Society.* **89**(2): 205-212.

Newman, E.I. (1985). Some factors affecting the abundance of mycorrhizas in grassland. In: *Nuclear techniques to study the role of mycorrhiza in increasing food crop production.* (Ed.S.K. Danso), IAEA-TECDOC-338. International Atomic Energy Agency. Vienna.

Newman, E.I., Heap, A.J. and Lawley, R.A. (1981). Abundance of mycorrhizas and root-surface micro-organisms of *Plantago lanceolata* in relation to soil and vegetation: A multivariate approach. *New Phytologist.* **89**: 95-108.

Ng, P.P., Cole, A.L.J., Jameson, P.E. and Mcwha, J.A.(1982). Cytokinin production by ectomycorrhizal fungi. *New Phytologist.* **91**: 57-62.

Nicolson, T.H. and Schenck, N.C. (1979). Endogonaceous mycorrhizal endophytes in Florida. *Mycologia.* **71**: 178-198.

Nirmala, C.B., Mohankumar, V. and Mahadevan, A. (1988). Viability of VAM spores in the fallow fields of Maduravoyal Tamil Nadu. In: *Mycorrhizae for Green Asia.* (Eds. A. Mahadevan, N. Raman and K. Natarajan), Madras University, Madras : 89-90.

Nirmala, R. and Vadivel, E. (1999). Effect of combined application of organic manure and biofertilizers on growth and productivity of cucumber. *South Indian Horticulture* **47** (1/6): 252-254.

Nirmala, R., Vadivel, E. and Azakiamanavalan, R.S. (1999). Influence of organic manures on fruit characters and yield of cucumber (*Cucumis sativus* Linn) cv. Local. *South Indian Horticulture* **47** (1/6): 65-68.

Oliveira, A.A.R. and Zambolim, L. (1986). Interaction between the endomycorrhizal fungus *Glomus etunicatum* and the gall nematode *Meloidogyne javanica* at different levels on bean (*Phaseolus vulgaris). Fitopatologia Brasileira.* **11**: 217-226.

Onkarayya, H. and Sukhada Mohandas. S. (1993). Studies on dependency of citrus rootstocks to VAM inoculation in Alfisol soil. *Advances in Horticulture and Forestry.* **3**:81-91.

Pacovsky, R.S. (1988). Influence of inoculation with *Azospirillum brasilense* and *Glomus fasciculatum* on sorghum nutrition. *Plant and Soil.* **110**: 283-287.

Padma, S.D. and Sullia, S.B. (1990). Development of vesicular arbuscular mycorrhizal fungi cultivars of Mulberry. In: *Mycorrhizal Symbiosis and Plant Growth.* (Eds. D.J. Bagyaraj and A. Manjunath), U.A.S., Bangalore : 31-33.

Padma, S.D., Sukumar, J., Kumar, T.D.N. and Bongale, U.D. (1999). Effect of vesicular arbuscular mycorrhiza, *Azotobacter* and Biophos on the growth of mulberry saplings. *International Journal of Tropical Agriculture* **17** (1/4): 183-187.

Padma, T.M.R. and Kandasamy, D. (1990). Effect of interaction between VA-mycorrhizae and graded levels of phosphorus on the growth of papaya (*Carica papaya*). In: *Current trends in mycorrhizal research. Proceedings of National Conference on Mycorrhiza*. (Eds. B.L. Jalali and H. Chand), H.A.U., Hisar: 133-34.

Padmavathi, T., Veeraswamy, J. and Venkateswarlu, K. (1990). Occurrence of *Sclerocystis* species in semi-arid soils of India. *Proceedings of the Indian Academy of Sciences. Plant Sciences*. **100**(4): 259-262.

Padmavathi, T., Veeraswamy, J. and Venkateswarlu, K. (1990a). Effect of pesticides on endomycorrhizal symbiosis in foxtail millet and sorghum. In: *Mycorrhizal Symbiosis and Plant Growth*. (Eds. D.J. Bagyaraj and A. Manjunath), U.A.S., Bangalore : 47-48.

Padmavathi, T., Veeraswamy, J. and Venkateswarlu, K. (1990b). Spore population of VAM fungi in foxtail millet fields of Anantapur district, Andhra Pradesh. In: *Current trends in mycorrhizal research. Proceedings of the National Conference on Mycorrhiza*. (Eds. B.L. Jalali and H. Chand), H.A.U., Hisar: 10.

Padmavathi, T., Veeraswamy, J., Venkateswarlu, K. and Rao, A.S. (1984). Occurrence of VA mycorrhizas on different legumes in a laterite soil. *Current Science*. **53**(23):1254-1255.

Pahwa, M.R. (1996). Effect of inoculation of *Rhizobium*, VAM fungi and phosphate solubilizing bacteria on nodulation and seedling vigour of *Leucaena leucocephala: Range Management & Agroforestry*. **16**:129-131.

Pai, G., Bagyaraj, D.J., Ravindra, T.P. and Prasad, T.G. (1994). Calcium uptake by cowpea as influenced by mycorrhizal colonization and water stress. *Current Science*. **66**(6): 444-445.

Pal, K.K. and Singh, C.S. (1995). Effect of different bacteria on vesicular-arbuscular mycorrhizal root colonization to pearl millet. In: *Mycorrhizae: Biofertilizers for the Future*. (Eds. Alok Adholeya and Sujan Singh), TERI, New Delhi:250-252.

Pal, S.B. (1978). (Ed.). *Enzyme-labelled Immunoassay of Hormones and Drugs*, Berlin, Walter de Guyter.

Palipane, J.B. (1999). Influence of vesicular-arbuscular mycorrhizal fungi on the growth of coffee plants. Abstracts: *National Conference on Mycorrhiza*, Barkatulla University, Bhopal: 11.

Pande, M., Tarafdar, J.C. and Gupta, G.N. (1999). Vesicular arbuscular mycorrhizal infection in arid zone trees of agroforestry systems. *Journal of the Indian Society of Soil Science*. **47**(1): 54-57.

Pandey, C.B. (1998). Plant parasitic nematodes reduce VAM infection in dry-tropical grasses. *Tropical Ecology* **39** (2): 285-287.

Pandey, D., Singh, J. P., Kashyap, N. and Dwivedi, G.K. (2003). Response of vesicular arbuscular mycorrhizae (VAM), rhizobium and phosphorus sources on nodulation, growth and yield of pea variety VL-7. *Crop Research (Hisar)* **25** (2): 333-336.

Pandey, M. and Tarafdar, J.C. (1999). Critical level of VAM fungal propagules in different arid crops. *Agrochimica* **43** (3/4): 187-192.

Pandey, M. and Gupta, G.N. (2000). Scope of biofertilizers in agroforestry. *Indian Journal of Agro forestry* **2:** 23-32.

Pandey, M. and Trafdar, J.C. (2002). Effect of phosphorus salinity and moisture on VAM fungal association in neem (*Azadirachta indica* Linn.). *Symbiosis (Rehovot)* **32** (3): 195-209.

Pandey, P. N., Verma, M.M. and Jain, R.K. (1998). Response of cowpea *(Vigna unguiculate* Linn. Walp.) to VAM fungi *Glomus fasciculatum* (Gf) and *Rhizobium* (Rh) inoculation under different sources and levels of phosphorus. *Proceedings of the National Academy of Sciences, India. Section B, Biological Sciences* **68** (3/4): 273-278.

Pandey, R., Gupta, M.L., Singh, H.B. and Kumar, S. (1999a). The influence of vesicular-arbuscular mycorrhizal fungi alone or in combination with *Meloidogyne incognita* on *Hyoscyamus niger* L. *Bioresource Technology*. **69**(3): 275-278.

Pandey, R., Singh, H.B., Gupta, M. L., Singh, H. N., Sharma, S. and Kumar, S. (1999b). Efficacy of VAM fungi, green manure, organic materials and pesticides in the management of root knot nematode in menthol mint. In: *Proceedings of national symposium on rational approaches in nematode management for sustainable agriculture.* Anand, India. (eds. S.C. Dhawan and K. K. Kaushal) New Delhi, India; Nematological Society of India: 47-50.

Pandey, S. and Singh, V.K. (1990a). Ecophysiological responses of litchi cultivars to vesicular arbuscular mycorrhizal colonization. In:*Mycorrhizal Symbiosis and Plant Growth.* (Eds. D.J. Bagyaraj and A. Manjunath), U.A.S., Bangalore : 49-50.

Pandey, S. and Singh, V.K. (1990b). Endophytic mycorrhiza in different varieties of *Litchi chinensis* Sonn. and its effect on rhizosphere-microbial population. In: *Current trends in mycorrhiza.* (Eds. B.L. Jalali and H. Chand), H.A.U., Hisar: 117.

Panja, B. and Chaudhuri, S. (1999). Mycorrhiza in *Cajanus* root association in alluvial soil. Abstracts : *National Conference on Mycorrhiza*, Barkatulla University, Bhopal: 50.

Pant, D.D., Nautiyal, D.D. and Misra, D.R. (1995). Studies in Endomycorrhiza of some Ophioglossaceous prothalli. In : *Microbes and Man.* (Eds. Sudhir Chandra, K.K. Khanna and H .K.Kehri), Bishen Singh Mahendra Pal Singh, Dehra Dun : 340-378.

Panwar, J. and Vyas, A. (2000). Changes in peroxidase and polyphenol oxidase activities in roots of *Tamarix aphylla* by different AM fungi *Journal of Eco-Physilogy* **3**(3/4): 103-106.

Panwar, J. and Vyas, A. (2002a). AM fungi: a biological approach towards conservation of endangered plants in Thar Desert, India. *Current Science* **82** (5): 576-578.

Panwar, J. and Vyas, A. (2002b). Relative efficiency of different AM fungi on biomass production and nutrient uptake in *Tamarix aphylla. Indian Forester* **128** (11): 1216-1220.

Panwar, J.D.S. (1993). Response of VAM and *Azospirillum* inoculation to water status and grain yield in wheat under water stress condition. *Indian Journal of Plant Physiology*. **36**(1):41-43.

Panwar, J.D.S. and Thakur, A.K. (1995). Physiological and biochemical studies in vesicular-arbuscular mycorrhizae, *Rhizobium* inoculated mungbean plants under field conditions. In: *Mycorrhizae: Biofertilizers for the Future.* (Eds. Alok Adholeya and Sujan Singh), TERI, New Delhi: 471-477.

Parameswaran, P. and Augustine, B. (1988). Distribution and ecology of a VAM in a scrub jungle. In: *Mycorrhizae for Green Asia.* (Eds. A. Mahadevan, N. Raman and K. Natrajan), Madras University, Madras : 91-94.

Paroha, S., Chandra, K.K. and Tiwari, K.P. (2000). Synergistic role of VAM and *Azotobacter* inoculation on growth and biomass production in forestry species. *Journal of Tropical Forestry* **16** (1): 13-21.

Paroha, S., Chandra, K.K. and Bhandari, P.S. (1999). Growth performance of *Bambusa arundinacea* in relation to VA mycorrhizal inoculation. *Vaniki Sandesh* **23** (1): 21-23.

Parvathi, K., Venkateswarlu, K. and Rao, A.S. (1985a). Influence of root-infecting fungi on development of *Glomus mosseae* in groundnut. *Current Science*. **54**(19): 1006-1007.

Parvathi, K., Venkateswarlu, K. and Rao, A.S. (1985b). Response of groundnut (*Arachis hypogea* L.) to combined inoculation with *Glomus mosseae* and *Rhizobium* sp. *Proceedings, Indian Academy of Sciences, Plant Sciences*. **95**(1): 35-40.

Parvathi, K.; Venkateswarlu, K.; and Rao, A.S. (1984). Occurrence of VA mycorrhizas on different legumes in a laterite soil. *Current Science*, India **52**(23): 1254-1255.

Pattanaik, R., Sahu, S., Padhi, G.S. and Mishra, A.K. (1995). Effect of inoculation of vesicular arbuscular mycorrhiza on horsegram (*Macrotyloma uniflorum*) grown in soil in iron-mine area. *Indian Journal of Agricultural Sciences*. **65**(3): 186-190.

Paulitz, T.C. and Linderman, R.G. (1991). Mycorrhizal interactions with soil organisms. In : *Handbook of Applied Mycology*, Vol. I. *Soil and Plants*. (Eds. D.K. Arora, Bharat Rai, K.G. Mukerji and G. Kundsen), Marcel Dekker Inc., New York : 77-129.

Petri, L. (1903). Riccrche sul significato morfologico dei prosporoidi (sporangioli di Janse) nelle micorrize endotrofiche, *Nuovo G. Bot. Ital.* **10**: 541.

Peyronel, B. (1924). Prime ricerche sulle micorrize endotrofiche e sulla micoflora radicicola normale delle fanerogame, *Rev. Biol.*, **6**:1.

Peyronel, B. (1924). *Specific di Endogone productrici di micorrize endotrofische. Bull. Staz. Patol. Veg.* Rome. **5**: 73-75.

Peyronel, B. (1937). *Le Endogone qualiproductrici di micorrize endotrofische nelle fanerogame alpestri. Nuova G. Bot. Ital.* **44**: 584-586.

Pfeiffer, C.M. and Bloss, H.E. (1988). Growth and nutrition of guayule (*Parthenium argentatum*) in a saline soil as influenced by vesicular arbuscular mycorrhiza and phosphorus fertilization. *New Phytologist*. **108**(3): 315-321.

Phirke, N.V., Chincholkar, S.B. and Kothari, R.M. (2002). Optimal exploitation of native arbuscular and vesicular-arbuscular mycorrhizae for improving the yield of banana through IPNM. *Indian Journal of Biotechnology* **1** (3): 280-285.

Pirozynski, K.A. and Dalpe, Y. (1989). Geological history of the Glomaceae, with particular reference to mycorrhizal symbiosis. *Symbiosis*. **7**: 1-36.

Plaut, Z. and Grieve, C.M. (1988). Photosynthesis of salt-stressed maize as influenced by Ca: Na ratios in the nutrient solution. *Plant and Soil*. **105**: 283-286.

Plenchette, C. (1982). Vesicular arbuscular endomycorrhizas: a potential tool to exploit in agriculture. *Phyto-protection*. **63**: 86

Poi, S.C., Ghosh, G. and Kabi, M.C. (1989). Response of chickpea (*Cicer arietinum* L.) to combined inoculation with *Rhizobium*, phosphobacteria and mycorrhizal organisms. *Zentralblatt fur Mikrobiologie*. **144**(4): 249-253.

Pokhriyal, T.C., Uniyal, K., Chaukiyal, S.P. and Thapar, H.S. (1992). Effect of VAM inoculation on growth and nitrogenase activity in *Albizia lebbek* (Linn.) Benth. *Van Vigyan*. **30**(3,4): 123-126.

Ponnambalam, A., Sugavanam, V., Rajendran, K. and Devraj, P. (1999). Growth and biomass production of *Casurina eqisetifolia* and Casurina junghuhniana in quartz sand dump. *Advances in Forestry Research in India*. **21**: 64-78.

Poss, J.A., Pond, E., Menge, J.A. and Jarrell, W.M. (1985). Effect of salinity on mycorrhizal onion and tomato in soil with and without additional phosphate. *Plant and Soil*. **88**: 307-319.

Potty, V.P. (1988a). Multiplication and inoculation of *Glomus macrocarpum* var. *Microcarpum* in vegetatively propagated tuber crops. In: *Mycorrhizae for Green Asia*. (Eds. A. Mahadevan, N. Raman and K. Natarajan), Madras University, Madras : 243-245.

Potty, V.P. (1988b). Response of cassava (*Manihot esculenta*) to VAM inoculation in acid laterite soil. In: *Mycorrhizae for green Asia*. (Eds. A. Mahadevan, N. Raman and K. Natarajan), Madras University, Madras : 246-249.

Potty, V.P. and Harikumar, V.S. (1995). Interaction of vesicular-arbuscular mycorrhizal fungi and phosphobacterium in sweet potato rhizosphere. In: *Mycorrhizae: Biofertilizers for the Future*. (Eds. Alok Adholeya and Sujan Singh), TERI, New Delhi: 240-244.

Potty, V.P. and Indira, P. (1990). Influence of vesicular arbuscular mycorrhizae on the photosynthesis and photorespiration of sweet potato (*Ipomoea batatas*). In: *Current trends in mycorrhizal research. Proceedings of the National Conference on Mycorrhiza*. (Eds. B.L. Jalali and H. Chand), H.A.U., Hisar: 73.

Powell, C. L.L. and Bagyaraj, D.J. (1984). *VA mycorrhiza*. C.R.C. Press Inc Boca Raton, Florida: 1-233.

Prabakaran, Arjunan, J.G., Ravi, K.B. and Ramamoorthi, N. (1995). Performance of sorghum cultivars to vesicular-arbuscular mycorrhizae inoculation in Alfisol. In: *Mycorrhizae: Biofertilizers for the Future*. (Eds. Alok Adholeya and Sujan Singh), TERI, New Delhi: 348-351.

Prabakaran, Ramasamy, J.A., Srinivasan, K. and Natarajan, N. (1995). Effect of endomycorrhizae and *Azospirillum* on growth and yield of CO-1 maize in acid soil. In: *Mycorrhizae: Biofertilizers for the Future*. (Eds. Alok Adholeya and Sujan Singh), TERI, New Delhi: 270-273.

Prabhakar, J.S.S., Murthy, P.K., Rao, M.S., Ramavatharam, N. and Reddy, K.S. (1992). Effect of vesicular-arbuscular mycorrhiza on growth, phosphorus nutrition and kernel yield of groundnut (*Arachis hypogaea* L.). *New Botanist*. **19**(1-4):51-58.

Pradhan, M., Sharma, G.D. amd Mishra, R.R. (1996). Isolation and characterization of vesicular-arbuscular (VAM) symbionts of actinorrhizal Himalayan alder (*Alnus nepalensis* D. Don) of Meghalaya. *Journal of Hill Research*. **9** (2) : 325-330.

Prakash, J. and Kasera, P.K. (2003). Response of nutritional treatments on growth and biomass of *Salvadora persica*, an important medicinal plant of Thar Desert, India. *Advances in Plant Sciences* **16** (1): 247-252.

Prasad, G.K., Singh, S.K., Das, P.K. and Natha, S. (1997). Influence of VAM, macro and micro-nutrients on vegetative propagation of *Dendrocalamus strictus. Indian Forester*. **123**(9):863-866.

Prasad, K. (1995). Physico-chemical characteristics of soil in relation to VAM (*Glumus fasciculatum*) colonization in *Sacchrarum officinarum* L. *Journal of Phytological Research* **8**(2):201-205.

Prasad, K. (1999a). Influence of indigenous VAM fungus (*Glomus fasciculatum*) and *Rhizobium* on growth, nutrient uptake and nodulation in *Acacia nilotica* plant. Abstracts: *National Conference on Mycorrhiza*, Barkatulla University, Bhopal:67.

Prasad, K. (1999b). Vesicular arbuscular mycorrhizal fungi (VAMF) associated with *Prosopis juliflora* (L.). *Journal of Tropical Forestry* **15** (3): 233-236.

Prasad, K. (2000). Growth responses in *Acacia nilotica (L.)* Del. inoculated with *Rhizobium* and *Glomus fasciculatum* VAM fungi. *Journal of Tropical Forestry* **16** (1): 22-27.

Prasad, K. (2002). Effect of arbuscular mycorrhizae on biomass yield, uptake and translocation of nitrogen, phosphorus and potassium in *Azadirachta indica* L. In*: Frontiers in Microbial Biotechnology and Plant Pathology* (Eds. C.Manoharachary, D.K. Purohit, S. Ram Reddy, M.A. Singara Charya, S. Girisham.): 187-191 ISBN 81-7233-291-2.

Prasad, K. and Bilgrami, R.S. (1995). Impact of vesicular-arbuscular mycorrhizae and phosphates on biomass yield and sucrose content of *Saccharum officinarum* L. In: *Mycorrhizae: Biofertilizers for the Future*. (Eds. Alok Adholeya and Sujan Singh), TERI, New Delhi:465-470.

Prasad, K.S. and Sailaja, K.Sri. (1995). Effect of vesicular-arbuscular mycorrhizae inoculation on periwinkle (*Catharanthus roseus* (L.) G. DON.) In: *Mycorrhizae: Biofertilizers for the Future.* (Eds. Alok Adholeya and Sujan Singh), TERI, New Delhi: 403-406.

Prasad, V., Manjunath, G.T.S., Redsy, C.N. (2000). Influence of *Glomus fasciculatum* inoculation on growth and phosphorus uptake in *Gladiolus* sp. *Mycorrhiza News* **11** (4): 17-18.

Prathibha, C.K., Alagawadi, A.R. and Sreenivasa, M.N. (1995) Establishment of inoculated organisms in rhizosphere and their influence on nutrient uptake and yield of cotton. *Karnataka Journal of Agricultural Sciences*. **8**(1): 22-27.

Pukakayastha, T. J. and Chhonkar, P. K. (2001). Influence of vesicular-arbuscular mycorrhizal fungi (*Glomus etunicatum* L.) on mobilization of zinc in wetland rice (*Oryza sativa* L.). *Biology and fertility of Soils* **33** (4): 323-327.

Punj, V., Gupta, R.P., Bhandari, S.C., Kaul, V.K. and Pandher, M.S. (1987). Dual inoculation with *Rhizobium* and *Glomus fasciculatum* enhances growth, nodulation and nitrogen fixation in *Leucaena leucocephala. Research and Development Reporter*. **4**(1): 1-9.

Punj, V., Gupta, R.P., Hittu Matta, N. (1992). Histopathological studies on mycorrhizal infected roots of *Leucaena leucocephala. Journal of Research, Punjab Agricultural University* . **29**(2): 225-229.

Puppi, G. and Riess, S. (1987). Role and ecology of VA mycorrhizae in sand dunes. *Angewandte Botanik*. **67**: 115-126.

Purcino, A.A.C., Lurlarp, C. and Lynd, J.Q. (1986). Mycorrhiza and soil fertility affects growth, nodulation and nitrogen fixation of *Leucaena* grown on a typic eutrustox. *Commun. In Soil Science Plant Anal*. **17**: 473-489.

Purkayastha, T.J., Singh, C.S. and Chhonkar, P.K. (1998). Growth and Iron nutrition of broccoli (*Brassica oleracea* L. var. *indica* Plenck). grown in a Typic Ustochrept, as influenced by vesicular-arbuscular mycorrhizal fungi in the presence of pyrite and farmyard manure. *Biology and Fertility of Soils*. **27**(1):35-38.

Pushpakumari, R. and Geethakumari, V.L. (1999). Economizing N and P through combined inoculation of mycorrhiza and *Azotobacter* in sweet potato. *Journal of Root Crops* 25 (1): 69-71.

Puther, J.T., Prasad, K.V.S.K., Sharmila, P. and Saradhi, P.P. (1998). Vesicular arbuscular mycorrhizal fungi improves establishment of micropropagated *Leucaena leucocephala* plantlets. *Plant Cell, Tissue and Organ Culture*. **53**(1): 41-47.

Rachel, E.K., Reddy, S.R. and Reddy, S.M. (1989). VA mycorrhizal colonization of different angiospermic plant species in the semi-arid soils of Andhra Pradesh. *Acta Botanica Indica*. **17**: 225-228.

Rachel, E.K., Reddy, S.R. and Reddy, S.M. (1990). Seasonal variation in the colonization and spore polulation of VA mycorrhizal fungi in semi-arid soil of A.P. *National Academy Science Letters* . **13**(6): 219-223.

Rachel, E.K., Reddy, S.R. and Reddy, S.M. (1991). Influence of soil liming on VA-mycorrhizal infection and growth response of cowpea. *Indian Botanical Reporter*. **10**(1-2): 24-27.

Rachel, E.K., Reddy, S.R. and Reddy, S.M. (1993a). Distribution of VA-mycorrhizal fungi in four different soils of Andhra Pradesh. *Indian Botanical Reporter*. **12**(1/2): 35-40.

Rachel, E.K., Reddy, S.R. and Reddy, S.M. (1993b). Performance of VAM fungi in comparison to super phosphate application in a sandy loam soil of Andhra Pradesh, *Indian Journal of Mycology and Plant Pathology*. **23**(2): 149-152.

Rachel, E.K., Shailaja, K.M., Reddy, S.R. and Reddy, S.M. (1996). Effects of some agrochemicals on VAM infection and growth of sunflower. *Journal of Indian Botanical Society*. **75**: 179-181.

Raghavendra, S. and Lakshman, H.C. (1995). Effect of different soil pH values on vesicular-arbuscular mycorrhiza associated tree species. In: *Mycorrhizae: Biofertilizers for the Future*. (Eds. Alok Adholeya and Sujan Singh), TERI, New Delhi: 510.518.

Ragupathy, S. and Mahadevan, A. (1993). Distribution of vesicular-arbuscular mycorrhizae in the plants and rhizosphere soils of the tropical plains, Tamil Nadu, India. *Mycorrhiza*. **3** (3) : 123-136.

Ragupathy, S. and Mahadevan, A. (1995). Effect of different mycorrhizal treatments on the growth of *Peltophorum pterocarpum*. In: *Mycorrhizae: Biofertilizers for the Future*. (Eds. Alok Adholeya and Sujan Singh), TERI, New Delhi: 390-393.

Ragupathy, S., Mohankumar, V. and Mahadevan, A. (1988). Distribution of VAM in Thanjavur district flora. In: *Mycorrhizae for Green Asia*. (Eds. A. Mahadevan, N. Raman and K. Natrajan), Madras University, Madras : 95-98.

Ragupathy, S., Nagarajan, G. and Mahadevan, A. (1998). Mycorrhizae in coastal sand dunes of Tuticorin, Tamil Nadu. *Journal of Environmental Biology* **19** (3): 281-284.

Rahangdale, R. and Gupta, N (1999). Vesicular-arbuscular mycorrhizal association of biomass tree species in the tropical forests of Madhya Pradesh. *Indian Journal of Forestry* **22** (1/2): 62-65.

Rai, R. (1988). Interaction response of *Glomus albidus* and *Cicer Rhizobium* strains on iron uptake and symbiotic N_2 fixation in calcareous soil. *Journal of Plant Nutrition*. **11**(6-11): 863-869.

Rai, R.P. (1990a). Dark septate hyphae as endophytes of potato roots. *Indian Phytopathology*. **43**(1): 107-108.

Rai, R.P. (1990b). Improved yields in potato through mycorrhizal inoculation. In *Trends in mycorrhizal research. Proceedings of the National Conference on Mycorrhiza*. (Eds. B.L. Jalali and H. Chand), H.A.U., Hisar : 148-149.

Raj, B. K. V. and Chandrashekar, K. R. (2003). Association of mycorrhizal fungi with the endemic dipterocarps of Sharavathi Valley forest (Karnataka) region. *Mycorrhiza News* **14** (4): 18-21.

Raj, J., Bagyaraj, D.J. and Manjunath, A. (1981). Influence of soil inoculation with vesicular-arbuscular mycorrhiza and phosphate-dissolving bacterium on plant growth and ^{32}P-uptake. *Soil Biology Biochemistry*. **13**: 105-108.

Raja, P. and Mahadevan, A. (1995). Mass multiplication of vesicular arbuscular mycorrhizae under controlled conditions. *Biology and Biotechnology of Mycorrhizae*. **56**:233-236.

Raja, P., Raghupathy, S. and Mahadevan, A. (1995). Mycorrhizal association of pteridophytes of Nilgiris and Kodaikanal Hills, South India. *Acta Botanica Indica*. **23**: 181-186.

Rajadurai, K.R. and Beaulah, A. (2000a). The effect of *Azospirillum* and VAM on nutrient uptake and available nutrients in post harvest soil of African marigold (*Tagetes erecta* L.). *Journal of Ecotoxicology &Environmental Monitoring* **10** (3/4): 173-176.

Rajadurai, K.R. and Beaulah, A. (2000b). The effect of inorganic fertilizers, *Azospirillum* and VAM on yield characters of African marigold (*Tagetes erecta*). *Journal of Ecotoxicology &Environmental monitoring* **10** (2): 101-105.

Rajadurai, K.R., Manivannan, K.,Jawaharlal, M. and Beaulah, A. (2000). Effect of *Azospirillum* and VAM on growth characters of African marigold (*Tagetes erecta* L.). *South Indian Horticulture* **48** (1/6): 83-87.

Rajagopal, D., Madhavendra, S.S. and Kaiser, J. (1989). Occurrence of vesicular-arbuscular mycorrhizal fungi in roots of *Morus alba*. *Current Science*. **58**(12): 687-689.

Rajalingam, G.V. and Kumar, N. (2001). Effect of digested coirpith compost (DCC) and biofertilizers on green leaf yield of tea. *In changing scenario in the production system of horticultural crops. Proceedings of a National Seminar, Coimbatore, Tamil Nadu, India. South Indian Horticulture* **49** (special): 142-144.

Rajan, S.K., Reddy, B.J.D. and Bagyaraj, D.J. (2000). Screening of arbuscular mycorrhizal fungi for their symbiotic efficiency with *Tectona grandis*. *Forest Ecology and Management*. **126** : 91-95.

Rajasekhar, A., Babu, G. P. and Reddy, T. K. K. (2002). Effect of dual inoculation of *Rhizobium* and vesicular arbuscular mycorrhiza on biomass of red sanders, *Pterocarpus santalinus* L. *Legume Research* **25** (3): 208-210.

Rajendran, G., Ramakrishanan, S. and Subramanian, S. (2001). Biomanagement of nematodes in horticultural crops. In *changing scenario in the production system of horticultural crops. Proceedings of a National Seminar, Coimbatore, Tamil Nadu, India. South Indian Horticulture* **49** (special): 227-230.

Rajendran, K., Sugavanam, V. and Devaraj, P. (2000). Influence of biofertilizers on the biomass production of *Casuarina equisetifolia* in farm forestry. *Bangladesh Journal of Forest Science* **29**(1): 26-36.

Rajeshwari, E., Latha, T.K.S., Vanangamudi, K., Selvan, K.A. and Narayanan, R. (2001). Effect of arbuscular mycorrhizae and phosphorus on seedling growth of *Casuarina equisetifolia. Indian Phytopathology* **54**(1): 85-87.

Rajeswari, E., Narayanan, R. and Vanangamudi, K. (1999,publ.2000). Survey of arbuscular mycorrhizal association of *Casuarina equisetifolia* Frost in Tamil Nadu. *Madras Agricultural Journal* **86** (4/6): 295-298.

Rajeswari, S. and Sankaranarayanan, C. (1995). Effect of nursery treated VAM on the nematode interaction in tomato. *International Journal of Tropical Plant Diseases*. **13**(1): 107-111.

Rajgopal, B. and Ramarethinam, S. (1997). Interaction between vesicular arbuscular mycorrhiza and tea roots. *Planter's Chronicle*. **92**(6): 263, 265, 267-268.

Rajput, R.L. (1999). Effect of bio-fertilizers application on yield of bajra (pearl millet).*Bhartiya Krishi Anusandhan Patrika* **14** (3/4): 43-46.

Rajshekharra, E., Sreenivasa, M.N. and Bhat, R.S. (1989). Effect of cropping system on vesicular-arbuscular mycorrhizal fungi in redgram and sunflower. *Karnataka Journal of Natural Sciences* **2**(3): 231-233.

Raju, P.S., Clark, R.B., Ellis, J.R. and Maranville, J.W. (1990). Effect of species of VA mycorrhizal fungi on growth and mineral uptake of Sorghum at different temperatures. *Plant and Soil.* **121**(2): 157-164.

Raju, R. and Reddy, M.N. (1999). Effect of biofertilizers and inorganic nutrition on black gram (*Phaseolus mungo*) grown in rice fallows. . *Indian Journal of Agronomy* **44** (4): 782-786.

Ramakrishnan, B., Johri, B.N. and Gupta, R.K. (1988a). Effect of vesicular arbuscular mycorrhizal fungus on photosynthesis and photorespiration in water stressed maize. *Photosynthetica.* **22**(3): 443-447.

Ramakrishnan, B., Johri, B.N. and Gupta, R.K. (1988b). Influence of VAM fungus *Glomus caledonius* on free proline accumulation in water-stressed maize. *Current Science*. **87**(19): 1082-1084.

Ramakrishnan, B., Johri, B.N. and Gupta, R.K. (1990). The response of mycorrhizal maize plants to variations in water potentials. In: *Current trends in mycorrhizal research. Proceedings of the National Conference on Mycorrhiza.* (Eds. B.L. Jalali and H. Chand), H.A.U., Hisar : 61-62.

Ramakrishnan, K., Thamizhiniyan, P., Sundaramoorthy, P. and Sriramachandrasekaran, M.V. (2001). Effect of sugar mill effluent polluted soil on the VAM fungi. *Journal of Ecobiology* **13**(3): 187-192.

Raman, N. and Sambandan, K. (2000). Axenic germination of *Scutellospora erythropa* and *Scutellospora nigra* in *in vitro* conditions. *Indian Journal of Experimental Biology* **38**(11): 1159-1163.

Raman, N. and Elumalai, S. (1995). Increase of indole-3 acetic acid in nodules of *Casuarina equisetifolia* inoculated with mycorrhizal fungi and *Frankia*. In: *Mycorrhizae: Biofertilizers for the Future.* (Eds. Alok Adholeya and Sujan Singh), TERI, New Delhi: 494-496.

Raman, N. and Elumalai, S. (1998). Gibberellic acid in mycorrhizal and actinorhizal inoculated *Casuarina equisetifolia. Indian Journal of Experimental Biology* **36**(7): 703-708.

Raman, N. and Gopinathan, S. (1992). Association and activity of vesicular arbuscular mycorrhizae of trees in tropical forest of South. *Journal of Tropical Forestry*. **8**(4): 966-973.

Raman, N. and Nagarajan (1995). Incidence of mycorrhizal association in a forest fire site in Servarayans Hills, Tamil Nadu. In: *Mycorrhizae: Biofertilizers for the Future*. (Eds. Alok Adholeya and Sujan Singh), TERI, New Delhi:104-106.

Raman, N. and Nagarajan, N. (1996). Effect of forest fire on VAM fungi in a tropical forest of southern India. *Commonwealth Forestry Review*. **75**(3): 247-252.

Raman, N. and Sambandan, K. (1998). Distribution of VAM fungi in tannery effluent polluted soils of Tamil Nadu, India. *Bulletin of Environmental Contamination and Toxicology*. **60**(1): 142-150.

Raman, N., Sambandan, K. , Sabadevan, C. and Selvaraj, T. (1995). Distribution of vesicular-arbuscular mycorrhizal fungi in tannery effluent polluted soils of Tamil Nadu, India. In: *Mycorrhizae: Biofertilizers for the Future*. (Eds. Alok Adholeya and Sujan Singh), TERI, New Delhi: 168-176.

Ramana, B.V. and Babu, R.S.H. (1999). Response of onion (*Allium cepa*) to inoculation of VAM fungi at different levels of phosphorus. Abstract: *National Conference on Mycorrhiza,* Barkatulla University, Bhopal: 54.

Ramananda, T.V. and Sreenivasa,M. N. (2000). Influence of different AM fungi on growth of onion. *Journal of Maharashtra Agricultural Universities* **25** (1): 93-94

Ramananda, T.V., Sreenivasa, M.N. and Gukumugthy, S.B. (2000). Response of onion (*Allium cepa* L.) to inoculation of arbuscular mycorrhizal (AM) fungi. *Environment and Ecology* **18** (2): 451-453.

Ramaraj, B. and Sharmugam, N. (1986). Growth responses of vesicular arbuscular mycorrhizae on pulses. *Madras Agricultural Journal*. **73**(1): 32-35.

Ramaraj, B. and Sharmugam, N. (1990). Interaction of vesicular-arbuscular mycorrhiza (*Glomus etunicatum*) and *Rhizobium* in cowpea. In: *Current trends in mycorrhizal research. Proceedings of the National Conference on Mycorrhiza*. (Eds. B.L. Jalali and H. Chand), H.A.U., Hisar: 107-108.

Ramaraj, B., Sharmugam, N. and Reddy, C.D. (1988). Biocontrol of *Macrophomina* root rot of cowpea and *Fusarium* wilt of tomato by using VAM fungi. *Mycorrhizae for Green Asia*. (Eds. A. Mahadevan, N. Raman and K. Natrajan), Madras University, Madras, India: 250-251.

Ramesh, C., Bojan, C.R. and Mahadevan, A. (1995). Light and electron microscopic observations of vesicular-arbuscular mycorrhizal symbiosis between *Pennisetum pedicellatum* and *Gigaspora*. *Biology and Biotechnology of Mycorrhizae*. **56**: 19-25.

Ramesh, C.,Chellappan, P. and Mahadevan. A. (2000). X-ray microanalysis of elements of VA mycorrhizal and non-mycorrhizal *Pennisetum pedicellatum* roots. *Indian Journal of Experimental Biology* **38** (4): 396-398.

Rana, S.K., Maiti, D., Barnwal, M.K., Singh R.K. and Variar, M. (2002). Effect of rice (*Oryza sativa*)-based intercropping systems on vesicular-arbuscular mycorrhial colonization, P uptake and yield. *Indian Journal of Agricultural Sciences* **72** (7): 400-403.

Rangarajan, M. and Santhanakrishnan, P. (1995). Plant growth promoting rhizobacteria and biofertilizers increase the fresh leaf yield and nutrient content in *Morus alba*. In: *Mycorrhizae: Biofertilizers for the Future*. (Eds. Alok Adholeya and Sujan Singh), TERI, New Delhi: 189-191.

Rangarajan, M., Kandasamy, D., Oblisami, G. and Narayanan, R. (1988). Synergistic effect of rhizobia and VAM on biomass production of certain tree legumes. In: *Mycorrhizae for Green Asia.* (Eds. A. Mahadevan, N. Raman and K. Natrajan), Madras University, Madras: 288-290.

Rangeshwaran, R., Rao, W.K., Ramaiah, P.K. (1990). Vesicular-arbuscular mycorrhiza in coffee. *Journal of Coffee Research.* **20**(1): 55-68.

Rani, A., Dhungana, H.N. and Sharma, G.S. (1995). Occurrence of vesicular-arbuscular mycorrhizal fungi in forest nursery seedlings in Assam. In: *Mycorrhizae: Biofertilizers for the Future.* (Eds. Alok Adholeya and Sujan Singh), TERI, New Delhi: 62-64.

Rani, P., Aggarwal, A. and Mehrotra, R.S. (1998). Establishment of nursery technology through *Glomus mosseae, Rhizobium* sp. and *Trichoderma harzianum* on better biomass yield of *Prosopis cinearia* Linn. *Proceedings of the National Academy of Sciences* **68** (3/4): 301-305.

Rani, P., Aggarwal, A. and Mehrotra, R.S. (1999). Growth responses in *Acacia nilotica* inoculated with VAM fungi (*Glomus mosseae*) *Rhizobium sp.* and *Trichoderma harzianum. Indian Phytopathology* **52** (2): 151-153.

Rani, P., Aggarwal, A. and Sharma, D. (2001). Improvement in biomass yield of *Prosopis cineraria* through VAM, *Rhizobium* sp. and *Trichoderma harzianum. Advances in Plant Sciences* **14** (2): 593-596.

Rani, R. and Mukerji, K.G. (1988). Taxonomy of Indian vesicular-arbuscular mycorrhizal fungi. In: *Mycorrhizae for Green Asia.* (Eds. A. Mahadevan, N. Raman and K. Natrajan) Madras University, Madras :103-105.

Rani, S.J. and Manoharachary, C. (1994). Occurrence and distribution of VAM fungi associated with safflower. *Indian Phytopathology.* **47**(3): 263-265.

Rani, V. and Bhaduria, S. (2001). Vesicular arbuscular mycorrhizas (VAM) association with some medicinal plants growing on alkaline soils of Mainpuri district, Uttar Pradesh. *Mycorrhiza News* **13** (2): 12-14.

Rano J. and Mukerji, K.G. (1988). VAM fungi - a tool for reforestation. *Journal of Phytological Research.* **1**(i):31-35.

Rao, A.V. and Tak, R. (2001). Influence of mycorrhizal fungi on the growth of different tree species and their nutrient uptake in gypsum mine soil in India. *Applied Soil Ecology* **17** (3): 279-284.

Rao, A.V. and Tarafdar, J.C. (1993). Role of VAM fungi in nutrient uptake and growth of clusterbean in an arid soil. *Arid Soil Research and Rehabilitation.* **7**(3): 275-280.

Rao, A.V. and Tarafdar, J.C. (1999). Soil solarization for mass scale production of arbuscular mycorrhizal fungal inoculum in Indian arid zone. *Indian Journal of Agricultural Sciences.* **60**(4): 271-274.

Rao, A.V., Tarafdar, J.C. and Tak, R. (1999a). Role of AM fungi in rehabilitation of mine spoils. Abstract: *National Conference on Mycorrhiza,* Barkatulla University, Bhopal: 51.

Rao, M. S.,Reddy, P.P. and Nagesh, M. (2000). Management of *Meloidogyne incognita* on tomato by integrating *Glomus mosseae* with *Pasteuria penetrans* under field conditions. *Pest Management in Horticultural Ecosystems* **6** (2): 130-134.

Rao, M.S. and Reddy, P.P. (2001). Control of *Meloidogyne incognita* on aubergines using *Glomus mosseae* interacted with *Paecilomyces lilacinus* and neem cake. *Nematologia Mediterranea* **29** (2): 153-157.

Rao, M.S., Reddy, P.P. and Mohandas, S. (1995a). Effect of integration of endomycorrhiza (*Glomus mosseae*) and neem cake on the control of root-knot nematode on tomato. *Zeitschrift fur Pflanzenkrankheiten und Pflanzenschutz*. **102**(5): 526-529.

Rao, M.S., Reddy, P.P. and Mohandas, S. (1998a). Bio-intensive management of *Meloidogyne incognita* on egg plant, by integrating *Paecilomyces lilacinus* and *Glomus mosseae. Nematologia Mediterranea*. **26**(2): 213-216.

Rao, M.S., Reddy, P.P., Somasekhar, N. and Nagesh, M. (1997). Management of root-knot nematodes, *Meloidogyne incognita* in tomato nursery by integration of endomycorrhiza, *Glomus fasciculatum* with castor cake. *Pest Management in Horticultural Ecosystem*. **3**(1): 31-35.

Rao, M.S., Reddy, P.P., Sukada, M., Nagesh, M. and Pankaj (1998b). Management of root-knot nematode on egg plant by integrating endomycorrhiza (*Glomus fasciculatum*) and castor (*Ricinus communis*) cake. *Nematologia Mediterranea*. **26**(2): 217-219.

Rao, M.S., Srinivas, P., Reddy, G.L. and Reddy, S.M. (1999b). VAM status of some plants growing naturally in open cast coal mine soils. Abstract: *National Conference on Mycorrhiza,* Barkatulla University, Bhopal: 38.

Rao, M.S., Srinivas, P., Reddy, G.L. and Reddy, S.R. (2002). Arbuscular mycorrhizal dependency of two forest tree species in coal mine disturbed soils. In: *Frontiers in Microbial Biotechnology and Plant Pathology* (Eds. C. Manoharachary, D.K. Purohit, S. Ram Reddy, M.A. Singara Charya, S. Girisham.) : 289-293.

Rao, P.V., Pawar, S.E. and Singh, S.N. (1990). Distribution and intensity of native VAM in Maharashtra region. *Indian Journal of Applied and Pure Biology*. **5**(2): 107-114.

Rao, V.K., Krishnappa, K. and Hegde, S.V. (1995b). Impact of VA mycorrhizal colonization on growth of chickpea infected with *Meloidogyne incognita* and *Fusarium oxysporum* f.sp. *ciceri* disease complex. *International Chickpea and Pigeonpea Newsletter*. No. 2, 32-34.

Rao, V.M. and Gupta, P. (1999). Occurrence of vesicular arbuscular mycorrhizal associations and selection of efficient VA mycorrhizal fungus in *Prosopis cineraria* (Linn.). *Journal of Phytological Research* **12** (1/2): 47-50.

Rao, V.P., Pawar, S.E. and Singh, S.N. (1995c). Production and application of vesicular-arbuscular mycorrhizal inocula for sustainable agriculture. In: *Mycorrhizae: Biofertilizers for the Future.* (Eds. Alok Adholeya and Sujan Singh), TERI, New Delhi:424-428.

Rao, V.U. and Rao, A.S. (1996). Response of blackgram (*Phaseolus mungo*) and greengram (*P. radiatus*) to inoculation with vesicular-arbuscular mycorrhizal fungus under different sources of phosphorus fertilizer. *Indian Journal of Agricultural Sciences*. **66**(10): 613-616.

Rao, Y.S.G., Suresh, C.K., Suresh, N.S. and Bagyaraj, D.J. (1987). Vesicular-arbuscular mycorrhizae in aromatic plants. *Current Research*. **16**(4): 55-57.

Rao, Y.S.G., Suresh, C.K., Suresh, N.S., Mallikarjunaiah, R.R. and Bagyaraj, D.J. (1989). Vesicular-arbuscular mycorrhizae in medicinal plants. *Indian Phytopathology*. **42**(3): 476-478.

Rathi, S.K. and Singh, L. (1990). Preliminary survey of mycorrhizal fungi in some weeds and cultivated plants in Meerut. In: *Current trends in mycorrhizal research. Proceedings of the National Conference on Mycorrhiza*. (Eds. B.L.Jalali and H. Chand), H.A.U., Hisar: 31-32.

Rathore, V.P. and Singh, H.P. (1995a). Influence of vesicular-arbuscular mycorrhizal fungi and phosphate on maize. *Journal of the Indian Society of Soil Science*. **43**(2): 207-210.

Rathore, V.P. and Singh, H.P. (1995b). Quantification and correlation of vesicular-arbuscular mycorrhizal propagules with soil properties of some mollisols of Northern India. *Mycorrhiza.* **5**(3): 201-203.

Rattan, A., Sharma, M.P., Chauhan, S.P. and Adholeya, A. (1999). Variation in the propagule density of vesicular arbuscular mycorrhizal fungi at rehabilitated waterlogged sites. Abstract: *National Conference on Mycorrhiza,* Barkatulla University, Bhopal: 52.

Ratti N., Kumar, S., Verma, H.N. and Gautam, S.P. (2001). Improvement in bioavailability of tricalcium phosphate to *Cymbopogon martinii* var. *motia* by rhizobacteria, AMF and *Azospirillum* inoculation. *Microbiological Research* ***156*** (2): 145-149.

Ratti, N. and Janardhanan, K.K. (1995). Vesicular-arbuscular mycorrhizal association in some alkaloid bearing plants. In: *Mycorrhizae: Biofertilizers for the Future.* (Eds. Alok Adholeya and Sujan Singh), TERI, New Delhi: 407-409.

Ratti, N. and Janardhanan, K.K. (1996a). Effect on growth, phosphorus and cytokinin contents of palmarosa (*Cymbopogon martinii* var *motia*) by *Glomus* inoculation. *Indian Journal of Experimental Biology.* **34**(11):1126-1128.

Ratti, N. and Janardhanan, K.K. (1996b). Response of dual inoculation with VAM and *Azospirillum* on the yield and oil content of Palmarosa (*Cymbopogon martinii* var. *motia*). *Microbiological Research.* **151**(3): 325-328.

Ratti, N., Abdul-Khaliq, Shukla, P.K., Akhtar Haseeb and Janardhanan, K.K. (2000). Effect of *Glomus mosseae* (Nicol. And Gerd.) Gerdemann and Trappe on root knot disease of menthol mint (*Mentha arvensis* sub sp. haplocalyx Briquiet) caused by *Meloidogyne incognita* (Kofoid and White) Chitwood. *Journal of Spices and Aromatic Crops* **9**(2): 129-132.

Ratti, N., Gautam, S.P. and Verma, H.N. (2002). Impact of four *Glomus* species on the growth, oil content, P content and phosphatase activity of *Vetiveria zizanoides. Indian Phytopathology* **55** (4): 434-437.

Raverkar, K.P. and Bhandari, A.R. (1995). Vesicular-arbuscular mycorrhizal fungi and health of tomato seedlings under nursery conditions. In: *Mycorrhizae: Biofertilizers for the Future.* (Eds. Alok Adholeya and Sujan Singh), TERI, New Delhi: 381-385.

Raverkar, K.P. and Tilak, K. V. B. R. (2002). Improved mycorrhizal colonization, nodulation status, biomass production and nutrient uptake in soybean due to inoculation with *Bradyrhizobium japonicum. Legume Research* **25** (1): 32-36.

Raverkar, K.P. and Tilak, K.V.B.R. (1988). Relative efficiency of different VAM on soybean (*Glycine max*). In: *Mycorrhizae for Green Asia.* (Eds. A. Mahadevan, N. Raman and K. Natrajan), Madras University, Madras: 162-165.

Raverkar, K.P., Mehta, K. and Bhandari, A. (1990). Occurrence of VA-mycorrhizal associations with fruit and ornamental plants. In: *Current trends in mycorrhizal research. Proceedings of the National Conference on Mycorrhiza.* (Eds. B.L.Jalali and H. Chand), H.A.U., Hisar: 47-48.

Ravi, K.B., Prabakaran, J. and Mariappan, S. (1995). Survey of vesicular-arbuscular mycorrhizae in agroforestry trees in alfisol. In: *Mycorrhizae: Biofertilizers for the Future.* (Eds. Alok Adholeya and Sujan Singh), TERI, New Delhi: 95-99.

Ravichandran, V.K. and Balasubramanian, T.N. (1999). Growth response of *Casuarina equisetifolia* seedlings to dual inoculation of *Frankia* and VAM (*Glomus fasciculatum*). *Journal of Ecobiology.* **11** (3): 175-179.

Rawat, A.K., Khare, A.K. and Vaishya, U.K. (1991). Dual inoculation effect of *Rhizobium* and *Mycorrhiza* on nodulation and yield in chickpea. *Legume Research.* **14**(3): 150-152.

Ray, S. and Dalei, B.K. (1998). VAM for root knot nematode management and increased productivity of grain legumes in Orissa. *Indian Journal of Nematology.* **28**(1): 41-47.

Read, D.J. (1991). Mycorrhizas in ecosystems. *Experientia.* **47**: 376-391.

Reddy, B., Bagyaraj, D.J. and Mallesha, B.C. (1996a). Selection of efficient VA mycorrhizal fungi for acid lime. *Indian Journal of Microbiology.* **36**(1): 13-16.

Reddy, B., Bagyaraj, D.J. and Mallesha, B.C. (1996b). Selection of efficient VA mycorrhizal fungi for papaya. *Biological Agriculture & Horticulture.* **13** (1) : 1-6.

Reddy, B.N. and Hindumathi, A. (1999). Biological control of charcoal rot of sorghum by using mycorrhiza. Abstract: *National Conference on Mycorrhiza,* Barkatulla University, Bhopal: 62.

Reddy, C.N. and Garampalli, H.R. (2002). Effect of fly ash on VAM formation and growth response of pulse crops infested with *Glomus aggregatum* in sterile soil. In*: Frontiers in Microbial Biotechnology and Plant Pathology* (Eds. C. Manoharachary, D.K. Purohit, S. Ram Reddy, M.A. Singara Charya, and S. Girisham.): 205-212.

Reddy, C.N. and Bais, R.S. (1990). Prevalence of native VAM fungi and their relative performance in infectivity of local crops. In: *Current trends in mycorrhizal research. Proceedings of the National Conference on Mycorrhiza.* (Eds. B.L. Jalali and H. Chand), H.A.U., Hisar : 103-104.

Reddy, C.N. and Goud, R. R. (1988). Occurrence of VAM association with *Cassia occidentalis.* In: *Mycorrhizae for Green Asia.* (Eds. A. Mahadevan, N. Raman and K. Natrajan), Madras University, Madras : 85-88.

Reddy, M.N., Devi, M.C. and Sridevi, N.V.(1999). VAM association in different varieties of turmeric (*Curcuma longa*). Abstract: *National Conference on Mycorrhiza,* Barkatulla University, Bhopal: 41.

Reddy, M.P., Rao, D.M.R., Verma, R.S., Srinath, B. and Katiyar, R.S. (2000). Effect of VAM inoculation and addition of phosphorus on the growth of S-13 mulberry saplings. . *Indian Journal of Sericulture* **39** (1): 12-15.

Reddy, M.P., Rao, D.M.R., Verma, R.S., Srinath, B. and Katiyar, R.S. (1998). Response of S-13 mulberry variety to VAM inoculum under semi-arid condition. *Indian Journal of Plant Physiology.* **3**(3): 194-196.

Reddy, M.S.B. and Bagyaraj, D.J. (1988). Selection of an efficient VAM fungus for inoculating *Leucaena* in oxisols and vertisols. In: *Mycorrhizae for Green Asia.* (Eds. A. Mahadevan, N. Raman and K. Natrajan), Madras University, Madras : 271-273.

Reddy, M.S.B. and Bagyaraj, D.J. (1990). Response of pigeonpea to vesicular arbuscular mycorrhizal colonization in an alfisol. In: *Mycorrhizal Symbiosis and Plant Growth.* (Eds. D.J. Bagyaraj and A. Manjunath),U.A.S., Bangalore:64-66.

Reddy, M.S.B. and Bagyaraj, D.J. (1991). The symbiotic efficiency of pigeonpea to VA mycorrhizal inoculation in an alfisol and vertisol. *Biological Agriculture & Horticulture*. **8**(2): 82.

Reddy, M.S.B., Bagyaraj, D.J. and Mallesha, B.C. (1990). Phosphate response curve of *Leucaena* inoculated with *Gigaspora margarita*. In: *Current trends in mycorrhizal research. Proceedings of the National Conference on Mycorrhiza*. (Eds. B.L. Jalali and H. Chand), H.A.U., Hisar: 68-69.

Reddy, M.V., Rao, J.N. and Krishna, K.R. (1988). Influence of mycorrhiza on chickpea *Fusarium* wilt. *International Chickpea Newsletter*. No. 19:16.

Reddy, P.P., Rao, M.S., Mohandas, S. and Nagesh, M. (1995a). Integrated management of the citrus nematode, *Tylenchulus semipenetrans* Cobb using VA mycorrhiza, *Glomus fasciculatum* (Thaxt.) Gerd. & Trappe and oil cakes. *Pest management in Horticultural Ecosystem*. **1**(1): 37-41.

Reddy, P.R.P., Reddy, P.J.M., Prakash, P. and Manoharachary, C. (1995b). Association of vesicular-arbuscular mycorrhizal fungi in soils polluted with industrial and sewage effluents. In: *Mycorrhizae: Biofetilizers for the Future*. (Eds. Alok Adholeya and Sujan Singh), TERI, New Delhi: 155-158.

Reddy, S. V. K. and Ahlawat, I. P. S. (2001). Dry matter accumulation and nutrient uptake in lentil (*Lens culinaris* Medikus) in relation to cultivar, phosphorus, zinc and biofertilizers. *Research on Crops* **2** (1): 21-24.

Reddy, S.M., Reddy, S.R., Giri, R.Y. and Bhanja, M.R. (1995c). Mycorrhizae of eucalyptus. *Applied Environmental Microbiology*: 11-13.

Reddy, S.R., Rachel, E.K. and Reddy, S.M. (1997). Effect of water stress on the VAM colonization and growth of sunflower. *Journal of Mycology and Plant Pathology*. **27**(3): 294-296.

Reddy, T.K.K. and Rajasekhar, A. (1988). Antibiotic induced VAM stimulation in peanut and microbial interaction. In: *Mycorrhizae for Green Asia*. (Eds. A. Mahadevan, N. Raman and K. Natrajan), Madras University, Madras : 252-254.

Redecker, D., Hijri, M., Dulieu, H., and Sanders, I. R. (1999). Phylogenetic analysis of a dataset of fungal 5.8S rDNA sequences shows that highly divergent copies of internal transcribed spacers reported from *Scutellospora castanea* are of ascomycete origin. *Fungal Genet. Biol.* **28**: 238-244

Redente, E.F. and Reeves, F.G. (1981). Interactions between vesicular arbuscular mycorrhiza and *Rhizobium* and their effect on sweetvetch growth. *Soil Science*. **132**:410-415.

Reena, J. and Bagyaraj, D.J. (1990). Growth stimulation of *Tamarindus indica* by selected VA mycorrhizal fungi. *World Journal of Microbiology and Biotechnology*. **6**(1): 59-63.

Rekha Rani and Mukerji, K.G. (1987). *National Symposium on Fungal Ecology in relation to Human welfare*, Hyderabad, 23.

Rekha Rani, Rano J. and Mukerji, K.G. (1987). Mycorrhizal status of some ornamental plants at their reproductive and vegetative stage. *Acta Botanica Indica*. **15**(2): 311-313.

Rhodes, L.H. and Gerdemann, J.W. (1980). Nutrient translocation in vesicular-arbuscular mycorrhizae. In: *Cellular interaction in symbiosis and parasitism*. (Eds. C.B. Cook, P.W. Pappas and E.D. Rudolph), The Ohio University Press, Columbus.: 173.

Rikhy, M. and K.G. Mukerji (1974). *Transactions of Mycological Society Japan*. **15**: 47-52.

Robert, C.P., Sivaprasad, P., Mammootty, K.P. and Lekha, K.S. (1995). Effect of fungicides on vesicular-arbuscular mycorrhizal fungi under natural conditions. In: *Mycorrhizae: Biofetilizers for the Future*. (Eds. Alok Adholeya and Sujan Singh), TERI, New Delhi:133-136.

Rodrigues, B. F. (1999). VAM association in seedlings of some plant species growing on iron ore mine spoils in Goa (India). *Indian Journal of Forestry* **22** (1/2): 136-140.

Rodrigues, B.F. (1995). Mycorrhizal status of naturally occurring and cultivated legumes of iron ore mine wastelands in Goa. In: *Mycorrhizae: Biofertilizers for the Future*. (Eds. Alok Adholeya and Sujan Singh), TERI, New Delhi: 42-44.

Rodrigues, B.F. (2000). Diversity of arbuscular mycorrhizal (AM) fungal species from iron ore mine wastelands in Goa. *Indian Forester* **126** (11): 1211-1216.

Rodrigues, B.F. and Jaiswal, V. (1999). Occurrence and distribution of AM fungi in the rhizospheric soils of coastal vegetation of Goa. Abstract: *National Conference on Mycorrhiza,* Barkatulla University, Bhopal: 6.

Rodrigues, B.F., Bukhari, M.S. (1997). Occurrence of VAMF colonization in herbaceous plant species growing on iron ore mine wastelands in Goa. In: *Microbial Biotechnology* (Eds. S.M. Reddy, H.P.Srivastava, D.K. Purohit and S. Ram Reddy) Jodhpur University, Jodhpur : 83-86.

Rose, S.L. (1980). Mycorrhizal association of some actinomycete nodulated nitrogen-fixing plants. *Canadian Journal of Botany*. **58**: 1449-1454.

Rose. S.L. and Yongberg, C.T. (1981). Tripartite associations in snowbush (*Ceanothus volutinus*). Effect of vesicular-arbuscular mycorrhizae on growth, nodulation and nitrogen fixation. *Canadian Journal of Botany*. **59**: 34-39.

Roseldahl, S. (1985). Interactions between the vesicular-arbuscular mycorrhizal fungus *Glomus fasciculatum* and *Aphanomyces euteiches* root rot of peas. *Journal of Phytopathology*. **114** : 31-40.

Routray, T.K., Patnaik, R., Padhi, G.S. and Mishra, A.K. (1997). Response of cowpea (*Vigna unguiculata*) to inoculation with coselected vesicular arbuscular mycorrhizal fungi and *Rhizobium* strains in iron mine waste soil. *Indian Journal of Experimental Biology*. **35**(1): 92-95.

Rovira, A.D. (1979). Biology of soil-root interface. In: *The Soil-Root Interface*. (Eds. J.L. Harley and S.R. Russell), Academic Press, London: 145.

Rovira, A.D. (1983). The significance of rhizosphere microflora and mycorrhiza in plant nutrition. *Encyclopedia of Plant Physiology* (New Series). **15**: 61-78.

Roy A.K., Rashmi Kumari, Chakraborty, B.N. and Chakraborty, U. (2002). VA Mycorrhizae in relation to growth of different tea varieties. *Mycorrhiza News* **14** (1): 9-11.

Rozema, J., Arp, W., Diggelen, J. Van, Esbroek,M. Van, Broekman, R. and Punte, H. (1986). Occurrence and ecological significance of vesicular arbuscular mycorrhiza in the salt marsh environment. *Acta Botanica Neerlandica*. **35**(4): 457-467.

Rupnawar, B.S. and Navale, A.M. (2000b). Effect of VAM inoculation to pomegranate layers with respect to N and P uptake. *Journal of Maharashtra Agricultural Universities* **25** (1): 47-49.

Rupnawar, B.S. and Navale, A.M. (2000a). Effect of VA-mycorrhizal inoculation on growth of pomegranate layers. *Journal of Maharashtra Agricultural Universities* **25** (1): 44-46

Sadasivan Nageswari and Babu, R.S. (1998). Interaction of *Glomus fasciculatum* (Thaxter sensu Gerd) Gerd and Trappe and *Heterodera cajani* Koshy, 1967 on cowpea (*Vigna unguiculata* (I.) Walp). In: *Nematology:Challenges and opportunities in 21st Century.* Proceedings of the Third International Symposium of Afro-Asian Society of Nematologists (TISAASN), Sugarcane Breeding Institute (ICAR), Coimbatore, India. (ed. U.K. Mehta) Luton, UK; Afro-Asian Society of Nematologists: 111-113.

Safir, G.R. and Nelson, C.E. (1981). Water and nutrient uptake by vesicular-arbuscular mycorrhizal plant. In: *Mycorrhizal Association and the Crop Production.* (Ed. R. Myers), Rutgers University Press, New Brunswick, N.J.

Safir, G.R., Boyer, J.S. and Gerdemann, J.W. (1972). Nutrient status and mycorrhizal enhancement of water transport in soyabean. *Plant Physiology*. **49**: 700-703.

Safir, G.R., Coley, S.C., Siqueira, J.O. and Carls, P.S. (1990). Improvement and synchronization of VA mycorrhizal fungal spore germination by short-term cold storage. *Soil Biology and Biochemistry*. **22**(1): 109-111.

Sagar, A., Minhas, M. and Lakhanpal, T.N. (1993). Preliminary observations on VAM associated with *Celtis australia*-an agroforestry tree. *Indian Journal of Mycology and Plant Pathology*. **23**(2): 145-148.

Saha, R., Saha, J., Bhattacharya, P.M., Maiti, D. and Chaudhuri, S. (1999). Arbuscular mycorrhizal responsiveness of two rice varieties in nutrient deficient laterite soil. Abstract: *National Conference on Mycorrhiza,* Barkatulla University, Bhopal: 10.

Sailo, G. L., Bagyaraj, D. J. and Joshi, S. (2002). Anatomical and histochemical changes in the roots and leaves of *Simarouba glauca* DC due to inoculation with an arbuscular mycorrhizal fungus and a mycorrhiza helper bacterium. *Journal of Plants Biology* **29** (3): 321-326.

Sailo, G.L. and Bagyaraj, D.J. (2003). Response of *Simarouba glauca* to inoculation with *Glomus mosseae, Bacillus coagulans* and *Azotobacter chroococcum. Biological Agriculture & Horticulture* **20** (4): 339-345.

Sajan, K.M., Gowda, K.K., Kumar, S.N. and Sreeramu, B.S. (2002). Effect of bio-fertilizers on growth and yield of chilli (*Capsicum annum* L.) cv. Byadagi Dabba at different levels of nitrogen and phosphorus. *Journal of Species and Aromatic Crops* **11** (1): 58-61.

Sambandan, K., Kannan, K. and Raman, N. (1992). Distribution of vesicular-arbuscular mycorrhizal fungi in heavy metal polluted soils of Tamil Nadu, India. *Journal of Environmental Biology*. **13**(2): 159-167.

Sampathkumar, G. and Ganeshkumar, A. (2003). Effect of AM fungi and *Rhizobium* on growth and nutrition of *Vigna mungo* L., and *Vigna unguiculata* L. *Mycorrhiza News* **14** (4): 15-18.

Sanders, F.E. and Tinker, P.B. (1973). Phosphate flow into mycorrhizal roots. *Pest Science*. **4**: 385.

Sanders, F.E., Mosse, B. and Tinker, P.B. (Eds.) (1975). *Endomycorrhiza*. Academic Press, London.

Sanders, R.I., Alt, M., Grappe, K., Boller, T. and Wiemken, A. (1995). *New Phytologist*. **130** : 419-427.

Sankaranarayanan, C. and Sundarababu, R. (2001). Influence of moisture and pH on the efficiency of VA-mycorrhiza, *Glomus mosseae* (Nicol & Gerd.) Gerd. & Trappe against *Meloidogyne incognita* L. (Kofoid and White) Chitw. on blackgram (*Vigna mungo* L.) Hepper. *Journal of Biological Control* **15**(1): 69-72.

Sankaranarayanan, C. and Sundarababu, R. (1994). Interaction of *Glomus fasciculatum* with *Meloidogyne incognita* inoculated at different timings on blackgram (*Vigna mungo*). *Nematologia Mediterranea.* **22**(1): 35-36.

Sankaranarayanan, C. and Sundarababu, R. (1997a). Effect of leaf extracts on growth of blackgram inoculated with vesicular arbuscular mycorrhiza (*Glomus fasciculatum*) and root-knot nematode (*Meloidogyne incognita*). *Indian Journal of Nematology*. **26**(2): 144-147.

Sankaranarayanan, C. and Sundarababu, R. (1997b). Effect of oil cakes and nematicides on the growth of blackgram (*Vigna mungo*) inoculated with VAM fungus (*Glomus fasciculatum*) and root-knot nematode (*Meloidogyne incognita*). *Indian Journal of Nematology*. **27**(1): 128-130.

Sankaranarayanan, C. and Sundarababu, R. (1998). Effect of *Rhizobium* on the interaction of vesicular-arbuscular mycorrhizae and root-knot nematode on blackgram *(Vigna mungo) Nematologia Mediterranea*. **26**(2): 195-198.

Sankaranarayanan, C. and Sundarababu, R. (1999). Role of phosphorus on the interaction of vesicular-arbuscular mycorrhiza *(Glomus mosseae)* and root knot nematode *(Meloidogyne incognita)* on blackgram *(Vigna mungo). Indian Journal of Nematology* **29** (1): 105-108

Sansamma George, Pillai, G.R. and Pushpakumari, R. (1998). Influence of biofertilizers on productivity of guinea grass inter cropped in coconut gardens. *Indian Journal of Agronomy* **43** (4): 622-627.

Santhaguru, K., Ponmalar, S.B.G. and Karunakaran, R. (1995). Vesicular-arbuscular mycorrhizae in tree-legumes and its rhizospheric soils in Alagar hills. *Indian Forester*. **121**(9): 817-823.

Santhanakrishnan, P. and Oblisami, G. (1990). Effect of interaction between *Rhizobium* and VA mycorrhizal fungi inoculation on the growth of groundnut applied with different levels of gypsum. In : *Current trends in mycorrhizal research. Proceedings of the National Conference on Mycorrhiza*. (Eds. B.L. Jalali and H. Chand), H.A.U., Hisar :155-156.

Santhi, A. and Sundarababu, R. (1995a). Effect of phosphorus on the interaction of vesicular-arbuscular mycorrhizal fungi with *Meloidogyne incognita* on cowpea. *Nematologia Mediterranea*. **23**(2): 263-265.

Santhi, A. and Sundarababu, R. (1995b). Effect of three species of VAM *viz., Glomus fasciculatum, G. versiforme, G. etunicatum* and root knot nematode *Meloidogyne incognita* on cowpea growing in different types of soil. *International Journal of Tropical Plant Diseases*. **13**(1): 63-68.

Santhi, R. and Kothandaraman, G.V. (1995). Effect of endomycorrhizal colonization and phosphorus sources on the phosphorus recovery, nutrient uptake and growth of greengram. *Indian Journal of Agricultural Research*. **29**(4): 209-214.

Santhi, R., Kothandaraman, G.V. and Rangarajan, M. (1988). Efficiency of VAM inoculation with different sources of phosphorus on the availability and uptake of nutrients and the yield of green gram. In: *Mycorrhizae for Green Asia.* (Eds. A. Mahadevan, N. Raman and K. Natrajan), Madras University, Madras : 166-168.

Saralabai, V.C. and Vivekanandan, M. (1995). Ability of endomycorrhizae to withstand extreme stress. In: *Mycorrhizae: Biofertilizers for the Future*. (Eds. Alok Adholeya and Sujan Singh), TERI, New Delhi:522-525.

Sarvanan, R.S. and Natarajan, K. (2000). Effect of ecto- and endomycorrhizal fungi along with *Bradyrhizobium* sp. on the growth and nitrogen fixation in *Acacia nilotica* seedlings in the nursery. *Journal of Tropical Forest Science* **12**(2): 348-356.

Sarwar, N. and Mukerji, K.G. (1995). Mycorrhizal technology in forestation. In: *Mycorrhizae: Biofertilizers for the Future.* (Eds. Alok Adholeya and Sujan Singh), TERI, New Delhi:342-344.

Sastry, M.S.R. and Johri, B.N. (1999). Arbuscular mycorrhizal fungal diversity of stressed soils of Bailadila iron ore sites in Bastar region of Madhya Pradesh. *Current Science.* **77**: 1095-1099.

Sastry, M.S.R., Sharma, A.K. and Johri, B.N. (2000). Effect of an AM fungal consortium and Pseudomonas on the growth and nutrient uptake of Eucalyptus hybrid. *Mycorrhiza* **10** (2): 55-61.

Saxena, A.K., Rathi, S.K. and Tilak, K.V.B.R. (1977). Differential effect of various endomycorrhizal fungi on nodulation ability of greengram by *Bradyrhizobium* sp. *(Vigna)* strain S-24. *Biology and Fertility of Soils.* **24**(2): 175-174.

Saxena, A.K., Rathi, S.K. and Tilak, K.V.B.R. (1995). Influence of vesicular-arbuscular mycorrhizae on competitive ability of *Bradyrhizobium* sp. (*Vigna*) strain S-24. In: *Mycorrhizae: Biofertilizers for the Future*. (Eds. Alok Adholeya and Sujan Singh), TERI, New Delhi:227-230.

Schenck, N.C. and Kellam, M.K. (1978). The influence of vesicular-arbuscular mycorrhiza on disease development. *Florida Agricultural Station, Technical Bulletin* : 798.

Schenck, N.C. and Kinloch, R.A. (1990). Incidence of mycorrhizal fungi on six field crops in monoculture on a newly cleared woodland site. *Mycologia.* **72**:445-456.

Schenck, N.C. and Perez, Y. (1987). *Manual of the Identification of VA mycorrhizal fungi.* University of Florida, Gainesville, FL.: 1-245.

Schenck, N.C. and Sequeira, J.O. (1987). Ecology of VA mycorrhizal fungi in temperate agroecosystem. In: *Mycorrhizae in the next decade: Practical applications and research priorities. Proceedingse. of 7th NACOM.* (Eds. D.M. Sylvia, L.L., Hung and J.H. Graham).

Schenck, N.C., Spain, J.L., Sieverding, E. and Howeler, R.H. (1984). Several new and unreported vesicular-arbuscular mycorrhizal fungi (Endogonaceae) from Columbia. *Mycologia.* **76**: 685-699.

Schisler, D.A. and Linderman, R.G. (1987). The influence of volatiles purged from soil around Douglas fir ectomycorrhizae on soil microbial populations. *7th North American Conference on Mycorrhizae*. University of Florida. Gaineville: 218.

Schlicht, A. (1889). Beitrag zur Kenntniss de Verbreitung under der Bedentung der Mykorrhizen, *Land. Wirtsch. Jahrb.,* **18**: 499.

Schoenbeck, F. and Dehne, H.W. (1979). *Unterschungen zum pilzliche spores parasiten. Olpidium brassicae*, T.M.V.Z. *Pflenzenkrankh. Pflanzenschutz*. **86**: 103.

Schubler, A., Gehrig, H., Schwarzott, D., and Walker, C. (2001). Analysis of partial Glomales SSU rRNA gene sequence: implications for primer design and phylogeny. *Mycological Research*. **105**: 5-15

Schuepp, H., Dehne, B. and Sticher, H. (1987). Interactions between VA mycorrhizas and heavy metal stress. *Angewandte Botanik*. **61**: 85-96.

Secilia, J. and Bagyaraj, D.J. (1987). Bacteria and actinomycetes associated with pot culture of vesicular-arbuscular mycorrhizas. *Canadian Journal of Microbiology*. **33**: 1069-1073.

Secilia, J. and Bagyaraj, D.J. (1988). Fungi associated with pot cultures of vesicular-arbuscular mycorrhizas. *Transactions of British Mycological Society*. **90**(1): 117-119.

Secilia, J. and Bagyaraj, D.J. (1990). Response of upland rice to vesicular arbuscular mycorrhizal fungi. In: *Mycorrhizal Symbiosis and Plant Growth*. (Eds. D.J. Bagyaraj and A. Manjunath), U.A.S., Bangalore :76-78.

Secilia, J. and Bagyaraj, D.J. (1994). Selection of efficient vesicular-arbuscular mycorrhizal fungi for wetland rice a preliminary screening. *Mycorrhiza*. **4**(6): 265-268.

Sekar, I., Vanangamudi, K. and Suresh, K.K. (1997). Influence of inoculation of biofertilizers on biomass production (growth) of seedlings of shola tree species. *Van Vigyan*. **35**(2): 57-62.

Selvaraj, M., Syamala, D., Arumugam, S. and Rao, M.V. (1996). Growth and productivity of AM inoculated tropical tree seedlings. *Indian Forester*. **112**(12): 1161-1167.

Selvaraj, R., Kannan, K., and Lakshminarasiman, C. (1986). Vesicular-arbuscular mycorrhizal fungi in roots and scale-like leaves of *Canna indica* L. (Cannaceae). *Current Science*. **55**(15): 728-730.

Selvaraj, S., Prabakaran, P., Mariappan, S. and Arjunan, G. (1995). Significance of rhizobia and endomycorrhizae on tree plantation nursery in Alfisol. In: *Mycorrhizae: Biofertilizers for the Future*. (Eds. Alok Adholeya and Sujan Singh), TERI, New Delhi : 274-280.

Selvaraj, T. and Bhaskaran, C. (1999). Some sporocarpic VA-mycorrhizal fungi associated with *Spinifex littoreus* and *Aeluropus lagopoides* in saline soils. Abstract: *National Conference on Mycorrhiza,* Barkatulla University, Bhopal: 40.

Selvaraj, T. and Subramanian, G. (1988). Light and Scanning electron microscopic studies of VAM in *Sesamum indicum* roots. In: *Mycorrhizae for Green Asia.* (Eds. A. Mahadevan, N. Raman and K. Natrajan), Madras University, Madras :106-110.

Selvaraj, T. and Subramanian, G. (1990a). Incidence of vesicular arbuscular mycorrhizal fungi in medicinal plants. In : *Mycorrhizal Symbiosis and Plant Growth.* (Eds. D.J. Bagyaraj and A. Manjunath), University of Agricultural Sciences, Bangalore : 34-35.

Selvaraj, T. and Subramanian, G. (1990b). Phenols and lipids in mycorrhizal and non-mycorrizal roots of *Sesamum indicum. Current Science*. **59**(9): 471-473.

Selvaraj, T., Murugan, R. and Bhaskaran, C. (2001). Arbuscular mycorrhizal association of Kashini (*Cichorium intybus* L.) in relation to physico-chemical characters. *Mycorrhiza News* **13** (2): 14-16.

Sempavalan, J., Kumutha, K. and Santhanakrishnan, P. (2002). Relative effect of vesicular arbuscular mycorrhizal fungi on growth and nutrient uptake of *Casuarina equisetifolia. Madras Agricultural Journal* **88** (10/12): 700-703.

Sempavalan, J., Santhanakrishnan, P. and Kumutha, K. (2001). Effect of *Frankia* strains associated with VA Mycorrhiza on *Casuarina equisetifolia. Madras Agricultural Journal* **88** (7/9): 519-521.

Sen, R. and Hepper, C.M. (1986). Characterization of vesicular-arbuscular mycorrhizal fungi (*Glomus* spp.) by selective enzyme staining following polyacrylamide gel electrophoresis. *Soil Biology and Biochemistry.* **18** : 29-34.

Senapati, H.K., Pal, A.K. and Swain, S.K. (1987). Effect of mycorrhiza on the growth characteristics of okra in laterite soil. *Indian Journal of Horticulture.* **44**(1-2): 93-95.

Sengupta, A. and Chaudhuri, S. (1990). Vesicular arbuscular mycorrhiza (VAM) in pioneer salt marsh plants of the Ganges delta in West Bengal (India). *Plant and Soil.* **122** : 113-115.

Sengupta, A. and Chaudhuri, S. (1994). A typical root endophytic fungi of mangrove plant community of Sundarban and their possible significance as mycorrhiza. *Journal of Mycopathological Research.* **32**(1): 29-39.

Sengupta, A. and Chaudhuri, S. (1995). Effect of dual inoculation of *Rhizobium* and mycorrhiza on growth response of *Sesbania grandiflora* L. in coastal saline and sand dune soil. *Indian Journal of Forestry.* **18**(1): 35-37.

Sengupta, A. and Chaudhuri, S. (2002). Arbuscular mycorrhizal relations of mangrove plant community at the Ganges river estuary in India. *Mycorrhiza* **12** (4): 169-174.

Senthilkumar, S., Saravanakumar, P. and Sathiyanathan, M. (1999). Role of Vesicular arbuscular mycorrhizae in nitrate reduction system and zinc uptake in *Zea mays*. *Ceylon Journal of Science, Biological Sciences* **27** (1): 27-32.

Setua, G. C., Kar, R., Maji, C. and Ghosh, J. K. (2000). Evaluation of quality of mycorrhizal mulberry leaves through silkworm rearing. *Uttar Pradesh Journal of Zoology* **20** (1): 73-77.

Setua, G.C., Das, K.K., Ghosh, J. K. and Saratchandra, B. (2001). Seasonal variation in the association of arbuscular mycorrhiazae with mulberry (*Morus alba* L.). *Journal of Mycopathological Research* **39** (1): 29-34.

Setua, G.C., Kar, R., Ghosh, J.K., Das, N.K. and Saratchandra, B. (1999a). Response of direct inoculation of VAM on growth, yield and phosphorus uptake in mulberry (*Morus alba*). *Indian Journal of Agricultural Sciences* **69** (6): 444-448.

Setua, G.C., Kar, R., Satpathy, B., Das, N.K., Ghosh, J.K. and Saratchandra, B. (1999b). Effect of vesicular arbuscular mycorrhiza on growth, leaf yield and phosphorus uptake in mulberry (*Morus alba*) under irrigated, alluvial soil conditions. *Indian Journal of Agricultural Sciences* **69** (12): 833-836.

Setua, G.C., Sudhakar, R., Kar, R., Das, N.K. and Ghosh, J.K. (1999c). Effect of vesicular arbuscular mycorrhizal association on 'S-1' mulberry (*Morus* sp.) at nursery stage. *Indian Journal of Agricultural Sciences.* **69**(3): 201-204.

Shailaja, K.M., Kumar, P.P. and Reddy, S.R. (1998). Influence of phyllody disease on VAM colonization in two cultivars of sesame. *Indian Phytopathology.* **51**(3): 273-274.

Shankar, A. and Varma, A. (1993). Isozyme analysis of VAM spores help in taxonomic enumeration. *Current Science.* **65**(2): 165-168.

Sharma, A.K., Singh, R. and Singh U.S. (1988). Effect of vesicular-arbuscular mycorrhiza on uptake of phosphorus and zinc in rice (*Oryza sativa* L.) *Current Science*. **57**(16): 901-902.

Sharma, A.K., Srivastava, P.C. and Johri, B.N. (1999a). Effect of vesicular-arbuscular mycorrhization on uptake and translocation of phosphorus in *Sesbania aculeata*. *Current Science*. 77(10):25.

Sharma, A.K., Srivastava, P.C. and Johri, B.N. (1999b). Multiphasic zinc uptake system in mycorrhizal and nonmycorrhizal roots of French bean (*Phaseolus vulgaris* L.). *Current Science*. **76**(2): 228-230.

Sharma, A.K., Srivastava, P.C., Johri, B.N. and Rathore, V.S. (1992). Kinetics of zinc uptake by mycorrhizal (VAM) and nonmycorrhizal corn (*Zea mays* L.) roots. *Biology and Fertility of Soils*. **13**(4): 206-210.

Sharma, D.D., Chandrashekar, D.S., Gunasekhar, V., Rekha, M. and Sarkar, A. (2001a). Comparative efficacy of different control measures against root knot nematode disease of mulberry. *Indian journal of Sericulture* **40**(2): 151-157.

Sharma, H., Kumar, V., Behl, R.K. and Narula, N. (2001b). Survival of *Azotobacter chroococcum* in the rhizosphere of three different wheat crosses: effect of AM fungi. In: *Plant Nutrition: food security and sustainability of agro-ecosystems through basic and applied research. Fourtine International Plant Nutrition Colloquium*, Hannover, Germany. (Eds. W.J. Horst, M.K. Schenk, A. Burkert, N. Claassen, H. Flessa, W.B. Frommer, H. Goldbach, H.W.Olfs, V. Romheld, *et.al.*) 634-635.

Sharma, J.K., Sankaran, K.V. and Balasundaran, M. (1995a). Screening of different mycorrhizal isolates for *Acacia, Casuarina* and *Pterocarpus* growing in degraded acid soils of Kerala. In: *Mycorrhizae: Biofertilizers for the Future.* (Eds. Alok Adholeya and Sujan Singh), TERI, New Delhi: 534-540.

Sharma, J.K., Sankaran, K.V., Balasundaran, M. and Sankar, S. (1996a). Use of mycorrhizal and nitrogen fixing symbionts in reforestation of degraded acid soils of Kerala, Peechi, India. (Kerala Forest Research Instt.) *KERI Research Report* No. 112, 11 + 51 pp.

Sharma, M. and Mukerji, K.G. (1995). Interaction between vesicular-arbuscular mycorrhizal fungi and free-living nitrogen fixers and its effect on growth of jute. In: *Mycorrhizae: Biofetilizers for the Future*. (Eds. Alok Adholeya and Sujan Singh), TERI, New Delhi: 253-257.

Sharma, M. and Roy, A.K. (1991). Vesicular arbuscular mycorrhizae in hypoglycaemic plants. *National Academy Science Letters*. **14**(2): 467-468.

Sharma, M., Bhargava, S., Verma, M.K. and Adholeya, A. (1994). Interaction between the endomycorrhizal fungus *Glomus fasciculatum* and the root-knot nematode, *Meloidogyne incognita* on tomato. *Indian Journal of Nematology.* **24**(2): 133-139.

Sharma, M.P. and Adholeya,A. (2000a). Response of *Eucalyptus tereticornis* to inoculation with indigenous AM fungi in a semiarid alfisol achieved with different concentrations of available soil P. *Microbiological Research* **154** (4): 349- 354.

Sharma, M.P. and Adholeya, A. (2000b). Enhanced growth and productivity following inoculation with indigenous AM fungi in four varieties of onion (*Allium cepa* L.) in an alfisol. *Biological Agriculture & Horticulture* **18** (1): 1-14.

Sharma, M.P., Atimanav, G., Bhatia, N.P. and Adholeya A. (1995b). Response of *Acacia nilotica* to indigenous vesicular-arbuscular mycorrhizal fungal inoculation and single superphosphate fertilization in degraded wasteland soils. In: *Mycorrhizae: Biofertilizers for the Future.* (Eds. Alok Adholeya and Sujan Singh), TERI, New Delhi: 331-335.

Sharma, M.P., Atimanav, G., Bhatia, N.P. and Adholeya, A. (1996b). Growth responses and dependence of *Acacia nilotica* var. *cupriciformis* on the indigenous arbuscular mycorrhizal consortium of a marginal wasteland soil. *Mycorrhiza.* **6**(5): 441-446.

Sharma, M.P., Bhargava, S., Varma, M.K. and Adholeya, A. (1995c). Effect of vesicular-arbuscular mycorrhizal fungus *Glomus fasciculatum* on survival and development of root-knot nematode *Meloidogyne incognita* in tomato. In: *Mycorrhizae: Biofertilizers for the Future.* (Eds. Alok Adholeya and Sujan Singh), TERI, New Delhi:115-114.

Sharma, M.P., Bhatia, N.P. and Adholeya, A. (2001c). Mycorrhizal dependency and growth response of *Acacia nilotica* and *Albizzia lebbeck* to inoculation by indigenous AM fungi as influenced by available soil P levels in a semi-arid Alfisol wasteland. *New Forests* **21**(1): 89-104.

Sharma, M.P., Chauhan, R.K.S. and Adholeya, A. (1999c). Mycorrhizal dependency of *Eucalyptus tereticornis* on arbuscular mycorrhizal fungi in a semi-arid alfisol. Abstract: *National Conference on Mycorrhiza,* Barkatulla University, Bhopal: 53.

Sharma, Pawan (1995). Effect of phosphorus fertilization on vesicular-arbuscular mycorrhiza in reclaimed sodic soils. In: *Mycorrhizae: Biofertilizers for the Future.* (Eds. Alok Adholeya and Sujan Singh), TERI, New Delhi:530-533.

Sharma, S. D. and Bhutani, V. P. (2000). Leaf nutrient status of apple seedlings as influenced by VAM, *Azotobacter* and inorganic fertilizers. *Journal of Hill Research* **13** (2): 63-66.

Sharma, S., Kashyap, S., Prasad, R. and Vasudevan, P. (2000). Role of mycorrhiza and secondary phytobiomass in ericulture. *Indian Journal of Sericulture* **39** (2): 152-159.

Sharma, S., Kashyap, S., Vasudevan, P and Prasad, R. (1999d). Effect of mycorrhiza on mushroom and silkworm cocoons yield. Abstract: *National Conference on Mycorrhiza,* Barkatulla University, Bhopal: 49.

Sharma, S.D. and Bhutani, V.P. (1995). Response to mycorrhizal inoculation and P-fertilization of apple-seedlings. In: *Mycorrhizae: Biofertilizers for the Future.* (Eds. Alok Adholeya and Sujan Singh), TERI, New Delhi:303-310.

Sharma, S.D., Bhutani, V.P. and Awasthi, R.P. (2002). Effect of vesicular- arbuscular Mycorrhizae and phosphorus on leaf and soil mineral nutrient status of apple seedling. *Indian Journal of Horticulture* **59**(2): 140-144.

Sharma, S.K., Sharma, G.D. and Mishra, R.R. (1986a). Records of *Acaulospora* spp. from India. *Current Science.* **55**(15): 724-726.

Sharma, S.K., Sharma, G.D. and Mishra, R.R. (1986b). Status of mycorrhizae in sub-tropical forest ecosystem of Meghalaya. *Acata Botanica Indica.* **14**(1): 87-92.

Sharma, S.K., Sharma, G.D. and Mishra, R.R. (1987). Endogonaceae in sub-tropical forest of north-east India. *Journal of the Indian Botanical Society.* **66**(3-4): 266-268.

Sharma, W. and Trivedi, P.C. (1997). Concomitant effect of *Paecilomyces lolacinus* and vesicular-arbuscular mycorrhizal fungi on root-knot nematide infected okra. *Annals of Plant Protection Sciences.* **5** :70-74.

Shashidhara, G.R. and Gopinath, G. (2002). Effect of nutrients and bio-inoculants on calendula. In *Floriculture research trend in India.* Proceedings of the national symposium *on Indian floriculture in the new millennium, Bangalore.* (Eds R.L Misra, and Sanyat Misra).

Shashikala, B.N., Reddy, B.J.D. and Bagyaraj, D.J. (1999a). Influence of *Glomus mosseae* on the growth of micropropagated *Syngonium* and *Spathiphyllum* at varied levels of fertilizer phosphorus. *Crop Research (Hisar)* **18** (3): 454-460.

Shashikala, B.N., Reddy, B.J.D. and Bagyaraj, D.J. (1999b). Response of micropropagated banana plantlets to *Glomus mosseae* at varied levels of fertilizer phosphorus. *Indian Journal of Experimental Biology.* **37**(5): 499-502.

Shetty, P.K. and Magu, S.P. (1997). Influence of metalaxyl on *Glomus fasciculatum* associated with wheat (*Triticum aestivum*). *Current Science.* **72**(4): 275-277.

Shinde, B.P. and Nair, L.N. (1995). Effect of VAM and *Rhizobium* on growth and nodulation of *Phaseolus radiatus* by using MPN method. *Bio. Ind.* **6**(1 & 2): 45-48.

Shinde, B.P. and Nair, L.N. (1996). Effect of *Rhizobium* and VAM on growth and nodulation of legumes: I-*Phaseolus. Recent Advances in Biotechnology*: 208-209.

Shinde, B.P. and Nair, L.N. (1998). Effect of continuous cropping on VAM association of Maize. *International Conference IIMACA Meeting* Panji, Goa. (In Press).

Shinde, B.P. and Nair, L.N. (1999). VAM fungi from earthworm casts of grape vine orchards. Abstracts: *National Conference on Mycorrhiza,* Barktaulla University, Bhopal: 41.

Shirsath, K.S., Wani, P.V. and Konde, B.K. (1998). Development and performance of mycorrhizal seedlings in *ber (Zizyphus mauritiana* L.). *Journal of Maharashtra Agricultural University.* **23**(1): 4-6.

Shivaram, S. and Rai, P.V. (1990). Influence of *Rhizobium* and VAM on nutrient uptake and competition of rhizobia in mixed culture. *Zentralblatt fur Mikrobiologie.* **145**(4): 313.

Shrihari, P.C. and Sreenivasa, M.N. (1999a). Host preference of arbuscular mycorrhizal fungi for tomato (*Lycopersicon esculentum* Mill.). *Karnataka Journal of Agricultural Sciences* **12** (1/4): 81-87.

Shrihari, P.C. and Sreenivasa, M.N. (1999b). Possible P savings in tomato through inoculation of *Glomus fasciculatum.* Abstract: *National Conference on Mycorrhiza,* Barkatulla University, Bhopal: 51.

Shukla, B.N. and Vanjare, N. (1990). Extraction of vesicular arbuscular mycorrhizae spores and effect of fertilizers on their population in oilseeds cropped soils. In *Current trends in mycorrhizal research. Proceedings of the National Conference on Mycorrhiza.* (Eds. B.L. Jalali and H. Chand), H.A.U., Hisar : 33-34.

Shukla, B.N. and Vanjare, N. (1992). Effect of available NPK on VAM spore population in oilseeds cropped soil. *Indian Journal of Mycology and Plant Pathology.* **22**(2): 187-189.

Siddiqui, Z.A. and Mahmood, I. (1995). Biological control of *Heterodera cajani* and *Fusarium udum* by *Bacillus subtilis, Bradyrhizobium japonicum* and *Glomus fasciculatum* on pigeonpea. *Fundamental Applied Nematology.* **18**(6): 559-566.

Siddiqui, Z.A. and Mahmood, I. (1996). Biological control of *Heterodera cajani* and *Fusarium udum* on pigeonpea by *Glomus mosseae, Trichoderma harzianum* and *Verticillium chlamydosporium. Israel Journal of Plant Sciences* . **44**: 49-56.

Siddiqui, Z.A. and Mahmood, I. (1997). Effect of plant growth promoting bacterium, an AM fungus and soil types on the morphometrics and reproduction of *Meloidogyne javanica* on tomato. *Applied Soil Ecology*. **8**: 77-84.

Siddiqui, Z.A. and Mahmood, I. (2000). Effect of *Bacillus subtilis, Glomus mosseae* and ammonium sulphate on the development of *Meloidogyne javanica* and on growth of tomato. *Thai Journal of Agricultural Science* **33** (1/2): 29-35.

Siddiqui, Z.A., Mahmood, I. and Hayat, S. (1998). Biocontrol of *Heterodera cajani* and *Fusarium udum* on pigeonpea using *Glomus mosseae, Paecilomyces lilacinus* and *Pseudomonas fluorescence*.

Sidhu, O.P. and Behl, H.M. (1995). Applications of vesicular- arbuscular mycorrhizae in salt-affected soils. In: *Mycorrhizae: Biofertilizers for the Future*. (Eds. Alok Adholeya and Sujan Singh), TERI, New Delhi: 526-529.

Sidhu, O.P. and Behl, H.M. (1997). Response of three *Glomus* species on growth of *Prosopis juliflora* Swartz at high pH levels. *Symbiosis (Rehobot)*. **23**(1): 23-34.

Sidhu, O.P., Behl, H.M., Gupta, M.L. and Janardhanan, K.K. (1990a). Occurrence of vesicular-arbuscular mycorrhiza on *Casuarina equisetifolia* L. *Current Science*. **59**(8): 422-423.

Sidhu, O.P., Misra, P.N. and Behl, H.M. (1990b). Clonal selectivity in *Populus deltoides* for vesicular-arbuscular mycorrhizal inoculation. In: *Current trends in mycorrhizal research, Proceedings of the National Conference on Mycorrhiza*. (Eds. B.L. Jalali and H. Chand), H.A.U., Hisar: 122-123.

Sieverding, E. and Leihner, D.E. (1984). Effect of herbicides on population dynamics of VA mycorrhiza with cassava. *Angewett Botanik*. **58**: 283-294.

Simmons, G.L. and Pope, P.E. (1988). Using VA mycorrhizae to enhance seedling root growth in compacted soil. *Northern Journal of Applied Forestry*. **5**(1): 65-68.

Simon, L., Levesque, R.C. and Lalonde, M. (1992). Rapid quantification by PCR of endomycorrhizal fungi colonizing roots. *PCR Methods and Applications*. **2** : 76-80.

Singh D. and Singh, R.S. (2001). Incidence of vesicular arbuscular mycorrhizal fungi in chickpea in Punjab. *Plant Disease Research* **16** (1): 107-109.

Singh R., Naveen Kumar and Rana, N.S. (2000a). Response of rainfed guinea grass (*Panicum maximum)* to biofertilizers inoculation and nitrogen. *Indian Journal of Agronomy* **45** (1): 205-209.

Singh, A., Singh, S. P. and Singh B. P. (2002b). Effect of VAM and inorganic fertilizers on growth and yield of union (*Allium cepa* L.). *Vegetable Science* **29** (1): 40-42.

Singh, A.K. and Mishra, R.R. (1995). Effect of vesicular-arbuscular mycorrhiza on growth and phosphorus-uptake of phosphorus deficiency tolerant and susceptible paddy varieties(RCPL 101 and RCPL 104) under different soil phosphorus levels. In: *Mycorrhizae: Biofertilizers for the Future*. (Eds. Alok Adholeya and Sujan Singh), TERI, New Delhi: 314-321.

Singh, C., Saxena, S. K., Goswami, A. M. and Sharma, R. R. (2000b). Effect of fertilizers on growth, yield and quality of sweet orange (*Citrus sinensis*) cv. Mosambi. *Indian Journal of Horticulture* **57** (2): 114-117.

Singh, C., Sharma, A. K. and Johri, B. N. (2002c). Host genotype determines the impact of soil phosphorus on arbuscular mycorrhizal symbiosis in maize (*Zea mays* L.). *Symbiosis (Rehovot)* **33** (2): 145- 164.

Singh, C.S. (1990). Vesicular-arbuscular (VA) mycorrhiza in presence of *Rhizobium* sp. enhances nodulation, N_2 fixation N utilization of pigeon pea (*Cajanus cajan*) as assessed with ^{15}N technique. In *Trends in mycorrhizal research. Proceedings of the National Conference on Mycorrhiza.* (Eds. B.L. Jalali and H. Chand), H.A.U., Hisar: 152-154.

Singh, C.S. (1992). Mass inoculum production of vesicular-arbuscular (VA) mycorrhizae: I. Selection of host in the presence of *Azospirillum brasilense. Zentalblatt fur Mikrobiologie*. **147**(7): 447-453.

Singh, C.S. (1995). Impact of N_2 fixing and phosphate solubilizing bacteria on root colonization and spore production. In: *Mycorrhizae: Biofertilizers for the Future.* (Eds. Alok Adholeya and Sujan Singh), TERI, New Delhi: 216-219.

Singh, C.S. and Jha, D. (1994). Mass inoculum production of vesicular-arbuscular (VA) mycorrhizae: effect of various bacteriological media and fertilizer solutions. *Microbiological Research.* **149**(1): 27-29.

Singh, C.S. and Kapoor, A. (1995). Root colonization of *Casuarina equisetifolia* in different soil samples in India. In: *Mycorrhizae: Biofertilizers for the Future.* (Eds. Alok Adholeya and Sujan Singh), TERI, New Delhi:86-87.

Singh, C.S. and Tyagi, S.P. (1989). Study on the occurrence of vesicular-arbuscular-mycorrhizal (VAM) fungi in the root of *Aeschynomene indica* under the influence of various ecological factors. *Zentralblatt fur Mikrobiologie.* **144**(4): 241-248.

Singh, C.S., Amawate, J.S., Tyagi, S.P. and Kapoor, A. (1990). Interaction effect of *Glomus fasciculatum* and *Azospirillum brasilense* on yields of various genotypes of wheat (*Triticum aestivum)* in pots. *Zentralblatt fur Mikrobiologie.* **145**(3): 203-208.

Singh, C.S., Kapoor, A. and Rana, J.P.S. (1998). Occurrence of ecto- and endo-mycorrhizae with the roots of *Casuarina equisetifolia* in various soil types. *Indian Forester.* **124**(7): 561-564.

Singh, C.S., Kapoor, A. and Wange, S.S. (1991). The enhancement of root colonization of legumes by vesicular-arbuscular mycorrhizal (VAM) fungi through the inoculation of the legume seed with commercial yeast (*Saccharomyces cerevisiae*). *Plant and Soil.* **131**(1): 129-133

Singh, D., Singh, R. S. and Singh, K. (2002 d). Study on the effect of *Glomus aggregatum* and phosphorus amendments on the colonization and spore population in chickpea. *Legume Research* **25** (2): 142-144.

Singh, G. and Tilak, K.V.B.R. (2001). Phosphorus nutrition through combined inoculation withphosphorus solubilizing rhizobacteriaand arbuscular mycorhizae. *Fertilizer News* **46** (9): 33-36.

Singh, G., Sekhon, H. S. and Sharma, P. (2001). Effect of *Rhizobium,* vesicular arbuscular mycorrhiza and phosphorus on the growth and yield of lentil (*Lens culinaris*) and field pea (*Pisum sativum*). *Environment and Ecology* **19** (1): 40-42.

Singh, H.P. (1994). Response to inoculation with *Bradyrhizobium*, vesicular-arbuscular mycorrhiza and phosphate solubilizing microbes on soybean in a Mollisol. *Indian Journal of Microbiology.* **34**(1):27-31.

Singh, H.P. and Singh, T.A. (1993). The interaction of rockphosphate, *Bradyrhizobium*, vesicular-arbuscular mycorrhizae and phosphate-solubilizing microbes on soybean growth in a sub-Himalayan Mollisol. *Mycorrhiza.* **4**(1): 37-43.

Singh, H.P., Singh, A. and Singh, S. (1995). Nodulation and growth of soyabeans as influenced by rock phosphate, *Bradyrhizobium*, mycorrhiza and phosphate-dissolving microbes. *Bhartiya Krishi Anusandhan Patrika.* **10**(3): 117-124.

Singh, H.P., Singh, T.A. and Singh, S. (1999). Uptake and concentration of nitrogen in soybean as influenced by rock phosphate, *Bradyrhizobium*, mycorrhiza and phosphate solubilizing microbes. In: *Proceedings of National Conference on Mycorrhiza.* Barkatulla University, Bhopal: 21

Singh, J., Gowda, S.T.K., Bagyaraj, D.J. and Rai, P.V. (1990). Effect of glyphosate on VA mycorrhizal association in groundnut. In : *Mycorrhizal Symbiosis and Plant Growth.* (Eds. D.J. Bagyaraj and A. Manjunath), U.A.S., Bangalore : 44-46.

Singh, K., Varma, A.K. and Mukerji, K.G. (1987). Vesicular-arbuscular mycorrhizal fungi in diseased and healthy plants of *Vicia faba. Acta Botanica Indica.* **15** (2) : 304-310.

Singh, K.K. and Mishra, J.P. (1998). Biofertilizers in pulses. *Fertiliser News.* **43**(12): 91-94.

Singh, L., Joshi N. and Singh, R. (2002a). Association of mycorrhizal fungi with wild and cultivated plants in Meerut College Campus including Abu drain. *Advances in Plant Sciences* **15** (1): 315-317.

Singh, M. and Tilak, K.V.B.R. (1989). Field response of chickpea to inoculation with *Glomus versiforme. Plant and Soil.* **119**(2): 281-284.

Singh, M. and Tilak, K.V.B.R. (1990a). Photoassimilate partitioning and translocation in mycorrhizal sorghum. In: *Current trends in mycorrhizal research. Proccedings of the National Conference on Mycorrhiza.* (Eds. B. L. Jalali and H. Chand), H.A.U. Hisar : 51-52.

Singh, M. and Tilak, K.V.B.R. (1995). Effect of *Glomus versiforme* on distribution of photoassimilates in sorghum (*Sorghum bicolor* L. Moench). In: *Mycorrhizae: Biofetilizers for the Future.* (Eds. Alok Adholeya and Sujan Singh), TERI, New Delhi: 462-464.

Singh, M.and Tilak, K.V.B.R. (1990b). Response of different cultivars of sorghum (*Sorghum vulgare*) to inoculation with *Glomus versiforme* in field. In: *Mycorrhizal Symbiosis and Plant Growth.* (Eds. D.J. Bagyaraj and A. Manjunath), U.A.S. Bangalore : 81.

Singh, R. and Adholeya, A. (2002). AMF biodiversity in wheat agro-systems of India. *Mycorrhiza News* **14** (3): 21-23.

Singh, R. and Adholeya, A. (2003a). Effect of crop rotation on diversity and functionality of arbuscular mycorrhizal fungi (AMF). *Mycorrhiza News* **15** (91): 25-26.

Singh, R. and Adholeya, A. (2003b). Management of mycorrhiza in agriculture. *Mycorrhiza News* **14** (4): 26-27.

Singh, R. and Pandya, R.K. (1995). The occurrence of vesicular-arbuscular mycorrhiza in pearl millet and other hosts. In: *Mycorrhizae: Biofertilizers for the Future* (Eds. Alok Adholeya and Sujan Singh), TERI, New Delhi: 56-58.

Singh, R.B. (1993). *Native VAM fungi and their performance on local crops under Gulbarga conditions.* Ph.D. thesis, Gulbarga Univ., Gulbarga, India.

Singh, R.S. and Singh, D. (1995). Effect of *Glomus aggregatum* inoculation and P-amendments on *Fusarium oxysporum* f. sp. *ciceri* causing chickpea wilt. In: *Mycorrhizae: Biofertilizers for the Future.* (Eds. Alok Adholeya and Sujan Singh), TERI, New Delhi:119-123.

Singh, R.S., Hardip and Kang, M.S. (1992). Effect of soil depth and seasonal changes on spore population and mycorrhizal colonization of kinnow and rough lemon seedlings. *Indian Phytopathology*. **43**(3): 337-243.

Singh, R.S., Singh, H. and Singh, O.S. (1993). Influence of VA mycorrhizal inoculation and phosphorus amendments on growth and nutrient uptake by rough lemon seedlings. *Plant Disease Research*. **8**(2): 115-120.

Singh, S. and Kapoor, K.K. (1998). Effects of inoculation of phosphate-solubilizing microorganisms and an arbuscular mycorrhizal fungus on mungbean grown under natural soil conditions. *Mycorrhiza*. **7**(5): 249-253.

Singh, S. and Kapoor, K.K. (1999). Inoculation with phosphate solubilizing microorganisms and a vesicular-arbuscular mycorrhizal fungus improves dry matter yield and nutrient uptake by wheat grown in a sandy soil. *Biology and Fertility of Soils*. **28**(2): 139-144.

Singh, S.N. (1990). Biotechnology for mass production of VA mycorrhiza inocula. In: *Curent trends in mycorrhizal research. Proceedings of the National Conference on Mycorrhiza.* (Eds. B.L. Jalali and H. Chand), H.A.U., Hisar : 80-86.

Singh, S.N., Rao, V.P. and Pawar, S.E. (1990). Studies on growth stimulation of onion (*Allium cepa* L.) with vesicular arbuscular mycorrhizas in Maharashtra region. *Indian Journal of Applied and Pure Biology*. **5**(1): 1-3.

Singh, S.S. and Tiwari, S.C. (1999). Impact of soil degradation on distribution of vesicular-arbuscular mycorrhizal (VAM) fungi population in forest soils. Abstract: *National Conference on Mycorrhiza,* Barkatulla University, Bhopal: 39.

Singh, S.S. and Tiwari, S.C. (2001). Modified wet sieving and decanting technique for enhanced recovery of vesicular-arbuscular mycorrhizal fungi from forest soils. *Mycorrhiza News* **12** (4): 12-13.

Singh, Y.P., Singh, R.S. and Sitaramaiah, K. (1990). Mechanism of resistance of mycorrhizal tomato against root-knot nematode. In: *Current trends in mycorrhizal research. Proceedings of the National Conference on Mycorrhiza.* (Eds. B.L. Jalali and H. Chand), H.A.U., Hisar: 96-97.

Singh. C., Singh, A.K. and Fohri, B.N. (2002a). Variation among maize lines for colonization and responsiveness to arbuscular mycorrhizal fungi. *Mycorrhiza News* **14** (3): 10-13.

Singha, K.D., Baruah, R. and Pathak, A.K. (2000). Occurrence and performance of VAM fungi on the growth of rice plant under rain fed lowland condition of Assam. *Annals of Agri Bio Research* **5** (2): 149-154.

Sirohi, S.S. and Singh, O.S. (1993). Relationship of endomycorrhizal association of unsterilized soils with available soil phosphorus, plant growth, phosphorus uptake and oil synthesis in peppermint. *Scientia Horticulturae*. **20**(2): 185-191.

Sitaramaiah, K. and Khanna, R. (1997). Effect of *Glomus fasciculatum* on growth and chemical composition of maize. *Journal of Mycology and Plant Pathology*. **27**(1): 21-24.

Sivaprasad, P. and Rai, P.V. (1991). Synergistic association between *Glomus fasciculatum* and *Rhizobium* species and its effect on pigeonpea (*Cajanus cajan*). *Indian Journal of Agricultural Sciences*. **61**(2): 97-101.

Sivaprasad, P. and Sheela, M.S. (1998). Distribution of phytonematode and nematode management. In *Nematology: Challenges and Opportunities in 21 st Century.* Proceedings of the Third International Symposium of Afro-Asian Society of Nematologists (TISAASN), Sugarcane Breeding Institute (ICAR), Combatore, India. (ed. U.K. Mehta) Luton, UK; Afro-Asian Society of nematologists: 102-105.

Sivaprasad, P., Hegde, S.V. and Rai, P.V. (1983). Effect of *Rhizobium* and *Mycorrhiza* inoculation on growth of *Leucaena*. *Leucaena Research Reports*. **4**, 42.

Sivaprasad, P., Jacob, A., Nair, S.K. and George, B. (1990). Influence of VA mycorrhizal colonization on root-knot nematode infestation in *Piper nigrum* L. In :*Current trends in mycorrhizal research. Proceedings of the National Conference on Mycorrhiza*. (Eds. B.L. Jalali and H. Chand), H.A.U., Hisar: 100-101.

Sivaprasad, P., Joseph, P.J. and Sulochana, K.K. (1999). Foot rot of black pepper and rhizome rot of ginger as influenced by arbuscular mycorrhizal fungi (AMF) in the field. Abstract: *National Conference on Mycorrhiza,* Barkatulla University, Bhopal: 61.

Sivaprasad, P., Rai, P.V. and Joshi, S.S. (1988). Ultrastructures of nodules of *Cajanus cajan* (L.) Millsp. inoculated with vesicular-arbuscular mycorrhizal (VAM) fungus. *Agricultural Research Journal of Kerala*. **26** : 198-203.

Sivaprasad, P., Ramesh, B., Mohanakumaran, N., Rajmohan, K. and Joseph, P.J. (1995a). Vesicular-arbuscular mycorrhizae for the *ex vitro* establishment of tissue culture plantlets. In: *Mycorrhizae: Biofetilizers for the Future*. (Eds. Alok Adholeya and Sujan Singh), TERI, New Delhi: 281-283.

Sivaprasad, P., Robert, C.P., Vijayan, M. and Joseph, P.J. (1995b). Vesicular-arbuscular mycorrhizal colonization in relation to foot-rot disease intensity in black pepper. In: *Mycorrhizae: Biofetilizers for the Future*. (Eds. Alok Adholeya and Sujan Singh), TERI, New Delhi:137-140.

Sivaprasad, P., Sulochana, K.K. and Nair, S.K. (1990). Comparative efficiency of different VA mycorrhizal fungi on cassava (*Manihot esculenta*, Crantz). *Journal of Root Crops*. **16**(1): 39-40.

Sivaprasad, P., Suneetha, S., Joseph, P.J., Sulochana, K.K. and Skujins, J., Allen, M.F. (1986). Use of mycorrhizae for land rehabilitation. MIRCEN, *Journal of Applied Microbiology and Biotechnology*. **2**(1): 161-176.

Sivaprasad, P.; Suneetha, S.; Joseph, P.J.; Sulochana, K.K. and Rajmohan, K. (1999). Influence of arbuscular mycorrhizal fungi (AMF) on establishment and growth of tissue culture plantlets of banana and alocasia. *National Conference on mycorrhiza*, Barkatulla University Bhopal: 12.

Skujins,J. and Allen, M. F.(1986).Use of mycorrhizae for land rehabilitation. *MIRCEN, Journal of Applied Microbiology and Biotechnology.* **2**(1): 161-176.

Smith, F.A. and Smith, S.E. (1986). Movements across membranes: physiology and genetical biochemistry. In: *Physiology and Genetical Aspect of Mycorrhizae*. (Eds. V. Gianinazzi-Pearson and S. Gianinazzi), Paris Institut National De La Recherche Agronomique, Paris.

Smith, S.E. (1993). Transport at the mycorrhizal interface. *Mycorrhiza*. **5**: 1-4.

Smith, S.E. and Gianinazzi-Pearson, V. (1988). Physiological interactions between symbionts in vesicular-arbuscular mycorrhizal plants. *Annual Review Plant Physiology and Molecular Biology*. **39**: 221-224.

Smith, S.E. and Smith, F.A. (1990). Structure and function of the interfaces in biotrophic symbiosis as they relate to nutrient transport. *New Phytologist*. **114**: 1-38.

Smith, S.E., St. John, B.J., Smith, F.A. and Bromley, J.L. (1986). Effect of mycorrhizal infection on plant growth, nitrogen and phosphorus nutrition of glasshouse grown *Allium cepa* L. *New Phytologist*. **103** : 359-373.

Smith, S.E., St. John, B.J., Smith, F.A. and Nicholas, D.J.D. (1985). Activity of glutamine synthetase and glutamase dehydrogenase in *Trifolium subterraneum L. Australian Journal of Plant Physiology*. **6**: 305-311.

Solaiman, M.Z. and Hirata, H. (1995). Effects of indigenous arbuscular mycorrhizal fungi in paddy fields on rice growth and N,P,K nutrition under different water regimes. *Soil Science and Plant Nutrition*. **41**(3): 505-514.

Sood, S.G. (2003). Chemo tactic response of plant-growth-promoting bacteria towards roots of vesicular-arbuscular mycorrhizal tomato plants. *FEMS Microbiology Ecology* **45** (3): 219-227.

Sosamma, V. K., Koshy, P. K. and Samuel, R. (1998). Effect of VAM, *Glomus mosseae* on nematodes infesting banana under coconut based high-density multispecies cropping system. In *Nematology: Challenges and Opportunities in 21 st Century. Proceedings of the Third International Symposium of Afro-Asian Society of Nematologists (TISAASN)*, Sugarcane Breeding Institute (ICAR), Combatore, India. (ed. U.K. Mehta) Luton, UK; Afro-Asian Society of Nematologists: 94-97.

Sosamma, V.K., Koshy, P.K., Samuel, R. and Bindu, S.M. (1999). Interaction of arbuscular mycorrhizal fungi (AMF) isolated from healthy high yielding coconut palms with burrowing nematode, *Radopholus similes* on coconut. In: *Proceedings of national symposium on rational approaches in nematode management for sustainable agriculture,* Anand, India. (Eds. S.C Dhawan and K.K. Kaushal) New Delhi, India; Nematological Society of India: 34-37.

Spanu, P., Boller, T., Ludwig, A., Wiemken,A., Faccio, A. and Bonfante-Fasolo, P. (1989). Chitinase in roots of mycorrhizal *Allium porrum* : Regulation and localization. *Planta*. **177** : 447-450.

Sreeamalu, K.R., Hanumanthappa, M., Andani Gowda, Murthy, K.N.K. and Jaysheela, N. (2000). Dual inoculation of *Azotobacter chroococcum* and *Glomus fasciculatum* improves growth and yield of sunflower under field conditions and saves N and P fertilizer application. *Environment and* Ecology **18** (2): 380-384.

Sreenivasa, M.N. (1992). Selection of an efficient vesicular-arbuscular mycorrhizal fungus for chilli (*Capsicum annuum* L.). *Scientia Horticulturae*. **50**(1-2): 53-58.

Sreenivasa, M.N. (1994a). Response of chilli (*Capsicum annuum*) to vesicular-arbuscular mycorrhiza at different phosphorus levels in field. *Indian Journal of Agricultural Sciences*. **64**(1) 47-49.

Sreenivasa, M.N. (1994b). VA mycorrhiza in conjunction with organic amendments improve growth and yield of chilli. *Environment and Ecology*. **12**(2): 312-314.

Sreenivasa, M.N. and Bagyaraj, D.J. (1988a). *Chloris gayana* (Rhodes grass), a better host for the mass production of *Glomus fasciculatum* inoculum. *Plant and Soil*. **106**(2): 289-290.

Sreenivasa, M.N. and Bagyaraj, D.J. (1988b). Effect of culture, age and pruning on mass production of the VA mycorrhizal fungus, *Glomus fasciculatum. Pertanika*. **11**(1): 143-145.

Sreenivasa, M.N. and Bagyaraj, D.J. (1988c). Optimum frequency of addition of nutrient solution for mass production of *Glomus fasciculatum* inoculum. *Current Science*. **57**(12): 665-667.

Sreenivasa, M.N. and Bagyaraj, D.J. (1988d). Selection of a suitable substrate for mass multiplication of *Glomus fasciculatum. Plant and Soil*. **109**(1): 125-127.

Sreenivasa, M.N. and Bagyaraj, D.J. (1988e). Suitable level of zinc, copper, and manganese for mass production of the vesicular-arbuscular mycorrhizal fungus, *Glomus fasciculatum. Proceedings of the Indian Academy of Science, Plant Sciences*. **98**(2): 135-138.

Sreenivasa, M.N. and Bagyaraj, D.J. (1989). Use of pesticides for mass production of vesicular-arbuscular mycorrhizal inoculum. *Plant and Soil*. **119**(1): 127-132.

Sreenivasa, M.N. and Bagyaraj, D.J. (1990). Suitable source and level of nitrogen for mass production of the VA mycorrhizal fungus *Glomus fasciculatum*. In *Current trends in mycorrhizal research. Proceedings of the National Conference on Mycorrhiza*. (Eds. B.L. Jalali and H. Chand), H.A.U., Hisar : 35-36.

Sreenivasa, M.N. and Gaddagimath, N.B. (1993). Mycorrhizal dependency of chilli cultivars. *Zentralblatt fur Mikrobiologie*. **148**(1): 55-59.

Sreenivasa, M.N. and Krishnaraj, P.U. (1992). Synergistic interaction between VA mycorrhizal fungi and a phosphate solubilizing bacterium in Chilli (*Capsicum annuum*). *Zentralblatt fur Mikrobiologie*. **147**(1-2): 126-130.

Sreenivasa, M.N. and Majjigudda, I.M. (1996). Residual influence of VA mycorrhizal inoculation on rhizosphere microflora and growth of subsequent crop. *Environment and Ecology*. **14**(4): 782-784.

Sreenivasa, M.N. and Rajashekhara, E. (1989). Influence of the stage of host development and genotypes on mycorrhizal colonization of field grown wheat. *Zentralblatt fur Mikrobiologie*. **144**(6): 381-384.

Sreenivasa, M.N. and Srihari, P.C. (1994). Influence of cropping system on native mycorrhiza. *Environment and Ecology*. **12**(2): 485-486.

Sreenivasa, M.N., Krishnaraj, P.U., Gangadhara, G.A. and Manjunathaiah, H.M. (1993). Response of chilli (*Capsicum annuum* L.) to the inoculation of an efficient vesicular-arbuscular mycorrhizal fungus. *Sicnentia Horticulturae*. **53**(1-2): 45-52.

Sreeramulu, K. R. (1998). Isolation and characterization of arbuscular Mycorrhizal fungi associated with different cultivars of cardamom. *Progressive Horticulture* **30** (1/2): 16-20.

Sreeramulu, K.R. and Bagyaraj, D.J. (1999). Arbuscular mycorrhizal fungi improve growth of Vazukka cultivar of cardamom. *Journal of Ecobiology* **11** (2): 125-129.

Sreeramulu, K.R. and Bagyaraj, D.J. (1998). Screening of arbuscular mycorrhizal fungi in cardamom cv. Mysore. *Journal of Plantation Crops* **27** (3): 207-211.

Sreeramulu, K.R., Gowda, V. N. and Bagyaraj, D.J. (1998). Influence of VA mycorrhiza on establishment and growth of Khirni (*Manilkara hexandra* (Roxb.) Dub.) seedlings. *Current Research- University of Agricultural Sciences (Bangalore)* **27** (3): 50-52.

Sreeramulu, K.R., Shetty, Y.V. and Setty, T.K.P. (1996). Effect of VA mycorrhiza on the growth of two important leafy vegetables. *Madras Agricultural Journal.* **83**(6): 362-364.

Srilakshmi, R., Sundaram, M.D. and Tholkappian, P. (1999). Effect of organic amendments on the mycorrhizal colonization, growth and yield of cotton (MCU-7). Abstract: *National Conference on Mycorrhiza,* Barkatulla University, Bhopal: 47.

Srinath, J., Bagyaraj, D.J. and Satyanarayana, B.N. (2003). Enhanced growth and nutrition of micropropagated *Ficus benjamina* to *Glomus mosseae* co-inoculated with *Trichoderma harzianum* and *Bacillus coagulans. World Journal of Microbiology & Biotechnology* **19** (1): 69-72.

Srinivas, K., Ramaraj, B. and Shanmugam, N. (1988a). Survey for the occurrence of native VAM fungi. In: *Mycorrhizae for Green Asia.* (Eds. A. Mahadevan, N. Raman and K. Natrajan), Madras University, Madras : 11-113.

Srinivas, K., Shanmugam, N. and Ramaraj, B. (1988b). Effect of VAM fungi on growth and nutrient uptake of forest tree seedlings. In: *Mycorrhizae for Green Asia.* (Eds. A. Mahadevan, N. Raman and K. Natrajan), Madras University, Madras :294-297.

Srinivas, P., Rao, M.S., Reddy, S.R. and Reddy, S.M. (1999a). Vesicular arbuscular mycorrhizal dependency of *Albizzia lebbeck* (L.) Benth.*Proceedings of the National Academy of Sciences India. Section B, Biological Sciences* **69** (3/4): 337-342.

Srinivas, P., Rao, M.S., Reddy, S.R. and Reddy, S.M. (2001). Dynamics of VAM fungi of Godavari belt forests. In: *Perspectives in Biotechnology.* Poceedings of a national symposium, Warangal, India. (Eds.Redy, S.M.; Rao, D.; Vidyavati), Jodhpur, India; Scientific Publisher: 23-30.

Srinivas, P., Reddy, G.S.L., Rao, M.S. and Reddy, S.M. (1999b). Mycorrhizal dependency of two multi-purpose tree species in coal mine soils. Abstract: *National Conference on Mycorrhiza,* Barkatulla University, Bhopal: 9.

Srinivas, P., Reddy, R. and Reddy, S.M. (1998). Dynamics of vesicular arbuscular mycorrhizae in association with *Albizzia lebbeck. Journal of Indian Botanical Society.* **77**: 245-246.

Srinivasan, D., Muthukumar, T. and Udaiyan, J. (1996). Influence of effluent irrigation on VA mycorrhizae. *Indian Journal of Microbiology.* **36**(1): 37-40.

Srinivasulu, Y. and Lahsman, H.C. (2002). Response of *Guizotia abyssinica* Cass. (Niger) and *Linum usitattissimim* (Linseed) to VA-mycorrhizal inoculation in an unsterile soil. *Karnataka Journal of Agriculture Science* **15** (2): 405-407.

Srivastava, D. and Mukerji, K.G. (1995). Field response of mycorrhizal and nonmycorrhizal *Medicago sativa* var. local in the F1 generation. *Mycorrhiza.* **5**(3): 219-221.

Srivastava, H.P., Chandel, S. and Kaushal, S. (1997). Arbuscular mycorrhizal fungi occurring in vegetated sand dunes of Indian Thar Desert. In :*Microbial Biotechnology.* (Eds. S.M. Reddy, H.P. Srivastava, D.K. Purohit and S. Ram Reddy), Jodhpur University, Jodhpur : 87-96.

Srivastava, J. N., Kumar, R., Mathur, M., Bhatnagar, V.P. and Kumar, A. (1999). Effect of VAM fungi on plant growth and increase mineral nutrient uptake in maize. *Journal of Living World* **6** (1): 1-3.

Srivastava, K.K., Mohan, V. and Verma, N. (2001). Impact of VAM inoculation on some semi-arid tree species. *Indian Forester* **127**(8): 936-940.

Srivastava, N. and Chandra, S. (1998). Mycorrhizal association in some floricultural crops. *Abstract : XXI Annual Meeting of Indian Botanical Society*, Vikram University, Ujjain :15.

Srivastava, N., Chandra, S. and Kehri, H. K. (2005). Selection of efficient VAM inoculant for *Chrysanthemum*. In: *Biological Diversity: Current Status* (Eds. S.P. Gautam, Y.K. Bansal and A, K, Pandey) Shree Publishers and Distributors, New Delhi: 315-322.

Srivastava, N.K. and Basu, M. (1995). Occurrence of vesicular-arbuscular mycorrhizal fungi in some medicinal plants. In: *Mycorrhizae: Biofertilizers for the Future*. (Eds. Alok Adholeya and Sujan Singh), TERI, New Delhi: 59-61.

Stahl, P.D., Williams, S.E. and Christense, M. (1988). Efficacy of native vesicular-arbuscular mycorrhizal fungi after severe soil disturbance. *New Phytologist*. **110**(3): 347-354.

Stevenson, F.J. (1986). *Cycles of Soil*. John Wiley and Sons. New York.

Subba Rao, N.S., Tilak, K.V.B.R. and Singh, C.S. (1981). Ecology of nonsymbiotic dinitrogen fixers. In *Progress in Microbial Ecology*. (Eds. K.G. Mukerji, V.P. Agnihotri and R.P.Singh), Lucknow Print House (India), Lucknow.

Subba Rao, N.S., Tilak, K.V.B.R. and Singh, C.S. (1985). Synergistic effect of vesicular-arbuscular mycorrhizas and *Azospirillum brasilense* on the growth of barley in pots. *Soil Biology and Biochemistry*. **17**: 119-121.

Subba Rao, N.S., Tilak, K.V.B.R. and Singh, C.S. (1986). Dual inoculation with *Rhizobium* sp. and *Glomus fasciculatum* enhances nodulation, yield and nitrogen fixation in chickpea (*Cicer arietinum* Linn.) *Plant and Soil*. **95**: 351-359.

Subbiah, K. and Narayanasamy, T. (2001). Interaction effect of phosphate and biofertilizers on P use efficiency, yield and quality of fruits in bhendi (*Abelmoschus escuentus* L. Moench.). In: *changing scenario in the production system of horticultural crops*. Proceedings of a National Seminar, Coimbatore, Tamil Nadu, India. *South Indian Horticulture* **49** (special): 168-170.

Subbiah, K. (1994). Effect of N, P and biofertilizers on yield and nutrient uptake in chilli and bellary onion. *Madras Agricultural Journal*. **81**(5): 277-279.

Subhan, S., Sharmila, P. and Saradhi, P.P. (1997). *Glomus fasciculatum* alleviates transplantation shock of micropropagated *Sesbania sesban*. *Plant Cell Reports*. **17**(4): 268-272.

Subhashini, D.V. (1990). The role of VA mycorrhiza in controlling certain root diseases of tobacco. In *Trends in mycorrhizal research. Proceedings of the National Conference on Mycorrhiza*. (Eds. B. L. Jalali and H. Chand), H.A.U., Hisar: 102.

Subhashini, D.V. and Krishnamurty, V. (1995). Influence of vesicular-arbuscular mycorrhiza on phosphorus economy, yield, and quality of fluecured Virginia tobacco in rain-fed Alfisols. In: *Mycorrhizae: Biofetilizers for the Future*. (Eds. Alok Adholeya and Sujan Singh), TERI, New Delhi:328-330.

Subhashini, D.V., Rana, B.S. and Pottat, V.P. (1988). Genotype dependency variation in mycorrhizal colonization and response to phosphorus in *Sorghum bicolor*. In: *Mycorrhizae for Green Asia.* (Eds. A. Mahadevan, N. Raman and K. Natrajan), Madras University, Madras : 255-259.

Subramanian, S., Santhanakrishnan, P., Arunachalam, N. and Thangaraj, T. (2001). Role of mycorrhizas in tomato production under drought stress conditions. In: *changing scenario in the production system of horticultural crops.* Proceedings of a National Seminar, Coimbatore, Tamil Nadu, India. *South Indian Horticulture* **49** (special): 268-270.

Sugavanam, V., Udaiyan, K. and Devaraj, P. (1999a). Influence of tripartite interactions of VAM fungi, *Frankia* and *Casuarina equisetifolia* on growth and nutrient accumulation in host under nursery conditions. *Bangladesh Journal of Forest Science* **28**(1): 25-31.

Sugavanam, V., Udaiyan, K. and Devraj, P. (1999b). Selection of an efficient vesicular mycorrhizal fungi and *Azospirillum* sp. for inoculating *Tectona grandis. Indian Journal of Forestry* **21** (4): 281-284.

Sugavanam, V., Udaiyan, K. and Manian, S. (1994). Effect of fungicides on vesicular-arbuscular mycorrhizal infection and nodulation in groundnut (*Arachis hypogaea* L.). *Agriculture, Ecosystems & Environment.* **48**(3): 285-293.

Sugavanam, V.,Udaiyan, K. and Devaraj. P. (2000). Effect of different arbuscular mycorrhizal fungi and *Azospirillum* on *Eucalyptus tereticornis* seedlings. *Journal of Tropical Forest Science* **12** (1): 37-43.

Sullia, S.B. and Sampath, P. (1990). Occurrence of vesicular arbuscular mycorrhizal fungi in medicinal plants. In: *Mycorrhizal Symbiosis and Plant Growth.* (Eds. D.J. Bagyaraj and A. Manjunath), U.A.S., Bangalore : 37-38.

Sulochana, K.K., Sivaprasad, P. and Vasanthakumar, K. (1995). Phosphorus nutrition and yield of cassava as influenced by vesicular-arbuscular mycorrhiza. In: *Mycorrhizae: Biofetilizers for the Future.* (Eds. Alok Adholeya and Sujan Singh), TERI, New Delhi: 397-399.

Sulochana, T. and Manoharachary, C. (1989). Vesicular-arbuscular mycorrhizal association of castor and safflower. *Current Science.* **58**(8): 459-461.

Sulochana, T., Manoharachary, C. (1990). Impact of season on the distribution of VAM fungi associated with sesame. In : *Current trends in mycorrhizal research. Proceedings on National Confer ence on Mycorrhiza.* (Eds. B.L. Jalali and H. Chand), H.A.U., Hisar : 18-19.

Sulochana, T., Manoharachary, C. and Rama Rao, P. (1988). VAM fungi associated with *Sesamum indicum* in relation to some ecological factors. In: *Mycorrhizae for Green Asia.* (Eds. A. Mahadevan, N. Raman and K. Natrajan), Madras University, Madras : 114-117.

Sulochana, T., Manoharachary, C., Ramarao, P. (1989). Growth response and root colonization in cultivars of sesame to VAM fungi. *Current Science.* **58**(9): 519-520.

Sumana, D.A. and Bagyaraj, D.J. (1998). Selection of efficient VA mycorrhizal fungi for *Dalbergia latifolia* Roxb. *Annals of Forestry.* **6**(2): 186-190.

Sumana, D.A. and Bagyaraj, D.J. (1999). Selection of efficient VA mycorrhizal fungi for inoculating neem. Abstract: *National Conference on Mycorrhiza,* Barkatulla University, Bhopal: 10.

Sundarababu, R., Mani, M.P. and Arulraj, P. (2001). Management of *Meloidogyne incognita* in chilli nursery with *Glomus mosseae.Annals of Plant Protection Sciences* **9** (1): 161-162.

Sundarababu, R., Sankaranarayanan, C. and Santhi, A. (1993). Interaction between vesicular-arbuscular mycorrhiza and *Meloidogyne javanica* on tomato as influenced by time of inoculation. *Indian Journal of Nematology*. **23**(1): 125-126.

Sundaram, M.D. and Arangarasan, V. (1995a). Effect of inoculation of vesicular-arbuscular mycorrhizal fungi on the yield and quality attributes in tomato (*Lycopersicon esculentum* M.) cv. Co-3. In: *Mycorrhizae: Biofetilizers for the Future*. (Eds. Alok Adholeya and Sujan Singh), TERI, New Delhi: 394-396.

Sundaram, M.D. and Arangarasan, V. (1995b). Role of vesicular-arbuscular mycorrhizal fungi on the growth, nematode (*Meloidogyne incognita* Kafoid and White) disease incidence and fruit yield of tomato (Co-3) grown in loamy soils of South Arcot District (Tamil Nadu). In: *Mycorrhizae: Biofertilizers for the Future*. (Eds. Alok Adholeya and Sujan Singh), TERI, New Delhi: 100-103.

Sundaresan, P., Raja, N.U., Gunasekaran, .; Lakshmanan, M. (1988). Studies on nitrate reduction by VAM fungal spores. *Current Science*. **57**(2):84-85.

Sundaresan, P., Raja, N.U., Gunasekaran, P. and Lakshmanan, M. (1993a). Studies on nitrate reduction by VAM fungal spores. *Current Science*. **57**(2): 84-85.

Sunderesan, P., Raja, Naa. V. and Gunasekaran, P. (1993b). *J. Biosciences*. **18** (2) : 291-301. cf. Dynamics of plant-pathogen-mycorrhiza interactions in legume crops (H. Chand and B.L. Jalali). In : *Trends in Microbial Exploitation*. (Eds. Bharat Rai, R.S. Upadhyay and N.K. Dubey), B.H.U., Varanasi : 71-77.

Suresh, C.K. and Bagyaraj, D.J. (1984). Interaction between a vesicular arbuscudlar mycorrhiza and root-knot nematodes and its effect on growth and chemical composition of tomatoes *Nematologia Mediteranea*. **12**:31-39.

Suresh, C.K., Bagyaraj, D.J. and Reddy, D.D.R. (1985). Effect of vesicular arbuscular mycorrhiza on survival, penetration and development of root-knot nematodes in tomato. *Plant and Soil*. **87**:305-308.

Suvercha and Mukerji, K.G. (1988). Influence of VAM fungi on the growth of two cultivars of *Capsicum annum*. In: *Mycorrhizae for Green Asia*. (Eds. A. Mahadevan, N. Raman and K. Natrajan), Madras University, Madras : 262-264.

Swaminathan, K. and Verma, B.C. (1977). Symbiotic effect of vesicular arbuscular mycorrhizal fungi on the phosphate nutrition of potatoes. *Proceedings of the Indian Academy of Sciences*. B **85**(5): 310-318.

Swaminathan, K. and Verma, B.C. (1979). Response of three crop species to vesicular-arbuscular mycorrhizal infection on zinc-deficient Indian soils. *New Phytologist*. **82**(2): 481-487.

Swarupa, S.G., Naik, T.B., Reddy, A.G.S., Raju, T. (1997). Effect of biofertilizers on germination of coffee seed. *Indian Coffee*. **61**(30): 3-5.

Sylvia, D.M. and Will, M.E. (1988). Establishment of a vesicular arbuscular mycorrhizal fungi and other micro-organisms on a beach replenishment site in Florida. *Applied Environmental Microbiology*. **54**: 348-352.

Talukdar, N.C. and Thakuria, D. (2001). Diversity and importance of vesicular-arbuscular mycorrhizal fungi in teak (*Tectona grandis*) and gomar (*Gmelina arborea*) plantations of Assam. In: *Tropical forestry research: Challenges in the new millennium. Proceedings of the International Symposium.* (Eds. R.V. Varma, K.V. Bhat, E.M. Murlidharan and J.K. Sharma.). Peechi, India; Kerala Forest Research Institute (KFRI): 177-182.

Tamuli, P. and Boruah, P. (2002). Vesicular-Arbuscular Mycorrhizal (VAM) association of agarwood tree in Jorhat District of the Brahmaputra Valley. *Indian Forester* **128** (9): 991-994.

Tandon, V. and Prakash, A. (1998). Influence of soil inoculation with VA mycorrhiza and phosphorus solubilising microorganisms on growth and phosphorus uptake in *Sesamum indicum. International Journal of Tropical Agriculture.* **16**(1/4): 201-209.

Tandon, V. and Prakash, A. (1999). Role of VAM fungi in distribution and uptake of heavy metal in *Sesamum indicum* seedlings. Abstract: *National Conference on Mycorrhiza*, Barkatulla University, Bhopal: 50.

Tarafdar, J.C. (1995). Visual demonstration of *in vitro* acid phosphatase activity of VA mycorrhizal fungi. *Current Science.* **69**(6):541-543.

Tarafdar, J.C. and Kumar, P. (1996). The role of vesicular arbuscular mycorrhizal fungi on crop, tree and grasses grown in an arid environment. *Journal of Arid Environments* . **34**(2): 197-203.

Tarafdar, J.C. and Rao, A.V. (1990). Survey of Indian arid zone tree species for occurrence of VAM infections. In: *Current trends in mycorrhizal research. Proceedings of the National Conference on Mycorrhiza.* (Eds. B.L. Jalali and H. Chand), H.A.U., Hisar: 42-43.

Tarafdar, J.C. and Rao, A.V. (1997a). Mycorrhizal colonization and nutrient concentration of naturally grown plants on gypsum mine spoils in India. *Agriculture, Ecosystems & Environment.* **61**(1): 13-18.

Tarafdar, J.C. and Rao, A.V. (1997b). Response of legumes to VAM fungal inoculation. *Symbiosis* (Rehobot). **22**(3): 265-274.

Tarafdar,J.C. and Rao,A.V. (2001). Response of cluster bean to *Glomus mosseae* and *Rhizobium* in an arid soil fertilized with nitrogen, phosphorus and farmyard manure *Journal of the Indian Society of Soil Science* **49**(4): 751-755.

Tewari, J.P., Skoropad, W.P., Mukerji, K.G. and Mishra, S. (1982). *Transactions of British Mycological Society.* **79**: 364-366.

Tewari, L., Johri, B.N. and Tandon, S.M. (1993). Host genotype dependency and growth enhancing ability of VA-mycorrhizal fungi for *Eleusine coracana* (finger millet). *World Journal of Microbiology and Biotechnology.* **9**: 191-195.

Thakur, A.K. and Panwar, J.D.S. (1995). Effect of *Rhizobium* VAM interactions on growth and yield in mungbean *(Vigna radiata* (L.) Wilczek) under field conditions. *Indian Journal of Plant Physiology.* **38**(1): 62-65.

Thakur, A.K. and Panwar, J.D.S. (1997). Response of *Rhizobium*-vesicular arbuscular mycorrhizal symbionts on photosynthesis, nitrogen metabolism and sucrose translocation in greengram (*Phaseolus radiatus). Indian Journal of Agricultural Sciences.* **67**(6): 245-248.

Thakur, M. and Lakhanpal, T.N.(1990). Studies on the mycorrhiza of apple (*Malus domestica* Borkh.). In *Current trends in mycorrhizal research. Proceedings of the National Conference on Mycorrhiza.* (Eds. B.L. Jalali and H. Chand), H.A.U., Hisar: 27-28.

Thammaiah, T.A., Bhattacharya, S., Shivaprakash, M.K. and Bagyaraj, D.J. (2001). Response of Robusta coffee (*Coffea canephora*) Sln. 3R [C x R] to VA mycorrhizal fungi. *Journal of Plant Biology* **28** (2): 213-215.

Thanuja, T.V., Hegde, R.V. and Sreenivasa, M.N. (2002). Induction of rooting and root growth in black pepper cuttings (*Piper nigrum* L.) with the inoculation of arbuscular mycorrhizae. *Scientia Horticulturae* **92**(3/4): 339-346.

Thapar, H. S., Sekhon, B.S., Atwal, A. and Singh, R. (1990). Effect of foliar application of urea on nitrogen metabolism of mycorrhizal moong plants under varying phosphorus levels. In: *Current trends in mycorrhizal research. Proceedings of the National Conference on Mycorrhiza.* (Eds. B.L. Jalali and H. Chand), H.A.U., Hisar: 74-75.

Thapar, H.S. and Khan, S.N. (1973). Studies on Endomycorrhiza in some forest species. *Proceedings of the Indian National Science Academy.* **39**(6): 687-694.

Thapar, H.S. and Khan, S.N. (1985a). Distribution of VA mycorrhizal fungi in forest soils in India. *Indian Journal of Forestry.* **8**(1): 5-7.

Thapar, H.S. and Khan, S.N. (1985b). Effect of VA mycorrhiza on the growth of Hoop Pine (*Araucaria cunninghamii* Sweet). *Journal of Tree Sciences.* **4**(1): 39-43.

Thapar, H.S. and Khan, S.N. (1988). Seasonal frequency of *Endogone* spores in new-forest soils. *Trends in Tree Sciences* (Eds. P.K. Khosla and R.N. Sehgal) : 161-162.

Thapar, H.S. and Uniyal, K. (1990). Survey of some VAM fungi of saline soils of Haryana State. In: *Current trends in mycorrhizal research. Proceedings of the National Conference on Mycorrhiza.* (Eds. B.L. Jalali and H. Chand), H.A.U., Hisar: 16-17.

Thapar, H.S. and Uniyal, K. (1996). Effect of VAM fungi and *Rhizobium* on growth of *Acacia nilotica* in sodic and New Forest soils. *Indian Forester.* **122**(11): 1033-1039.

Thapar, H.S. and Vijayan, A.K. (1990). Vesicular-arbuscular mycorrhizal associations and root colonization in some important tree species. In: *Current trends in mycorrhizal research. Proceedings of the National Conference on Mycorrhiza.* (Eds. B.L. Jalali and H. Chand), H.A.U., Hisar : 111-112.

Thapar, H.S., Khan, S.N. and Uniyal, K. (1985). Endogonaceae of Forest Soils - New Records. *Journal of Tree Sciences.* **4**(2): 47-52.

Thapar, H.S., Khan, S.N., Ram, M.A.R. and Bakshi, B.K. (1984). Effect of fertilizers on development of endomycorrhiza in Jauri Pine (*Agathis robusta*). *The Indian Forester.* **110**(11):1093-1097.

Thapar, H.S., Uniyal, K. and Verma, R.K. (1991). Survey of native VAM fungi of sodic soils of Haryana State. *Indian Forester.* **117**(12): 1059-1069.

Thapar, H.S., Vijyan, A.K. and Uniyal, K. (1992). Vesicular-arbuscular mycorrhizal associations and root colonization in some important tree species. *The Indian Forester.* **118**: 207-212.

Thatoi, H., Parida, D., Padhi, G.S. and Mishra, A. (1995). Effect of *Rhizobium* and AM fungi inoculation on growth and rhizosphere microbial population of *Acacia nilotica* in iron mine waste soil. *Naturalis (Sao Paulo).* **20**: 177-190.

Thatoi, H.N., Padhi, G.S. and Misra, A.K. (1999). Selection of efficient VAM fungi for improvement of growth and nodulation of *Acacia nilotica* growth in chromite mine waste soil. Abstract: *National Conference on Mycorrhiza*, Barkatulla University, Bhopal: 54.

Thatoi, H.N., Sahu, S., Misra, A.K. and Padhi, G.S. (1993). Comparative effect of VAM inoculation on growth, nodulation and *Rhizobium* population of sababul (*Leucaena leucocephala* (Lam.) de Wit.) grown in iron mine waste soil. *Indian Forester*. **119**(6): 481-489.

Thippeswamy, R. and Sreenivasa, M.N. (1999, publ. 1999). Physiological, nutritional and yield responses of tomato to the inoculation of *Glomus fasciculatum* under water stress condition. *Journal of Maharashtra Agricultural Universities* **23** (3): 265-268.

Tholkappian, P., Prakash, M. and Sundaram, M.D. (2001). Effect of AM-fungi on proline, nitrogen and pod number of soybean under moisture stress. *Indian Journal of Plant Physiology* **6**(1): 98-99.

Thomas, G.V., Sundararaju, P., Ali, S.S. and Ghai, S.K. (1989). Individual and interactive effects of VA mycorrhizal fungi and root-knot nematode, *Meloidogyne incognita* on cardamum. *Tropical Agriculture, U.K.* **66**(1): 21-24.

Thomas, R.S., Dakessian, S., Ames, R.N., Brown, M.S. and Bethlenfalvay, G.J. (1986). Aggregation of a silty clay loam soil by mycorrhizal onion roots. *Soil Science Society of American Journal* **50**: 1494-1499.

Thompson, J.P. (1987). Decline of vesicular-arbuscular mycorrhizae in long fallow disorder of field crops and its expression in phosphorus deficiency of sunflower. *Australian Journal of Agricultural Research* . **38**(5): 847-867.

Tien, T.M., Gaskins, M.H. and Hobbell, D.H. (1979). Plant growth substances produced by *Azospirillum brasilense* and their effect on the growth of pearl millet (*Pennisetum americanum* L.) *Applied Environmental Microbiology*. **37**: 1016-1024.

Tilak, K.V.B.R. (1985a). Interactions of vesicular-arbuscular mycorrhizae and nitrogen fixers. *Proc. Soil Biology Symposium*. H.A.U., Hisar: 219-226.

Tilak, K.V.B.R. (1985b). Recent developments in the field of *Azospirillum* and its interaction with vesicular-arbuscular mycorrhizal (VAM) fungi in increasing the crop yields. In:*Crop Productivity*. (Eds. H.L. Srivastava, S. Bhaskaman, and Menon), Oxford & IBH Book Publ. Co., New Delhi: 287-305.

Tilak, K.V.B.R. (1990). Interaction of VA-mycorrhizae with beneficial soil microorganisms. In: *Current trends in mycorrhizal research. Proceedings of the National Conference on Mycorrhiza*. (Eds. B.L. Jalali and H. Chand), H.A.U., Hisar : 87-90.

Tilak, K.V.B.R. (1995). Vesicular-arbuscular mycorrhiza and *Azospirillum brasilense* rhizocoenosis in pearl millet in semi-arid tropics. In: *Mycorrhizae: Biofetilizers for the Future*. (Eds. Alok Adholeya and Sujan Singh), TERI, New Delhi: 177-179.

Tilak, K.V.B.R. and Dwivedi, A. (1990a). Enhancement of spore germination of *Glomus fasciculatum* by bacterial cell free extracts. *Indian Journal of Experimental Biology*. **28**(4): 373-375.

Tilak, K.V.B.R. and Dwivedi, A. (1990b). Nitrate reductase activity of vesicular-arbuscular mycorrhizal fungi. In: *Current trends in Mycorrhizal Research. Proceedings of the National Conference on Mycorrhiza*. (Eds. B.L. Jalali and H. Chand), H.A.U., Hisar : 59-60.

Tilak, K.V.B.R. and Dwivedi, A. (1991). Effect of *Glomus versiforme* inoculation on the yield of chickpea at varying N levels. *International Chickpea Newsletter*. No. 24, 44-45.

Tilak, K.V.B.R. and Murthy, B.N. (1987). Association of vesicular-arbuscular mycorrhizal fungi with the roots of different cultivars of barley (*Hordeum vulgare). Current Science*. **56**(21): 1114-1116.

Tilak, K.V.B.R. and Singh, C.S. (1988). Response of pearl millet (*Pennisetum americanum*) to inoculation with vesicular-arbuscular mycorrhizae and *Azospirillum brasiliense* with different sources of phosphorus. *Current Science*. **57**(1): 43-44.

Tilak, K.V.B.R., Li, C.Y. and Ho, I. (1989). Occurrence of nitrogen-fixing *Azospirillum* in vesicular-arbuscular mycorrhizal fungi. *Plant and Soil*. **116**(2): 286-288.

Tilak, K.V.B.R., Li, C.Y., Trappe, J.M. (1988). Characterization of nitrogen-fixing *Azospirillum* isolated from within sporocarps of ectomycorrhizal fungi associated with Douglas-fir {*Pseudotsuga menziesii* (Mirb.) Franco}. *Indian Journal of Microbiology*. **28**(4): 315-319

Tilak, K.V.B.R., Rathi, S.K. and Saxena, A.K. (1995a). Field evaluation of different arbuscular mycorrhizae on soyabean. In: *Mycorrhizae: Biofertilizers for the Future*. (Eds. Alok Adholeya and Sujan Singh), TERI, New Delhi: 339-341.

Tilak, K.V.B.R., Saxena, A.K. and Sadasivam, K.V. (1995b). Synergistic effects of phosphate-solubilizing bacterium (*Pseudomonas striata*) and arbuscular mycorrhizae on soybean. In: *Mycorrhizae: Biofertilizers for the Future*. (Eds. Alok Adholeya and Sujan Singh), TERI, New Delhi: 224-226..

Tinker, P.B. (1980). Role of rhizosphere microorganisms in phosphorus uptake by plants. In: *The Role of Phosphorus in Agriculture*. (Eds. F.E. Khasawnch., E.C., Sample and E.J. Kamprath), *American Society of Agronomy*. Madison. Wisc. : 615-654.

Tinker, P.B. (1982). Mycorrhizas: The present position. In: *Whither Soil Research, Trans. 12th Int. Congr. Soil Science*. New Delhi: 150.

Tiwari, N., Chauhan, U.K. and Tiwari, D. (1995). Response of vesicular-arbuscular mycorrhizal colonization in the rhizosphere soil of *Cajanus cajan* L. and *Zea mays* L. at different growth stages. In: *Mycorrhizae: Biofertilizers for the Future*. (Eds. Alok Adholeya and Sujan Singh), TERI, New Delhi: 352-355.

Tiwari, P. and Adholeya, A. (2002). In vitro co-culture of two AMF isolates *Gigaspora margarita* and *Glomus intraradices* on Ri T-DNA transformed roots. *FEMS Microbiology Letter* **206** (1): 39-43.

Tiwari, P.K. and Shukla, B.N. (1991). Influence of pesticides on *Glomus fasciculatum*, and growth, development and nutrient uptake of soybean. *Indian Journal of Mycology and Plant Pathology*. **21**(1): 49-55.

Tiwari, S.C. (1995). *Alnus nepalensis* D. Don biomass production and growth response to inoculation with *Frankia* and vesicular- arbuscular mycorrhiza. In : *Mycorrhizae: Biofertilizers for the Future*. (Eds. Alok Adholeya and Sujan Singh), TERI, New Delhi:184-188.

Tobar, R.M., Azcon, R. and Barea, J.M. (1994). The improvement of plants N acquisition from an ammonium-treated, drought stressed soil by the fungal symbiont in arbuscular mycorrhizae. *Mycorrhiza*. **4**: 105-108.

Tomar, A., Kumar, N., Pareek, R.P. and Chaube, A.K. (2001). Synergism among VA mycorrhizal, phosphate solubilizing bacteria and *Rhizobium* for symbiosis with black gram (*Vigna mungo* L.) under field conditions. *Pedosphere*. 11(4): 327-332

Trappe, J.M. (1987). Phylogenetic and ecological aspects of mycotrophy in the Magnoliophyta (Angiosperms) from an evolutionary stand point. In: *Ecophysiology of vesicular arbuscular mycorrhizal plants*. (Ed. G. Safir), C.R.C. Press. Boca. Raton. Fl.: 5-25.

Trappe, J.M. and Schenck, N.C. (1982). Taxonomy of the fungi forming endomycorrhizae. A. Vesicular arbuscular mycorrhizal fungi (Endogonales). In: *Methods and Principles of Mycorrhizal Research*. (Ed. N.C. Schenck), American Phytopathological Publication, St. Paul, MN.: 1-10.

Trappe, J.M., Molina, R. and Castellano, M. (1984). Reactions of mycorrhizal fungi and mycorrhiza formation of pesticides. *Annual Review of Phytopathology*. **22**: 331-359.

Trate V. and Srivasuki, K.P. (2001). Influence of *Glomus fasciculatum* and *Pisolithus tinctorius* on growth and drought tolerance of some tropical tree species. *Biologia* (*Bratislava*) **56**(4): 411-416.

Udaiyan, K. and Muthukumar, T. (1999). Growth improvement of acacias by indigenous arbuscular mycorrhizal fungi, *Rhizobium* and phosphate solubilizing bacteria in non-sterile field soil. Abstract: *National Conference on Mycorrhiza*, Barkatulla University, Bhopal: 22.

Udaiyan, K. and Sugavanam, V. (1995). Interaction of vesicular-arbuscular mycorrhizal fungi and *Frankia* sp. on growth and nutrient uptake in *Casuarina equisetifolia* Forst. In : *Mycorrhizae: Biofertilizers for the Future*. (Eds. Alok Adholeya and Sujan Singh), TERI, New Delhi: 231-237.

Udaiyan, K., Devi, A.P.G., Chitra, A. and Greep, S. (1997). Possible role of arbuscular mycorrhizal (AM) fungi on drought tolerance in *Vigna unguiculata* sub sp. *unguiculata* (L.) Walp and *Leucaena latisiliqua* L. *Pertanika Journal of Tropical Agricultural Science*. **20**(2/3): 135-146.

Udaiyan, K., Greep, S., Muthukumar, T. and Chitra, A. (1999). Effect of fumigation and pesticide drenches on VAM status and growth in cereals. *Journal of Environmental Biology* **20** (2): 167-175.

Udaiyan, K., Manian, S., Muthukumar, T. and Greep, S. (1995). Biostatic effect of fumigation and pesticide drenches on an endomycorrhizal-*Rizobium*-legume tripartite association under field conditions. *Biology and Fertility of Soils*. **20**(4): 275-283.

Udaiyan, K., Muthukumar, T., Vasantha, K., Greep, S. and Bai, V.N. (2001). Effect of fumigant and pesticides on the mycorrhization and nodulation of tree legume seedlings. *Journal of Tropical Forest Science* **13** (1): 19-30.

Udaiyan, K., Sugavanam, V. and Manian, S. (1996). Effects of formaldehyde fumigation and Fytalan drench on VAM fungi and nodulation in some leguminous forest tree seedlings in India. *Pertanika Journal of Tropical Agricultural Science*. **19**(1): 33-41.

Umesh, K.C., Krishnappa, K. and Bagyaraj, D.J. (1989). Interaction of burrowing nematode, *Radopholus similis* (Cabb, 1893) Thorne 1949 and VA mycorrhiza, *Glomus fasciculatum* (Thaxt). Gerd. and Trappe in banana (*Musa acuminata* Colla.). *Indian Journal of Nematology*. **18**(1): 6-11.

Uniyal, K. and Thapar, H.S. (1995). Growth responses in *Prosopis juliflora* inoculated with VA mycorrhizal fungi and *Rhizobium* in Sodic Soil. *Van Vigyan*. **33**(3-4): 182-184.

Uniyal, K. and Thapar, H.S. (1997). Status of VAM colonization in *Acacia nilotica* seedlings of different seed origin. *The Indian Forester*. **123**: 256-259.

Uniyal, K. and Uniyal, D.P. (2000). Population dynamics of arbuscular mycorrhizal fungi in *Dalbergia sissoo* Roxb. *Indian Forester* **126**(7): 782-787.

Varma, A. (1988). Growth and some aspects of nitrogen metabolism of *Gigaspora gigantea*. In: *Mycorrhizae for Green Asia*. (Eds. A. Mahadevan, N. Raman and K. Natrajan), Madras University, Madras: 120-124.

Varma, A. and Adholeya, A. (1995). Selection of suitable host for *in vitro* inoculation by vesicular-arbuscular mycorrhizal fungi *Gigaspora margarita.* In: *Mycorrhizae: Biofertilizers for the Future.* (Eds. Alok Adholeya and Sujan Singh), TERI, New Delhi: 478-481.

Varma, A. and Adholeya, A. (1996). Cost-economics of existing methodologies for inoculum production of vesicular-arbuscular mycorrhizal fungi. In: *Concepts of Mycorrhizal Research.* (Ed. K.G. Mukerji), Kluwer Academic Publishers, Printed in the Netherlands: 179-194.

Varma, A. and Schuepp, H. (1994). Positive influence of arbuscular mycorrhizal fungus on *in vitro* raised hortensis plantlets *Angewandte Botanik.* **64**(3/4) : 108-115.

Varma, A. and Schuepp, H. (1995). Mycorrhization of micropropagated plantlets. In : *Mycorrhizae: Biofertilizers for the Future.* (Eds. Alok Adholeya and Sujan Singh), TERI, New Delhi: 322-327.

Varshney, A., Sharma, M.P., Adholeya, A., Dhawan, V. and Srivastava, P.S. (2002). Enhanced growth of micro propagated bulb lets of *Lilium sp.* inoculated with arbuscular mycorrhizal fungi at different P fertility levels in an alfisol. *Journal of Horticultural Science and Biotechnology* **77** (3): 258-263.

Vasanthakrishna, M., Bagyaraj, D.J. (1993). Selection of efficient VA mycorrhizal fungi for inoculating *Casuarina equisetifolia. Arid Soil Research and Rehabilitation.* **7**(4): 377-380.

Vasanthakrishna, M., Bagyaraj, D.J. and Muthanna, M.B. (1994a). VA mycorrhizal fungi associated with *Casuarina equisetifolia. Microbiological Research.* **194**(1): 23-25.

Vasanthakrishna, M., Bagyaraj, D.J. and Nirmalnath, P.J. (1994b). Response of *Casuarina equisentifolia* to inoculation with *Glomus fasciculatum* and/or *Frankia. Forest Ecology and Management.* **68**(2/3): 399-402.

Vasanthakrishna, M.; Vinayak, K. and Bagyaraj, D.J. (1990). Occurrence of vesicular arbuscular mycorrhizal fungi in tropical ornamental plants. In: *Mycorrhizal Symbiosis and Plant Growth.* (Eds. D.J. Bagyaraj and A. Manjunath), U.A.S., Bangalore : 39-40.

Veeraswamy, J., Padmavathi, T. and Venkateswarlu, K. (1988). Development of VAM on *Sorghum* under field conditions. In: *Mycorrhizae for Green Asia.* (Eds. A. Mahadevan, N. Raman and K. Natarajan), Madras University, Madras: 265-267.

Veeraswamy, J., Padmavathi, T. and Venkateswarlu, K. (1990). Occurrence of endomycorrhizal fungi in rhizosphere soils of sorghum and foxtail millet. In: *Mycorrhizal Symbiosis and Plant Growth.* (Eds. D.J. Bagyaraj and A. Manjunath), U.A.S., Bangalore : 41-42.

Veeraswamy, J., Padmavathi, T., Venkateswarlu, K. and Daft M.J. (1992). Species of *Glomus* from semi-arid soils of Andhra Pradesh. *Environment and Ecology.* **10**(4): 804-807.

Venkataraman, M.N., Borthakur, R. and Singh, H.D. (1985a). Endogonaceae of north eastern India with special reference to *Gigaspora nigra. Acta Botanica Indica.* **13**L 73-76.

Venkataramanan, M.N., Borthakur, R. and Singh, H.D. (1985b). Occurrence of endotrophic mycorrhizal fungus in agarwood plant *Aquillaria agallocha* (Roxb). *Current Science.* **54**(18): 928.

Venkataramanan, M.N., Borthakur, R., Borthakur, B. and Singh, H.D. (1986). Survey of vesicular arbuscular mycorrhizal association in some weeds in North Eastern India. *Acta Botanica Indica.* **14**:202-204.

Venkataramanan, M.N., Kotoky, R. and Singh, H. (1990). Occurrence and distribution of vesicular-arbuscular mycorrhizal fungi in acid soils of North Eastern India. In: *Current trends in Mycorrhizal Research. Proceedings of the National Conference on Mycorrhiza.* (Eds. B.L. Jalali and H. Chand), H.A.U., Hisar : 29-30.

Venkateswarlu, K. and Daft, M.J. (1995). Interaction effects of carbofuran and *Glomus clarum* on growth of groundnut. In: *Mycorrhizae: Biofetilizers for the Future*. (Eds. Alok Adholeya and Sujan Singh), TERI, New Delhi: 124-127.

Verma, N.K. (1994). Shifting cultivation and mycorrhizal loss in the forest soils of north-eastern hills of India. In: *Ecology and man in the Himalayas* (Eds. A.K. Kapoor and S. Kapoor), New Delhi, India, M D Publications Pvt. : 193-205.

Verma, N.S., Kaur, R. and Verma, A.K. (1990). Effect of soil inoculation with endomycorrhizal fungi on growth of *Cyamopsis tetragonaloba*. In: *Mycorrhizal symbiosis and plant growth.* (Eds. D.J. Bagyaraj and A. Manjunath), U.A.S., Bangalore : 82-83.

Verma, R.K. and Jamaluddin (1995). Association and activity of arbuscular mycorrhizae of teak (*Tectona grandis*) in central India. In *Special issue: focus on teak, Indian Forester*. **121**(6): 533-539.

Verma, R.K., Jamaluddin and Gupta, B.N. (1994). Effect of inoculation of VAM fungi and *Rhizobium* on growth and biomass production in *Acacia nilotica* in nursery. *Indian Forester*. **120**(12): 1089-1094.

Vestberg, M. and Sukainen, M.U. (1992). *The Mycologist* **6** : 38.

Vietti, A.J., Staden, J. V. (1990). Histology of VA mycorrhizal development in guayule seedlings. *Mycorrhizal Research* **94**(6):831-834.

Vijaya, T. and Srivasuki, K.P. (2001). Growth response of *Azadirachta indica* seedlings to dual inoculation with *Glomus fasciculatum* and *Aspergillus fumigatus. Biologia (bratislava)* **56** (1): 125-129.

Vijaya, T., Kumar, R.V. Reddy, B.V.P., Sastry, P.S.S. and Srivastav, A.K. (1995). Studies on occurrence of endomycorrhiza in some forest soils of Andhra Pradesh. In: *Mycorrhizae: Biofertilizers for the Future*. (Eds. Alok Adholeya and Sujan Singh), TERI, New Delhi: 45-47.

Vijaya, T., Srivasuki, K.P. and Sastry, P.S. (1996). Role of gibberellic acid in teak and seed germination and the effect of *Glomus macrocarpus* on growth and sodic soil tolerance. *Annals of Forestry*. **4**(2): 211-212.

Vijayakumar, B.S., Bhiravamurthy, P.V. and Anand, M.G. (2000). VAM fungal association in *Lycopersicon esculentum* L. grown in semi-arid tropical soils of Puttaparthy, A.P. *Journal of Ecobiology* **12**(1): 73-74.

Vijayakumar, B.S.; Bhiravamurthy, P. V.; and Anand, M G. (1999). Studies on VAM and micro fungi in soils of the Prasanthinilayam Hills. *Journal of Hill Research* **12** (1): 81-82 .

Vijayakumar, K. and Abraham, T.K. (2001). Effect of pesticides on AM colonization and spore production in some endemic trees of Western Ghats. *Journal of Mycopathological Research* **39** (1): 9-14.

Vijayalakshmi, M. and Rao, A.S. (1988a). Vesicular-arbuscular mycorrhizal associations of some Asteraceae and Amaranthaceae. *Acta Botanica Indica*. **16**(2): 168-174.

Vijayalakshmi, M. and Rao, A.S. (1988b). Vesicular-arbuscular mycorrhizal associations of *Sesamum*. *Proceedings of the Indian Academy of Sciences, Plant Sciences* . **98**(1): 55-59.

Vijayalakshmi, M. and Rao, A.S. (1993). Effects of six insecticides and one fungicide on the development of VAM fungi in peanut (*Arachis hypogaea* L.) *Zentralblatt fur Microbiologie.* **148**(1): 60-65.

Vijaykumar and Abraham, T. K. (2002). Mycorrhizal and rhizosphere fungi of some endemic trees of Western Ghats. *Journal of Mycopathological Research* **40** (1): 25-28.

Vijaykumar, B.S. and Bhiravamurthy, P.V. (1999). Selection of efficient indigenous local VAM fungi for improving the groundnut crop in semi-arid tropical soils of Anantapur district, Andhra Pradesh. Abstract: *National Conference on Mycorrhiza*, Barkatulla University, Bhopal: **48**.

Vijaykumar, K. and Abraham, T.K. (1998). Arbuscular mycorrhizal association of *Bentinckia condapanna*. Berry ex Roxb., an endemic palm of Western Ghats. *Journal of Mycopathological Research*. **36**(2): 67-72.

Vijaykumari, B. and Janardhanan, K. (2003). Effect of biofertilizers on seed germination, seedling growth and biochemical change in silk cotton [*Ceiba pentandra* (Linn.) Gaertn.]. *Crop Research* (Hisar) **25** (2): 328-332.

Vilarino, A. and Arines, J. (1992). The influence of aqueous extracts of burnt or heated soil on the activity of vesicular arbuscular mycorrhizal fungi propagules. *Mycorrhiza*. **1**(2): 79-82.

Vinayak, K. and Bagyaraj, D.J. (1990a). Selection of efficient VA mycorrhizal fungi for Trifoliate orange. *Biological Agriculture & Horticulture*. **6**(4): 305-311.

Vinayak, K. and Bagyaraj, D.J. (1990b). Vesicular-arbuscular mycorrhizae screened for Troyer citrange. *Biology and Fertility of Soils*. **9**(4): 311-314.

Vineeta and Singh, O.S. (1997). Growth responses of mint plants infected by vesicular arbuscular mycorrhizal (VAM) fungus. In: *Achievements and prospects in mycology and plant pathology*. (Eds. S.S. Chahal, I.B. Parashar, H.S. Randhawa and S. Arya), International Book Distributors, Dehra Dun : 55-58.

Vyas, A. (1990). Occurrence and distribution of endomycorrhizal fungi in soyabean and chickpea in a black soil. In: *Mycoprrhizal Symbiosis and Plant Growth* (Eds. D. J. Bagyaraj and A. Manjunath), UAS, Bangalore: 12.

Vyas, R. and Srivastava, H.P. (1988). Studies on VAM of moth bean with special reference to their ecological implications. In: *Mycorrhizae for Green Asia*. (Eds. A. Mahadevan, N. Raman and K. Natrajan), Madras University, Madras: 118-119.

Vyas, R. and Srivastava, H.P. (1995). Effect of dual inoculation with *Rhizobium japonicum* and *Glomus fasciculatum* on yield and growth of cluster bean. In : *Mycorrhizae: Biofertilizers for the Future*. (Eds. Alok Adholeya and Sujan Singh), TERI, New Delhi: 338-339.

Vyas, S.C. and Vyas, S. (1995). Influence of pesticides on my corrhizal infection and chlamydospores formation. In: *Mycorrhizae: Biofertilizers for the Future*. (Eds. Alok Adholeya and Sujan Singh), TERI, New Delhi: 147-154.

Walker, C. (1983). Taxonomic concepts in the Endogonaceae: spore wall characteristics in species descriptions. *Mycotaxon*. **18**: 443-455.

Walker, C. (1987). Current concepts in the taxonomy of the Endogonaceae. In: *Mycorrhizae in the Next Decade: Practical Applications and Research Priorities*. (Eds. D.M. Slyvia, L.L. Hung and J.H. Graham), University of Florida Publication, Gainesville, FL.: 300-302.

Walker, J.C. (1950). *Plant Pathology*, McGraw Hill, New York.

Walsh, P.S., Metzger, D.A. and Higuchi, R. (1991). Chelex 100 as a medium for simple extraction of DNA for PCR-based typing from forensic material. *Biotechniques*. **10** : 506-513.

Wani, P.V. and Konde, B.K. (1996). Vesicular arbuscular mycorrhizal association in garlic. *Journal of Maharashtra Agricultural Universities* . **21**(3): 420-423.

Wani, P.V. and Konde, B.K. (1998a). Response of garlic to inoculation with vesicular arbuscular mycorrhizae. *Journal of Maharashtra Agricultural Universities* **23** (1): 35-38.

Wani, P.V. and Konde, B.K. (1998b). Effects of *Glomus mosseae* inoculation using different P-sources on garlic. *Journal of Maharashtra Agricultural Universities* **23** (1): 39-42.

Wani, S.P. and Lee, K.K. (1995). Effect of crop and soil management practices on mycorrhizal association with crops in the semi-arid tropics. In: *Mycorrhizae: Biofertilizers for the future*. (Eds. Alok Adholeya and Sujan Singh), TERI, New Delhi: 384-390.

Warner, A. and Mosse, B. (1979). cf. A Review of serological techniques for the identification of mycorrhizal fungi. In : *Mycorrhiza News*, **3** (3) 1991. In : *Rothamsted Report for 1977*. Harpenden, Rothamsted Experimental Station : 239.

Wath, M.R. and Deotare, P.W. (1999). VA-mycorrhizal status of *Alysicarpus species* dominant in natural tropical grassland ecosystem. Abstract: *National Conference on Mycorrhiza*, Barkatulla University, Bhopal: 35.

White, J.A., Munn, L.C. and Williams, S.E. (1989). Edaphic reclamation aspects of vesicular-arbuscular mycorrhizae in Red desert soils. *Soil Science Society of America*. **53**(1): 86-90.

White, P.R. 1954. In : *Cultivation of Animal and Plant cells.* 2nd Ed. Ronald Press, New York.

Williams, A.J., Chandra, K.K., Dugaya, D., Mishra, A.K., and Banerjee, S.K. (1995). Population status of VAM fungi in rehabilitated ash disposal area. *Environment and Ecology*, **13**(2): 284-387.

Williams, S.E. and Allen, M.F. (1984). VA mycorrhiza and reclamation of arid and semiarid lands. In: *Proc. Conf. held at Dubois, Wyoming*, 1982. Wyoming Agriculture Experiment Station, Wyoming.

Wilson, J.M., Trinik, M.J. and Parker, C.A. (1983). cf. A Review of serological techniques for the identification of mycorrhizal fungi. In : *Mycorrhiza News*, **3** (3) 1991. *Soil Biology and Biochemistry*. **15** : 439-445.

Wyss, P. and Bonfonte, P. (1993). Amplification of genomic DNA of arbuscular-mycorrhizal (AM) fungi by PCR using short arbitrary primers. *Mycological Research*. **95**: 1351-1357.

Yaragoppa, S.D., Desai, B.K., Pujari, B.T., Halepyati, A.S. and Alloli, T.B. (2003a). Dry matter and its accumulation pattern in *Sesbania aculeate* as influenced by plant densities and phosphorus management. *Karnataka Journal of Agricultural Sciences* **16** (2): 294-296.

Yaragoppa, S.D., Desia, B.K., Halepyati, A.S. and Pujari, B.T. (2003b). Influence of plant densities and phosphorus management of growth and seed yield of *Sesbania aculeate* (Wills.) Poir. *Karnataka Journal of Agricultural Sciences* **16** (2): 297-299.

Zaidi, R. and Mukerji, K.G. (1983). Incidence of vesicular-arbuscular mycorrhizas (VAM) in diseased and healthy plants. *Indian Journal of Plant Pathology*. **1**(1): 24-26.

Zajicek, J.M., Hetrick, B.A.D. and Alberecht, M.L. (1987). Influence of drought stress and mycorrhizae on growth of two native herbs. *Journal of American Society for Horticulture Science*. **63**: 362-369.

Zak, B. (1976). Pure culture synthesis of pacific madrone ectendomycorrhizae. *Mycologia*. **63**: 362-369.

Zambolim, L. (1986). *In ANAIS da Reunais Brasieira Sobre Microrrhizas*, iia 14 de November de (1985). Larvas, M.G.(1986): 76-00.

Zeuske, D., Schaat, A. and Weber, H. C. (1999). Progressive series related to mycotrophy in Araliaceae-Ginseng (Panax spec.) as a final stage. In: *Morphology, anatomy and systematics at the centenary of Wilhelm Troll's birth.* Procedings of the 13th Sympopsium on Morphology, Anatomy. Leuven, Belgium April 1997 (eds. L.P. Ronse Decrance, E. Robbrecht, E. Smets), *Systematics and Geography of Plants* **68** (1/2): 85-93.